Lecture Notes in Artificial Intelligence 7250

Subseries of Lecture Notes in Computer Science

Michael R. Berthold (Ed.)

Bisociative Knowledge Discovery

An Introduction to Concept, Algorithms, Tools, and Applications

Series Editors

Randy Goebel, University of Alberta, Edmonton, Canada
Jörg Siekmann, University of Saarland, Saarbrücken, Germany
Wolfgang Wahlster, DFKI and University of Saarland, Saarbrücken, Germany

Volume Editor

Michael R. Berthold
University of Konstanz
Department of Computer and Information Science
Konstanz, Germany
E-mail: michael.berthold@uni-konstanz.de

Acknowledgement and Disclaimer
The work reported in this book was funded by the European Commission in the 7th Framework Programme (FP7-ICT-2007-C FET-Open, contract no. BISON-211898).

ISSN 0302-9743 e-ISSN 1611-3349
ISBN 978-3-642-31829-0 e-ISBN 978-3-642-31830-6
DOI 10.1007/978-3-642-31830-6
Springer Heidelberg Dordrecht London New York

Library of Congress Control Number: 2012941862

CR Subject Classification (1998): I.2, H.3, H.2.8, H.4, C.2, F.1

LNCS Sublibrary: SL 7 – Artificial Intelligence

Typesetting: Camera-ready by author, data conversion by Scientific Publishing Services, Chennai, India

Printed on acid-free paper

Springer is part of Springer Science+Business Media (www.springer.com)

Foreword

We have all heard of the success story of the discovery of a link between the mental problems of children and the chemical pollutants in their drinking water. Similarly, we have heard of the 1854 Broad Street cholera outbreak in London, and the linking of it to a contaminated public water pump. These are two high-profile examples of bisociation, the combination of information from two different sources.

This is exactly the focus of the BISON project and this book. Instead of attempting to keep up with the meaningful annotation of the data floods we are facing, the BISON group pursued a network-based integration of various types of data repositories and the development of new ways to analyze and explore the resulting gigantic information networks. Instead of finding well-defined global or local patterns they wanted to find domain-bridging associations which are, by definition, not well defined since they will be especially interesting if they are sparse and have not been encountered before.

The present volume now collects the highlights of the BISON project. Not only did the consortium succeed in formalizing the concept of bisociation and proposing a number of types of bisociation and measures to rank their "bisociativeness," but they also developed a series of new algorithms, and extended several of the existing algorithms, to find bisociation in large bisociative information networks.

From a personal point of view, I was delighted to see that some of our own work on finding structurally similar pieces in large networks actually fit into that framework very well: Random walks, and related diffusion-based methods, can help find correlated nodes in bisociative networks. The concept of bisociative knowledge discovery formalizes an aspect of data mining that people have been aware of to some degree but were unable to formally pin down. The present volume serves as a great basis for future work in this direction.

May 2012 Christos Faloutsos

Foreword

Table of Contents

Part I: Bisociation

Part II: Representation and Network Creation

Part III: Network Analysis

Part V: Applications and Evaluation

Towards Bisociative Knowledge Discovery*

Michael R. Berthold

Nycomed Chair for Bioinformatics and Information Mining,
Department of Computer and Information Science,
University of Konstanz, Germany
Michael.Berthold@Uni-Konstanz.DE

Abstract. Knowledge discovery generally focuses on finding patterns within a reasonably well connected domain of interest. In this article we outline a framework for the discovery of new connections between domains (so called *bisociations*), supporting the creative discovery process in a more powerful way. We motivate this approach, show the difference to classical data analysis and conclude by describing a number of different types of domain-crossing connections.

1 Motivation

Modern knowledge discovery methods enable users to discover complex patterns of various types in large information repositories. Together with some of the data mining schema, such as CRISP-DM and SEMMA, the user participates in a cycle of data preparation, model selection, training, and knowledge inspection. Many variations on this theme have emerged in the past, such as Explorative Data Mining and Visual Analytics to name just two, however the underlying assumption has always been that the data to which the methods are applied to originates from one (often rather complex) domain. Note that by *domain* we do not want to indicate a single feature space but instead we use this term to emphasize the fact that the data under analysis represents objects that are all regarded as representing properties under one more or less specific aspect. Multi View Learning [19] or Parallel Universes [24] are two prominent types of learning paradigms that operate on several spaces at the same time but still operate within one domain.

Even though learning in multiple feature spaces (or views) has recently gained attention, methods that support the discovery of connections across previously unconnected (or only loosely coupled) domains have not received much attention in the past. However, methods to detect these types of connections promise tremendous potential for the support of the discovery of new insights. Research on (computational) creativity strongly suggests that this type of out-of-the-box thinking is an important part of the human ability to be truly creative. Discoveries such as Archimedes' connection between weight and (water) displacement and the – more recent – accidental ("serendipitous") discovery of Viagra are two illustrative examples of such domain-crossing creative processes.

* Extended version of [1].

M.R. Berthold (Ed.): Bisociative Knowledge Discovery, LNAI 7250, pp. 1–10, 2012.

In this introductory chapter we summarise some recent work focusing on establishing a framework supporting the discovery of domain-crossing connections continuing earlier work [3]. In order to highlight the contrast of finding patterns within a domain (usually associations of some type) with finding relations across domains, we refer to the term *bisociation*, first coined by Arthur Koestler in [13]. We argue that *Bisociative Knowledge Discovery* represents an important challenge in the quest to build truly creative discovery support systems. Finding predefined patterns in large data repositories will always remain an important aspect, but these methods will increasingly only scratch the surface of the hidden knowledge. Systems that trigger new ideas and help to uncover new insights will enable the support of much deeper discoveries.

2 Bisociation

Defining bisociation formally is, of course, a challenge. An extensive overview of related work, links to computational creativity and related areas in AI, as well as a more thorough formalisation can be found in [7]. Here we will concentrate on the motivational parts and only intuitively introduce the necessary background.

Boden [4] distinguishes among three different types of creative discoveries: Combinatorial, Exploratory, and Transformational Creativity. Where the second and third category can be mapped on (explorative) data analysis or at least the discovery process within a given domain, Combinatorial Creativity nicely represents what we are interested in here: the combination of different domains and the creative discovery stemming from new connections between those domains.

Informally, bisociation can be defined as (sets of) concepts that bridge two otherwise not –or only very sparsely– connected domains whereas an association bridges concepts within a given domain. Of course, not all bisociation candidates are equally interesting and in analogy to how Boden assesses the interestingness of a creative idea as being new, surprising, and valuable [4], a similar measure for interestingness can be specified when the underlying set of domains and their concepts are known. Going back to Koestler we can summarise this setup as follows:

> *"The creative act is not an act of creation in the sense of the Old Testament. It does not create something out of nothing; it uncovers, selects, re-shuffles, combines, synthesises already existing facts, ideas, faculties, skills. The more familiar the parts, the more striking the new whole."*

Transferred to the data analysis scenario, this puts the emphasis on finding patterns across domains whereas finding patterns in the individual domains themselves is a problem that has been tackled already for quite some time. Put differently, he distinguishes associations that work within a given domain (called *matrix* by Koestler) and are limited to repetitiveness (here: finding other/new occurrences of already identified patterns) and bisociations representing novel connections crossing independent domains (matrices).

3 Types of Bisociation

Obviously the above still remains relatively vague and for concrete implementations the type of bisociative patterns that are sought needs to be specified better. In the past years a number of bisociation types emerged in the context of Bisociative Knowledge Discovery: Bridging Concepts, Bridging Graphs, and Bridging by Structural Similarity, see [14] for a more detailed analysis. Since these ideas are also addressed in other areas of research, additional types most likely exist in those fields as well.

3.1 Bridging Concepts

The most natural type of bisociation is represented by a concept linking two domains, Figure 1 illustrates this.

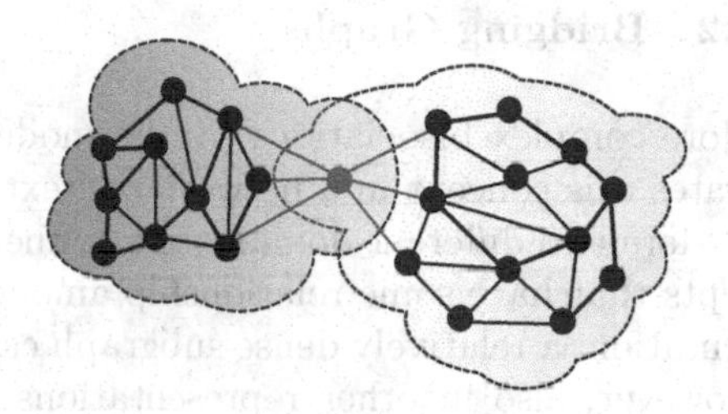

Fig. 1. Bridging concept (from [14])

Such bridging concepts do not need to exist in the context of a network based representation, as suggested by the figure, but can also be found in other representations. In [21], for instance, different textual domains were analysed to find bisociative terms that link different concepts from the two domains.

An example of a few bridging concepts is shown in Figure 2. Here a well known data set containing articles from two domains (migraine and magnesium) was searched for bridging terms (see [21] for more details). Note that this example reproduces an actual discovery in medicine.

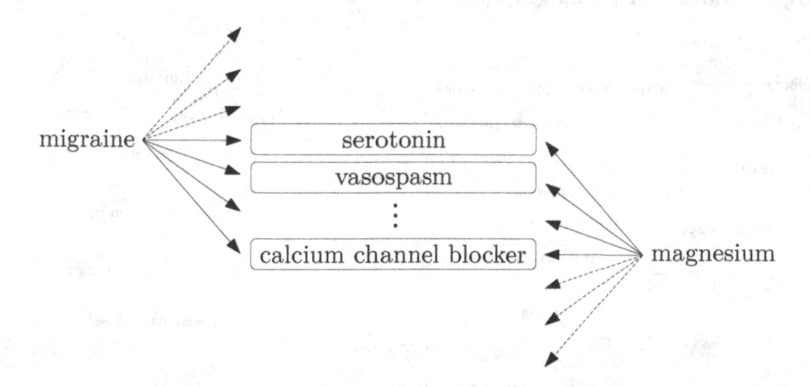

Fig. 2. Bridging concepts - an example reproducing the Swanson discovery (from [21])

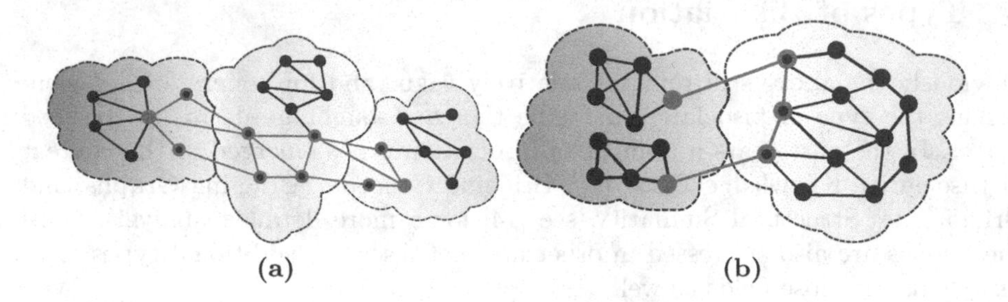

Fig. 3. Bridging graphs (from [14])

3.2 Bridging Graphs

More complex bisociations can be modelled by bridging graphs, Figure 3 illustrates this concept in a network context.

Here two different domains are connected by a (usually small) subset of concepts that have some relationship among themselves. In a network-based representation, a relatively dense subgraph can be identified connecting two domains. However, also in other representations, such "chains of evidence" can be formalised, connecting seperate domains.

Two examples for bridging graphs are shown in Figure 4 (the data stems from Schools-Wikipedia, see [17] for details). These demonstrate well how the two concepts "probability space" and "arithmetic mean" connect the domain of movies with a number of more detailed concepts in the statistics domain. This is at first glance surprising but finds its explanation in the (in both cases also somewhat "creative") use of those concepts in the two films or the series of films dominated by one actor. The second example nicely bridges physical properties and usage scenarios of phonographs.

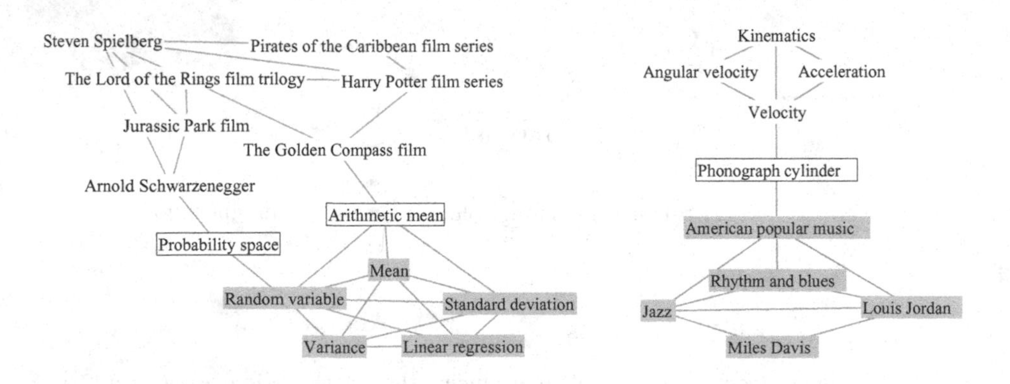

Fig. 4. Bridging graphs - two examples (from [17])

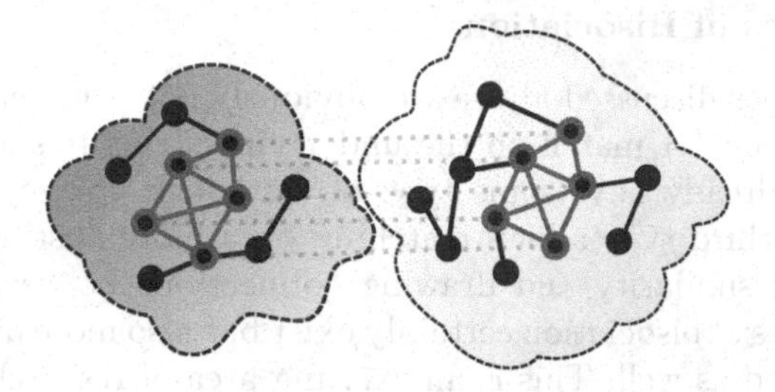

Fig. 5. Bridging by graph similarity (from [14])

3.3 Bridging by Structural Similarity

The third, so far most complex type of bisociation does not rely on some straightforward type of link connecting two domains but models such connections on a higher level. In both domains two subsets of concepts can be identified that share a structural similarity. Figure 5 illustrates this – again in a network-based representation; also here other types of structural similarity can exist.

An interesting example of such structural similarities can be seen in Figure 6. The demonstration data set based on Schools-Wikipedia was used in this example again. The two nodes slightly off centre ("Euclid" on the left and "Plato" on the right) are farther apart in the original network but share structural properties such as being closely connected to the hub of a subnetwork ("mathematics" vs. "philosophy"). Note that "Aristotle" also fills a similar role in the philosophy domain.

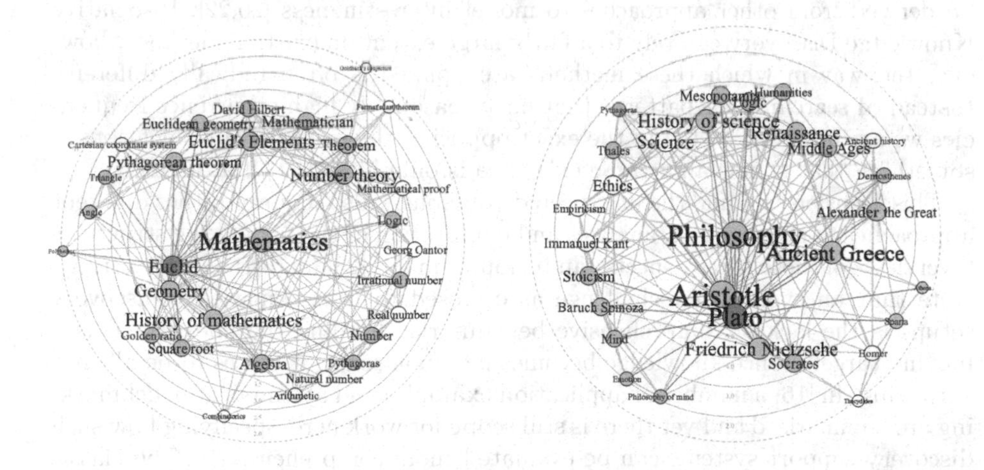

Fig. 6. Bridging by graph similarity - example (from [22])

3.4 Other Types of Bisociation

The bisociation types discussed above are obviously not complete. The first two types are limited to a 1:1 match on the underlying structures and require that the two domains already have some type of (although sparse) neighbourhood relation. Only the third type allows matching on a more abstract level, finding areas of structural similarity and drawing connections between those. Other, more abstract, types of bisociation certainly exist but also more direct bisociation types can be defined as well. This is an exciting area of research and one could also imagine systems that observe user interaction and learn new complex types of bisociation from user feedback and successful discoveries.

4 Bisociation Discovery Methods

In order to formalise the types of bisociations and develop methods for finding them, a more detailed model of the knowledge space needs to be available. When dealing with various types of information and the desire to find patterns in those information repositories a network-based model is often an appropriate choice due to its inherent flexibility. A number of methods can be found in this volume [2]. We hasten to add, however, that this is not the only way to model domains and bisociations, some contributions finding bisociation in non-network type domains can be found here as well, see for example the text-based bisociative term discoveries in [12,21].

It is interesting to note that quite a few of the existing methods in the machine learning and data analysis areas can be used, frequently with only minor modifications. For instance, methods for item set mining can be applied to the detection of concept graphs [15] and measures of bisociation strength can also be derived from other approaches to model interestingness [20,22]. Bisociative Knowledge Discovery can rely to a fairly large extent on existing methods, however the way in which these methods are applied is often radically different. Instead of searching for patterns that have reasonably high occurrence frequencies we are often interested in the exact opposite: the essenace of bisociations is something that is new and whose existence is only hinted at, if at all so far.

This focus on "finding the unexpected" obviously also requires rather different approaches to the creation, analysis and exploration of the underlying structure. Overviews of these three aspects can be found in [5], [23], and [18,9] respectively. Note that an even bigger challenge as opposed to usual knowledge discovery setups is the lack of comprehensive benchmarks. Finding the unexpected is a moving target – once knowledge becomes common sense, it ceases to be all that surprising. In [16] a number of application examples and attempts at benchmarking are summarised and yet there is still scope for work here, specifying how such discovery support systems can be evaluated more comprehensively. The classic setup of benchmark repositories is unlikely to be sufficient, as pure numerical performance does not really quantify a method's potential for creativity support – in fact individual methods will be hard to be evaluated properly, as they only become useful in concert with a larger system enabling truly explorative use.

5 Future Directions

The work briefly outlined in this paper is only the start, of course. Much more needs to be done to fully understand and make use of bisociative knowledge discovery systems. For once, the nature of bisociative patterns is far from complete – so far we have mainly addressed more classical approaches of finding numerical ways to assess the potential for bisociative insights of fairly simple patterns. The true potential lies in finding more abstract techniques to discover bisociations, similar to the methods described in [22] or [11]. Using abstract features to describe neighbourhoods – quite similar to the fingerprint similarity measures used in molecular searches for a long time already – shows enormous promise. Finding structurally similar patterns in different domains allows more complex knowledge to be transferred among the involved domains than only pointing to an existing (or missing) link.

However, in order to support the exploration of these more complex patterns it will be paramount to develop methods that allow smooth transitions among the associated levels of abstraction. Formal groundings for view transformations, similar to the methods described in [6] will be required. This will need to be accompanied by other powerful visual tools, of course, in order to actually give users access to these view transformations. BioMine [8] or CET [10] have been used successfully but even more flexible methods will be needed to integrate various views within the same structure. An interesting additional challenge will be the integration of user feedback not only in terms of guiding the search but also with respect to actually learning from the users' feedback to avoid proposing uninteresting patterns over and over again. Unfortunately, as discussed above, "(un)interesting" is a moving target and heavily depends on the current scope of analysis. Active Learning approaches offer interesting mechanisms to quickly update internal models of interest to make those systems respond in a useful way. An interesting side effect could be that such learning systems observe the users, learn patterns of bisociation that were of interest in the past and actually transfer those patterns among different analyses, thus forming meta level bisociations over time.

6 Conclusions

Bisociative Knowledge Discovery promises great impact especially in those areas of scientific research where data gathering still outpaces model understanding. Once the mechanisms are well understood the task of data analysis tends to change and the focus on (statistically) significant and validated patterns is much stronger. However, in the early phase of research, the ability to collect data outperforms by far the experts' ability to make sense out of those gigantic data repositories and use them to form new hypotheses. This trend can be seen in the life sciences where data analysis barely scratches the surface of the wealth of generated data. Current methods not only fall short of offering true, explorative access to patterns within domains, but are also considerably lacking when it comes to offering this kind of access across domains. The framework sketched here (and more substantially founded in [7]) can help to address this shortcoming. Much work still needs to be done, however, as many more types of bisociations can be formalised and many of the existing methods in the machine learning and data analysis/mining community are waiting to be applied to these problems.

One very interesting development here can be seen in the network-based bisociation discovery methods which are beginning to bridge the gap nicely between solidly understood graph theoretical algorithms and overly heuristic, poorly controllable methods. Putting those together can lead to the discovery of better understood bisociative (and other) patterns in large networks.

The data mining community has been looking for an exciting "Grand Challenge" for a number of years now. Bisociative Knowledge Discovery could offer just that: inventing methods and building systems that support the discovery of truly new knowledge across different domains will have an immense impact on how research in many fields can be computer supported in the future.

Acknowledgements. The thoughts presented in this paper would not have emerged without countless constructive and very fruitful discussions with the members of the EU FP7 BISON Project and the BISON Group at Konstanz University. In particular, I would like to thank Tobias Kötter, Kilian Thiel, Uwe Nagel, Ulrik Brandes, and our frequent guest bison, Werner Dubitzky, for many discussions pertaining the nature of bisociation and creative processes. Thanks are also due to Nada Lavrač for constructive criticism and enthusiastic support, Luc De Raedt and Hannu Toivonen for their different but always positive views on the topic, and Christian Borgelt, Andreas Nürnberger, and Trevor Martin for their support for this project from the start. Throughout the project itself our reviewers Agnar Aamodt, Wlodzislaw Duch, and David Pierce as well as Paul Hearn, the project officer, have always provided positive, practical feedback, which often gave new, inspiring stimuli to the work reported here. Rumours have it that the BISON project owes particular thanks to one of the project application reviewers, who strongly supported its funding!

This work was funded by the European Commission in the 7th Framework Programme (FP7-ICT-2007-C FET-Open, contract no. BISON-211898). Heather Fyson deserves much gratitude for hiding most of the infamous EU project coordination complexity from my view.

References

1. Berthold, M.R.: Bisociative Knowledge Discovery. In: Gama, J., Bradley, E., Hollmén, J. (eds.) IDA 2011. LNCS, vol. 7014, pp. 1–7. Springer, Heidelberg (2011)
2. Berthold, M.R. (ed.): Bisociative Knowledge Discovery. LNCS (LNAI), vol. 7250. Springer, Heidelberg (2012)

3. Berthold, M.R., Dill, F., Kötter, T., Thiel, K.: Supporting Creativity: Towards Associative Discovery of New Insights. In: Washio, T., Suzuki, E., Ting, K.M., Inokuchi, A. (eds.) PAKDD 2008. LNCS (LNAI), vol. 5012, pp. 14–25. Springer, Heidelberg (2008)
4. Boden, M.A.: Précis of the creative mind: Myths and mechanisms. Behavioural and Brain Sciences 17, 519–570 (1994)
5. Borgelt, C.: Network Creation: Overview. In: Berthold, M.R. (ed.) Bisociative Knowledge Discovery. LNCS (LNAI), vol. 7250, pp. 51–53. Springer, Heidelberg (2012)
6. Dries, A., Nijssen, S., De Raedt, L.: BiQL: A Query Language for Analyzing Information Networks. In: Berthold, M.R. (ed.) Bisociative Knowledge Discovery. LNCS (LNAI), vol. 7250, pp. 147–165. Springer, Heidelberg (2012)
7. Dubitzky, W., Kötter, T., Schmidt, O., Berthold, M.R.: Towards Creative Information Exploration Based on Koestler's Concept of Bisociation. In: Berthold, M.R. (ed.) Bisociative Knowledge Discovery. LNCS (LNAI), vol. 7250, pp. 11–32. Springer, Heidelberg (2012)
8. Eronen, L., Toivonen, H.: Biomine: Predicting links between biological entities using network models of heterogeneous database. BMC Bioinformatics (accepted for publication, 2012)
9. Gossen, T., Nitsche, M., Haun, S., Nürnberger, A.: Data Exploration for Knowledge Discovery: A Brief Overview of Tools and Evaluation Methods. In: Berthold, M.R. (ed.) Bisociative Knowledge Discovery. LNCS (LNAI), vol. 7250, pp. 287–300. Springer, Heidelberg (2012)
10. Haun, S., Gossen, T., Nürnberger, A., Kötter, T., Thiel, K., Berthold, M.R.: On the Integration of Graph Exploration and Data Analysis: The Creative Exploration Toolkit. In: Berthold, M.R. (ed.) Bisociative Knowledge Discovery. LNCS (LNAI), vol. 7250, pp. 301–312. Springer, Heidelberg (2012)
11. Henderson, K., Gallagher, B., Li, L., Akoglu, L., Eliassi-Rad, T., Tong, H., Faloutsos, C.: It's who you know: graph mining using recursive structural features. In: KDD, pp. 663–671 (2011)
12. Hynönen, T., Mahler, S., Toivonen, H.: Discovery of Novel Term Associations in a Document Collection. In: Berthold, M.R. (ed.) Bisociative Knowledge Discovery. LNCS (LNAI), vol. 7250, pp. 91–103. Springer, Heidelberg (2012)
13. Koestler, A.: The Act of Creation. Macmillan (1964)
14. Kötter, T., Berthold, M.R.: From Information Networks to Bisociative Information Networks. In: Berthold, M.R. (ed.) Bisociative Knowledge Discovery. LNCS (LNAI), vol. 7250, pp. 33–50. Springer, Heidelberg (2012)
15. Kötter, T., Berthold, M.R. (Missing) Concept Discovery in Heterogeneous Information Networks. In: Berthold, M.R. (ed.) Bisociative Knowledge Discovery. LNCS (LNAI), vol. 7250, pp. 230–245. Springer, Heidelberg (2012)
16. Mozetič, I., Lavrač, N.: Applications and Evaluation: Overview. In: Berthold, M.R. (ed.) Bisociative Knowledge Discovery. LNCS (LNAI), vol. 7250, pp. 359–363. Springer, Heidelberg (2012)
17. Nagel, U., Thiel, K., Kötter, T., Piatek, D., Berthold, M.R.: Towards Discovery of Subgraph Bisociations. In: Berthold, M.R. (ed.) Bisociative Knowledge Discovery. LNCS (LNAI), vol. 7250, pp. 263–284. Springer, Heidelberg (2012)
18. Nürnberger, A.: Exploration: Overview. In: Berthold, M.R. (ed.) Bisociative Knowledge Discovery. LNCS (LNAI), vol. 7250, pp. 285–286. Springer, Heidelberg (2012)
19. Rüping, S., Scheffer, T. (eds.): Learning with Multiple Views. Processings of the ICML 2005 Workshop (2005)

20. Segond, M., Borgelt, C.: Cover Similarity Based Item Set Mining. In: Berthold, M.R. (ed.) Bisociative Knowledge Discovery. LNCS (LNAI), vol. 7250, pp. 104–121. Springer, Heidelberg (2012)
21. Sluban, B., Juršič, M., Cestnik, B., Lavrač, N.: Exploring the Power of Outliers for Cross-domain Literature Mining. In: Berthold, M.R. (ed.) Bisociative Knowledge Discovery. LNCS (LNAI), vol. 7250, pp. 325–337. Springer, Heidelberg (2012)
22. Thiel, K., Berthold, M.R.: Node Similarities from Spreading Activation. In: Berthold, M.R. (ed.) Bisociative Knowledge Discovery. LNCS (LNAI), vol. 7250, pp. 246–262. Springer, Heidelberg (2012)
23. Toivonen, H.: Network Analysis: Overview. In: Berthold, M.R. (ed.) Bisociative Knowledge Discovery. LNCS (LNAI), vol. 7250, pp. 144–146. Springer, Heidelberg (2012)
24. Wiswedel, B., Hoeppner, F., Berthold, M.R.: Learning in parallel universes. Data Mining and Knowledge Discovery 21, 130–150 (2010)

Towards Creative Information Exploration Based on Koestler's Concept of Bisociation

Werner Dubitzky[1], Tobias Kötter[2], Oliver Schmidt[1], and Michael R. Berthold[2]

[1] University of Ulster, Coleraine, Northern Ireland, UK
w.dubitzky@ulster.ac.uk, schmidt-o1@email.ulster.ac.uk
[2] University of Konstanz, Konstanz, Germany
{tobias.koetter,michael.berthold}@uni-konstanz.de

Abstract. *Creative information exploration* refers to a novel framework for exploring large volumes of heterogeneous information. In particular, creative information exploration seeks to discover new, surprising and valuable relationships in data that would not be revealed by conventional information retrieval, data mining and data analysis technologies. While our approach is inspired by work in the field of computational creativity, we are particularly interested in a model of creativity proposed by Arthur Koestler in the 1960s. Koestler's model of creativity rests on the concept of *bisociation*. Bisociative thinking occurs when a problem, idea, event or situation is perceived simultaneously in two or more "matrices of thought" or domains. When two matrices of thought interact with each other, the result is either their *fusion* in a novel intellectual synthesis or their *confrontation* in a new aesthetic experience. This article discusses some of the foundational issues of computational creativity and bisociation in the context of creative information exploration.

"Creativity is the defeat of habit by originality." – Arthur Koestler

1 Introduction

According to Higgins, *creativity* is the process of generating something new that has value [19]. Along with other essentially human abilities, such as intelligence, creativity has long been viewed as one of the unassailable bastions of the human condition. Since the advent of the computer age this monopoly has been challenged. A new scientific discipline called *computational creativity* aims to model, simulate or replicate creativity with a computer [7]. This article explores the concept of *bisociation* [20] in the context of computational creativity. While our discussion may be relevant to a large number of domains in which creativity plays a central role, we emphasize domains with clear practical applications, such as science and engineering. We start our discourse on bisociation with the familiar concept of *association*.

The concept of association is at the heart of many of today's powerful computer technologies such as information retrieval and data mining. These technologies typically employ "association by similarity or co-occurrence" to locate or discover

M.R. Berthold (Ed.): Bisociative Knowledge Discovery, LNAI 7250, pp. 11–32, 2012.

information relevant to a user's tasks. A typical feature of these approaches is that the underlying information pool (document corpora, databases, Web sites, etc.) contains information that has been pre-selected in some way to focus and simplify the discovery process. For example, a biological study would pre-select scientific papers from relevant life science journals or abstracts before applying a particular text mining task. Pre-selecting information in this way already introduces certain limits on how creative these conventional approaches can be. This means that under normal circumstances such resources would not be combined to facilitate creative insights and solutions. A novel information exploration paradigm that aims to facilitate the generation of *creative* insight or solutions could be referred to as *creative information exploration* (*CIE*). Domains were CIE is critical include design and engineering, the arts (e.g., painting, sculpture, architecture, music and poetry) as well as scientific discovery disciplines.

In the remainder of this article we use the terms creative domains and creative disciplines to designate domains and disciplines in which creative information discovery plays an important role.

People working in creative domains employ creative thinking to connect seemingly unrelated information, for example, by using metaphors, analogy and other ways of thinking and reasoning [6]. Creative styles of thought allow the mixing of conceptual categories and contexts that are normally separated. Our goal is to develop computer-based solutions that support creative thinking. Inspired by Koestler's notion of *bisociation* [20], our particular aim is to develop concepts and solutions that facilitate bisociative CIE tasks in creative domains. Intuitively, bisociative CIE could be viewed as an approach that seeks to combine elements from two or more "incompatible" concept or information spaces (domains) to generate creative solutions and insight.

The remainder of this article is organized as follow: Sections 2 and 3 introduce a working definition of creativity with a view to its computational realization. In Sections 4 to 6 we review Kostler's notion of bisociation and offer an initial formal definition of this concept. Before we reflect on the work presented in this article and offer some concluding remarks (Section 8), we present a short review of related work in Section 7.

2 Creativity

2.1 What Is Creativity?

Human creativity, like other human capabilities, is difficult to define and formalize. In this article we adopt the following working definition of *creativity* based on the work by Margaret Boden [6].

Definition 1 (creativity). *Creativity is the ability to come up with ideas or artifacts that are new, surprising, and valuable.*

In this working definition of creativity the notions of *idea* and *artifact* refer to concepts and creations from art as well as science and engineering and other areas. Here we view creativity as an ability which is an intrinsic part of an intelligent agent (human, machine-based or otherwise). In the following discussion we elaborate the meaning of the concepts *new*, *surprising* and *valuable* in the definition of creativity.

The word *new* in our working definition of creativity may refer to two dimensions: *historic creativity* or *personal creativity.* By historic creativity we mean ideas or artifacts that are original in the sense that they represent the first occurrence of a particular idea or artefact in human history. The history of science and modern legal practice tell us that sometimes it may not be straightforward to determine precisely the first occurrence of a scientific or engineering idea. Examples of disputes over historic creativity include the theory of evolution, the invention of gun powder, and the social Web site Facebook. Personal creativity, on the other hand, means that someone comes up with an idea or invention independently from someone else who had already conceived of the same thing *before*. From the perspective of the "re-inventor" this still constitutes "true" creativity.

An important factor in our working definition of creativity concerns the notion of *surprise* – for a new idea to be considered creative there has to be an element of surprise. An idea or artefact may be surprising because it is unlikely (has a low probability of occurring) or unfamiliar. When a new idea unexpectedly falls into an already familiar conceptual framework (or thinking style) one is intrigued to not have realized it before. For example, in 1996 Akihiro Yokoi invented a "digital pet" called Tamagotchi which soon became a best seller. While the concept of looking after plants, pet animals and soft toy pets has been around for a long time, no one had dared to think that this idea could be applied to devices that resemble digital pocket calculators. A different type of surprise occurs when we encounter an apparently *impossible* concept or artefact. For instance, in 1905 Einstein shocked the scientific establishment by suggesting that energy is being *transmitted* in finite "packets" called quanta [11]. Max Planck, the originator of quantum theory, initially rejected Einstein's proposal even though his own theory suggested that energy *transfer* to and from matter is not continuous but discrete.

The last element in our working definition of creativity is the notion of *value* – a new concept or artefact must be valuable in some non-trivial way to qualify as creative. In the fine arts aesthetic values are difficult to recognize or agree about: what makes a painting by one artist hundred times more expensive than a painting by another? To formally define aesthetic values is even harder. Furthermore, values vary over time and within and across cultures. Even in science there is often considerable disagreement over the "simplicity", "elegance" or "beauty" of a theory or scientific argument. Einstein and Bohr, for instance, had argued over decades about the value (correctness and completeness) of the two prevailing models of the atom (the probabilistic and discrete model, favored by Bohr, and the deterministic and continuous model, which was preferred by Einstein) [22]. Whether a particular hypothesis is interesting or valuable may depend on scientific, social, economic, political and other factors. So even when we agree on novelty and the factor of surprise, there may still be a considerable disagreement over how valuable a new idea or artefact is, hence over the degree of creativity.

This brief discussion about the nature of creativity and the difficulty to recognize and agree on what creativity actually is serves as a context for the development of computational creativity techniques. Ultimately, what constitutes human or machine creativity is difficult to judge and needs to be assessed on a case-by-case basis.

2.2 Three Roads to Creativity

Following Boden [6] we distinguish three processes of creativity; these relate to the three forms of surprises discussed above.

Combinatorial Creativity. Arthur Koestler[1] is credited with the following characterization of creativity:

> The creative act is not an act of creation in the sense of the Old Testament. It does not create something out of nothing; it uncovers, selects, reshuffles, combines, synthesizes already existing facts, ideas, faculties, skills. The more familiar the parts, the more striking the new whole.

This idea is very much in line with the first process of creativity identified by Boden, which generates unfamiliar combinations of familiar concepts and constructs. In humans, analogy is a fundamental cognitive process in which familiar elements appear in an unfamiliar arrangement. A typical example of analogy establishes an analogical relationship between Niels Bohr's model of the atom with the basic structure of the heliocentric solar system. Facilitating this kind of creative process requires a rich knowledge structure and flexible ways of manipulating this structure. Clearly, the novel combination of elements must have a point or a meaning. Therefore, purely random shuffling and re-combination of elements will not be sufficient to generate creativity.

Exploratory Creativity. Margaret Boden defines *conceptual spaces* as a "structured style of thought". In her definition, a key characteristic of conceptual spaces is that they are not originated by an individual but are a structure adopted from the cultures and peer groups within which people live [6]. Conceptual spaces include ways of writing prose, styles of architecture and art, theories of nature, as well as approaches to design and engineering. So any systematic way of thinking which is valued by a certain group or culture could be thought of as a conceptual space.

In Boden's framework, a conceptual space defines a space of possible combinations of its elements, where each combination represents a particular thought, idea or artifact. While the number of possible thoughts within a conceptual space may be very large, only a fraction of these may have actually been realized. Consider, for instance, the games of chess and checkers. In chess the number of possible legal positions or "configurations" has been estimated at $10^{15\,790}$ and for checkers the number is 10^{18} [16,29]. Clearly, even with the long history of chess playing, only a very small number of possible "combinations" could have

[1] Prolific writer and author of *The act of creation* [20].

been explored so far. Clearly, Boden's concept of a conceptual space is much broader. For the game of chess, for example, it would not only include all possible chess board positions but also all knowledge structures employed by chess players to play the game as well as other facts and information about chess.

No matter what the actual size of a given conceptual space, someone who comes up with a new combination within that space is considered to be creative in the exploratory sense (provided the combination "has a point"). Boden likens the exploration of conceptual spaces to the exploration of a territory with a map. The map encompasses all possibilities, but to discover a particular and valuable possibility one needs to go out and explore the actual territory. *Exploratory creativity* is important as it facilitates the discovery of so far unknown possibilities. Once such novel possibilities come to light, the explorers may even be able to reflect deeper on the limits and potentials of a particular conceptual space.

Transformational Creativity. Exploratory creativity is limited by the possibilities defined within a conceptual space or thinking style (or "map"). Essentially, each conceptual space restricts the kind of thoughts that can be thought. To overcome this limitation, and to attempt to think what is unthinkable within a given conceptual space, it is necessary to change or *transform* the conceptual space. It must be transformed so that thoughts that were inconceivable within the previous version of the space now become possible. Such transformations may be subtle or radical. *Transformational creativity* constitutes the deepest form of creative processes in Boden's model of creativity.

3 Computational Creativity

Teaching humans to be creative is a flourishing business and the number of creativity techniques available is large [19]. Teaching or programming a computer to be creative or appear to be creative is another matter altogether. *Computational creativity* refers to an active scientific discipline that aims to model, simulate or replicate creativity using a computer [7].

Computational creativity draws on many concepts developed within the field of *artificial intelligence* (*AI*). Analogously to computational creativity, AI could be defined as a discipline aiming to model, simulate or replicate (human) *intelligence*. Boden suggests that AI concepts could be used to define and construct artificial conceptual spaces which could then be studied and eventually be used to combine elements from the spaces, and to explore and transform such spaces with the aim of generating creative insight and solutions. Boden describes concrete AI-based approaches to computational creativity [6,7].

4 Koestler's Concept of Bisociation

People working in creative domains employ creative thinking to connect seemingly unrelated information (true negatives under the association paradigm), for example, by using a metaphoric or analogical way of thinking. Analogical and metaphoric styles of thought allow the mixing of conceptual categories and contexts that are normally separated. In the 1960s Arthur Koestler developed a model of creative thinking referred to as *bisociation* [20]. Bisociation facilitates the mixture in one human mind of concepts from two contexts or categories of objects that are normally considered separate by the literal processes of the mind.

Koestler proposed the bisociation concept to distinguish the type of metaphoric thinking that leads to the acts of great creativity from the more "pedestrian" associative style of thinking, with which we are so familiar in our everyday lives and which pervades many of todays computing approaches. Associative thinking is based on the "habits" or set of routines that have been established over a period of time. Associative processes combine elements from the same "matrix" of thought. The associative mode of thinking differs from the bisociative mode that underlies the creative act. Bisociation, according to Koestler, means to join unrelated, often conflicting, information in a new way. It is being "double minded" or able to think simultaneously on more than one plane or matrix of thought (see Figure 1). "When two independent matrices of perception or reasoning interact with each other the result ... is a ... fusion in a new intellectual synthesis ..." [20]. Frank Barron reinforces this idea and characterizes bisociation as "the ability to tolerate chaos or seemingly opposite information" [3]. Koestler makes a clear distinction between more routine or habitual thinking (association) operating within a single plane or matrix of thought, and the more creative bisociative mode of thinking which connects independent autonomous matrices.

Koestler's basic concept of bisociation is illustrated in Figure 1. The diagram depicts two matrices of thought (domains or knowledge bases in our terminology), M_1 and M_2, as orthogonal planes. M_1 and M_2 represent two self-contained but "habitually incompatible" matrices of thought. An event, idea, situation, concept or problem, π, which is perceived simultaneously in both matrices is not merely linked to one associative context (M_1 or M_2) but *bisociated* with two associative contexts (M_1 and M_2). In the diagram, π is illustrated by the thick line cutting across M_1 and M_2. The diagram illustrates six concepts labeled $c_1, \ldots, c_6$. The concepts c_1, c_2, c_3 and c_6 are perceivable in matrix M_2 and c_1, c_2, c_3, c_4 and c_5 are perceivable in M_1. The concepts c_1, c_2, c_3 are associated with the problem π – because c_1, c_2, c_3 are perceivable in both matrices, it is possible to "see" the problem simultaneously from two frames of mind.

Central to Koestler's concept of bisociation are the notions of a *matrix* and a *code* Koestler [20]; we quote from page 38:

> ... to introduce a pair of related concepts which play a central role in this book and are indispensable to all that follows. ... I shall use the word '*matrix*' to denote any ability, habit, or skill, any pattern of ordered behavior governed by a '*code*' of fixed rules.

A matrix[2] in Koestler's framework denotes any ability, skill, habit or pattern of ordered behavior. Matrices shape our perceptions, thoughts, and activities; they

[2] Other terms Koestler uses for the concept of a matrix include the following: matrix of thought, matrix of behavior, matrix of experience, matrix of perception, associative context, frame of reference, universe of discourse, type of logic, code of behavior.

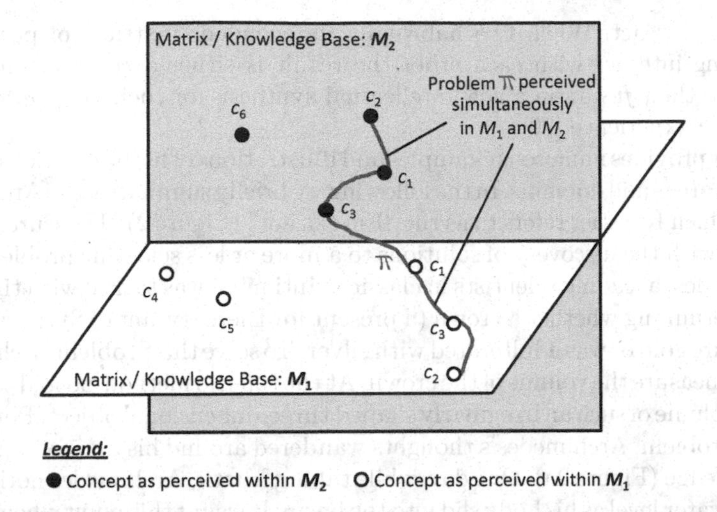

Fig. 1. Illustration of Koestler's concept of bisociation (adapted from Koestler [20])

could be viewed as condensations of learning into "habit". For example, a spider has an innate skill that enables it to build webs, a mathematician possesses the ability of mathematical reasoning, and a chess grandmaster has a knowledge base which allows him to play chess at a very high level. The abilities and skills represented by a matrix may be applied to concrete problems and tasks in a flexible way. For example, depending on the environment a spider finds itself in, it may choose three, four or more points of attachment to suspend its web.

Each matrix in Koestler's model of bisociation is governed by a set of fixed *codes* or rules. The rules could be innate or acquired. For example, in the game of chess, the rules of the game are fixed, while the patterns of knowledge (allowing one to play well or not so well) vary across players[3]. In mathematics, operations such as multiplication, differentiation, integration, etc. constitute fixed rules that govern mathematical reasoning. Another example of a code are the assumptions, concepts, notions, etc. that underly religious, political, economic, philosophical and similar debates and arguments. For instance, a debate on abortion may be held "in terms of" religious morality or social responsibility. Often the rules that govern a matrix of skill (ability, habit) function on a lower level of awareness than the actual performance itself (playing the piano, carrying out a conversation, formulating a strategy).

Once people have reached adulthood they have formed more or less rigid, automated patterns of behavior and thinking ("habits" or knowledge bases). Sometimes these patterns are interrupted by spontaneous sparks of insight which presents a familiar concept or situation in a new light. This happens when we connect previously unconnected matrices of perception or experience in a creative act of bisociation. Considering the field of humor, science and engineering as well as the arts, Koestler's conjecture was that bisociation is a general mechanism

[3] Certain ways of playing chess are also relatively frequent or nearly constant. For example, certain moves in chess openings, or certain endgame patterns.

for the creative act. When two habitually independent matrices of perception or reasoning interact with each other the result is either a *collision* ending in laughter, or their *fusion* in a new intellectual synthesis, or their *confrontation* in an aesthetic experience [20].

Koestler provides numerous examples and illustrations of his bisociation concept in different areas and domains. In the following we briefly summarize the Archimedes example which Koestler refers to as the "Eureka act" (Figure 2). The Eureka act is concerned with the discovery of solutions to a more or less scientific problem.

Archimedes, a leading scientists in classical antiquity, was tasked with the problem of determining whether a crown (a present for Hiero, tyrant of Syracuse) consisted of pure gold or was adulterated with silver. To solve this problem Archimedes needed to measure the volume of the crown. At the time no method existed to determine the volume of such an irregularly shaped three-dimensional object. Pondering over this problem, Archimedes's thoughts wandered around his matrix of geometrical knowledge (Figure 2a). One day, while taking a bath, Archimedes noticed the rise of the water level as his body slid into the basin. It was at this point when he connected the matrix of and experience associated with taking a bath with the matrix of his knowledge of geometry. He realized that the volume of water displaced was equal to the volume of the immersed parts of his own body. This Eureka moment is illustrated in Figure 2b. When Archimedes found the solution to this problem both matrices (associations of taking a bath and knowledge of geometry) were simultaneously active. In a sense Archimedes was looking at the same problem from two different perspectives of knowledge or experience at the same time. This "double-mindedness" allowed him to see the solution which was obscured under the view of either of the two individual perspectives.

Consider the diagram in Figure 2a. The dashed line illustrates Archimedes's search through the conceptual space to find a solution for his problem. While the search path traverses both knowledge bases (M_1 and M_2), the reasoning of Archimedes is initially confined to perceiving only one knowledge base at a time. Thinking about the problem in this "habitual" way, Archimedes fails to "see" the solution, because he does not simultaneously perceive the concepts describing the solution (c_1 and c_6) and the problem (c_1, c_2 and c_3).

Now consider the diagram in Figure 2b. At some point Archimedes is able to perceive the concepts describing both the problem (P) *and* the solution (S) simultaneously from the perspective of both knowledge bases. This is depicted by the line connecting the corresponding concepts across both knowledge bases. It is at this point when Archimedes experiences the Eureka moment which is created by the bisociative view spanning two matrices of thought.

The example of Archimedes and the crown illustrates how a familiar but unnoticed aspect of a phenomenon (rise of water level as a result of the immersion of an object) is suddenly perceived at an unfamiliar and significant angle (determining the purity of substance of an irregularly shaped object). Koestler refers to this as the "bisociative shock" often associated with discoveries when we suddenly see familiar objects and events in a strangely new and revealing light.

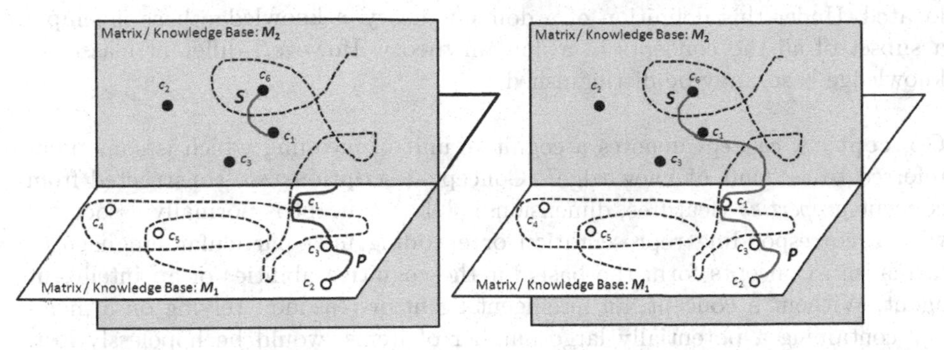

Fig. 2. Illustration of the Eureka act (adapted from Koestler [20]). The matrix or knowledge base M_1 represents concepts or associations of *geometrical knowledge*, and M_2 those of *taking a bath*. The dashed lines represent the search or exploration of the matrices as part of the problem-solving process. (a) Diagram on the left: The line connecting the concepts c_1, c_2 and c_3 represents the problem, P, as perceived by Archimedes based on his geometric knowledge base M_1. The arc connecting the concepts c_1 and c_6 in M_2 represents the solution, S. (b) Diagram on the right: The concepts associated with the problem *and* solution when perceived *simultaneously* in both knowledge bases.

The distinguishing characteristics of associative and bisociative thought are summarized in Table 1.

Table 1. Comparison of characteristics of bisociation and association based on Koestler [20]

Habit (Associative)	Originality (Bisociative)
association within a given matrix	bisociation of independent matrices
rigid to flexible variations on a theme	super-flexibility
repetitiveness	novelty
conservative	destructive-constructive

5 Elements of Bisociative Computational Creativity

Before we formally define bisociation, we analyze and compare the concepts and models of creativity proposed by Boden and Koestler. We do this by adopting an AI perspective of the notions involved and, on this basis, attempt a synthesis. In essence, we define creativity and bisociation in terms of *domain theories* and *knowledge bases*. Simply put, a domain theory consists of all knowledge (concepts) relevant to a given domain (at a given point in time), regardless of the type of knowledge, how it is encoded (formalism, substrate) or where it is located. Under this definition of a domain theory, a knowledge base is simply a subset of all the concepts in a domain theory. However, different classes of knowledge bases may be distinguished.

Concept. A *concept* denotes a cognitive unit of meaning which is sometimes referred to as "unit of knowledge". Concept descriptions are constructed from concept *properties* (features, dimensions) [30]. A concept is normally associated with a corresponding representation or encoding in a particular language or formalism. Concepts form the basis for the cognitive abilities of an intelligent agent. Without a concept, an intelligent agent or reasoner, relying on a memory containing a potentially large number of items, would be hopelessly lost. If a reasoner perceived each entity as unique, it would be overwhelmed by the enormous diversity of what it experiences, unable to remember but a fraction of what it encounters. A concept captures the notion that many objects, ideas or events are alike in some important respects, and can therefore be thought about in the same or similar ways. Once an entity has been assigned to a concept on the basis of its perceptible properties, a concept may also be used to infer some of the entity's non-perceptible attributes. Having, for example, chosen perceptible attributes like size, shape and material to decide an object is a book, it can be inferred that the object contains pages and textual information. This idea of inferability is based on the assumption that all instances of a given concept are subject to the same, or similar, underlying mechanisms (e.g., cause-effect relationships) which may or may not be completely known. Such mechanisms may be simple, in the case of books, or complex in chess positions.

Different views and models of concepts have been proposed [30]; these vary in a number of aspects, in particular, in the degree to which they are deterministic/probabilistic and intensional/extensional. In this article, concepts form the basic units from which domain theories and knowledge bases are constructed. Here, concepts include all forms of knowledge, including the three kinds of knowledge normally distinguished in epistemology: "knowledge that" (propositional, declarative knowledge), "knowledge how" (procedural knowledge) and "acquaintance knowledge" (about places, situations, cases, experiences) [1]. The knowledge that concepts represent may be tacit or explicit, it may be implemented on living tissue, electronic structures, paper or any other substrate. Critical for our discussion on domain theories and knowledge bases is that concepts are normally associated with one or more domains.

Notice that here we do not differentiate the representation languages or formalisms used to specify concrete knowledge structures (frames, rules, trees, networks, heuristics, case bases, etc.).

Domain Theory. For the purpose of this discussion, a *domain* is viewed as a formal or common sense topic, subject area, field of interest, for example, a scientific discipline (e.g., biology), a game (e.g., chess), social, cultural, economic or political topics (e.g., religious morality), common patterns of activity (e.g., taking a bath), and so on. Based on this view of a domain, we define a domain theory as follows:

Definition 2 (domain theory). *A domain theory D_i defines a set of concepts (knowledge units) that are associated with a particular domain i.*

Notice that a particular concept may belong to more than one domain theory at the same time.

In this view of a domain theory it is easy to see that most domain theories would be formed from an heterogeneous and distributed pool of knowledge "sources", including humans, documents, electronic systems, and so on. For example, the domain theory of chess would be "encoded" in books, reports of tournaments, databases, chess programs, and the minds of a large number of chess players. While many of the concepts within the domain theory of chess would be shared across many chess players, other concepts may be unique to and accessible by individual players only (or by groups of players)[4].

A domain theory is shared across a peer group. One consequence of the distributed and heterogeneous nature of most non-trivial domain theories is that they are usually associated with a particular peer group, culture, society, etc., rather than with an individual or a very small group of people. Notice, a domain theory, as it is defined here, usually includes elements that are not accessible by the entire peer group associated with it. For example, the subjective case base (acquaintance-knowledge) a particular chess master has accumulated over his career is not likely to be accessible by other chess masters (members of the peer group). Likewise, certain documents or electronic resources about chess knowledge may be accessible only to a limited group of peer members.

A domain theory is fixed or changing only very slowly. An established domain theory would normally not change radically but remain relatively stable and undergo mostly minor modifications over time. Radical changes of a domain theory would be related to changes in fundamental concepts of a domain theory. For example, nowadays in chess it rarely happens that a "standard" move in a particular game would be shown to be unsound.

A domain theory incorporates "hidden" concepts. At a given point in time, a large database holds facts and patterns that have already been explicitly reported or are known by at least one intelligent agent. However, at the same time there may be many "hidden" facts or patterns contained in the same database which have not been discovered yet. Analogously, a domain theory captures concepts which are explicitly documented or known by an intelligent agent. At the same time, a domain theory harbors concepts which are yet to be discovered. Notice that while the total number of hidden and explicit/known concepts within a domain theory may be very large (or even infinite), not every conceivable concept may be expressible under the constraints of a particular domain theory.

Knowledge Base. A *knowledge base* is constructed from the concepts of a domain theory; we define a knowledge base as a subset of a domain theory as follows:

[4] Detailed psychological studies suggest that, for example, the number of symptom-illness correspondences known by a medical specialist, or the number of board positions memorized by a chess master, appear to be in the range of 30 000 to 100 000 [23].

Definition 3 (knowledge base). *A knowledge base K_i is defined as a subset of a domain theory D_i, i.e., $K_i \subseteq D_i$.*

This means that in the extreme case a knowledge base and a domain theory could be identical. This is of course only a theoretical possibility, because for real-world domain theories, a knowledge base is normally a highly selective subset of the domain theory. In particular, a knowledge base would tend to have the following characteristics.

A knowledge base is domain-specific. As a consequence of how a knowledge base is defined, it is always defined with respect to a particular domain. Hence, a knowledge base contains only concepts from the underlying domain theory.

A knowledge base is focused, selective, goal-oriented, biased ... A knowledge base is normally not formed by a random process which selects elements from a domain theory and puts them together to make up a knowledge base. Instead, a knowledge base is either intentionally constructed or it is evolved, and as a consequence a knowledge base normally represents a focused, selective, goal-oriented, biased, subjective, etc. subset of the domain theory. When a knowledge base is designed, its construction is guided by the function it is supposed to fulfill, by other design constraints and requirements, and by the set of biases, skills, abilities, etc. of its designers. In this process particular choices are made in terms of which concepts from the underlying domain theory will be included in the knowledge base. When an intelligent agent acquires knowledge (learning, evolution) it normally does so under a set of constraints, including the goals it pursues, its prior experience, abilities, skills, the environment it operates in, and so on. A knowledge base which is thus constructed or evolved has selected (or acquired) a set of domain concepts in a very biased or "habitual" way. Notice, as an intelligent agent evolves a knowledge base, it does not only assimilate knowledge from the domain theory that is shared by other peer members, but it also creates a part of the domain theory space that is normally not accessible to other peer members of the domain.

Agent-specific knowledge bases. In our definition of a knowledge base, a book on a particular variant of the Sicilian Defence could be considered as a knowledge base in the domain of chess. Often, however, in this discussion we are concerned with knowledge bases that are tied to or integrated within a specific intelligent agent[5]. In this case, we are talking about the type of knowledge base which is highly subjective, containing domain concepts which are not shared with the domain's peer members. It is precisely the non-shared concepts in such an agent-specific knowledge base that form a kind of "inertial system" or "reference system" against which the common or shared parts of the knowledge base are viewed and interpreted. What is important to understand is that an intelligent agent has *exactly one* knowledge base for a given domain! This knowledge base may be

[5] Here we use the term "intelligent agent" to denote a uniquely identifiable entity with cognitive abilities such as reasoning, planning, hypothesizing, etc. It is irrelevant on which physical substrate such an entity is implemented or whether or not it is highly localized in physical space.

empty, if the agent knows no concept in that domain, or it may be a non-empty knowledge base consisting of shared and non-shared concepts of the domain theory. The non-shared domain concepts impose a unique, biased perspective of the agent on the domain. The fact that an agent captures part of the domain theory which is normally not shared with other agents in the domain, makes such an agent-specific knowledge base special.

Agent-specific knowledge bases are habitually incompatible. Another critical aspect of the concept of an agent-specific knowledge base is that, given a concrete problem, normally (or "habitually") only a single knowledge base would be active at a given time. This is what Koestler refers to as "habitually incompatible" matrices.

Models of Creativity. Both Boden and Koestler base their models on a corpus of domain-specific knowledge or concepts called *conceptual space* by Boden and *code* by Koestler. In our conceptualization both a code and a conceptual space are viewed as a *domain theory.*

With respects to Boden's model of creativity, domain theories are equivalent to the notion of *conceptual spaces.* They satisfy the characteristic of not being tied to an individual as well as being relatively stable over time. Indeed, a domain theory encompasses all the knowledge (or concepts) known about a domain at a given point in time. Furthermore, a domain theory represents Boden's "generative structure" [6] that contains the "possibilities" of hitherto unknown knowledge which may be discovered in the creative process (combinatorial or exploratory creativity). Essentially, these are all possible concepts within a domain theory that have not been made explicit in any form (documented) or are not known by any agent of the domain's peer group. Boden's transformational creativity is facilitated by a change or transformation of the underlying domain theory. Such a change would typically be realized by a modification or addition of concepts in a given domain theory.

In Koestler's framework of creativity the notion of a *code* is equivalent to our concept of a domain theory. Like Koestler's concept of a code, a domain theory constitutes a relatively fixed system of rules (or concepts) which governs the processes of creativity.

Unlike Koestler's model, which incorporates the notion of a *matrix*, Boden does not make a distinction between a matrix and a conceptual space. Comparing her model with that of Koestler, Boden states: "Matrices appear in my terminology as conceptual spaces, and different forms of bisociation as association, analogy, exploration, or transformation." [6]. This is where Koestler's model appears to be more differentiated. With the notion of a matrix, Koestler puts the subjective perspective of the entity that engages in creative thought in the center of his model. Indeed, the matrix notion provides this degree of individuality that appears to be associated with many creative ideas and inventions. In our model, Koestler's matrix concept is reflected in the concept of a knowledge base. A knowledge base, like a matrix in Kostler's framework, is uniquely linked to a particular reasoner or intelligent agent. Indeed, a knowledge base carries the characteristics that Koestler associates with his matrices:

1. There is exactly one knowledge base per agent for each domain.
2. A knowledge base reflects the subjective personal, prejudiced and unobjective views and patterns of thinking and behavior – i.e., a *habitual frame of thought* – that provide a unique (albeit biased) perspective of the domain. Usually, when pondering over a task or problem, only the concepts of a single knowledge base would be active. This is why Koestler calls his matrices "habitually" incompatible. This notion does not seem to be reflected in Boden's model.
3. Because each agent or reasoner incorporates a set of (partially overlapping) knowledge bases in a *highly integrated* fashion (with in a single "mind"), such an agent is equipped with the unequaled potential to discern patterns of bisociation by bringing together or superimposing multiple knowledge bases simultaneously. It is this structure that allows an agent to "see" or perceive a problem, situation or idea simultaneously from different frames of mind (knowledge bases).

Viewing Koestler's matrix as a knowledge base appears to be a more realistic model for combinatorial, exploratory and transformational creativity, because it takes into account the fact that an entity's (agent) view of the world is normally limited by the set of knowledge bases it has. One can assume that agents operating on the basis of Boden's conceptual spaces are also limited to a subset of the conceptual space, but this is not so clear in the model of Boden.

Boden argues that bisociation can be incorporated in her model. However, in the absence of a clear account of the "habitual" dimension (represented by matrices in Koestler's framework and by knowledge bases in our model) involved in bisociation, Boden's model seems less convincing.

6 Towards a Formal Definition of Bisociation

Based on above considerations we now attempt to provide a formal definition of bisociation. In our definition we employ the following symbols:

Let U denote the *universe of discourse*, which consists of all concepts.

Let $c \in U$ denote a *concept* in U.

Within the universe of discourse, a problem, idea, situation or event π is associated with the concepts $X \subset U$. Typically, in a concrete setting, a subset $P \subset X$ is used to describe and reason about π.

D_i denotes a *domain theory* which represents the total knowledge (concepts) within a domain. Notice that the union of all domain theories represents the universe of discourse: $\cup_i D_i = U$. Furthermore, $\exists i, j : D_i \cap D_j \neq \emptyset$. This means that many domain theories overlap.

R denotes a *reference system* or *intelligent agent* which possesses exactly one knowledge base (empty or non-empty) per domain theory D_i.

$K_i^R \subset D_i$ denotes the *knowledge base* with respect to the reference system or intelligent agent R and domain theory D_i. Notice, an intelligent agent R has exactly a single knowledge base K_i^R (empty or non-empty) per domain theory i. For example, the knowledge base K_{chess}^R defines the chess knowledge base an intelligent R has.

$K^R = \cup_i K_i^R$ denotes the entire *set of knowledge bases* incorporated in the reference system or intelligent agent R. K^R represents the total knowledge that R has in all the domains. For example, an intelligent agent R may possess non-empty knowledge bases for the domains of chess, biology and religious morality, and an empty knowledge base for the domain of geometry.

Definition 4 (habitually incompatible knowledge bases). *Two agent-specific knowledge bases K_i^R and K_i^R $(i \neq j)$ are said to be habitually incompatible if, at a given point in time t, there is no concept $c : c \in K_i^R \wedge c \in K_j^R$ that is active or perceived simultaneously in K_i^R and K_i^R.*

In other words, an intelligent agent usually employs a single frame of mind (knowledge base) at a given moment in time to think about a problem. One could compare this "pedestrian" way of thinking to a "sequential" mode of reasoning in which a reasoner switches between the matrices (knowledge bases) but only uses one matrix at the time.

Definition 5 (bisociation). *Let π denote a concrete problem, situation or event and let $X \subset U$ denote the concepts associated with π. Further, let K_i^R and K_j^R denote two habitually incompatible agent-specific knowledge bases $(i \neq j)$. Bisociation occurs when elements of X are active or perceived simultaneously in both K_i^R and K_j^R at a given point in time t.*

This refers to the situation where a problem is perceived simultaneously in two frames of reference or matrices of thought (Figure 1).

For example, at time t the concepts $B = \{c_1, c_2, c_3\}$ may be active or perceived simultaneously in K_i^R and K_j^R. In this case we say that the concepts in A are *bisociated.*

Definition 6 (association). *Let π denote a concrete problem, situation or event and let $X \subset U$ denote the concepts associated with π. Further, let K_i^R denote an agent-specific knowledge base. Association occurs when elements of X are active or perceived in K_i^R at time t only.*

For example, at time t the concepts $A = \{c_1, c_2, c_3\}$ may be active in K_i^R only. In this case we say that the concepts in A are *associated* (with each other).

7 Related Work

The key notion of bisociation is a knowledge structure that is defined on the concepts originating from multiple domains. Below we briefly look at some of the literature which is closely related to bisociation. This short review does not claim to be exhaustive. A more comprehensive literature review should include areas such as data and information fusion, heterogenous information networks, interchange of knowledge bases and ontologies, multi-agent systems, hybrid intelligent systems, metaphor-based reasoning (conceptual/cognitive metaphors), conceptual blending, discourse reasoning, and others.

Analogical Reasoning. Analogy is a powerful form of logical inference which allows to make assertions about an entity or concept, X, based on its similarity with another entity or concept, Y. For example, we use our knowledge about water flow to determine properties of electrical circuits. The underlying assumption of analogical reasoning is that if two entities or concepts are similar in some respects, then they are probably alike in other respects as well. Like inductive reasoning, which proceeds from the particular to the general, analogical reasoning does not guarantee the truth of the conclusion given a true premise. Despite this similarity with inductive reasoning, analogical reasoning is often viewed as a form of reasoning which is distinct from inductive reasoning. For instance, Sowa and Majumdar view analogical reasoning as a two-step reasoning process which first inductively creates a theory from a set of cases, and then deductively generates an answer to a specific question or problem on the basis of the theory [32]. In AI, analogical reasoning is often described as a representational or *analogical mapping* from a known "source" domain to a (novel) "target" domain [17].

A key element in analogy is the mechanism of selection. Not all commonalities between two concepts are equally important when we compare the concepts and make predictions based on similarities. Therefore, a central issue in analogical mapping is to determine the selection constraints that guide our assessment of similarity and dissimilarity[6]. Two broad classes of selection constraints have been investigated in AI: goal-relevance and structure-relevance. The former is used to focus analogical mapping on information that is considered critical to the problem or goal at hand. The latter is used to guide analogical mapping based on the structural commonalities between two entities or concepts.

[6] Similarity should consider the common and distinctive features of the entities under investigation. For example, let x and y denote two entities, and X and Y the sets of their characterizing features. Then the similarity, $sim(x, y)$, between x and y is a function of their common and distinctive features as follows:

$$sim(x, y) = \theta f(X \cap Y) - \alpha g(X \setminus Y) - \beta h(Y \setminus X),$$

where $f(X \cap Y)$ expresses the *similarity* based on common features in x and y, $g(X \setminus Y)$ the *dissimilarity* based on properties x has but y does not, and $h(Y \setminus X)$ the *dissimilarity* based on properties y has but x does not. θ, α and β influence how the various components affect the overall score, with θ, α, $\beta \in [0, 1]$.

Investigating the mechanisms of analogical reasoning in humans, Gentner and co-workers developed the *structure-mapping theory* of analogy [13]. The underlying assumptions in the structure-mapping theory are that (a) connected knowledge (concepts) is preferred over independent facts; this assumption is known as the systematicity principle, and (b) analogical mappings are based on structure-relevance selection constraints. The structure-mapping theory has been used to create a computational model called the *structure-mapping engine* [12]. The structure-mapping engine can find a mapping between the appropriate relations (between concepts in the considered domains) given a properly constructed representation of the domains of interest. Chalmers and co-workers [9] proposed a different approach to explain and model analogical reasoning. They view analogical reasoning as a product of a more general cognitive function called *high-level perception*. Morrison et al. interpret high-level perception and the structure-mapping theory as two aspects of analogy, rather than viewing them as mechanisms on two distinct cognitive organizational levels [27].

Human cognition is continually establishing *potential* mappings between knowledge domains or contexts. Analogical mapping occurs in a richly interconnected conceptual space in long-term memory. Attribute/category information plays a crucial role for the discovery of analogies across the conceptual spaces in long-term memory. Based on such a model of human memory, the following (simplified) analogical reasoning processes could be distinguished [14]:

1. **Retrieval:** In respond to some input case, an analogous or similar case is retrieved from long-term memory transferred to working memory.
2. **Mapping:** The two cases (the input case and the retrieved analogous case) are "aligned" in terms of their analogous features. This enables the identification of their common and distinctive properties and the inference of unknown properties of the input case based on the properties of the retrieved case.

Clearly, one of the problems of the above procedure is that mapping should already be part of the retrieval process.

Arguably, analogical reasoning is closely related to bisociative reasoning, in particular its domain-crossing conceptual space (long-term memory) bears the hallmarks of bisociation. Furthermore, the concept of "richly interconnected conceptual space in long-term memory" is very similar to the assumption in our formulation of bisociation that there needs to be an overlap of concepts in two domains to facilitate bisociation.

Bisociation is different to analogical reasoning in a number of ways. First, while analogy may be a mechanism in some forms of biosociation, bisociation is not about analogy per se. Perceiving a problem *simultaneously* from the perspective of two distinct knowledge bases, does not mean that one views the entire problem from one knowledge base and then from the other. In a sense, when bisociation occurs, a fraction of both knowledge bases becomes unified into a single knowledge base in the context of the problem at hand. Also, when one considers some of the examples Koestler describes in the context of humor, it is clear that some of these do no rely on the concept of analogy [20]. The

Eureka act described in Figure 2 does not seem to be an example of analogical reasoning. Second, in contrast to bisociation, analogical reasoning seems to suggest a similar (analogous) structure of the long-term memory entity that is retrieved and the input case prompting the retrieval. Bisociation is more akin to Minsky's concept of *knowledge lines* [26], which are a kind of "scaffold" attached to the "mental agencies" (facts, concepts, routines, habits, associations) that were active in creating a certain idea or solving a particular problem in the past. The knowledge lines later work as a way to re-activate the same structures in the context of a new problem. Bisociation could be view in similar terms, except that bisociation explicitly models knowledge lines that cut across knowledge bases embodying domain-specific mental agencies. Thus, when bisociation occurs, mental agencies usually (habitually) active in the context of a specific domain, are activated together with mental agencies usually active in another domain. There are also other perhaps more subtle difference between analogical reasoning and bisociation that are not discussed here.

Swanson's Theory. *Swanson's theory* [33], also known as to as "Swanson linking", is based on the assumption that new knowledge and insight may be discovered by connecting knowledge sources which are thought to be previously unrelated. By "unrelated" Swanson originally meant that there is no co-authorship, no citation and no officially stated relationship among the considered knowledge sources. Swanson coined the phrase "undiscovered public knowledge" to refer to published knowledge that is effectively hidden in disjoint topical domains because researchers working in different domains are unaware of each others' work and scientific publications. He demonstrated his ideas by discovering new relationships in the context of biology and other areas. The field of *literature-related discovery* has emerged from Swanson's work. It aims at discovering new and interesting knowledge by associating two or more concepts described in the literature that have not been linked before [21]. *Conceptual biology* is another line of research in this direction – here the idea is to complement empirical biology by generating testable and falsifiable hypotheses from digital biological information using data mining, text mining and other techniques [4,28]. The methodologies from literature-related discovery and Swanson's theory have already been incorporated in conceptual biology. In combination with systems biology, automatic hypothesis generation is being investigated to facilitate automated modeling and simulation of biological systems [2].

The work by Swanson, literature-related discovery and conceptual biology are related to bisociative information exploration in their attempt to discover information across normally disjoint information spaces. Perhaps one aspect that is strikingly different between the Swanson's approach and bisociation is the notion of unrelatedness and topical disjointedness in Swanson. This assumption separates conceptual spaces on the basis of the originators of knowledge. In our definition of bisociation we do not make this distinction. Nevertheless, the Swanson's theory, while being currently focused on literature as its main source of knowledge, is interesting in the context of bisociation. Further investigations are needed to determine how bisociation and Swanson's approach could complement each other.

Computational Creativity in Science. Computational creativity [7] in art, music and poetry has been around for some time. A recent development is computational creativity applied to the fields of science and engineering. For example, the aim of the Symposium on Computational Approaches to Creativity in Science[7] (Stanford, US, 2008) was to explore (among other things) (a) the role creativity plays in various scientific areas and how ICT-based tools could contribute to scientific tasks and processes, (b) the nature of creativity in search through a problem space and the representation of the search space and the problem description, (c) the role background knowledge plays in aiding and possibly interfering with creative processes in science, and (d) the interactions among scientists that increase creativity and how computational tools could support these interactions.

There was a wide range of contributions at the Symposium which covered themes such as the design of discovery systems; inter-disciplinary science and communication; abstraction, analogy, classification; spatial transformations and comparisons; conceptual simulation; strategies for searching a problem space; the question of how discovery and creativity differs; knowledge acquisition/refinement approaches and systems; knowledge-based and knowledge management systems, and "knowledge trading zones"; and explanations, models and mechanisms of creative cognition.

Computational creativity in science is a fruitful area and also an area in which large amounts of data, information and knowledge are readily available in computer-readable format. Given the specialization of science on the one hand, and the need for inter-disciplinary science to tackle highly challenging problems on the other hand, it seems that computational creativity in science offers a formidable platform to further investigate biosociative information exploration.

8 Discussion and Conclusion

Computational creativity, in particular computational creativity in non-art applications, is a relatively new computing paradigm [15,8]. For example, computational creativity in science and engineering means that a scientist or an engineer cedes part of her control over the discovery or design process to a computer system that operates with a degree of autonomy, and contributes to the results. In this article we have outlined a rationale or framework for computational creativity based on Koestler's concept of bisociation [20]. The framework presented here facilitates bisociation by "connecting" the knowledge bases of an intelligent agent in the context of a concrete problem, situation or event (Figure 1).

Koestler's treatise and other accounts of bisociation often illustrate bisociation by either bisociating two common or general knowledge domains, or by bisociating one more specialized subject matter domain with a commonsense knowledge domain. For example, the Eureka act (Section 4) bisociates the commonsense domain of taking a bath with the domain of geometry. If we want to reflect this kind of *structure* in a computational creativity solution for non-art

[7] `http://cll.stanford.edu/symposia/creativity/`

applications, this would mean that we need to develop a knowledge base reflecting the application domain *and* a knowledge base containing commonsense or general knowledge. A commonsense knowledge base contains the knowledge that most people possess and use to make inferences about the ordinary world [24]. Information in a commonsense knowledge base includes things like ontologies of classes and individuals; properties, functions, locations and uses of objects; locations, duration, preconditions and effects of events; human goals and needs; and so on. A commonsense knowledge base must be able to facilitate spatial, event and temporal reasoning. Tasks that require a commonsense knowledge base are considered "AI-complete" or "AI-hard", meaning that it would require a computer to be as intelligent as people to solve the task.

Another approach to bisociation-based computational creativity would require the bisociating of knowledge bases from different non-commonsense domains, for example, biology and quantum mechanics. Here we have a two-fold challenge:

First, we need to somehow provide some form of interoperability of the involved knowledge bases; this is a topic of active research [10]. Our approach to integrating the concepts from different domains is by creating a *heterogeneous information networks* (called BisoNet in this case) from underlying information sources. The topic of mining of heterogeneous information networks and linked data has become an area of very active research in recent years [18,5].

Second, when the content of bisociated concepts are presented to the user, there may be a considerable problem for the user to recognize potentially useful information from the other domain. For example, a life scientist investigating a detailed mechanism in relation to gene regulation and nuclear receptors may be presented with a scientific article in the field of quantum theory that discusses metric tensors in the context of entanglement entropy. Even if the bisociated article is potentially useful, the life scientist may not be able to "see" the usefulness because he does not have the necessary domain knowledge in field of quantum mechanics.

Another issue – that is shared with all approaches to computational creativity – of the presented framework concerns the assessment of whether or not a discovered item, relationship or bisociation is indeed creative in the sense of being *new*, *surprising* and *valuable* (see Definition 1). This problem is analogous to the issue of determining the degree of interestingness or usefulness[8] of patterns discovered by means of data mining or machine learning techniques [25]. Sosa and Gero [31] argue that creativity is a social construct based on individual-generative and group-evaluative processes. This suggests that the assessment of creativeness needs to incorporate social aspects that transcend the within-individual cognitive dimension. This points to a rather complex challenge for computational creativity and is something that future studies of computational bisociation need to take on board.

[8] In addition to these, the discovered patterns are usually also required to be non-trivial, valid, novel and comprehensible. Depending on the technique used and the application area, an automated assessment of these additional dimensions may also pose a considerable challenge.

With the increasing power of ICT and the growing amounts of data, information and knowledge sources, there is a new wave of efforts aiming to construct computing solutions that exhibit creative behavior in the context of challenging applications such as science and engineering [8]. This article presents a framework for computational creativity based on the concept of bisociation [20]. As a pioneering effort in this field, the BISON project[9] has been exploring bisociation networks for creative information discovery. This article presents some of the rationale, ideas and concepts we have explored in an effort to formally define the concept of bisociation and bisociative information exploration. Clearly, more work is needed to develop a more comprehensive formal understanding of bisociation and how this concept can be used to create novel ICT methods and tools.

References

1. Adams, M.P.: Empirical evidence and the knowledge-that/knowledge-how distinction. Synthese 170, 97–114 (2009)
2. Ananiadou, S., Kell, D.B., Tsujii, J.-I.: Text mining and its potential applications in systems biology. Trends in biotechnology 24(12), 571–579 (2006)
3. Barron, F.: Putting creativity to work. In: Sternberg, R.J. (ed.) The Nature of Creativity, pp. 76–98. Cambridge University Press, Cambridge (1988)
4. Bekhuis, T.: Conceptual biology, hypothesis discovery, and text mining: Swanson's legacy. Biomedical Digital Libraries 3, 2 (January 2006)
5. Bizer, C., Heath, T., Berners-Lee, T.: Linked data – the story so far. International Journal on Semantic Web and Information Systems (IJSWIS) 5(3), 1–22 (2009)
6. Boden, M.A.: Précis of the creative mind: Myths and mechanisms. Behavioural and Brain Sciences 17, 519–570 (1994)
7. Boden, M.A.: Computer models of creativity. In: Sternberg, R.J. (ed.) Handbook of Creativity, pp. 351–372. Cambridge University Press, Cambridge (1999)
8. Bridewell, W., Langley, P.: Symposium on computational aproaches to creativity in science. Final report for NSF grant IIS-0819656. Technical report, Institute for the Study of Learning and Expertise (2008), `http://cll.stanford.edu/symposia/creativity/text/scacs08.report.pdf`
9. Chalmers, D.J., French, R.M., Hofstadter, D.R.: High-level perception, representation, and analogy: A critique of artificial intelligence methodology. Journal of Experimental and Theoretical Artificial Intelligence 4, 185–211 (1992)
10. Chaudhri, V.K., Farquhar, A., Fikes, R., Karp, P.D., Rice, J.: OKBC: A programmatic foundation for knowledge base interoperability. In: Proceedings of the 15th National Conference on Artificial Intelligence (AAAI 1998), Madison, Wisconsin, USA, pp. 600–607 (1998)
11. Einstein, A.: Über einen die Erzeugung und Verwandlung des Lichtes betreffenden heuristischen Gesichtspunkt. Annalen der Physik 17, 132–148 (1905)

[9] Website of the BISON (Bisociation Networks for Creative Information Discovery) project: `http://www.bisonet.eu`.

12. Falkenhainer, B., Forbus, K.D., Gentner, D.: The structure-mapping engine. Artificial Intelligence 41, 1–63 (1989)
13. Gentner, D.: Structure-mapping: A theoretical framework for analogy. Cognitive Science 7(2), 155–170 (1983)
14. Gentner, D., Colhoun, J.: Analogical process of human thinking and learning. In: Glatzeder, B.M., Goel, V., von Muller, A. (eds.) Towards a Theory of Thinking. On Thinking, pp. 35–48. Springer, Heidelberg (2010)
15. Gero, J.S.: Computational models of innovative and creative design processes. Technological Forecasting and Social Change 64(2-3), 183–196 (2000)
16. Good, J.: A five-year plan for automatic chess. Machine Intelligence 2, 110–115 (1968)
17. Hall, R.P.: Computational approaches to analogical reasoning: A comparative analysis. Artificial Intelligence 39, 39–120 (1989)
18. Han, J.: Mining Heterogeneous Information Networks by Exploring the Power of Links. In: Gama, J., Costa, V.S., Jorge, A.M., Brazdil, P.B. (eds.) DS 2009. LNCS (LNAI), vol. 5808, pp. 13–30. Springer, Heidelberg (2009)
19. Higgins, J.M.: 101 Creative problem solving techniques. New Management Publishing Company (1994)
20. Koestler, A.: The act of creation. Penguin Books, New York (1964)
21. Kostoff, R.N.: Literature-Related Discovery (LRD): Introduction and background. Technological Forecasting and Social Change 75(2), 165–185 (2008)
22. Kumar, M.: Quantum: Einstein, Bohr and the great debate about the nature of reality. Icon Books Ltd. (2008)
23. Kurzweil, R.: The age of intelligent machines. Massachusetts Institute of Technology (1990)
24. Lenat, D.B., Guha, R.V., Pittman, K., Pratt, D., Shepherd, M.: Cyc: Toward programs with common sense. Communications of the ACM 33, 30–49 (1990)
25. McGarry, K.: A survey of interestingness measures for knowledge discovery. The Knowledge Engineering Review 20(1), 39–61 (2005)
26. Minsky, M.: K-lines: A theory of memory. Cognitive Science 4, 117–133 (1980)
27. Morrison, C.T., Dietrich, E.: Structure-mapping vs. high-level perception: The mistaken fight over the explanation of analogy. In: Proceedings of the 17th Annual Conference of the Cognitive Science Society, pp. 678–682 (1995)
28. Natarajan, J., Berrar, D., Hack, C.J., Dubitzky, W.: Knowledge discovery in biology and biotechnology texts: A review of techniques. Critical Reviews in Biotechnology 25(1/2), 31–52 (2005)
29. Schaeffer, J., Culberson, J., Treloar, N., Knight, B., Lu, P., Szafron, D.: A world championship caliber checkers program. Artificial Intelligence 53(2-3), 273–290–115 (1992)
30. Smith, E.E., Medin, D.L.: Categories and concepts. Harvard University Press, Cambridge (1981)
31. Sosa, R., Gero, J.S.: A computational framework for the study of creativity and innovation in design: Effects of social ties. In: Gero, J.S. (ed.) Design Computing and Cognition 2004, pp. 499–517. Kluwer Academic Publishers, Dordrecht (2004)
32. Sowa, J., Majumdar, A.K.: Analogical Reasoning. In: Ganter, B., de Moor, A., Lex, W. (eds.) ICCS 2003. LNCS (LNAI), vol. 2746, pp. 16–36. Springer, Heidelberg (2003)
33. Swanson, D.R.: Fish oil, Raynaud's syndrome, and undiscovered public knowledge. Perspectives in Biology and Medicine 30(1), 7–18 (1986)

From Information Networks to Bisociative Information Networks

Tobias Kötter and Michael R. Berthold

Nycomed-Chair for Bioinformatics and Information Mining,
University of Konstanz, 78484 Konstanz, Germany
Tobias.Koetter@uni-Konstanz.de

Abstract. The integration of heterogeneous data from various domains without the need for prefiltering prepares the ground for bisociative knowledge discoveries where attempts are made to find unexpected relations across seemingly unrelated domains. Information networks, due to their flexible data structure, lend themselves perfectly to the integration of these heterogeneous data sources. This chapter provides an overview of different types of information networks and categorizes them by identifying several key properties of information units and relations which reflect the expressiveness and thus ability of an information network to model heterogeneous data from diverse domains. The chapter progresses by describing a new type of information network known as bisociative information networks. This kind of network combines the key properties of existing networks in order to provide the foundation for bisociative knowledge discoveries. Finally based on this data structure three different patterns are described that fulfill the requirements of a bisociation by connecting concepts from seemingly unrelated domains.

1 Introduction

Applications of bisociative creative information exploration derive their potential to produce creative discoveries, insight and solutions from exploring bisociations across large volumes of information originating from two or more domain theories. To facilitate such applications it is necessary to integrate these domain theories (or associated knowledge bases) in such a way that the integrated pool can be processed coherently. Integration of such data is a considerable challenge not only because of the data volumes, but also because of the semantic (ontologies of different domains) and syntactic (data and knowledge formats) heterogeneity involved.

An obvious approach to integrate these large volumes of information from various domains with varying quality is a flexible representation in terms of an information network. A number of different types of information networks have been proposed in the last few years [38] particularly in the area of biomedical domains. This area of research is known for its diverse information sources that need to be considered, for example, in the drug discovery process [12]. The integrated sources

M.R. Berthold (Ed.): Bisociative Knowledge Discovery, LNAI 7250, pp. 33–50, 2012.

range from experimental data, such as gene expression results, through to highly curated ontologies, such as the ontology of Medical Subject Headings[1].

Information networks are commonly composed of information units representing physical objects as well as immaterial objects such as ideas or events and relations representing semantic or solely correlational connections between information units. They are almost always based on a graph structure with vertices and edges, where vertices represent units of information, e.g. genes, proteins or diseases, and the relations between these units of information are usually represented by edges. In some information networks relations are represented by vertices as well, and therefore apply a bi-partite graph representation. This type of representation has the added advantage that relations between more than two information units can be easily supported. Furthermore an edge can be directed or undirected depending on the relationship it represents. Most networks also allow additional attributes or properties to be attached to vertices and edges, such as a vertex type, e.g. gene or protein, describing the nature of the information unit. Such information networks that connect multi-typed vertices are also known as heterogeneous information networks [28].

In order to integrate not only structured and well annotated repositories but also other types of information such as experimental data or results from text mining, some information networks support weighted edges. Therefore interactions in biological systems, which can be noisy and erroneous, are often modeled by Bayesian networks [22,24,31]. In these approaches the edge weight represents the probability of the existence of the connection. However, the edge weight of networks used by information retrieval techniques, such as knowledge or Hopfield networks [14], represents the relatedness of terms. Usually the weights in these approaches are computed only once. In contrast to these approaches, Belew enables each user of an adaptive information retrieval (AIR) model [6] to adapt the weights according to their relevance feedback. The disadvantage of this approach is that over time the network will be strongly biased by the opinions of the majority of the users. Another weighted-graph method constructs a weighted graph based on information extracted from available databases [49]. In doing so the edge weight represents the quality of the relation and is based on three factors: edge reliability, relevance and rarity. They assume that each edge type has a natural inverse, such as "coded by" and "is referred by". Similarly, there is one inverse edge for each edge, leading to an undirected graph with directed edge labels.

Once the data is represented in an information network this well-defined structure can be used to discover patterns of interest, extract network summarizations or abstractions and develop tools for the visual exploration of the underlying relations. A general analysis of the structure of complex networks stemming from real-world applications has been conducted by Albert and Barabasi [2]. They have discovered that these networks often share a number of common properties such as the small-world property, clustering coefficient or degree distribution. A survey on link mining has been conducted by Getoor and Diehl [27].

[1] http://www.nlm.nih.gov/mesh/meshhome.html

They classified the link mining task into three categories: object-related tasks, link-related tasks and graph-related tasks.

Network summarizations representing different levels of detail can be visualized to gain insight into the structure of the integrated data. A general introduction to network analysis can be found in [11]. An overview of existing graph clustering methods can be found in [48] and a review of graph visualization tools for biological networks can be found in [45]. The paper compares the functionality, limitation and specific strength of these tools.

Approaches from the semantic Web community include formalization of general semantic networks where the most popular variants have resulted in the RDF standard [40] and for formalism of topic maps [23]. Both techniques imply the construction of various formalizations in the form of different graph constructs. A highly complex example is the formalization of topic maps via *shifted hypergraphs* [3]. In this approach a hypergraph model for topic maps is defined in which the standard hypergraph is extended to a multi-level hypergraph via a shift function. RDF models were proposed in the form of different graph structures: graph [29], bipartite graph [30] and hypergraph [42]. Standard graphs allow the modeling of relations between two nodes, whereas bipartite graphs and hypergraphs permit the integration of relations among any number of members.

In order to visually analyze large networks with several million vertices and many more edges, visualization has to focus on a sub-graph or at least summarize the network to match the user's interest or provide some kind of overview of existing concepts. Various visualization and graph summarization techniques have been developed to address this problem. Examples can be seen in the generalized fisheye views [25], the interactive navigation through different levels of abstraction [1], the extraction of sub-graphs that contain most of the relevant information by querying [21] or by spreading-activation [18]. Other approaches summarize the graph by clustering or pruning it based on the topology [57] or additional information such as a given ontology [50].

The next section describes different types of information networks and characterizes them based on the features they support, which are relevant to the integration of heterogeneous data types. We subsequently introduce bisociative information networks, which have been tailored to support the integration of heterogeneous data sources. Before we move on to the conclusion, we discuss patterns of bisociation in this type of network that support creative thinking by connecting seemingly unrelated domains.

2 Different Categories of Information Network

In order to differentiate among information networks, distinctions can be made between different properties of information units and relations. These properties are, of course, not exclusive. The properties of an information network define its expressiveness and thus its ability to model data of a diverse nature, e.g. ontologies or experimental data.

2.1 Properties of Information Units

The basic information unit does not posses any additional semantical information. However, they will at least include a label attached to them in order to identify the object or concept they represent. Additional properties are the following:

Attributed. units of information can have additional attributes attached to them. An attribute might be a link to the original data it stems from, or a translation of a user-readable label. These attributes might be considered while reasoning or analyzing the network but do not carry general semantic information, such as the following properties.

Typed. information units carry an additional label that is used to distinguish between different semantics of information units, e.g. gene or protein. These types can additionally be organized in a hierarchy or an ontology.

Hierarchical. information units represent a sub-graph composed of any number of information units and relations that can be used to condense parts of the network or to represent more complex concepts such as cellular processes.

2.2 Properties of Relations

The basic connection between information units represents a relationship between the corresponding members. They are not required to carry a label.

Attributed. relations have attributes attached to them and also fall into this category. Similar to attributed information units, they can be considered during the reasoning process, but do not carry a general semantic information.

Typed. relations are similar to typed information units and can carry a label identifying their type. This attribute is used to distinguish between different semantics of relations such as activates or encodes. These types, as well as typed information units, can be organized in a hierarchy or an ontology.

Weighted. relations carry a special type of label - the weight - which represents the strength of a relation, e.g. a number reflecting the probability or strength of a correlation or some other measure of reliability that allows the integration of facts and pieces of evidence.

Directed. relations can be used to explicitly model relationships that are only valid in one direction, such as parent child dependency in a hierarchy.

Multi-relation. relations are generally represented as edges supporting only two members. Topic maps (see Section 3.3) in contrast represent relations as multi edges supporting any number of members. This allows a more flexible modeling of relationships with any number of members, e.g. co-expressed genes of an experiment or co-authors of a paper. Furthermore connections among relations themselves can be represented. Note that it is complicated to combine this property with the directed property mentioned above. Additional information would need to be provided, such as an embedding graph to identify sources and targets in a relation with more than two members.

3 Prominent Types of Information Networks

This section describes prominent types of information networks and characterizes them based on the previously discussed properties (see section 2) they support.

3.1 Ontologies

Ontologies are based on typed and directed relations using a controlled vocabulary for information units and relations dedicated to a certain domain. The creation of the curated vocabulary leads in general to a manual or semi-automatic creation of an ontology, requiring a comprehensive knowledge of the area to be described.

Figure 1 depicts a simple ontology where information units are represented as nodes and relations are represented as labeled arrows.

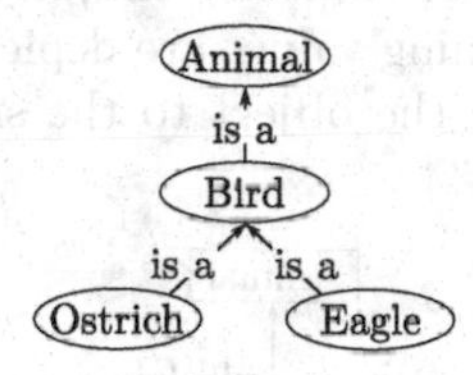

Fig. 1. Example of an ontology

In the area of life sciences particularly, many ontologies have been developed to share data from diverse research areas such as chemistry, biology or pharmacokinetics. One of the probably best known and most integrated ontologies in the biological field is the Gene Ontology (GO) [17]. The GO consists of three main ontologies describing the molecular function, biological process and cellular component of genes.

An attempt to integrate diverse ontologies has been made by the Open Biomedical Ontologies (OBO) consortium [52]. They have created a file exchange format and over 60 ontologies for different domains defining a general vocabulary that can be used by other systems.

A classification of biomedical ontologies has been completed by Bodenreider [10]. He classified these ontologies into three major categories: knowledge management; data integration, exchange and semantic interoperability; decision support and reasoning.

An ontology-based data integration platform is described in [33]. The authors describe a system that extends the existing text-mining framework ONDEX. ONDEX uses a core set of ontologies, which are aligned by several automated methods to integrate biological databases. The existing system is extended to support not only the alignment and integration of texts but heterogeneous data sources. The data is represented as a graph with attributed edges.

Tzitzikas et al. [56] describe a system that is based on the hierarchical integration of ontologies from different data sources. The system uses a mediator ontology, which bridges the heterogeneity of the different data source ontologies.

3.2 Semantic Networks

Semantic networks use typed relations to model the semantic of the integrated information units and their relations. Information units in semantic networks, in contrast to ontologies, are not represented by a curated vocabulary but rather described by attaching any number of attributes to them whose semantic is defined by the type of the relation.

Most of the semantic networks rely on Semantic Web [8] technologies such as the Resource Description Framework (RDF) [40], RDF Vocabulary Description Language (RDF Schema) and the Web Ontology Language (OWL) defined by the W3C consortium[2]. RDF is a knowledge representation and storage framework that uses triples. A triple consists of a subject, predicate and object. The subject and object are information units that are connected by a directed relation defined by the predicate.

In Figure 2 subjects and objects that are uniquely identifiable are depicted in ellipses, whereas objects containing values are depicted in boxes. Predicates are shown as arrows pointing from the object to the subject with the type of the relation as an annotation.

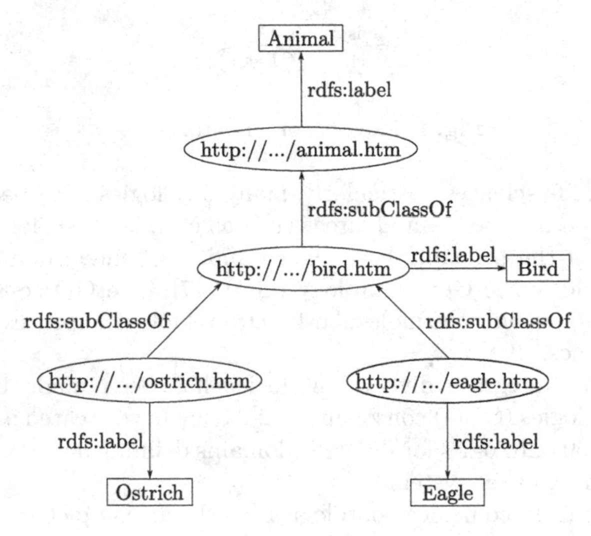

Fig. 2. Graph representation of a Semantic Web

The RDF Schema defines a core vocabulary that can be used to describe properties and classes. These properties and classes can be used to describe the members of a triple. OWL extends the RDF Schema by providing a set of additional standard terms to describe properties and classes in more detail such as relations between classes. It also defines the behavior of properties, e.g. symmetry or transitivity. OWL as well as the RDF Schema extend RDF by providing the means to model the semantics of the integrated data therefore enabling machines to make sense of the data. They are both described using the RDF.

[2] `http://www.w3.org/2001/sw/`

Bales and Johnson [5] analyzed large semantic networks created from 1998-2005 that involve both a graph theoretic perspective and semantic information. The results indicate that networks derived from natural language share common topological properties, such as scale-free and small-world characteristics.

Chen et al. [13] provide an introduction to semantic networks and semantic graph mining. In four case studies, they demonstrate the usage of semantic web technologies to analyze disease-causal genes, GO category cross-talks, drug efficacy and herb-drug interactions.

Belleau et al.[7] propose the Bio2RDF project to integrate data from different biological sources. Bio2RDF is used to integrate data from more than twenty different public bioinformatic sources by converting them into the RDF format.

YeastHub [15] another RDF-based data integration approach likewise integrates the data from heterogeneous sources into a RDF-based data warehouse. In addition they propose a standard RDF format for tabular data integration. The format can be used to convert any data table into a standardized RDF format.

A loosely coupled integration of semantic networks is proposed by Smith et al. [51] in the form of the LinkHub system. The system consists of smaller networks that can be connected by sharing a common hub. Thus independently maintained networks can be connected to the whole system by connecting them to one of the already integrated sub networks.

Biozon [9] combines the flexible graph structure with an ontology for vertex and edge types similar to the semantic web approach. This combined approach allows a more detailed description of a biological entity by either imposing more constraints on its nature in the hierarchy or on the structure of its relations to other entities in the graph. All vertices within Biozon are direct analogs to physical entities and sets of entities. Proteins, for example, are identified by their sequence of amino acids. In contrast to pure semantic networks Biozon allows any number of attributes to be attached to information units as well as to relations.

3.3 Topic Maps

Topic maps [23,47] use typed information units and relations. Furthermore topic maps support the modeling of multi relations with any number of members. The semantic of a topic is described by attaching any number of attributes to it.

Figure 3 depicts the three major elements of a topic map: topics (ellipses), associations (solid lines) and occurrences (boxes). Association and occurrence types are connected by the dashed lines whereas occurrences are connected by the dotted line.

A topic can generally be anything, for example a person, a concept or an idea. Topics can be assigned zero or more topic types, which are, in turn, defined as topics describing the semantics of the topic such as gene or protein.

Relations between any number of topics are represented by so-called associations. Associations are assigned a type that describes the association in more detail. Members of associations play a certain role defined by the association role. As with topic and occurrence types, association types and association roles

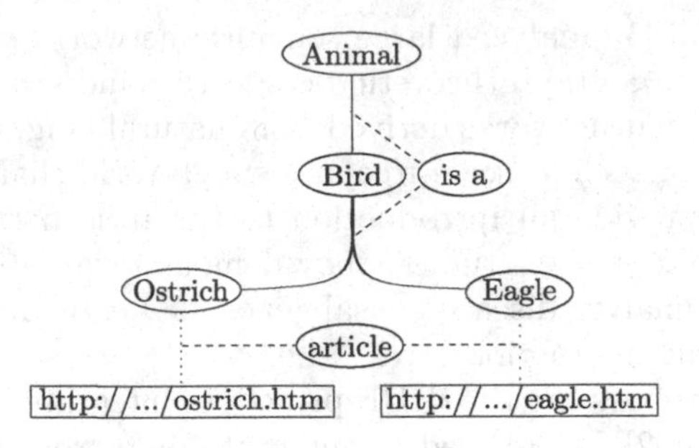

Fig. 3. Example of a topic map

are defined as topics themselves. In order to attach attributes to an association it needs to be converted into a topic by the act of reification.

Information resources that represent a topic or describe it in more detail are linked to topics by so-called occurrences. Occurrences are not generally stored in the topic map itself but are referenced using mechanisms supported by the system, e.g. Uniform Resource Identifiers (URI). Occurrences can have any number of different types, so-called occurrence types, that describe their semantics. These types are also defined as topics. Topic maps are self-documenting due to the fact that virtually everything in topic maps is a topic in the map itself, forming the ontology of the used topics and relation types.

An example of a topic-map-like data integration approach is PathSys [4]. In PathSys a relation is also represented as a vertex. This approach models relationships between relations themselves. To distinguish between information units and relations they introduce vertex types. Besides primary vertices representing information units and connector vertices representing relationships, they also introduce graph vertices. By introducing graph vertices, PathSys combines the multi relation property of topic maps with the hierarchical information unit property allowing the sub-graph representation to describe more complex objects such as protein complexes or cellular processes.

3.4 Weighted Networks

In most weighted networks the edge weight represents the strength of a relation such as reliability or probability. Weighted networks often exhibit additional properties such as types in order to be more expressive by modeling the semantic of the integrated data sources. They generally only support relationships with two members represented by the edges of the graph.

Figure 4 depicts a weighted network modeling the probability of a bird to be either a bird of prey or a flightless bird.

Probabilistic Weights. Probabilistic networks model the probability of the existence of a relationship. They are mostly used in the biological field to model interaction networks, e.g. gene-gene or protein-protein interaction networks. In order to model the probability of the relations the networks often depend

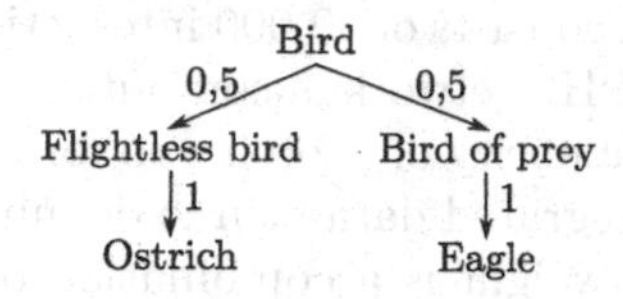

Fig. 4. Example of a weighted network

on a specific network structure or weight distribution. Bayesian networks, for example, depend on a directed acyclic graph, whose vertices model the random variables an its relations indicate their conditional dependencies [46].

Franke et al. [24] use three steps to fuse the information from the GO with microarray co-expression results and protein-protein interaction data using naive Bayesian networks. The resulting network called Genenetwork can be used to detect genes that are related to a disease based on genetic mutation.

Li et al. [41] use a two-layered approach to integrate gene relations from heterogeneous data sources. The first layer creates a fully connected Bayesian network for each integrated source, which represents the gene functional relations. The second layer combines these relations from the different data sources into one integrated network using a naive Bayesian method.

Jansen et al. [31] likewise propose a combination of naive Bayesian networks and fully connected Bayesian networks to create a protein-protein interaction network. They use the fully connected Bayesian networks to integrate experimental interaction data and naive Bayesian networks to incorporate other genomic features such as the biological process from the GO. To combine all results they use a naive Bayesian network as well.

In [55], Troyanskaya et al. introduce MAGIC (Multisource Association of Genes by Integration of Clusters). For each integrated data source, MAGIC creates a gene-gene relationship matrix to predict the functional relationship of two given genes. The matrices are generated from diverse high-throughput techniques such as gene expression microarrays. These gene-gene relationship matrices are weighted by the confidence in the integrated source and combined into a single matrix. This approach allows genes to be members of more than one group, which subsequently allows fuzzy clustering.

Heuristic Weights. Heuristic weights are mostly used to model the reliability or relevance of a given relation, thus allowing the integration of well-curated sources such as ontologies and pieces of evidence such as noisy experimental data in a single network.

In order to integrate data from diverse biological sources for protein function prediction, Chua et al. [16] propose Integrated Weighted Averaging (IWA). This combines local prediction methods with a global weighting strategy. Each data source is transformed into an undirected graph with proteins as vertices and relationships between proteins as edges. Each source graph has a score reflecting its reliability. Finally, all source graphs are combined in a single graph using IWA.

Kiemer et al. [32] use a weighted network to integrate yeast protein information from different data sources forming a protein-protein interaction network called WI-PHI. The network consists of 50,000 interactions from all data sources. The edge weight of the WI-PHI network is computed using the socio-affinity index [26], quantifying the propensity of proteins to form partnerships, multiplied by a weight constant per integrated data source defining its accuracy.

In Biomine [49] the edge weight is a combination of three different weights: reliability, relevance and rarity. Reliability reflects the reliability of the source the edge stems from. By changing the relevance of different node or edge types, e.g. proteins, genes, a user can focus on the types he or she is most interested in. Finally rarity is computed using the degree of the incident vertices. Edges that connect vertices with a low degree have a higher rarity score than edges that connect vertices with a high degree. Vertices and edges have a type assigned describing their nature. Each edge has its inverse edge with a natural inverse type such as "coded by" and "is referred by". Thus forming a weighted undirected graph with directed edge types.

In the next section we describe bisociative information networks that combine the properties of the existing network types in order to support the integration of heterogeneous data sources.

4 BisoNets: Bisociative Information Networks

Bisociative information networks (BisoNets) provide the flexibility to integrate relations from semantically meaningful information as well as loosely coupled information fragments with any number of members by adopting a weighted k-partite graph structure (see Figure 5).

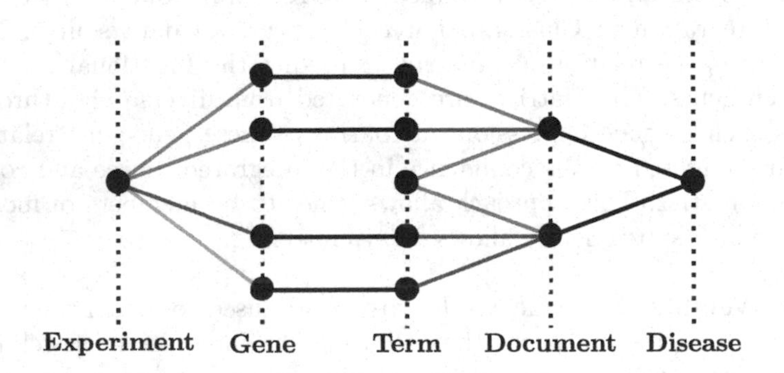

Fig. 5. Example of a 5-partite BisoNet

Vertices in BisoNets represent arbitrary units of information, e.g., a gene, protein, specific molecule, index term, or document, or abstract concepts such as ideas, acts or events. Vertices of the same type are grouped into vertex partitions such as documents, authors, genes or experiments. Since a vertex can play diverse roles it can be assigned to several partitions.

Depending on a certain view, the vertices of a partition can act as relations or information units. Let us consider a document author network to illustrate this concept. In one view the documents can describe the relationship between co-authors. Whereas in another view the authors describe the relationship between documents that have been written by the same authors. Thus the role of a vertex partition depends on the current view on the data.

Connections between vertices are represented by edges. An edge can only exist between vertices of diverse partitions; this leads to the k-partite graph structure. Hence a relation between two information units (e.g., authors) is described by a third information unit (e.g., document). A BisoNet therefore consists of at least two partitions, the first partition representing the information units and the second partition describing the relations between the information units.

The certainty of a connection is represented by the weight of the edge. A stronger weight represents a higher certainty in the existence of the connection. Thus, a connection derived from a reliable data source (e.g., a manually curated ontology) is assigned a stronger weight than a connection derived from an automated method (e.g., text mining method).

BisoNets model the main characteristics of the integrated information repositories without storing all the detailed data from which these characteristics are derived. By focusing on the concepts and their relations alone, BisoNets therefore allow very large amounts of data to be integrated.

Definition 1 (BisoNet). *A BisoNet* $B = (V_1, ..., V_k, E, \lambda, \omega)$ *is an attributed graph, where* $V = \bigcup_{i \leq k} V_i$ *represents the union of all vertex partitions and* $k \geq 2$ *denotes the number of existing partitions. Every vertex* $v \in V$ *represents a unit of information and can be a member of multiple partitions.*

The set of edges $E = \{\{u, v\} : u \in V_i; v \in V_j; j \neq i\}$ *connects vertices from two different vertex partitions, whereas an edge* $e = \{u, v\} \in E$ *represents a connection between the two vertices* $u \in V_i$ *and* $v \in V_j$ *where* $i \neq j$ *and* $2 \leq i, j \leq k$.

The function $\lambda : V \rightarrow \Sigma^*$ *assigns each vertex* $v \in V$ *an unique label from* Σ^**. This allows for the identification of a vertex by its unique label.*

The certainty of a relation is represented by the weight of an edge $e \in E$*, which is assigned by the function* $\omega : E \rightarrow [0, 1]$ *and where a weight of 1 represents the highest certainty.*

4.1 Summary

Table 1 compares the prominent types of information networks from section 3 with BisoNets based on the properties they support. The table shows that most of the networks support typed relations whereas topic maps and BisoNets also support typed information units. The types enable us to distinguish between different types of information units and relations, leading to to a better understanding of the integrated data. In addition the type information allows semantical information to be processed by a computer system. But the usage of type information requires detailed knowledge about the information that should be integrated into the network. The creation of a suitable type collection that allows the integration of data from diverse sources is thus an elaborated task which

Table 1. Properties matrix of prominent types of information network in conjunction with BisoNets (A=Attributed, T=Typed, H=Hierarchical, W=Weighted, D=Directed and M=Multi relation)

	Information Units			Relations				
	A	T	H	A	T	W	D	M
Ontologies					X		X	
Semantic Networks	X				X		X	
Topic Maps	X	X		X	X			X
Weighted Networks						X		
BisoNets	X	X	X	X	X	X	X	X

often has to be done manually. Moreover, not all data sources do possess the required semantical information to assign the right type and therefore manual annotations of the integrated information units and relations might be required. If information units and relation types are abandoned, the integration of data from heterogeneous sources is much easier but it might make the comprehension of the integrated data more difficult. As a result, BisoNets support typed information units and relations and allow their usage if the integrated data sources provide this information, however they are not mandatory. In contrast to topic maps, BisoNets also support weighted relations, thus allowing not only the integration of facts but also pieces of evidence. BisoNets combine the properties of the existing network types in order to provide a well-defined and powerful data structure that provides the flexibility to integrate relations from heterogeneous data sources.

5 Patterns of Bisociation in BisoNets

Once the information has been integrated into a BisoNet, it can be analyzed in order to find interesting patterns in the integrated data. One class of pattern is bisociation. So far, we have identified three different kinds of bisociations [37], which are described in more detail below.

5.1 Bridging Concept

Bridging concepts connect dense sub-graphs from different domains (see Figure 6). Bridging concepts employ ambiguous concepts or metaphors and are often used in humor [34] and riddles [19]. While ambiguity is useful for making jokes or telling stories, it is less popular in serious scientific or engineering applications. For example, the concept of a "jaguar" is ambiguous since it may refer to either an animal or a car. Metaphors, on the other hand, describe a form of understanding or reasoning in which a concept or idea in one domain is understood or viewed in terms of concepts or ideas from another domain. The statement "You are wasting my time", for instance, can be seen as a metaphor that connects the time with the financial domain. Metaphors play a major role in our everyday life as they afford a degree of flexibility that facilitates discoveries by connecting seemingly unrelated subjects [39].

A first approach to detect bridging concepts is the discovery of concept graphs [35,36] in the integrated data. Concept graphs can be used to identify existing and missing concepts in a network by searching for densely connected quasi bi-partite sub-graphs. Once a concept graph has been detected the domains, its aspect and member vertices stem from, can be analyzed in order to find concepts graphs, e.g. concepts that connect information units from different domains.

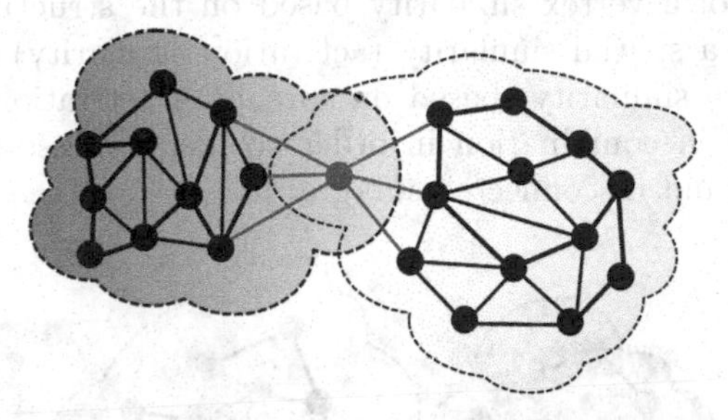

Fig. 6. Bridging concept

5.2 Bridging Graphs

Bridging graphs are sub-graphs that connect concepts from different domains (see Figure 7). They may lead to surprising information arising from different domains since they are able to link seemingly unrelated domains (see Figure 7a). An example of where bridging graph could be used to realize bisociation is the Eureka act of the Archimedes example [20]. A bridging graph may also lead to the linking of two disconnected concepts from the same domain via a connection through and unrelated domain (see Figure 7b).

A first step in the direction of the discovery of bridging graphs is the formalization and detection of such domain-crossing sub-graphs [43,44]. The discovered sub-graphs can be further ranked according to their potential interestingness. Therefore the interestingness is measured by a so called b-score that takes into account the size of the connected domains, the sparsity of the connections between the different domains and the distribution of the neighbors of the bridging vertices.

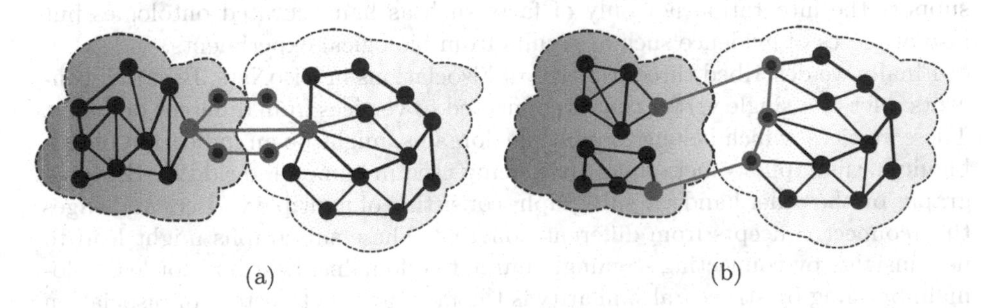

Fig. 7. Bridging graphs

5.3 Bridging by Graph Similarity

Bisociations based on graph similarity are represented by sub-graphs of two different domains that are structurally similar (see Figure 8). This is the most abstract pattern of bisociation that has the potential to lead to new discoveries by linking domains that do not have any connection except for the similar interaction of the bridging concepts and their neighbors.

These structurally similar but disconnected regions in a BisoNet can be discovered by means of a vertex similarity based on the structural properties of vertices. In [53,54] a spatial similarity (activation similarity) and a structural similarity (signature similarity) based on spreading activation are introduced, which can be used in combination in order to identify bisociations based on structurally similar but disconnected sub-graphs.

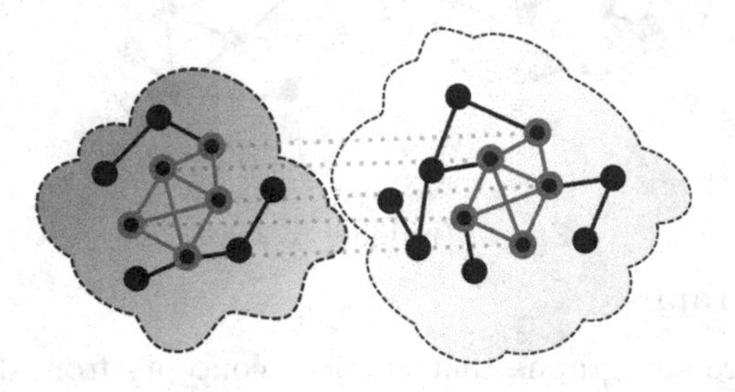

Fig. 8. Bridging by graph similarity

6 Conclusion

In this chapter we identified several key properties of information units and relations used in information networks. We provided an overview of different types of information networks and categorized them based on the identified properties. These properties reflect the expressiveness and thus the ability of an information network to model data of a diverse nature.

We further describe BisoNets as a new type of information network that is tailored to the integration of heterogeneous data sources from diverse domains. They possess the main properties required to integrate large amounts of data from a variety of information sources. By supporting weighted edges BisoNets support the integration not only of facts such as hand curated ontologies but also of pieces of evidence such as results from biological experiments.

Finally we described three patterns of bisociations in BisoNets. Bridging concepts refer to a single vertex that is connected to vertices from different domains. These vertices, which belong to multiple domains, might be an indication of ambiguity or metaphor - metaphors often being used in humor and riddles. Bridging graphs on the other hand are sub-graphs consisting of multiple vertices and edges that connect concepts from different domains. These sub-graphs might lead to new insights by connecting seemingly unrelated domains. Last but not least, domain bridging by structural similarity is the most abstract pattern of bisociation with the potential to lead to truly new discoveries by linking domains that are otherwise unconnected, except for the similar structure of their corresponding sub-graphs.

References

1. Abello, J., Korn, J.: Mgv: a system for visualizing massive multidigraphs. Transactions on Visualization and Computer Graphics 8(1), 21–38 (2002)
2. Albert, R., Barabasi, A.-L.: Statistical mechanics of complex networks. Reviews of Modern Physics 74, 47–97 (2002)
3. Auillans, P., de Mendez, P.O., Rosenstiehl, P., Vatant, B.: A Formal Model for Topic Maps. In: Horrocks, I., Hendler, J. (eds.) ISWC 2002. LNCS, vol. 2342, pp. 69–83. Springer, Heidelberg (2002)
4. Baitaluk, M., Qian, X., Godbole, S., Raval, A., Ray, A., Gupta, A.: Pathsys: integrating molecular interaction graphs for systems biology. BMC Bioinformatics 7, 55 (2006)
5. Bales, M.E., Johnson, S.B.: Graph theoretic modeling of large-scale semantic networks. Journal of Biomedical Informatics 39, 451–464 (2006)
6. Belew, R.: Adaptive information retrieval: using a connectionist representation to retrieve and learn about documents. In: Proceedings of the 12th Annual International ACM SIGIR Conference on Research and Development in Information Retrieval, pp. 11–20 (1989)
7. Belleau, F., Nolin, M.-A., Tourigny, N., Rigault, P., Morissette, J.: Bio2rdf: towards a mashup to build bioinformatics knowledge systems. Journal of Biomedical Informatics 41, 706–716 (2008)
8. Berners-Lee, T., Hendler, J., Lassila, O.: The semantic web. Scientific American 5, 34–43 (2001)
9. Birkland, A., Yona, G.: Biozon: a system for unification, management and analysis of heterogeneous biological data. BMC Bioinformatics 7, 70 (2006)
10. Bodenreider, O.: Biomedical ontologies in action: role in knowledge management, data integration and decision support. IMIA Yearbook of Medical Informatics 1, 67–79 (2008)
11. Brandes, U., Erlebach, T.: Network Analysis: Methodological Foundations. Springer (2005)
12. Burgun, A., Bodenreider, O.: Accessing and integrating data and knowledge for biomedical research. IMIA Yearbook of Medical Informatics 1, 91–101 (2008)
13. Chen, H., Ding, L., Wu, Z., Yu, T., Dhanapalan, L., Chen, J.Y.: Semantic web for integrated network analysis in biomedicine. Briefings in Bioinformatics 10, 177–192 (2009)
14. Chen, H., Ng, T.: An algorithmic approach to concept exploration in a large knowledge network (automatic thesaurus consultation): Symbolic branch-and-bound search vs. connectionist hopfield net activation. Journal of the American Society for Information Science 46(5), 348–369 (1995)
15. Cheung, K.-H., Yip, K.Y., Smith, A., Deknikker, R., Masiar, A., Gerstein, M.: Yeasthub: a semantic web use case for integrating data in the life sciences domain. Bioinformatics 21(suppl.1), i85–i96 (2005)

16. Chua, H.N., Sung, W.-K., Wong, L.: An efficient strategy for extensive integration of diverse biological data for protein function prediction. Bioinformatics 23, 3364–3373 (2007)
17. Consortium, G.O.: Creating the gene ontology resource: design and implementation. Genome Research 11, 1425–1433 (2001)
18. Crestani, F.: Application of spreading activation techniques in information retrieval. Artificial Intelligence Review 11, 453–482, 12 (1997)
19. Dienhart, J.M.: A linguistic look at riddles. Journal of Pragmatics 31(1), 95–125 (1999)
20. Dubitzky, W., Kötter, T., Schmidt, O., Berthold, M.R.: Towards Creative Information Exploration Based on Koestler's Concept of Bisociation. In: Berthold, M.R. (ed.) Bisociative Knowledge Discovery. LNCS (LNAI), vol. 7250, pp. 11–32. Springer, Heidelberg (2012)
21. Durand, P., Labarre, L., Meil, A., Divol, J.-L., Vandenbrouck, Y., Viari, A., Wojcik, J.: Genolink: a graph-based querying and browsing system for investigating the function of genes and proteins. BMC Bioinformatics 7(1), 21 (2006)
22. Figeys, D.: Combining different 'omics' technologies to map and validate protein-protein interactions in humans. Briefings in Functional Genomics and Proteomics 2, 357–365 (2004)
23. I.O. for Standardization. Information Technology – Document Description and Processing Languages – Topic Maps – Data Model. ISO, Geneva, Switzerland (2006)
24. Franke, L., van Bakel, H., Fokkens, L., de Jong, E.D., Egmont-Petersen, M., Wijmenga, C.: Reconstruction of a functional human gene network, with an application for prioritizing positional candidate genes. The American Journal of Human Genetics 78, 1011–1025 (2006)
25. Furnas, G.W.: Generalized fisheye views. In: Proceedings of the SIGCHI Conference on Human Factors in Computing Systems, vol. 17(4), pp. 16–23 (1986)
26. Gavin, A.-C., Aloy, P., Grandi, P., Krause, R., Boesche, M., Marzioch, M., Rau, C., Jensen, L.J., Bastuck, S., Dümpelfeld, B., Edelmann, A., Heurtier, M.-A., Hoffman, V., Hoefert, C., Klein, K., Hudak, M., Michon, A.-M., Schelder, M., Schirle, M., Remor, M., Rudi, T., Hooper, S., Bauer, A., Bouwmeester, T., Casari, G., Drewes, G., Neubauer, G., Rick, J.M., Kuster, B., Bork, P., Russell, R.B., Superti-Furga, G.: Proteome survey reveals modularity of the yeast cell machinery. Nature 440, 631–636 (2006)
27. Getoor, L., Diehl, C.: Link mining: a survey. ACM SIGKDD Explorations Newsletter 7(2), 3–12 (2005)
28. Han, J.: Mining Heterogeneous Information Networks by Exploring the Power of Links. In: Gama, J., Costa, V.S., Jorge, A.M., Brazdil, P.B. (eds.) DS 2009. LNCS, vol. 5808, pp. 13–30. Springer, Heidelberg (2009)
29. Hayes, J.: A graph model for RDF. Master's thesis, Technische Universität Darmstadt, Dept. of Computer Science, Darmstadt, Germany. In: Collaboration with the Computer Science Dept., University of Chile, Santiago de Chile (2004)
30. Hayes, J., Gutierrez, C.: Bipartite Graphs as Intermediate Model for RDF. In: McIlraith, S.A., Plexousakis, D., van Harmelen, F. (eds.) ISWC 2004. LNCS, vol. 3298, pp. 47–61. Springer, Heidelberg (2004)
31. Jansen, R., Yu, H., Greenbaum, D., Kluger, Y., Krogan, N.J., Chung, S., Emili, A., Snyder, M., Greenblatt, J.F., Gerstein, M.: A bayesian networks approach for predicting protein-protein interactions from genomic data. Science 302, 449–453 (2003)
32. Kiemer, L., Costa, S., Ueffing, M., Cesareni, G.: Wi-phi: A weighted yeast interactome enriched for direct physical interactions. Proteomics 7, 932–943 (2007)

33. Koehler, J., Rawlings, C., Verrier, P., Mitchell, R., Skusa, A., Ruegg, A., Philippi, S.: Linking experimental results, biological networks and sequence analysis methods using ontologies and generalised data structures. Silico Biology 5, 33–44 (2005)
34. Koestler, A.: The Act of Creation. Macmillan (1964)
35. Kötter, T., Berthold, M.R.: (Missing) concept discovery in heterogeneous information networks. In: Proceedings of the 2nd International Conference on Computational Creativity, pp. 135–140 (2011)
36. Kötter, T., Berthold, M.R.: (Missing) Concept Discovery in Heterogeneous Information Networks. In: Berthold, M.R. (ed.) Bisociative Knowledge Discovery. LNCS (LNAI), vol. 7250, pp. 230–245. Springer, Heidelberg (2012)
37. Kötter, T., Thiel, K., Berthold, M.R.: Domain bridging associations support creativity. In: Proceedings of the International Conference on Computational Creativity, pp. 200–204 (2010)
38. Kwoh, C.K., Ng, P.Y.: Network analysis approach for biology. Cellular and Molecular Life Sciences 64, 1739–1751 (2007)
39. Lakoff, G., Johnson, M.: Metaphors We Live by. University of Chicago Press (1980)
40. Lassila, O., Swick, R.R.: Resource Description Framework (RDF) model and syntax specification. W3C Working Draft (February 2002)
41. Li, J., Li, X., Su, H., Chen, H., Galbraith, D.W.: A framework of integrating gene relations from heterogeneous data sources: an experiment on arabidopsis thaliana. Bioinformatics 22(16), 2037–2043 (2006)
42. Martinez Morales, A.A.: A directed hypergraph model for RDF. In: Simperl, E., Diederich, J., Schreiber, G. (eds.) Proceedings of the KWEPSY 2007, vol. 275 (2007)
43. Nagel, U., Thiel, K., Kötter, T., Piątek, D., Berthold, M.R.: Bisociative Discovery of Interesting Relations between Domains. In: Gama, J., Bradley, E., Hollmén, J. (eds.) IDA 2011. LNCS, vol. 7014, pp. 306–317. Springer, Heidelberg (2011)
44. Nagel, U., Thiel, K., Kötter, T., Piatek, D., Berthold, M.R.: Towards Discovery of Subgraph Bisociations. In: Berthold, M.R. (ed.) Bisociative Knowledge Discovery. LNCS (LNAI), vol. 7250, pp. 263–284. Springer, Heidelberg (2012)
45. Pavlopoulos, G., Wegener, A.-L., Schneider, R.: A survey of visualization tools for biological network analysis. BioData Mining 1(1), 1–12 (2008)
46. Pearl, J.: Probabilistic Reasoning in Intelligent Systems: Networks of Plausible Inference. Morgan Kaufmann Publishers (1988)
47. Pepper, S.: The tao of topic maps: finding the way in the age of infoglut. In: Proceedings of XML Europe (2000)
48. Schaeffer, S.E.: Graph clustering. Computer Science Review 1, 27–64 (2007)
49. Sevon, P., Eronen, L., Hintsanen, P., Kulovesi, K., Toivonen, H.: Link Discovery in Graphs Derived from Biological Databases. In: Leser, U., Naumann, F., Eckman, B. (eds.) DILS 2006. LNCS (LNBI), vol. 4075, pp. 35–49. Springer, Heidelberg (2006)
50. Shen, Z., Ma, K.-L., Eliassi-Rad, T.: Visual analysis of large heterogeneous social networks by semantic and structural abstraction. IEEE Transactions on Visualization and Computer Graphics 12(6), 1427–1439 (2006)
51. Smith, A.K., Cheung, K.-H., Yip, K.Y., Schultz, M., Gerstein, M.K.: Linkhub: a semantic web system that facilitates cross-database queries and information retrieval in proteomics. BMC Bioinformatics 8, S5 (2007)
52. Smith, B., Ashburner, M., Rosse, C., Bard, J., Bug, W., Ceusters, W., Goldberg, L.J., Eilbeck, K., Ireland, A., Mungall, C.J., Consortium, O.B.I., Leontis, N., Rocca-Serra, P., Ruttenberg, A., Sansone, S.-A., Scheuermann, R.H., Shah, N.,

Whetzel, P.L., Lewis, S.: The obo foundry: coordinated evolution of ontologies to support biomedical data integration. Nature Biotechnology 25, 1251–1255 (2007)
53. Thiel, K., Berthold, M.R.: Node similarities from spreading activation. In: Proceedings of the IEEE International Conference on Data Mining (2010)
54. Thiel, K., Berthold, M.R.: Node Similarities from Spreading Activation. In: Berthold, M.R. (ed.) Bisociative Knowledge Discovery. LNCS (LNAI), vol. 7250, pp. 246–262. Springer, Heidelberg (2012)
55. Troyanskaya, O.G., Dolinski, K., Owen, A.B., Altman, R.B., Botstein, D.: A bayesian framework for combining heterogeneous data sources for gene function prediction (in saccharomyces cerevisiae). Proceedings of the National Academy of Sciences 100, 8348–8353 (2003)
56. Tzitzikas, Y., Constantopoulos, P., Spyratos, N.: Mediators over ontology-based information sources. In: Second International Conference on Web Information Systems Engineering, pp. 31–40 (2001)
57. van Ham, F., van Wijk, J.: Interactive visualization of small world graphs. In: van Wijk, J. (ed.) Proc. IEEE Symposium on Information Visualization INFOVIS 2004, pp. 199–206 (2004)

Network Creation: Overview

Christian Borgelt

European Centre for Soft Computing,
Calle Gonzalo Gutiérrez Quirós s/n, E-33600 Mieres (Asturias), Spain
christian.borgelt@softcomputing.es

Although networks are a very natural and straightforward way of organizing heterogeneous data, as argued in the introductory chapters, few data sources are in this form. We rather find the data we want to fuse, connect, analyze and thus exploit for creative discoveries, stored in flat files, (relational) databases, text document collections and the like. As a consequence, we need, as an initial step, methods that construct a network representation by analyzing tabular and textual data, in order to identify entities that can serve as nodes and to extract relevant relationships that should be represented by edges.

Rather than simply connect all (named) entities for which there is evidence that they may be related in some way, it is clearly desirable that these methods should try to select edges that have a higher chance of being part of a bisociation (or should at least try to endow such edges with higher weights) and should try to identify nodes that have a higher chance of being a bridging concept. In this way the created networks will be better geared towards the goal of creative information discovery. In addition, we need a representational formalism that allows us to reason about graph relationships, in order to support the network analysis and exploration methods described in Parts III and IV, respectively.

Contributions

Most of the following chapters deal with constructing BisoNets from text document collections, like web pages, (scientific) abstracts and papers, or news clippings. In order to process such data sources, the authors all start with standard text mining techniques for keyword extraction in order to obtain an initial set of node candidates. These candidates may then be filtered in order to identify potential bridging concepts or at least to rank them higher than other terms.

In more detail, the first chapter by Segond and Borgelt [1] simply takes the extracted keywords as the node set of the BisoNet that is to be constructed and focuses on the task of selecting appropriate edges. Since standard measures for the association strength of terms turn out to be of fairly limited value, the authors suggest a new measure, which has become known as "Bison measure" or "bisociation index". This measure is based on the insight that for selecting appropriate edges the similarity of the term weights is at least as important as, if not more important than, the magnitude of these weights.

In contrast to this, the chapter by Juric *et al.* [2] concentrates on the selection and ranking of terms and keywords in order to identify bridging concepts. Starting with a more detailed description of the used text mining techniques and

M.R. Berthold (Ed.): Bisociative Knowledge Discovery, LNAI 7250, pp. 51–53, 2012.

document representations, the authors provide a thorough overview of a variety of approaches to compute term weights and of several distance measures between vectors that represent documents in a bag-of-words vector space model. From these the authors derive heuristics to rank terms based on their occurrence in two or more document collections. Included here are heuristics relying on classifiers that are trained to distinguish between the document collections, and for which the misclassified terms are interpreted as potential bridging concepts. They show that in this way a significantly higher number of bridging concepts appear at the top of the ranking list than can be expected in a chance ranking.

The chapter by Hynönen *et al.* [3] again emphasizes the relation between terms, but rather than selecting edges for a BisoNet the authors try to identify terms that are connected in a document even though they are usually not in the underlying document collection as a whole. The core idea is that such unusually correlated terms can indicate a new development or a new insight that is described in the corresponding document(s). In order to measure the connection strength, the authors introduce two new aspects: the first consists in measures for the term pair frequency to assess the strength of correlation in a document and the term pair uncorrelation to describe the background of the document collection to which it is compared. The second aspect is that they take the document apart into sentences in order to achieve more fine-grained assessments.

The second chapter by Segond and Borgelt [4] presents a new item set mining technique, which may be applied to text document mining by seeing each term as an item and each document as a transaction of the terms that occur in it. The core idea of the approach is to go beyond terms pairs and to find correlations between multiple terms, which correspond to possible hyperedges in a BisoNet. However, since the standard measure for the selection of item sets, the support (number of transactions containing all items) is not well suited to assess the association of terms, the authors introduce an approach based on the similarity of item covers (sets of transactions containing the items) and develop an efficient algorithm to mine item sets with several such similarity measures.

Finally, the chapter by Kimmig *et al.* [5] discusses a representation and reasoning framework for graphs with probabilistically weighted edges that relies on the ProbLog language. The authors demonstrate how both graphs and graph patterns can conveniently be described in a logical framework and how deductive, abductive and inductive reasoning are supported, as is shown with several precise examples. In addition, modifications of the knowledge base can easily be expressed, including graph simplification, subgraph extraction, abstraction etc. Finally, the authors demonstrate how probabilistic edge weights (interpreted as the probability that an edge is present) can be incorporated and how all discussed logical concepts can be transferred and extended to probabilistic graphs.

Conclusions

Whether graphs are described by explicit graph data structures or in a logical framework (endowed with probabilistic edge weights or not), they are a powerful framework for knowledge representation. However, creating them from heterogeneous and, in particular, from unstructured data like documents, is a challenging task, especially if one wants to support creative information discovery. Even though standard text mining techniques form the starting point for all of the approaches discussed in this part, they are not sufficient for creating useful BisoNets. As the following chapters demonstrate, several enhancements of the selection of both nodes and edges can increase the chance of obtaining edges that support bisociative discoveries and of identifying nodes that are potential bridging concepts. It has to be conceded, though, that the methods are not perfect yet and that there is a lot of room for improvement. However, the described methods are highly promising and they could be shown to produce significantly better results than known techniques.

References

1. Segond, M., Borgelt, C.: Selecting the Links in BisoNets Generated from Document Collections. In: Berthold, M.R. (ed.) Bisociative Knowledge Discovery. LNCS (LNAI), vol. 7250, pp. 54–65. Springer, Heidelberg (2012)
2. Juršič, M., Sluban, B., Cestnik, B., Grčar, M., Lavrač, N.: Bridging Concept Identification for Constructing Information Networks from Text Documents. In: Berthold, M.R. (ed.) Bisociative Knowledge Discovery. LNCS (LNAI), vol. 7250, pp. 66–90. Springer, Heidelberg (2012)
3. Hynönen, T., Mahler, S., Toivonen, H.: Discovery of Novel Term Associations in a Document Collection. In: Berthold, M.R. (ed.) Bisociative Knowledge Discovery. LNCS (LNAI), vol. 7250, pp. 91–103. Springer, Heidelberg (2012)
4. Segond, M., Borgelt, C.: Cover Similarity Based Item Set Mining. In: Berthold, M.R. (ed.) Bisociative Knowledge Discovery. LNCS (LNAI), vol. 7250, pp. 104–121. Springer, Heidelberg (2012)
5. Kimmig, A., Galbrun, E., Toivonen, H., De Raedt, L.: Patterns and Logic for Reasoning with Networks. In: Berthold, M.R. (ed.) Bisociative Knowledge Discovery. LNCS (LNAI), vol. 7250, pp. 122–143. Springer, Heidelberg (2012)

Selecting the Links in BisoNets Generated from Document Collections

Marc Segond and Christian Borgelt

European Center for Soft Computing,
Calle Gonzalo Gutiérrez Quirós s/n, E-33600 Mieres (Asturias), Spain
{marc.segond,christian.borgelt}@softcomputing.es

Abstract. According to Koestler, the notion of a bisociation denotes a connection between pieces of information from habitually separated domains or categories. In this chapter, we consider a methodology to find such bisociations using a BisoNet as a representation of knowledge. In a first step, we consider how to create BisoNets from several textual databases taken from different domains using simple text-mining techniques. To achieve this, we introduce a procedure to link nodes of a BisoNet and to endow such links with weights, which is based on a new measure for comparing text frequency vectors. In a second step, we try to rediscover known bisociations, which were originally found by a human domain expert, namely indirect relations between migraine and magnesium as they are hidden in medical research articles published before 1987. We observe that these bisociations are easily rediscovered by simply following the strongest links.

1 Introduction

The concept of association is at the heart of many of today's powerful ICT technologies such as information retrieval and data mining. These technologies typically employ "association by similarity or co-occurrence" in order to discover new information that is relevant to the evidence already known to a user.

However, domains that are characterized by the need to develop innovative solutions require a form of creative information discovery from increasingly complex, heterogeneous and geographically distributed information sources. These domains, including design and engineering (drugs, materials, processes, devices), areas involving art (fashion and entertainment), and scientific discovery disciplines, require a different ICT paradigm that can help users to uncover, select, re-shuffle, and combine diverse contents to synthesize new features and properties leading to creative solutions. People working in these areas employ creative thinking to connect seemingly unrelated information, for example, by using metaphors or analogical reasoning. These modes of thinking allow the mixing of conceptual categories and contexts, which are normally separated. The functional basis for these modes is a mechanism called *bisociation* (see [1]).

According to Arthur Koestler, who coined this term, *bisociation* means to join unrelated, and often even conflicting, information in a new way. It means

M.R. Berthold (Ed.): Bisociative Knowledge Discovery, LNAI 7250, pp. 54–65, 2012.

being "double minded" or able to think on more than one plane of thought simultaneously. Similarly, Frank Barron [2] says that the ability to tolerate chaos or seemingly opposite information is characteristic of creative individuals.

Several famous scientific discoveries are good examples of bisociations, for instance Isaac Newton's theory of gravitation and James C. Maxwell's theory of electromagnetic waves. Before Newton, a clear distinction was made between *sub-lunar* (below the moon) and *super-lunar physics* (above the moon), since it was commonly believed that these two spheres where governed by entirely different sets of physical laws. Newton's insight that the trajectories of planets and comets can be interpreted in the same way as the course of a falling body joined these habitually separated domains. Maxwell, by realizing that light is an electromagnetic wave, joined the domains of optics and electromagnetism, which, at his time, were also treated as unrelated areas of physical phenomena.

Although the concept of bisociation is frequently discussed in cognitive science, psychology and related areas (see, for example, [1,2,3]), there does not seem to exist a serious attempt at trying to formalize and computerize this concept. In terms of ICT implementations, much more widely researched areas include association rule learning (for instance, [4]), analogical reasoning (for example, [5,6]), metaphoric reasoning (for example, [7]), and related areas such as case-based reasoning (for instance, [8]) and hybrid approaches (for example, [9]).

In order to fill this gap in current research efforts, the BISON project[1] was created. This project focuses on a knowledge representation approach with the help of networks of named entities, in which bisociations may be revealed by link discovery and graph mining methods, but also by computer-aided interactive navigation. In this chapter we report first results obtained in this project.

The rest of this chapter is structured as follows: in Section 2 we provide a definition of the core notion of a *bisociation*, which guides our considerations. Based on this definition, we justify why a network representation—a so-called *BisoNet*—is a proper basis for computer-aided bisociation discovery. Methods for generating BisoNets from heterogeneous data sources are discussed in Section 3, including procedures for selecting the named entities that form its nodes and principles for linking them based on the information extracted from the data sources. In particular, we present a new measure for the strength of a link between concepts that are derived from textual data. Such link weights are important in order to assess the strength of indirect connections like bisociations.

Afterwards, in Section 5 we report results on a benchmark data set (consisting of titles and abstracts of medical research articles), in which a human domain expert already discovered hidden bisociations. By showing that with our system we can create a plausible BisoNet from this data source, in which we can rediscover these bisociations, we provide evidence that the computer-aided search for bisociations is a highly promising technology.

Finally, in Section 6 we draw conclusions from our discussion.

---

[1] See `http://www.bisonet.eu/` for more information on this EU FP7 funded project.

2 Reminder: Bisociation and BisoNets

Since the core notion of our efforts is *bisociation*, we start by trying to provide a sufficiently clear definition, which can guide us in our attempts to create a system able to support a user in finding bisociations. A first definition within the BISON project[2] characterizes *bisociation* as follows:

> A *bisociation* is a link L that connects two domains D_1 and D_2 that are unconnected given a specific context or view V by which the domains are defined. The link L is defined by a connection between two concepts c_1 and c_2 of the respective domains.

Although the focus on a connection between two habitually (that is, in the context a user is working in) separated domains is understandable, this definition seems somewhat too narrow. Linking two concepts from the same domain, which are unconnected within the domain, but become connected by employing indirect relations that pass through another domain, may just as well be seen as bisociations. The principle should rather be that the connection is not fully contained in one domain (which would merely be an association), but needs access to a separate domain. Taking this into account, we generalize the definition:

> A *bisociation* is a link L between two concepts c_1 and c_2, which are unconnected given a specific context or view V. The concepts c_1 and c_2 may be unconnected, because they reside in different domains D_1 and D_2 (which are seen as unrelated in the view V), or because they reside in the same domain D_1, in which they are unconnected, and their relation is revealed only through a *bridging concept* c_3 residing in some other domain D_2 (which is not considered in the view V).

In both of these characterizations we define domains formally as sets of concepts. Note that a *bridging concept* c_3 is usually also required if the two concepts c_1 and c_2 reside in different domains, since direct connections between them, even if they cross the border between two domains, can be expected to be known and thus will not be interesting or relevant for a user.

Starting from the above characterization of *bisociation*, a network representation, called a *BisoNet*, of the available knowledge suggests itself: each concept (or, more generally, any named entity) gives rise to a node. Concepts that are associated (according to the classical paradigm of similarity or co-occurrence) are connected by an edge. Bisociations are then indirect connections (technically paths) between concepts, which cross the border between two domains.

Note that this fits both forms of bisociations outlined above. If the concepts c_1 and c_2 reside in different domains, the boundary between these two domains necessarily has to be crossed. If they reside in the same domain, one first has to leave this domain and then come back in order to find a bisociation.

---

[2] See `http://www.inf.uni-konstanz.de/bisonwiki/index.php5`, which, however, is not publicly accessible at this time.

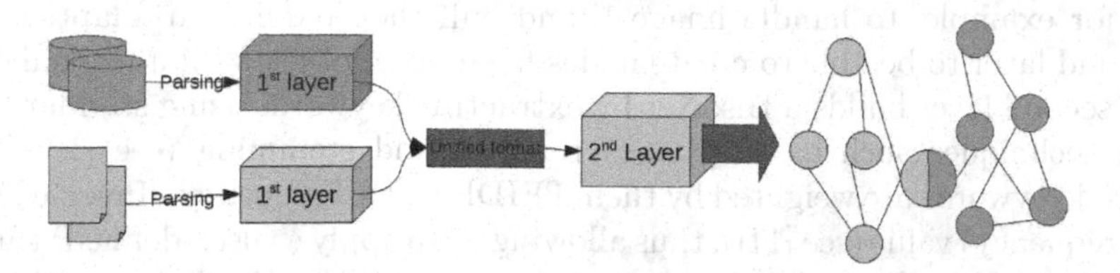

Fig. 1. Illustration of the structure of the BisoNet generator

3 BisoNet Generation

A system for generating BisoNets requires three ingredients: (1) A component to access the original, usually heterogeneous data sources. In order to cope with different data formats, we suggest, in Section 3.1, a two-layer architecture. (2) A method for choosing the named entities that are to form the nodes of the BisoNet. Here we rely on standard keyword extraction techniques, as discussed in Section 3.2. (3) A procedure for linking the nodes of a BisoNet and for endowing them with weights that indicate the association strength. For this we suggest, in Section 4, a new association measure for keywords.

3.1 Data Access and Pre-processing

As explained above, a BisoNet is a network that promises to contain bisociations. In order to generate such networks, we first have to consider two things: we must be able to read different and heterogeneous data sources, and we have to be able to merge the information derived from them in one BisoNet. Data sources can be databases (relational or of any other type), text collections, raw text, or any data that provide information about a domain. Due to the wide variety of formats a data source can have, the choice we made here is not to provide an interface of maximal flexibility that can be made to read any data source type, but to structure our creation framework into two separate steps.

In the first step, we directly accesses the data source and therefore a parser has to be newly developed for or at least adapted to the specific format of the data source. The second step is actual the BisoNet generation part. It takes its information from the first step, always in the same format, and therefore can generate a BisoNet from any data source, as far as it is parsed and exported in the form provided by the first step process (see Figure 1 for a sketch).

The way data should be provided to the second layer is fairly simple, because in this chapter we confine our considerations to textual data. As a consequence, the second layer creates nodes from data that are passed as records containing textual fields. These textual fields can contain, for now, either words or authors names. This procedure and data format is well adapted to textual databases or text collections, but is meant to evolve in future development in order to be able to take other types of data sources into account. However, since most of the data sources that we have used so far were textual data sources, this protocol seems simple and efficient. Future extensions could consist in including raw data fields (for example, to handle images), and will then require an adaptation of the second layer to be able to create nodes from other objects than textual data.

The second layer builds a BisoNet by extracting keywords using standard text mining techniques such as stop word removal and stemming (see [10]). The extracted keywords are weighted by their TFIDF (Text Frequency - Inverse Document Frequency) value (see [11]), thus allowing us to apply a (user-defined) threshold in order to filter the most important keywords, as will be detailed in Section 3.2. Links between nodes are created according to the presence of co-occurrences of the corresponding keywords in the same documents, and are weighted using a similarity measure adapted to the specific requirements of our case, which will be presented in Section 4. In the case that author lists are provided with each text string, extracted keywords are also linked to the related authors. These links are weighted according to the number of times a keyword occurs in a given author's work.

3.2 Creating Nodes

In our BisoNets nodes represent concepts. As we only talk about textual databases, we made the choice to characterize concepts by keywords that are extracted from the textual records taken from the data sources. In the second layer of our framework, each textual record j is processed with a stop word removal algorithm. Then the text frequency values are computed for each remaining term i as follows: $\mathrm{tf}_{i,j} = \frac{n_{i,j}}{\sum_k n_{k,j}}$, where $n_{i,j}$ is the number of occurrences of the considered term in textual record j and $\sum_k n_{k,j}$ is the sum of number of occurrences of all terms in textual record j.

Naturally, this procedure of keyword extraction is limited in its power to capture the contents of the text fields. The reason is that we are ignoring synonyms (which should be handled by one node rather than two or more), hyper- and hyponyms, pronouns (which may refer to a relevant keyword and thus may have to be counted for the occurrence of this keyword) etc. However, such linguistic properties are very difficult to take into account and need sophisticated tools (like thesauri etc.). Since such advanced text mining is not the main goal of our work (which rather focuses on BisoNet creation), keeping the processing simple seemed a feasible option. Nevertheless, advanced implementations may require such advanced processing, because ignoring, for example, synonyms and pronouns can distorts the statistics underlying, for instance, the term frequency value: ignoring pronouns that refer to a keyword, or not merging two synonyms makes the term frequency lower than it should actually be.

After all records have been processed, the inverse document frequency of each keyword i is computed the following way: $\mathrm{idf}_i = \log \frac{|D|}{|\{d \in D | t_i \in d\}|}$, where $|D|$ is the total number of records in the database and $|\{d \in D \mid t_i \in d\}|$ is the number of records in which the term t_i appears.

Each node is then weighted with its corresponding average TFIDF value: $\mathrm{tfidf}_i = \frac{1}{|D|} \sum_{j=1}^{|D|} \mathrm{tf}_{i,j} \cdot \mathrm{idf}_i$

This TFIDF approach is a very well known approach in text mining that is easy to implement and makes one able to easily apply a threshold, thus selecting only the most important nodes (keywords). A node then contains, as an attribute, a list of the term frequency values of its associated term in the different documents of the collection. This allows us to compute similarity measures presented in Section 4 in order to create links.

According to the definition of a bisociation presented in Section 2, two concepts have to be linked by other concepts that are not in their proper domain (so-called *bridging concepts*). This leads us to introduce the notion of domains, into which the nodes are grouped, so that we can determine when borders between domains are crossed. In order to be able to classify nodes according to their membership in different domains, it is important that they keep, also as an attribute, the domains the data sources belong to, from which they have been extracted. Since the same keyword can occur in several data sources, taken from different domains, one has to be able (for example, for graph mining and link discovery purposes) to know whether a certain keyword has to be considered from a certain domain's point of view. The nodes therefore keep this information as vector of domains their associated keyword belongs to.

This can be interesting, for example, to mine or navigate the BisoNet, keeping in mind that a user may be looking for ideas related to a certain keyword belonging to a domain A. The results of a search for bisociations might also belong to domain A, because it is the domain of interest of the user. However, these results should be reached following paths using keywords from other domains, that is to say bisociations. This procedure provides related keywords of interest for the user, as they belong to its research domain, but they might be also original and new connections as they are the result of a bisociation process.

4 Linking Nodes: Different Metrics

As explained in Section 3.2, nodes are associated with a keyword and a set of documents in which this keyword occurs with a certain term frequency. Practically, this is represented using a vector of real values containing, for each document, the term frequency of the node's keyword. In order to determine whether a link should be created between two nodes or not, and if there is to be a link, to assign it a weight, we have to use a similarity measure to compare two nodes (that is to say: the two vectors of term frequency values).

Links in our BisoNets are weighted using similarity measures shown below. This approach allows us to use several different kinds of graph mining algorithms, such as simply thresholding the values to select a subset of the edges, or more complex ones, like calculating, for example, shortest paths.

4.1 Cosine and Tanimoto Measures

One basic metric that directly suggests itself is an adaptation of the Jaccard index (see [12]): $J(A,B) = \frac{|A \cap B|}{|A \cup B|}$.

Here $|A \cap B|$ represents the number of elements at the same index that both have a positive value in the two vectors and $|A \cup B|$ the total number of elements in the two vectors.

It can also be interpreted as a probability, namely the probability that both elements are positive, given that at least one is positive (contain a given term i, i.e., $\mathrm{tf}_i > 0$).

Cosine similarity is a measure of similarity between two vectors of n dimensions by finding the angle between them. Given two vectors of attributes, A and B, the cosine similarity, $\cos(\theta)$, is represented using a dot product and magnitude as $\cos(\theta) = \frac{A \cdot B}{\|A\|\|B\|}$, where, in the case of text matching, the attribute vectors A and B are usually the tf-idf vectors of the documents.

This cosine similarity metric may be extended such that it yields the Jaccard index in the case of binary attributes. This is the Tanimoto coefficient $T(A, B)$, represented as $T(A, B) = \frac{A \cdot B}{\|A\|^2 + \|B\|^2 - A \cdot B}$.

These measures allow us to compare two nodes according to the number of similar elements they contain, but do not take into account the importance of the text frequency values.

4.2 The Bison Measure

In the Jaccard measure, as applied above, we would consider only whether a term frequency is zero or positive and thus neglect the actual value (if it is positive). However, considering two elements at the same index i in two vectors, one way of taking their values into account would be to use their absolute difference (that is, in our case, the absolute difference of the term frequency values for two terms, but the same document). With this approach, it is easy to compare two vectors (of term frequency values) by simply summing these values and dividing by the total number of values (or the total number of elements that are positive in at least one vector).

However, this procedure does not properly take into account that both values have to be strictly positive, because a vanishing term frequency value means that the two keywords do not co-occur in the corresponding document. In addition, we have to keep in mind that having two elements, both of which have a term frequency value of 0.2, should be less important than having two elements with a term frequency value of 0.9. In the first case, the keywords associated with the two nodes we are comparing appear only rarely in the considered document. On the other hand, in the latter case these keywords appear very frequently in this document, which means that they are strongly linked according to this document.

A possibility of taking the term frequency values itself (and not only their difference) into account is to use the product of the two term frequency values as a coefficient to the (absolute) difference between the term frequency values. This takes care of the fact that the two term frequency values have to be positive, and that the similarity value should be the greater, the larger the term frequency values are (and, of course, the smaller their absolute difference is). However, in our case, we also want to take into account that it is better to have two similar term frequency values of 0.35 (which means that the two keywords both appear rather infrequently in the document) than to have term frequency values of 0.3 and 0.7 (which means the first keywords appears rarely, while the other quite frequently).

In order to adapt the product to this consideration, we use the expression in Equation 1, in which k can be adjusted according to the importance one is willing to give to low term frequency values.

$$B(A,B) = (\mathrm{tf}_i^A \cdot \mathrm{tf}_i^B)^k \cdot (1 - |\,\mathrm{tf}_i^A - \mathrm{tf}_i^B\,|), \quad \mathrm{tf}_i^A, \mathrm{tf}_i^B \in [0,1] \tag{1}$$

Still another thing that we have to take into account in our case is that the same difference between tf_i^A and tf_i^B can have a different impact depending on whether tf_i^A and tf_i^B are large or small. To tackle this issue, we combine Equation 1 with the use of the arctan function, and thus obtain the similarity measure shown in Equation 2, which we call the Bison measure. This form has the advantage that it takes into account that two term frequency values for the same index have to be positive, that the similarity should be the greater, the larger the term frequency values are, and that the same difference between tf_i^A and tf_i^B should have a different impact according to the values of tf_i^A and tf_i^B.

$$B(A,B) = (\mathrm{tf}_i^A \cdot \mathrm{tf}_i^B)^k \cdot \left(1 - \frac{|\arctan(\mathrm{tf}_i^A) - \arctan(\mathrm{tf}_i^B)|}{\arctan(1)}\right), \quad \mathrm{tf}_i^A, \mathrm{tf}_i^B \in [0,1] \tag{2}$$

4.3 The Probabilistic Measure

Another way of measuring the similarity between two nodes is based on a probabilistic view. Considering two terms, it is possible to compute, for each document they appear into, the probability of randomly selecting this document by randomly choosing an occurrence of the considered term, all of which are seen as equally likely. This value is given by the law of conditional probabilities shown in Equation 3

$$P(d_i/t_j) = \frac{P(t_j/d_i) \cdot P(d_i)}{P(t_j)} \tag{3}$$
$$\text{with } P(t_j) = \sum_d P(t_j/d) \cdot P(d)$$

This leads us to represent a node by a vector of all the conditional probabilities of the documents they appear in instead of a vector of text frequencies.

Having this representation, we can compare two nodes using the similarity measure shown in Equation 4.

$$S(A,B) = \sqrt{\frac{1}{n} \cdot \sum_n (P(d_n/t_A) - P(d_n/t_B))^2} \tag{4}$$

We can add that $P(d_i/t_j)$ in Equation 3 is equivalent to the term frequency if $P(d_i)$ is constant, which is the case in most of the textual data sources. We can however use this $P(d_i)$ to give arbitrary weights to certain documents.

5 Benchmarks

Having shown how BisoNets can be built from textual data sources, we present benchmark applications in this section. The idea is to provide a proof of principle, that this approach of creating a BisoNet can help a user to discover bisociations.

In order to assess how effective the different similarity measures are, we count how many domain crossing links there are in the generated BisoNets, then we use different threshold values on the links in order to keep only the "strongest" edges according to the similarity measure used.

5.1 The Swanson Benchmark

Swanson's approach [13] to literature-based discovery of hidden relations between concepts A and C via intermediate B-terms is the following: if there is no known direct relation A-C, but there are published relations A-B and B-C one can hypothesize that there is a plausible, novel, yet unpublished indirect relation A-C. In this case the B-terms take the role of *bridging concepts*. In his paper [13], Swanson investigated plausible connections between migraine (A) and magnesium (C), based on the titles of papers published before 1987. He found eleven indirect relations (via bridging concepts B) suggesting that magnesium deficiency may be causing migraine.

We tried our approach on the Swansons data source which consists of 8000 paper titles, taken from the PubMed database, published before 1987 and talking about either migraine or magnesium, to see if it was possible to find again these relations between migraine and magnesium. In order to generate a BisoNet, we implemented a parser for text files containing the data from PubMed able to export them in the format understandable by the second layer of our framework. Then, this second layer performed the keywords extraction, using these keywords as nodes and linking these nodes in the way described in Section 3.

By ranking and filtering the edges we then produced BisoNets that contained the "strongest" edges and their associated nodes. The left graphic of Figure 2 shows how many domain crossing links that are kept using different threshold values on the edges. On this graphic, we can observe that the Bison measure is the one able to keep the most crossing-domain links even if only the very strongest edges are kept (threshold set to keep only the best 5% of the edges). These tests demonstrate that the Bison measure is very well suited for bisociation discovery, since with it the strongest links are the bisociative ones.

We can observe this also in Figure 3 where the difference between the Tanimoto and the Bison measure is graphically highlighted, showing that if we keep only the 5% best edges, the Tanimoto measure loses any relation between magnesium and migraine whereas the Bison measure manages to keep at least some.

5.2 The Biology and Music Benchmark

As we aim to discover bisociations, that is associations between concepts that appear unrelated from a certain, habitual point of view, an interesting benchmark

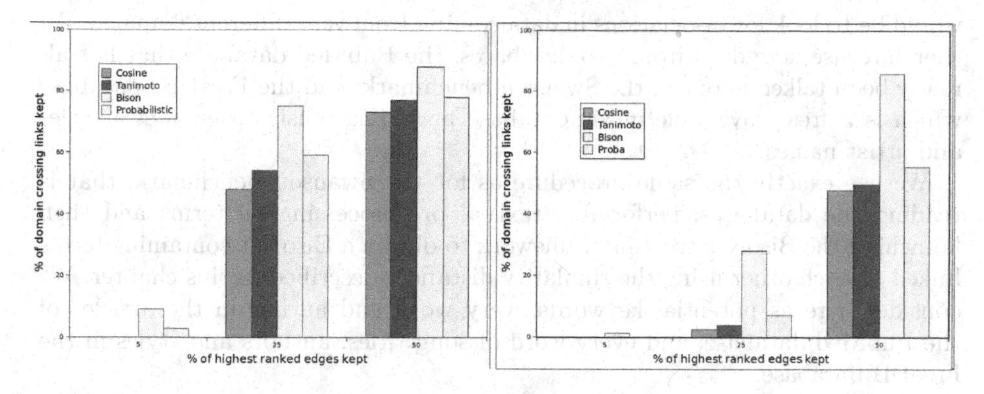

Fig. 2. Comparison between different similarity measures on the Swanson benchmark on the left and on the biology-music benchmark on the right

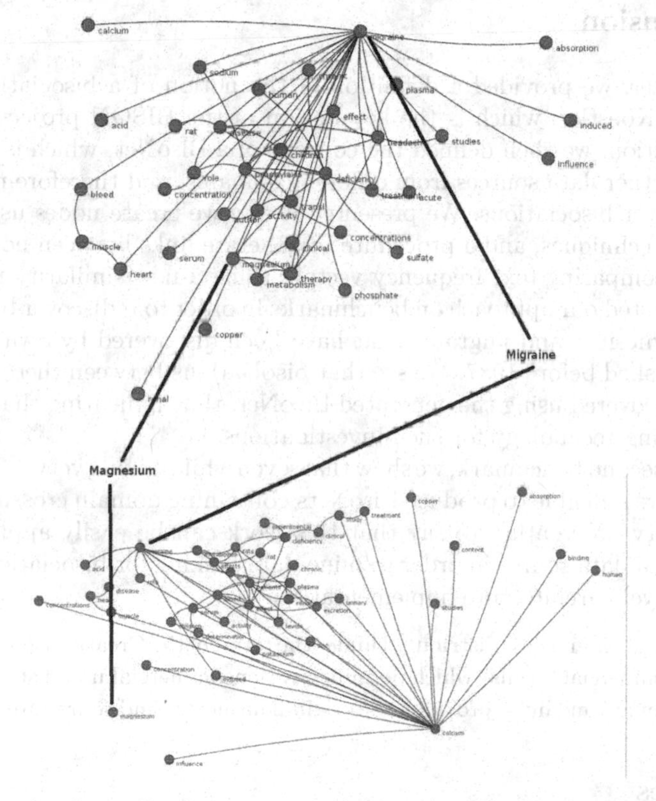

Fig. 3. Example of two BisoNets generated from the Swanson benchmark using the Bison similarity measure and the probabilistic similarity measure

would be to look for bisociations in data coming from very different domains. We therefore use here data from two databases: the PubMed database that has already been talked about in the Swanson benchmark, and the FreeDB[3] database which is a freely available music database providing music titles, music styles and artist names.

We use exactly the same procedure as for the Swanson benchmark, that is reading the databases, performing textual pre-processing on terms and then launching the BisoNet creation framework to obtain a BisoNet containing terms linked to each other using the similarity distances described in this chapter. We consider here as potential keywords every word and author in the articles of the PubMed database, and every word of song titles, authors and styles in the FreeDB database.

The right graphic of Figure 2 shows how many domain crossing links that are kept using different threshold values on the edges.

6 Conclusion

In this chapter, we provided a definition of the notion of a bisociation, as understood by Koestler, which is the key notion of the BISON project. Building on this definition, we then defined the concept of a BisoNet, which is a network bringing together data sources from different domains, and therefore may help a user to discover bisociations. We presented a way we create nodes using simple text-mining techniques, and a procedure to generate links between nodes, which is based on comparing text frequency vectors using a new similarity measure.

We then tested our approach on benchmarks in order to rediscover bisociations between magnesium and migraine that have been discovered by Swanson using articles published before 1987. We see that bisociations between these two terms are easily discovered using the generated BisoNet, thus indicating that BisoNets are a promising technology for such investigations.

Using the second benchmark, we show that, even while mixing very different data sources, we are still able to produce BisoNets containing domain crossing links.

In summary, we venture to say that this work can be easily applied to any kind of textual data source in order to mine data looking for bisociations, thanks to the two layers architecture implementation.

References

1. Koestler, A.: The act of creation. London Hutchinson (1964)
2. Barron, F.: Putting creativity to work. In: The Nature of Creativity. Cambridge Univ. Press (1988)

[3] `http://www.freedb.org/`

3. Cormac, E.M.: A cognitive theory of metaphor. MIT Press (1985)
4. Agrawal, R., Imielinski, T., Swami, A.: Mining association rules between sets of items in large databases. In: Proceedings of the ACM SIGMOD Internation Conference on Management of Data, pp. 207–216 (1993)
5. Chalmers, D.J., French, R.M., Hofstadter, D.R.: High-level perception, representation and analogy: a critique of artificial intelligence methodology. Journal of Experimental and Theoretical Artificial Intelligence 4, 185–211 (1992)
6. Falkenhainer, B., Forbus, K.D., Gentner, D.: The structure mapping engine: algorithm and examples. Artificial Intelligence 41, 1–63 (1989)
7. Barnden, J.A.: An Implemented System for Metaphor-Based Reasoning - With Special Application to Reasoning about Agents. In: Nehaniv, C.L. (ed.) CMAA 1998. LNCS (LNAI), vol. 1562, pp. 143–153. Springer, Heidelberg (1999)
8. Aamodt, A., Plaza, E.: Case-based reasoning: foundational issues, methodological variations and system approaches. Artificial Intelligence Communications 7(1), 39–59 (1994)
9. Cardoso, A., Costa, E., P.M.F.P.P.G.: An architecture for hybrid creative reasoning. In: Soft Computing in Case Based Reasoning. Springer, Heidelberg (2000)
10. van Rijsbergen, C.J., Robertson, S.E., Porter, M.F.: New models in probabilistic information retrieval. In: British Library Research and Development Report. Number 5587. London British Library (1980)
11. Gerard Salton, M.M.G.: Introduction to modern information retrieval. McGraw-Hill (1983)
12. Jaccard, P.: Étude comparative de la distribution florale dans une portion des alpes et du jura. Bulletin de la Société Vaudoise des Sciences Naturelles 37, 547–579 (1901)
13. Swanson, D.R., Smalheiser, N.R., Torvik, V.I.: Ranking indirect connections in literature-based discovery: The role of medical subject headings. Journal of the American Society for Information Science and Technology (JASIST) 57(11) (September 2006)

Bridging Concept Identification for Constructing Information Networks from Text Documents

Matjaž Juršič[1], Borut Sluban[1], Bojan Cestnik[1, 2], Miha Grčar[1], and Nada Lavrač[1, 3]

[1] Jožef Stefan Institute, Ljubljana, Slovenia
[2] Temida d.o.o., Ljubljana, Slovenia
[3] University of Nova Gorica, Nova Gorica, Slovenia
{matjaz.jursic,borut.sluban,bojan.cestnik,
miha.grcar,nada.lavrac}@ijs.si

Abstract. A major challenge for next generation data mining systems is creative knowledge discovery from diverse and distributed data sources. In this task an important challenge is information fusion of diverse mainly unstructured representations into a unique knowledge format. This chapter focuses on merging information available in text documents into an information network – a graph representation of knowledge. The problem addressed is how to efficiently and effectively produce an information network from large text corpora from at least two diverse, seemingly unrelated, domains. The goal is to produce a network that has the highest potential for providing yet unexplored cross-domain links which could lead to new scientific discoveries. The focus of this work is better identification of important domain-bridging concepts that are promoted as core nodes around which the rest of the network is formed. The evaluation is performed by repeating a discovery made on medical articles in the migraine-magnesium domain.

Keywords: Knowledge Discovery, Text Mining, Bridging Concept Identification, Information Networks, PubMed, Migraine, Magnesium.

1 Introduction

Information fusion can be defined as the study of efficient methods for automatically or semi-automatically transforming information from different sources and different points in time into a representation that provides effective support for human and automated decision making [5]. Creative knowledge discovery can only be performed on the basis of a sufficiently large and sufficiently diverse underlying corpus of information. The larger the corpus, the more likely it is to contain interesting, still unexplored relationships.

The diversity of data and knowledge sources demands a solution that is able to represent and process highly heterogeneous information in a uniform way. This means that unstructured, semi-structured and highly structured content needs to be integrated. Information fusion approaches are diverse and domain dependent. For instance, there are recent investigations [7, 19] in using information fusion to support

M.R. Berthold (Ed.): Bisociative Knowledge Discovery, LNAI 7250, pp. 66–90, 2012.

scientific decision making within bioinformatics. Smirnov et al. [22] exploit the idea of formulating an ontology-based model of the problem to be solved by the user and interpreting it as a constraint satisfaction problem taking into account information from a dynamic environment.

In this chapter we explore a graph-theoretic approach [1, 2] which appears to provide the best framework to accommodate the two dimensions of information source complexity – type diversity as well as volume size. Efficient management and processing of very large graph structures can be realized in distributed computing environments, such as grids, peer-to-peer networks or service-oriented architectures on the basis of modern database management systems, object-oriented or graph-oriented database management systems. The still unresolved challenge of graph-theoretic approaches is the creation, maintenance and update of the graph elements in the case of very large and diverse data and knowledge sources.

The core notion that guided our research presented in this chapter is based on the concept of *bisociation*, as defined by Koestler [11] and refined in our context by Dubitzky et al. [6]. Furthermore, Petrič et al. [15] explore the analogy between Koestler's creativity model and comparable cross-domain knowledge discovery approaches from the field of literature mining. In the field of biomedical literature-mining, Swanson [24] designed the *ABC model* approach, which investigates whether agent *A* is connected with phenomenon *C* by discovering complementary structures via interconnecting phenomena *B*. The process of discovery when domains A and C are known in advance and the goal is to find interconnecting concepts from B is called a *closed discovery process*. On the other hand, if only domain A is known then this is an *open discovery process* since also domain C has to be discovered.

Our research deals only with the closed discovery setting and is to some extent similar to the work of Smalheiser and Swanson [21] where they developed an online system ARROWSMITH, which takes as input two sets of titles from disjoint domains *A* and *C* and lists bridging terms (*b*-terms) that are common to literature *A* and *C*; the resulting b-terms are used to generate novel scientific hypotheses. Other related works in the domain of biomedical literature mining are work of Weeber et al. [28] where authors partly automate Swanson's discovery and work of Srinivasan et al. [23] where they develop an algorithm for bridging term identification with even less expert interaction needed.

This work extensively uses the concepts of bisociation, bridging concept, b-term identification, closed discovery, cross-context and A-C domains presented in the previous paragraph. Furthermore, we have based the evaluation techniques mostly on the results reported by Swanson et al. [26] and Urbančič et al. [27].

The chapter is structured as follows. The second section explains the initial problem we are solving into much more detail, defines the terminology used in this work and outlines the structure of the solution proposed in this chapter. The next section is more technical and it lays ground for some basic procedures for retrieving and preprocessing a collection of documents. It also introduces the standard text-mining procedures and terminology which is essential for understanding the subsequent sections. The fourth section presents the core contribution of this work, i.e., bisociative bridging concept identification techniques which are used to extract key network concepts (nodes). Evaluation of these core ideas on a previously well studied domain is presented in the following section. The sixth section builds upon the results from concept identification part (Sections 4 and 5) and shows how the final information networks are constructed.

2 Problem Description

This section describes the problem addressed in this work. The initial goal is straightforward: to construct an information network from text documents. The input to the procedure consists of text documents (e.g., titles and abstract of scientific documents) from two disparate domains. The output of the procedure is an information network which could, for example, look like the graph shown in Fig. 1. However, the strong bias towards bisociations leads us to using advanced bridging term identification techniques for detecting important network nodes and relations. The following paragraphs define in detail the input, the output, open issues and sketch the proposed solution.

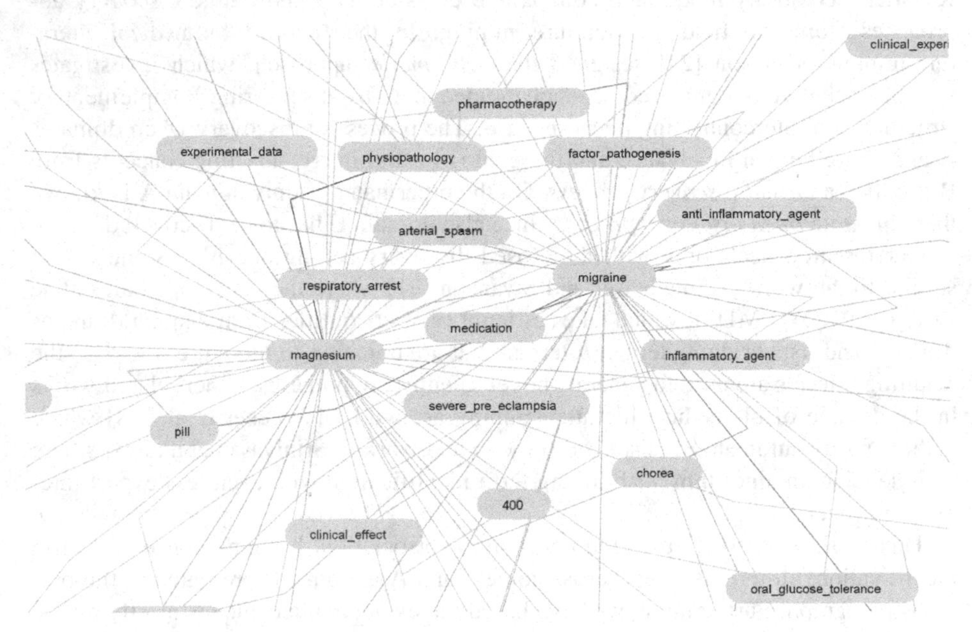

Fig. 1. Part of a network created from PubMed articles on migraine and magnesium

This chapter focuses – similarly as related work from the literature-mining field – on text documents as the primary data source. Texts are in general considered to be one of the most unstructured data sources available, thus, constructing a meaningful graph of data and knowledge (also named an information network) is even more of a challenge.

We are solving the closed discovery problem, which is the topic of research of this chapter and one of the basic assumptions of our methodology. The selected source text documents are originating from at least two dissimilar domains (M1 and M2 contexts by Koestler's naming or A and C domains according to Swanson and his followers). In this chapter, we always describe the methodology using exactly two domains even though it could be generalised to three or more domains.

In this work, the selected knowledge representation formalism is the so-called *bisociative information network*, called *BisoNet*. The BisoNet representation, as investigated in the BISON[1] project and discussed by Kötter and Berthold [12] is a graph representation, consisting of labeled nodes and edges (see Fig. 1). The original idea underlying the BISON project was to have a node for every relevant concept of an application domain, captured by terms denoting these concepts, that is, by *named entities*. For example, if the application domain is drug discovery, the relevant (named) entities are diseases, genes, proteins, hormones, chemical compounds etc. The nodes representing these entities are connected if there is evidence that they are related in some way. Reasons for connecting two terms/concepts can be linguistic, logical, causal, empirical, a conjecture by a human expert, or a co-occurrence observed in documents dealing with considered domains. E.g., an edge between two nodes may refer to a document (for example, a research paper) that includes the represented entities. Unlike semantic nets and ontologies, a BisoNet carries little semantics and to a large extend encodes just circumstantial evidence that concepts are somehow related through edges with some probability.

Open issues in BisoNet creation are how to identify entities and relationships in data, especially from unstructured data like text documents; i.e., which nodes should be created from text documents, what edges should be created, what are the attributes with which they are endowed and how should element weights be computed. Among a variety of solutions, this chapter presents the one that answers such questions by optimizing the main criterion of generated BisoNets: maximizing their bisociation potential. Bisociation potential is a feature of a network that informally states the probability that the network contains a bisociation. Thus, we want to be able to generate such BisoNets that contain as many bisociations as possible using the given data sources. In other words, maximizing the bisociation potential of the generated BisoNet is our main guidance in developing the methodology for creating BisoNets from text documents.

When creating large BisoNets from texts, we have to address the same two issues as in network creation from any other source: define a procedure for identifying key nodes, and define a procedure for discovering relations among the nodes. However, in practice, a workflow for converting a set of documents into a BisoNet is much more complex than just identifying entities and relations. We have to be able to preprocess text and filter out noise, to generate a large number of entities, evaluate their bisociation potential and effectively calculate various distance measures between the entities. As these tasks are not just conceptually difficult, but also computationally very intensive, great care is needed when designing and implementing algorithms for BisoNet construction.

---

[1] Bisociation Networks for Creative Information Discovery: `http://www.BisoNet.eu/`

Our approach to confront the network construction problem is based on developing the following ingredients:

1. Provide basic procedures for automatic text acquisition from different sources of interest on the Web.
2. Employ the state of the art approaches for text preprocessing to extract as much information as available in raw text for the needs of succeeding procedures.
3. Incorporate as much as possible available background knowledge into the stages of text preprocessing and candidate concept detection.
4. Define a candidate concept detection method.
5. Develop a method for relevant bisociative concept extraction from identified concept candidates and perform its evaluation.
6. Select a set of relevant extracted bisociative concepts to form the nodes of a BisoNet.
7. Construct relations between nodes and set their weights according to the Bisociation Index measure published and evaluated by Segond and Borgelt [4].

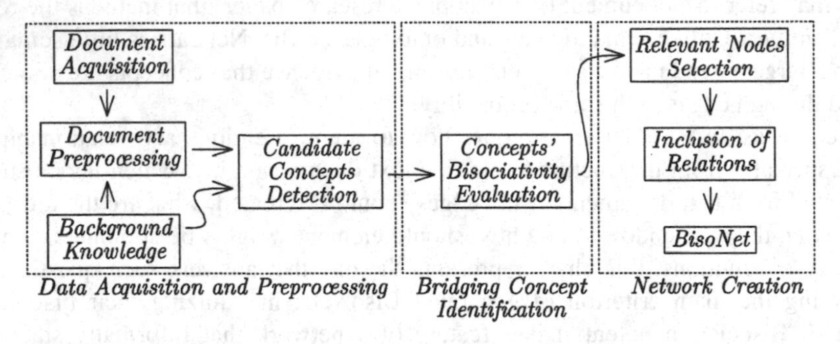

Fig. 2. Conceptual workflow of the proposed solution for BisoNet creation

Fig. 2 illustrates the steps of the methodology proposed by our work. This chapter concentrates mostly on the part of the new methodology for bridging concept evaluation (frame in the middle Fig. 2). As this is an important scientific contribution we provide an evaluation that justifies the design choices in our methodology conception. An evaluation of the final results – BisoNets – is not provided since an experimental evaluation is hard, if not impossible, to construct according to the data we currently possess and work on. We argue that by providing evaluation for high-quality bridging concept identification and evaluation (done in this work) and using the proven bisociative relation measure (defined by Segond and Borgelt [4]), the resulting BisoNets are also of high quality according to the loos defined measure of bisociation potential.

3 Document Acquisition and Preprocessing

This section describes the data preparation part (leftmost frame in Fig. 2) and is written from a technical perspective as it sets grounds for the reproducibility of the subsequent scientifically more interesting steps. Alongside the reproducibility, it addresses also the introduction of some essential text-mining concepts, which are crucial for understanding specific parts of our methodology. A top-level overview of the methodology, discussed along with a description of the actual working system, defines the preprocessing steps supporting the main goal addressed by this work – bisociative concept detection.

The system for text processing proposed and implemented in this work, named TexAs (Text Assistant), was used to produce the results presented in this chapter. The described TexAs implementation is built on top of the LATINO[2] library (Link analysis and text mining toolbox). This library contains a majority of elementary text mining procedures, but, as the creation of BisoNet is a very specific task (in the field of text mining), a lot of modules had to be implemented from scratch or at least optimized considerably.

3.1 Document Acquisition

For the study, we use only one data source, i.e., PubMed[3], which was used to retrieve the datasets (migraine-magnesium) used in the following sections. However, when experimenting with other domains, we identified and partly supported in TexAs the following text acquisition scenarios:

- Using locally stored files in various application dependent formats – this is the traditional setting in data mining; however, it usually requires large amounts of partly manual work for transforming the data between different formats.
- Acquiring documents using the SOAP web services (e.g. PubMed uses SOAP web service interface to access their database).
- Selecting documents from the SQL databases – it is a fast and efficient but rarely available option.
- Crawling the internet gathering documents from web pages (e.g. Wikipedia).
- Collecting documents from snippets returned from web search engines.

3.2 Document Preprocessing

In addition to explaining various aspects of preprocessing, this section also briefly describes basic text mining concepts and terminology, some of which are taken from the work of Feldman and Sanger [8]. Preprocessing is an important part of network extraction from text documents. Its main task is the transformation of unstructured data from text documents into a predefined well-structured data representation. As shown below, preprocessing is inevitability very tightly connected to the extraction of network entities. In our case, actual bisociative concept candidates are defined already when preprocessing is finished. The subsequent processing step 'only' ranks the entities and to remove the majority of lower ranked entities from the set.

---

[2] LATINO library: `http://sourceforge.net/projects/latino/`

[3] PubMed: A service of U.S. National Library of Medicine, which comprises more than 20 million citations for biomedical literature: `http://www.ncbi.nlm.nih.gov/pubmed`

In general, the task of preprocessing consists of the extraction of *features* from text documents. The set of all features selected for a given document collection is called a *representational model*. Each document is represented by a vector of numerical quantities – one for each aligned feature of the selected representational model. Using this construction, we get the most standard text mining document representation called feature vectors where each numerical component of a vector is related to a feature and represents a form of weight related to the importance of the feature in the selected document. Usually the majority of weights in a vector are equal to zero showing that one of the characteristics of feature vectors is their sparseness – they are often referred to as sparse vectors. The goal of preprocessing is to extract a feature vector for each document from a given document collection.

Commonly used document features are characters, words, terms and concepts [8]. Characters and words carry little semantic information and are therefore not interesting to consider. Terms and concepts on the contrary carry much more semantic information. Terms are usually considered as single or multiword phrases selected from the corpus by means of term-extraction mechanisms (e.g. because of their high frequency) or are present in an external lexicon of a controlled vocabulary. Concepts or keywords are features generated for documents employing the categorization or annotation of documents. Common concepts are derived from manually annotating a document with some predefined keywords or by inserting a document into some predefined hierarchy. When we refer to document features, we mean the terms and the concepts that we were able to extract from the documents. In the rest of this chapter, we do not distinguish between terms or concepts. In the case if a document set contains both, we merge them and pretend that we have only one type of document features, i.e. terms/concepts.

A standard collection of preprocessing techniques [8] is listed below, together with a set of functionalities implemented in our system TexAs:

- *Tokenization*: continuous character stream must be broken up into meaningful sub-tokens, usually words or terms in the case where a controlled vocabulary is present. Our system uses a standard Unicode tokenizer: it mainly follows the Unicode Standard Annex #29 for Unicode Text Segmentation[4]. The alternative is a more advanced tokenizer, which tokenizes strings according to a predefined controlled vocabulary and discards all the other words/terms.
- *Stopword removal*: stopwords are predefined words from a language that usually carry no relevant information (e.g. articles, prepositions, conjunctions etc.); the usual practice is to ignore them when building a feature set. Our implementation uses a predefined list of stopwords – some common lists that are already included in the library are taken from Snowball[5].
- *Stemming or lemmatization*: the process that converts each word/token into the morphologically neutral form. The following alternatives have been made

[4] Unicode Standard Annex #29:
`http://www.unicode.org/reports/tr29/#Word_Boundaries`

[5] Snowball – A small string processing language designed for creating stemming algorithms:
`http://snowball.tartarus.org`

available: Snowball stemmers, the Porter stemmer [17], and the one that we prefer, the LemmaGen lemmatization system [10].

- *Part-of-speech (POS) tagging*: the annotation of words with the appropriate POS tags based on the context in which they appear.
- *Syntactic parsing*: performs a full syntactical analysis of sentences according to a certain grammar. Usually shallow (not full) parsing is used since it can be efficiently applied to large text corpora.
- *Entity extraction*: methods that identify which terms should be promoted to entities and which not. Entity extraction by grouping words into terms using n-gram extraction mechanisms (an n-gram is a sequence of n items from a given sequence) has been implemented in TexAs.

3.3 Background Knowledge

Since high-quality features are hard to acquire, all possible methods that could improve this process should be used at this point. The general approach that usually helps the most consists in incorporating background knowledge about the documents and their domain. The most elegant technique to incorporate background knowledge is to use a controlled vocabulary. A controlled vocabulary is a lexicon of all terms that are relevant in a given domain. Here we can see a major difference when processing general documents as compared to scientific documents. For many scientific domains there exists not only a controlled vocabulary, but also a pre-annotation for a lot of scientific articles. In this case we can quite easily create feature vectors since we have terms as well as concepts already pre-defined. Other interesting approaches to identifying concepts include methods such as KeyGraph [13], which extract terms and concepts with minimal assumptions or background knowledge, even from individual documents. Other alternatives are using domain ontologies which could be, for example, semi-automatically retrieved by a combination of tools such as OntoGen and TermExtractor [9].

3.4 Candidate Concept Detection

The design choice of our approach is that the entities of the BisoNets will be the features of documents, i.e., the terms and concepts defined in the previous section. The subsequent steps are independent of term and concept detection procedure.

Entities need to be represented in a way which enables efficient calculation of different distance measures between the entities. We chose a representation in which an entity is described by a set (vector) of documents in which it appears. In the same way as documents are represented as sparse vectors of features (entities), the entities can also be represented as sparse vectors of documents. This is illustrated in Example 1 where entity ent_1 is present in documents doc_1, doc_3 and doc_4 and hence its feature vector consists of all these documents (with appropriate weights). By analogy to the original vector space – the *feature space* – the newly created vector space is named a *document space*.

Documents	Extracted entities
doc_1	ent_1, ent_2, ent_3
doc_2	ent_3, ent_4, ent_4
doc_3	ent_1, ent_2, ent_2, ent_5
doc_4	ent_1, ent_1, ent_1, ent_3, ent_4, ent_4

Original documents and extracted entities

Feature space	ent_1	ent_2	ent_3	ent_4	ent_5
doc_1	$w^f_{1:1}$	$w^f_{1:2}$	$w^f_{1:3}$		
doc_2			$w^f_{2:3}$	$w^f_{2:4}$	
doc_3	$w^f_{3:1}$	$w^f_{3:2}$			$w^f_{3:5}$
doc_4	$w^f_{4:1}$		$w^f_{4:3}$	$w^f_{4:4}$	

Sparse matrix of documents: $w^f_{x:y}$ denotes the weight (in the feature space) of entity y in the feature vector of document x

Document space	doc_1	doc_2	doc_3	doc_4
ent_1	$w^d_{1:1}$		$w^d_{1:3}$	$w^d_{1:4}$
ent_2	$w^d_{2:1}$		$w^d_{2:3}$	
ent_3	$w^d_{3:1}$	$w^d_{3:2}$		$w^d_{3:4}$
ent_4		$w^d_{4:2}$		$w^d_{4:4}$
ent_5			$w^d_{5:3}$	

Sparse matrix of entities: $w^d_{x:y}$ denotes the weight (in the document space) of document y in the document vector of entity x

Example 1: Conversion between the feature and the document space

Note that if we write document vectors in the form of a matrix, then the conversion between the feature space and the document space is performed by simply transposing the matrix (see Example 1). The only question that remains open for now is what to do with the weights? Is weight $w^f_{x:y}$ identical to weight $w^d_{y:x}$? This depends on various aspects, but mostly on how we define weights of the entities in the first place when defining document vectors.

There are four most common weighting models for assigning weights to features:

- *Binary*: a feature weight is either one, if the corresponding feature is present in the document, or zero otherwise.
- *Term occurrence*: a feature weight is equal to the number of occurrences of this feature. This weight might be sometimes better than a simple binary since frequently occurring features are likely more relevant as repetitions indicate that the text is strongly concerned with them.
- *Term frequency*: a weight is derived from the term occurrence by dividing the vector by the sum of all vector's weights. The reasoning of the quality of such weight is similar to term occurrence with the additional normalization that equalizes each document importance – regardless of its length.
- *TF-IDF*: Term Frequency-Inverse Document Frequency is the most common scheme for weighting features. It is usually defined as:

$$w^{TFIDF}_{x:y} = \text{TermFreq}(ent_x, doc_y) \cdot \log(N/DocFreq(ent_x)),$$

where $\text{TermFreq}(ent_x, doc_y)$ is the frequency of feature ent_x inside document doc_y (equivalent to term frequency defined in bullet point above), N is the number of all documents and $DocFreq(ent_x)$ is the number of documents that contain ent_x. The idea behind the TF-IDF measure is to lower the weight of features that appear in many documents as this is usually an indication of them being less important (e.g. stopwords). The quality of this approach has also been quantitatively proven by numerous usages in solutions to various problems in text-mining.

These four methods can be further modified by vector normalization (dividing each vector so that the length – usually the Euclidian or Manhattan length – of the vector is 1). If and when this should be done depends on several factors: one of them is the decision which distance measure will be used in the next, the relation construction step. If the cosine similarity is used, a pre-normalization of the vectors is irrelevant, as this is also done during the distance calculation. Example 2 shows the four measures in practice – documents are taken from the first table in Example 1. Weights are calculated for the feature space and are not normalized.

It is worthwhile to note again the analogy between the feature space and the document space. Although we have developed the methodology for entities network extraction, the developed approach can be used also for document network extraction. Moreover, both approaches can be used to extract a unified network representation where documents and entities are nodes, connected using some special relations.

	ent_1	ent_2	ent_3	ent_4	ent_5
doc_1	1	1	1		
doc_2			1	1	
doc_3	1	1			1
doc_4	1		1	1	

Binary weight

	ent_1	ent_2	ent_3	ent_4	ent_5
doc_1	1	1	1		
doc_2			1	2	
doc_3	1	2			1
doc_4	3		1	2	

Term occurrence

	ent_1	ent_2	ent_3	ent_4	ent_5
doc_1	$^1/_3$	$^1/_3$	$^1/_3$		
doc_2			$^1/_3$	$^2/_3$	
doc_3	$^1/_4$	$^2/_4$			$^1/_4$
doc_4	$^3/_6$		$^1/_6$	$^2/_6$	

Term frequency

	ent_1	ent_2	ent_3	ent_4	ent_5
doc_1	$(^1/_3)\cdot\log(^4/_3)$	$(^1/_3)\cdot\log(^4/_2)$	$(^1/_3)\cdot\log(^4/_3)$		
doc_2			$(^1/_3)\cdot\log(^4/_3)$	$(^2/_3)\cdot\log(^4/_2)$	
doc_3	$(^1/_4)\cdot\log(^4/_3)$	$(^2/_4)\cdot\log(^4/_2)$			$(^1/_4)\cdot\log(^4/_1)$
doc_4	$(^3/_6)\cdot\log(^4/_3)$		$(^1/_6)\cdot\log(^4/_3)$	$(^2/_6)\cdot\log(^4/_2)$	

TF-IDF: term frequency – inversed document frequency

Example 2: Weighting models of features in document vectors (from Example 1)

3.5 Distance Measures between Vectors

Although distance calculation addressed in this section is not used in the document preprocessing step, it is explained at this point since the content is directly related to the Section 3.4, and since the distance measures are extensively used in the two following sections about bridging concept identification as well as network creation.

The most common measures in vector spaces, which are also implemented in our system TexAs, are the following:

- Dot product: $\text{DotProd}(vec_x, vec_y)$.
- Cosine similarity: $\text{CosSim}(vec_x, vec_y) = \frac{\text{DotProd}(vec_x, vec_y)}{|vec_x| \cdot |vec_y|}$.
  is the dot product normalized by the length of the two vectors. In the cases where the vectors are already normalized, the cosine similarity is identical to the dot product.
- Jaccard index: this similarity coefficient measures the similarity between sample sets. It is defined as the cardinality of the intersection of the sample sets:
  $$\text{JaccInx}(vec_x, vec_y) = \frac{|vec_x \cap vec_y|}{|vec_x \cup vec_y|} = \frac{\text{DotProd}(vec_x, vec_y)}{|vec_x| + |vec_y| - \text{DotProd}(vec_x, vec_y)},$$
  where lengths $|vec_x|$ and $|vec_y|$ are Manhattan lengths of these vectors.
- Bisociation index: it is the similarity measure defined for the purpose of bisociation discovery in the BISON project. It is explained in more detail in [4]. This measure cannot be expressed by the dot product. Therefore, the following definition uses the notation from Example 1:
  $$\text{BisInx}(vec_x, vec_y) = \sum_{i=0}^{M} \left(\sqrt[k]{w_{x:i} \cdot w_{y:i}} \cdot \left(1 - \frac{|\tan^{-1}(w_{x:i}) - \tan^{-1}(w_{y:i})|}{\tan^{-1}(1)}\right)\right),$$
  where M is the number of all the entities.

In general, the choice of a suitable distance measure should be tightly connected to the choice of the weighting model. Some of the combinations are very suitable and have understandable interpretations or were experimentally evaluated as useful, while others are less appropriate. We list the most commonly used pairs of weighting model and distance measure below:

- *TF-IDF weighting and cosine similarity*: this is the standard combination for computing the similarity in the feature space.
- *Binary weighting and dot product distance*: if this is used in the document space the result is the co-occurrence measure, which counts the number of documents where two entities appear together.
- *Term occurrence weighting and dot product distance*: this is another measure of co-occurrence of entities in the same documents. Compared to the previous measure, this one considers also multiple co-occurrences of two entities inside a document and gives them a greater weight in comparison with the case were each appears only once inside the same document.
- *Binary weighting and Jaccard index distance*: Jaccard index was primary defined on sets, therefore the most suitable weighting model to use with it is the binary weighting model (since every vector then represents a set of features).
- *Term frequency weighting and the Bisociation Index distance*: the Bisociation Index was designed with the term frequency weighting in mind, thus it is reasonable to use this combination when determining a weighting model for the Bisociation index.

4 Identifying Bridging Concept Candidates for High Quality Network Entities Extraction

This section presents the key part of our methodology for bisociative bridging terms identification. We propose a set of heuristics which are promising for b-term discovery. In Section 5 we use them to rank all the terms from a document collection and thus obtain some terms which have a higher probability of being b-terms than a randomly selected term.

4.1 Heuristics Description

Heuristics are functions that numerically evaluate the term's quality by assigning a *bisociation score* (tendency that a term is a b-term) to it. For the definition of an appropriate set of heuristics we define a set of special (mainly statistical) properties of terms which will separate b-terms from regular terms. Thus, these heuristics can also be viewed as advanced term statistics.

All heuristics operate on the data retrieved from the documents in preprocessing or obtained from the background knowledge. Using an ideal heuristic and sorting all the terms by the its calculated bisociation scores should result in finding all the b-terms at the top of a list. However, sorting by actual heuristic bisociation scores (either ascending or descending) should still bring much more b-terms than non b-terms to the top of the term list.

Formally, a heuristic is a function with two inputs, i.e., a set of domain labeled documents D and a term t appearing in these documents, and one output, i.e., a number that correlates with the term's bisociation score.

In this chapter we use the following notation: to say that the bisociation score b is equal to the result of a heuristic named *heurX*, we can write it as $b = heurX\ (D,t)$. However, since the set of input documents is static when dealing with a concrete dataset, we can – for the sake of simplicity – omit the set of input documents from a heuristic notation and use only $b = heurX\ (t)$. Whenever we need to explicitly specify the set of documents on which the function works (never needed for a heuristic, but sometimes needed for auxiliary functions used in a formula for a heuristic), we write it as $funcX_D(t)$. For specifying an auxiliary function's document set we have two options: either we use D_u that stands for the (union) set of all the documents from all the domains, or we use $D_n\colon n \in \{1..N\}$, which stands for a set of documents from the domain n. In general the following statement holds: $D_u = \bigcup_{n=1}^{N} D_n$ where N is the number of domains. In the most common scenario, where we have exactly two distinct domains, we also use the notation D_A for D_1 and D_C for D_2, since we introduced A and C as representatives of the initial and the target domain in the closed discovery setting introduced in Section 1. Due to a large number of heuristics and auxiliary functions we use a multi word naming scheme for easier distinction; names are formed by word concatenation and capitalization of all non-first words (e.g.: *freqProdRel* and *tfidfProduct*).

It is valuable to note that all the designed heuristics are symmetric in the domains, as switching the order of domains (which domain is the initial domain and which is the target) should not affect the outcome of a heuristic. By allowing asymmetric heuristics the approach would lose generality and also the possibility to generalize it to more than two domains.

We divided the heuristics into different sets for easier explanation; however, most of the described heuristics work fundamentally in a similar way – they all manipulate solely the data present in document vectors and derive the terms' bisociation score. The only exceptions to this are the outlier based heuristics which firstly calculate outlier documents and only later use the information from the document vectors.

The heuristics can be logically divided into four sets which are based on: frequency, tf-idf, similarity, and, outliers. Besides those sets we define also two special heuristics which are used as a baseline for other heuristics.

4.2 Frequency Based Heuristics

For easier definition of frequency based heuristics we need two auxiliary sub-functions:

- $countTerm_D(t)$: counts the number of occurrences of term t in a document set D (called term frequency in tf-idf related contexts),
- $countDoc_D(t)$: counts the number of documents in which term t appears in a document set D, (called document frequency in tf-idf related contexts).

We define the following basic heuristics:

(1) $freqTerm(t) = countTerm_{D_u}(t)$: term frequency across both domains,

(2) $freqDoc(t) = countDoc_{D_u}(t)$: document frequency across both domains,

(3) $freqRatio(t) = \frac{countTerm_{D_u}(t)}{countDoc_{D_u}(t)}$: term to document frequency ratio,

(4) $freqDomnRatioMin(t) = \min\left(\frac{countTerm_{D_1}(t)}{countTerm_{D_2}(t)}, \frac{countTerm_{D_2}(t)}{countTerm_{D_1}(t)}\right)$: minimum of term frequencies ratio between both domains,

(5) $freqDomnProd(t) = countTerm_{D_1}(t) \cdot countTerm_{D_2}(t)$: product of term frequencies in both domains,

(6) $freqDomnProdRel(t) = \frac{countTerm_{D_1}(t) \cdot fcountTerm_{D_2}(t)}{countTerm_{D_u}(t)}$: product of term frequencies in both domains relative to the term frequency in all domains.

4.3 Tf-idf Based Heuristics

Tf-idf is the standard measure of term's importance in a document which is used heavily in text mining research. In the following heuristic definitions we use the following auxiliary functions:

- $tfidf_d(t)$ stands for tf-idf of a term t in a document d, and,
- $tfidf_D(t)$ represents tf-idf of a term in the centroid vector of all the documents $d: d \in D$. The centroid vector is defined as an average of all document vectors and thus presents an average document from the document collection D

Heuristics based on tf-idf are listed below:

(7) $tfidfSum(t) = \sum_{d \in D_u} tfidf_d(t)$: sum of all tf-idf weights of a term across both domains – analogy to $freqTerm(t)$,

(8) $tfidfAvg(t) = \frac{\sum_{d \in D_u} tfidf_d(t)}{freq\_doc_{D_u}(t)}$: average tf-idf of a term,

(9) $tfidfDomnProd(t) = tfidf_{D_1}(t) \cdot tfidf_{D_2}(t)$: product of a term's importance in both domains.

(10) $tfidfDomnSum(t) = tfidf_{D_1}(t) + tfidf_{D_2}(t)$: sum of a term's importance in both domains.

4.4 Similarity Based Heuristics

Another approach to construct a relevant heuristic measures is to use the cosine similarity measure. We start by creating a representational model as a document space and by converting terms (entities) into document vectors (see section 3.4). Next, we get the centroid vectors for both domains in the document space representation. Furthermore, we apply tf-idf weighting on top of all the newly constructed vectors and centroids. Finally we use the following auxiliary function to construct the heuristics:

- $simCos_D(t)$: calculates the cosine similarity of the document vector of term t and the document vector of a centroid of documents $d \in D$.

Constructed heuristics:

(11) $simAvgTerm(t) = simCos_{D_u}(t)$: similarity to an average term – the distance from the center of the cluster of all terms,

(12) $simDomnProd(t) = simCos_{D_1}(t) \cdot simCos_{D_2}(t)$: product of a term's similarity to the centroids of both domains,

(13) $simDomnRatioMin(t) = \min\left(\frac{simCos_{D_1}(t)}{simCos_{D_2}(t)}, \frac{simCos_{D_2}(t)}{simCos_{D_1}(t)}\right)$: minimum of a term's frequencies ratio between both domains.

4.5 Outlier Based Heuristics

Conceptually, an outlier is an unexpected event, entity or – in our case – document. We are especially interested in outlier documents since they frequently embody new information that is often hard to explain in the context of existing knowledge. Moreover, in data mining, an outlier is frequently a primary object of study as it can potentially lead to the discovery of new knowledge. These assumptions are well aligned with the bisociation potential that we are optimizing, thus, we have constructed a couple of heuristics that harvest the information possibly residing in outlier documents.

We concentrate on a specific type of outliers, i.e., domain outliers, which are the documents that tend to be more similar to the documents of the opposite domain than to those of their own domain. The procedures that we use to detect outlier documents build a classification model for each domain and afterwards classify all the documents using the trained classifier. The documents that are misclassified are declared as outlier documents, since according to the classification model they do not belong to their domain of origin.

We defined three different outlier sets based on three classification models used. These outlier sets are:

- D_{CS}: retrieved by Centroid Similarity (CS) classifier,
- D_{RF}: retrieved by Random Forest (RF) classifier,
- D_{SVM}: retrieved by Support Vector Machine (SVM) classifier.

Centroid similarity is a basic classifier model and is also implemented in the TexAs system. It classifies each document to the domain whose centroid's tf-idf vector is the most similar to the document's tf-idf vector. The description of the other two classification models is beyond the scope of this chapter, as we used external procedures to retrieve these outlier document sets. The detailed description is provided by Sluban et al. [20].

For each outlier set we defined two heuristics: the first counts the frequency of a term in an outlier set and the second computes the relative frequency of a term in an outlier set compared to the relative frequency of a term in the whole dataset. The resulting heuristics are listed below:

(14) $outFreqCS(t) = countTerm_{D_{CS}}(t)$: term frequency in CS outlier set,
(15) $outFreqRF(t) = countTerm_{D_{RF}}(t)$: term frequency in RF outlier set,
(16) $outFreqSVM(t) = countTerm_{D_{SVM}}(t)$: term frequency in SVM outlier set,
(17) $outFreqSum(t) = countTerm_{D_{CS}}(t) + countTerm_{D_{RF}}(t) + countTerm_{D_{SVM}}(t)$: sum of term frequencies in all three outlier sets,
(18) $outFreqRelCS(t) = \frac{countTerm_{D_{CS}}(t)}{countTerm_{D_u}(t)}$: relative frequency in CS outlier set,
(19) $outFreqRelRF(t) = \frac{countTerm_{D_{RF}}(t)}{countTerm_{D_u}(t)}$: relative frequency in RF outlier set,
(20) $outFreqRelSVM(t) = \frac{countTerm_{D_{SVM}}(t)}{countTerm_{D_u}(t)}$: relative frequency in SVM outlier set,
(21) $outFreqRelSum(t) = \frac{countTerm_{D_{CS}}(t) + countTerm_{D_{RF}}(t) + countTerm_{D_{SVM}}(t)}{countTerm_{D_u}(t)}$: sum of relative term frequencies in all three outlier sets.

4.6 Baseline Heuristics

We have two other heuristics which are supplementary and serve as a baseline for the others. The auxiliary functions used in their calculation are:

- $randNum()$: returns random number from the interval (0,1) regardless of the term under investigation,
- $inBoth(t)$: 1 if a term t appears in both domains and 0 otherwise.

The two baseline heuristics are:

(22) $random(t) = randNum()$: random baseline heuristic,
(23) $appearInAllDomn(t) = inBoth(t) + (randNum())/2$: it is a better baseline heuristic which can separate two classes of terms – the ones that appear in both domains and the ones that appear only in one. The terms that appear only in one domain have a strictly lower heuristic score than those that appear in both. The score inside of these two classes is still random.

5 Heuristics Evaluation

This section presents the evaluation of the heuristics defined in the previous section. First we describe the evaluation procedure, then the domain on which we evaluate the heuristics is presented, and finally the results of the evaluation along with the discussion of the results.

5.1 Evaluation Procedure

In the experimental setting used in this chapter we are given the following: a set of documents from two domains and a "gold standard" list of b-terms. Consequently, we are able to mark the true b-terms and evaluate how well our constructed heuristics are able to promote these b-terms compared to the rest of the terms.

We compare the heuristics using ROC (Receiver Operating Characteristic) curve and AUC (Area Under ROC) analysis. Some ideas on using the ROC for our evaluation were taken from Foster et al. [18]. ROC curves are constructed in the following way:

- Sort all the terms by their descending heuristic score.
- Starting from the beginning of the term list, do the following for each term: if a term is a b-term, then draw one vertical line segment (up) on the ROC curve, else draw one horizontal line segment (right) on the ROC curve.
- Sometimes, a heuristic outputs the same score for many terms and therefore we cannot sort them uniquely. Among terms with the same bisociation score b, let b_b be the number of terms that are b-terms and nb_b the number of non-b-terms. We then draw a line from the current point p to the point $p + (nb_b, b_b)$. In this way we may produce slanted lines, if such an equal scoring term set contains both b-terms and non b-terms.

Using the stated procedure, we get one ROC curve for each heuristic. The ROC space is defined by its two axes. The ROC's vertical axis scale goes from zero to the number of b-terms and the horizontal goes from zero to the number of non b-terms. AUC is defined as the percentage of the area under curve – the area under the curve is divided by the area of the whole ROC space. If a heuristic is perfect (it detects all the b-terms and ranks them at the top of the ordered list), we get a curve that goes first just up and then just right with an AUC of 100%. The worst possible heuristic sorts all the terms randomly regardless of being a b-term or not and achieves AUC of 50%. This random heuristic is represented by the diagonal in the ROC space.

The fact that some heuristics output the same score for many terms can produce different sorted lists and thus different performance estimates for the same heuristic on the same dataset. In the case of such equal scoring term sets, the inner sorting is random (which indeed produces different performance estimates). However, the ROCs that are provided (and constructed by the instructions in the paragraph above) correspond to the average ROC over all possible such random inner sortings. Besides AUC, we list also the interval of AUC which tells how much each heuristic varies among the best and the worst sorting of a possibly existing equal scoring term set. Preferable are the heuristics with a smaller interval which implies that they produce smaller and fewer equal scoring sets.

5.2 Migraine-Magnesium Dataset

This section describes the dataset used to evaluate the heuristics' potential of successful b-term identification. The dataset that we used is the well-researched *migraine-magnesium* domain pair which was introduced by Swanson [24] and later explored by several authors in several studies [25, 28, 26, 14]. In the literature-based discovery process Swanson managed to find more than 60 pairs of articles connecting the migraine domain with the magnesium deficiency via 43 b-terms. In our evaluation we are trying to rediscover these b-terms stated by Swanson to connect the two domains (see Table 1).

Table 1. B-terms identified by Swanson et al. in [26]

1 5 ht	16 convulsive	31 prostaglandin
2 5 hydroxytryptamine	17 coronary spasm	32 prostaglandin e1
3 5 hydroxytryptamine receptor	18 cortical spread depression	33 prostaglandin synthesis
4 anti aggregation	19 diltiazem	34 reactivity
5 anti inflammatory	20 epilepsy	35 seizure
6 anticonvulsant	21 epileptic	36 serotonin
7 antimigraine	22 epileptiform	37 spasm
8 arterial spasm	23 hypoxia	38 spread
9 brain serotonin	24 indomethacin	39 spread depression
10 calcium antagonist	25 inflammatory	40 stress
11 calcium blocker	26 nifedipine	41 substance p
12 calcium channel	27 paroxysmal	42 vasospasm
13 calcium channel blocker	28 platelet aggregation	43 verapamil
14 cerebral vasospasm	29 platelet function	
15 convulsion	30 prostacyclin	

The dataset contains scientific paper titles which were retrieved by querying the PubMed database with the keyword *"migraine"* for the migraine domain and with the keyword *"magnesium"* for the magnesium domain. Additional condition to the query was the publishing date which was limited to before the year 1988, since Swanson's original experiment – which we want to reproduce – also considered only articles published before that year. The query resulted in 8,058 titles (2,425 from the migraine domain and 5,633 from the magnesium domain) of the average length of 11 words. We preprocessed the dataset using the standard procedures described in Section 3.2 and by additionally specifying terms as n-grams of maximum length 3 (max. three words were combined to form a term) with minimum occurrence 2 (each n-gram had to appear at least twice to be promoted to a term). Using this preferences we produced a dataset containing 13,525 distinct terms or 1,847 distinct terms that appear at least once in each domain; both numbers include also all the 43 terms that Swanson marked as b-terms. An average document in the dataset consists of 12 terms and 394 (4,89%) documents contain at least one b-term.

5.3 Comparison of the Heuristics

This section presents the results of the comparison of the heuristics on the magnesium-migraine dataset using ROC analysis. The experimental setting was presented in detail in the previous sections. Nevertheless, for the purpose of this evaluation, it was slightly extended, due to additional knowledge about b-terms in this domain (this may be a general observation for any future domain). We realized that all the 43 b-terms appear in both domains; therefore, it is more fair for the comparison that the heuristics are also aware of this fact. Therefore, we made sure that every heuristic ordered all the terms that appear in both datasets (1,847 terms) before all the other terms (11,678 terms), however, every heuristic used its own score for ordering within these two sets of terms. In this way, we incorporated the stated background knowledge about b-terms in this domain into all the heuristics.

Table 2. Comparison of the results of all the defined heuristics ordered by the quality – AUC. The first column states the name of the heuristic; the second displays a percentage of the area under the ROC curve; and the last is the nterval of AUC.

Heuristic	AUC	Interval
(21) *outFreqRelSum*	95,33%	0,35%
(19) outFreqRelRF	95,24%	0,55%
(20) outFreqRelSVM	95,06%	1,26%
(18) outFreqRelCS	94,96%	1,30%
(17) outFreqSum	94,96%	0,70%
(8) tfidfAvg	94,87%	0,00%
(15) outFreqRF	94,73%	1,53%
(16) outFreqSVM	94,70%	2,06%
(14) outFreqCS	94,67%	1,80%
(4) freqDomnRatioMin	94,36%	0,62%
(10) tfidfDomnSum	93,85%	0,35%
(6) freqDomnProdRel	93,71%	0,40%
(13) simDomnRatioMin	93,58%	0,00%
(7) tfidfSum	93,58%	0,00%
(9) tfidfDomnProd	93,47%	0,39%
(5) freqDomnProd	93,42%	0,44%
(3) freqRatio	93,35%	5,23%
(23) *appearInAllDomn*	93,31%	6,69%
(12) simDomnProd	93,27%	0,00%
(1) freqTerm	93,20%	0,50%
(2) freqDoc	93,19%	0,50%
(11) simAvgTerm	92,71%	0,00%
(22) *random*	50,00%	50,00%

The first look at numerical result comparison (Table 2) reveals the following:

- The overall AUC results of all heuristics, except for the (22)random baseline, are relatively good and in the range of from approx. 93% to 95%.
- The difference among AUC results is small (only 2.5% between the worst and the best performing heuristic).
- The improved baseline heuristic (23)appearInAllDomn performs well and is not worse than some other heuristics.
- Outlier based heuristics seem to perform the best.
- Some heuristics, including the best performing ones, have a relatively high AUC interval which means that they output the same score for many terms.

Observing the results in Table 2, followed by the detailed ROC analysis described below, we selected the best heuristic that will be used as the heuristic for network node weighting, which is the final result of this work. The chosen heuristic is simply the first from the list in Table 2 – (21)outFreqRelSum – due to the fact that it has highest AUC and especially since it shows a low uncertainty. In other words, it has

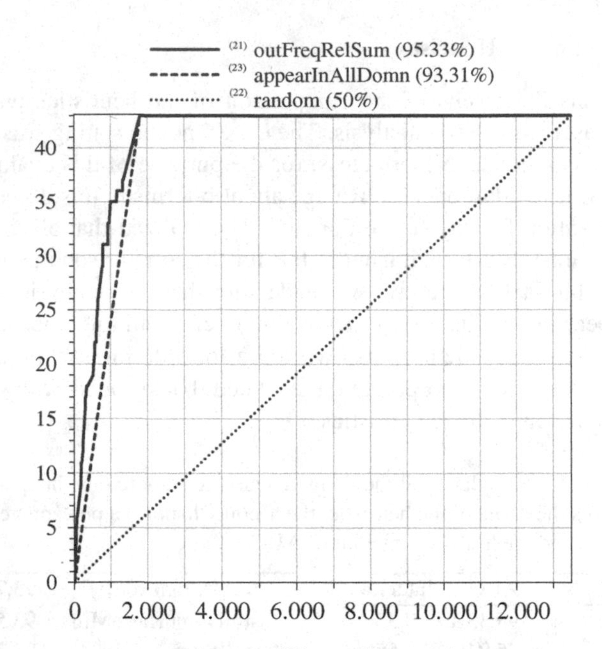

Fig. 3. ROC curve of the selected heuristic (21)outFreqRelSum along with the baseline heuristic (22)random and improved baseline heuristic (23)appearInAllDomn on detecting the 43 b-terms among all 13,525 candidate concepts.

small AUC interval, which means that it better defines the position of b-terms and we do not need to rely so much on random sorting of potential equal scoring term sets. We also assume it to be less volatile across domains since it actually represents cooperation (sum) of three other well performing heuristics: (19)outFreqRelRF, (20)outFreqRelSVM, and, (18)outFreqRelCS.

Detailed ROC curve analysis of the chosen heuristic (see Fig. 3) shows that our heuristic is only slightly better than the improved baseline heuristic, which is evident also from Table 2. However, when examined carefully we perceive the property of the heuristic which is the initial assumption of this research, i.e., extremely steep incline at the beginning of the curve which is much steeper than the incline of the baseline heuristics. This means that the chosen heuristic is able to detect b-terms at the beginning of the ordered list much faster than the baseline. The steep incline is even more evident in Fig. 4.

Fig. 4 shows the zoom-in perspective on the ROC curves of the selected outlier based heuristics – enumerated from (18) to (21) – along with the baselines. The zoom-in (applied also in Fig. 5) refers to the axis x since we show only 1,804 terms which is the point where all the heuristics (except (22)random) reach the top point (43 found b-terms). In Fig. 4 we can see the steep incline property of the (21)outFreqRelSum even more clearly. At the position of the first tick on the axis x (by the term 50 in the ordered list of terms) the chosen heuristic is able to detect already 5-6 b-terms while the baseline heuristic only approximately one. Similarly, we notice at the 200th term the baseline heuristics detects 5 b-terms while (21)outFreqRelSum detects already 11.

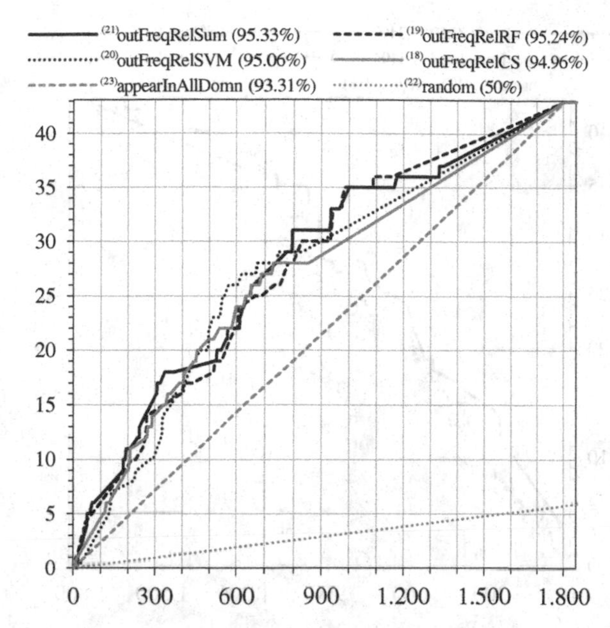

Fig. 4. ROC curves of the best-performing set of heuristic – relative frequency of a term in outlier sets – along with both baseline heuristics on detecting the b-terms among only 1,847 candidate concepts (only the concepts that appear in both domains)

If we follow the curve further we see a decrease in relative difference; nevertheless, at the 1000th term the ratio is still 24:35, even though the performance here is not of such importance as the performance at the beginning of the curve. The presented behavior at the beginning of the curve is highly appreciated especially from the expert's point of view who needs to go through such an ordered list of terms and detect potential b-terms. In such a setting we would really want to present some valuable b-terms at the very beginning of the list, even if other b-terms are dispersed evenly across it.

Even though we chose the heuristic from the outlier set we are still interested how the heuristics from the other sets performed. This comparison is presented in Fig. 5 where we show one (the best performing one) heuristic from each set of heuristics. Notice the outlier heuristic (19)outFreqRelRF which undoubtedly wins. It is harder to establish an order between the other three heuristics. The undesired property is exposed by (13)simDomnRatioMin where the ROC curve shows performance worse than (23)appearInAllDomn at the right side of the curve; however, even this would be tolerable if there is outperformance at the beginning of the curve. The conclusion for the other sets (besides the outlier one) is that even though they are slightly better than the baseline heuristic we are not able to infer their significant outperformance over it.

Overall, the results of the evaluation are beneficial for the insight into heuristic performance on the examined migraine-magnesium dataset. The conclusion is that it is extremely hard to promote b-terms in an ordered list of terms by observing only the terms' statistical properties in the documents. However, we managed to construct a well performing heuristic which is based on relative frequency of a term in three outlier sets of all the documents. The outlier sets of documents are retrieved using

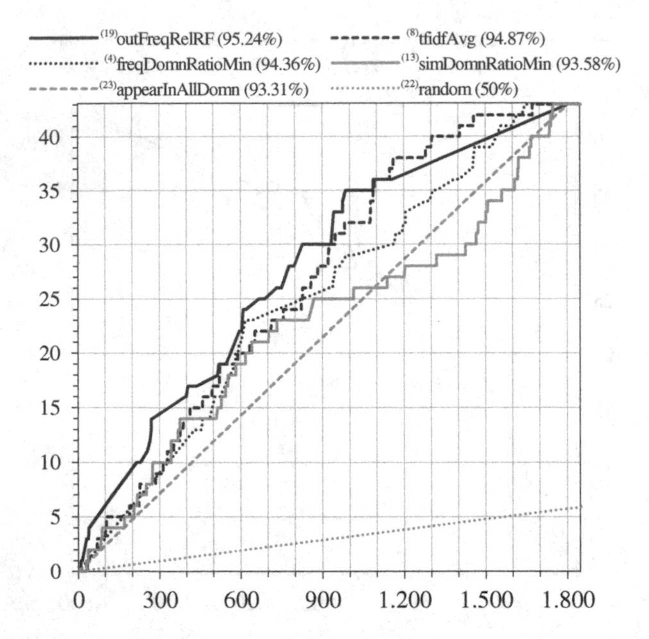

Fig. 5. ROC curves of the best-performing heuristics – one from each set (based on: frequency, tf-idf, similarity, outliers) along with both baseline heuristics on detecting the b-terms among only 1,847 candidate

three types of classifiers: Centroid Similarity, Random Forest, and, Support Vector Machine. The conclusion of our evaluation is well aligned with the results presented by Sluban et al. [20] and Petrič et al. [16].

The presented chapter motivated our future work in several directions of which we will first proceed with the following:

- Reevaluate the findings on a new independent test domains. We have already done some initial tests on the autism-calcineurin domain pair presented by Urbančič et al. [27], which show similar results as the presented evaluation.
- Try to do some further research on heuristics based on statistical properties of the terms. If no heuristics which outperform (23)appearInAllDomn is found, we will consider completely abandoning this type of heuristics.
- Add some new, fundamentally different classes of heuristics to rank the terms. We have a couple of ideas to try, including using SVM keywords (SVM trained to separate between domains) as potential b-terms with high score.
- Implement the findings of this research as a web application where the user (a domain expert) will be able to perform an experimentation and b-term retrieval on his own domains of interest.

6 Network Creation

This section briefly presents the ideas behind the creation of a BisoNet – an information network of concepts identified and weighted by the presented methodology.

The initial plan for BisoNet construction is first to take all the terms/concepts identified in the preprocessing step, next to weight them using the bisociation score of the $^{(21)}$outFreqRelSum heuristic and finally to add links among concepts according to the Bisociation Index measure defined by Segond and Borgelt [4].

Table 3. The 40 highest ranked terms using the preferred heuristic $^{(21)}$outFreqRelSum along with the weights (bisociation score) retrieved by the same heuristic. There are 5 gold standard b-terms in this list and they are all marked with asterisks.

#	Term	Score	#	Term	Score
1	sturge	3.50	26	cerebral artery	2.50
2	sturge weber	3.50	27	medication	2.50
3	weber	3.50	28	animal human	2.50
4	inflammatory agent	3.00	29	trial treatment	2.50
5	double blind clinical	3.00	30	*brain serotonin* *	2.50
6	migraine therapy magnesium	3.00	31	comparative double blind	2.50
7	ophthalmologic	3.00	32	comparative double	2.50
8	clinical aspect therapy	3.00	33	400	2.50
9	anti inflammatory agent	3.00	34	hyperventilation	2.50
10	therapy magnesium glutamate	3.00	35	cortical spread	2.50
11	bruxism	3.00	36	concentration serotonin	2.50
12	magnesium glutamate	3.00	37	pill	2.50
13	blind clinical	3.00	38	physiopathological	2.50
14	aspect therapy	3.00	39	vasospastic	2.50
15	physiopathology	2.83	40	respiratory arrest	2.50
16	hypotension	2.66	41	peripheral artery	2.50
17	treatment spontaneous	2.66	42	*spread depression* *	2.43
18	oral glucose tolerance	2.50	43	pharmacotherapy	2.33
19	*cerebral vasospasm* *	2.50	44	*arterial spasm* *	2.33
20	response serum	2.50	45	acid metabolism	2.33
21	factor pathogenesis	2.50	46	clinical experimental study	2.33
22	*cortical spread depression* *	2.50	47	chorea	2.33
23	severe pre	2.50	48	lactase	2.33
24	severe pre eclampsia	2.50	49	arginine	2.33
25	experimental data	2.50	50	clinical effect	2.33

We will explain BisoNet construction by creating an example network from the migraine-magnesium domain pair. Table 3 states first 50 terms which are the output of the first two steps of the procedure: candidate concept detection and $^{(21)}$outFreqRelSum heuristic scoring. How many terms do we consider for inclusion in the final BisoNet depends on the use-case of the created network. In the case when the network is an input of the following automatic procedures for bisociation detection, we want to keep as many nodes as possible, i.e., all candidate concepts nodes (13,525 in the migraine-magnesium domain). There may be a need to trim the number of nodes down either due to the computational complexity of the subsequent bisociation discovery procedures or due to the fact that the network is meant to be explored by a human. In such a case we have two primary options to consider: the first is to remove all the nodes that do not appear in both domains since those are less probable to contain interesting bisociations (we are left with 1,847 nodes in the migraine-magnesium domain). The second option is to use the scores of $^{(21)}$outFreqRelSum to cut the nodes under the specified threshold limit.

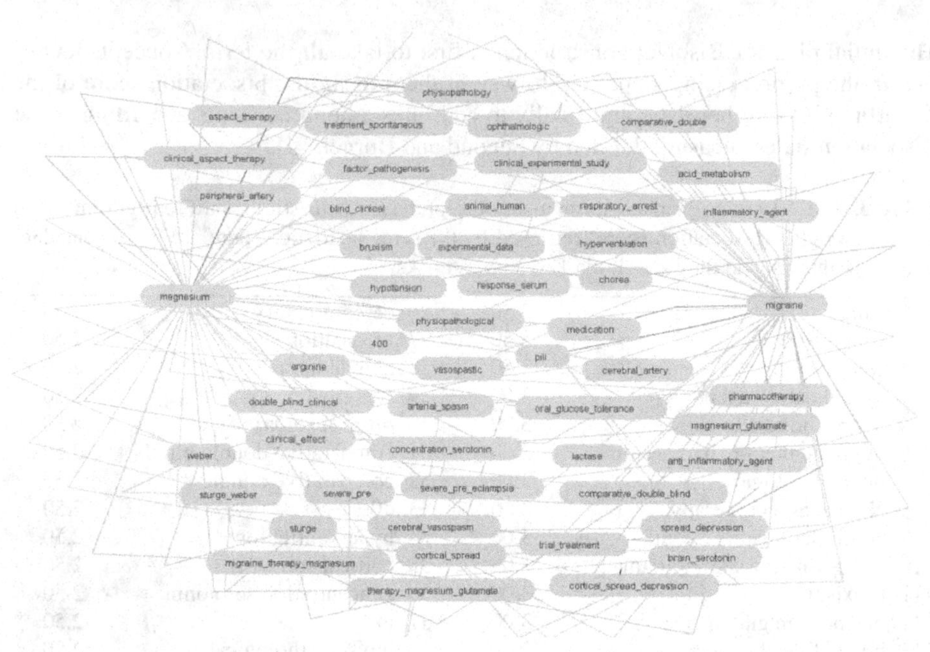

Fig. 6. Part of the network constructed from the migraine-magnesium database using [(21)]outFreqRelSum heuristic for weighting the nodes and Bisociation Index for weighting the links

The only step remaining to finalize a BisoNet construction is to calculate the links. If we have a reasonably large number of nodes (e.g. 1,000 or more) then it is infeasible to calculate all the links since there are $(n \cdot (n-1))/2$ of them if n is the number of nodes. Therefore, we again use thresholding to cut away lower weighted links. In extreme cases where there is a really vast number of nodes (e.g. 100,000 or more) there are special approaches needed to calculate all the links – even before thresholding is applied and the nodes are stored. However, these algorithms are beyond the scope of this work.

Fig. 6 shows a section of the final BisoNet constructed by the methodology described in this work. A section contains all the highest-ranking nodes retrieved using a threshold on the concepts' [(21)]outFreqRelSum heuristic score (see Table 3) and the two – in this domain – special nodes: migraine and magnesium. The links among nodes were calculated as described and were not thresholded. Weights on the links and nodes are not shown due to clarity; however, the node weights are stated in Table 3 while link weights can be inferred from the strength – darkness of the links.

With the presentation of this example we conclude this chapter. We addressed the problem of producing an information network, named BisoNet, from a large text corpus consisting of at least two diverse domains. The goal was to produce a BisoNet that has a high potential for providing yet unexplored cross-domain links which could lead to new scientific discoveries. We devoted most of this chapter to the sub-problem: *how to better identify important domain-bridging concepts* which become core nodes of the resulting network. We also provided a detailed description of all the preprocessing steps required to reproduce this work. The evaluation of bridging concept identification was performed by repeating a discovery made on medical articles in the migraine-magnesium domain. Further work is tightly related to the main focus of this chapter – heuristics for b-term identification and their evaluation – therefore, we stated the ideas for further work at the end of Section 5.

Acknowledgements. The work presented in this chapter was supported by the European Commission under the 7th Framework Programme FP7-ICT-2007-C FET-Open project BISON-211898, and the Slovenian Research Agency grant Knowledge Technologies (P2-0103)

References

1. Albert, R., Barabasi, A.L.: Statistical mechanics of complex networks. Rev. Mod. Phys. 74(1), 47–97 (2002)
2. Bales, M.E., Johnson, S.B.: Graph theoretic modeling of large scale semantic networks. Journal of Biomedical Informatics 39(4), 451–464 (2006)
3. Berthold, M.R., Dill, F., Kötter, T., Thiel, K.: Supporting Creativity: Towards Associative Discovery of New Insights. In: Washio, T., Suzuki, E., Ting, K.M., Inokuchi, A. (eds.) PAKDD 2008. LNCS (LNAI), vol. 5012, pp. 14–25. Springer, Heidelberg (2008)
4. Segond, M., Borgelt, C.: "BisoNet" Generation using Textual Data. In: Proceedings of Workshop on Explorative Analytics of Information Networks at ECML PKDD (2009)
5. Boström, H., Andler, S.F., Brohede, M., Johansson, R., Karlsson, A., van Laere, J., Niklasson, L., Nilsson, M., Persson, A., Ziemke, T.: On the definition of information fusion as a field of research. Technical report, University of Skovde, School of Hum.and Inf., Skovde, Sweden (2007)
6. Dubitzky, W., Kötter, T., Schmidt, O., Berthold, M.R.: Towards Creative Information Exploration Based on Koestler's Concept of Bisociation. In: Berthold, M.R. (ed.) Bisociative Knowledge Discovery. LNCS (LNAI), vol. 7250, pp. 11–32. Springer, Heidelberg (2012)
7. Dura, E., Gawronska, B., Olsson, B., Erlendsson, B.: Towards Information Fusion in Pathway Evaluation: Encoding Relations in Biomedical Texts. In: Proceedings of the 9th International Conference on Information Fusion (2006)
8. Feldman, R., Sanger, J.: The Text Mining Handbook: Advanced Approaches in Analyzing Unstructured Data. Cambridge University Press (2007)
9. Fortuna, B., Lavrač, N., Velardi, P.: Advancing Topic Ontology Learning through Term Extraction. In: Ho, T.-B., Zhou, Z.-H. (eds.) PRICAI 2008. LNCS (LNAI), vol. 5351, pp. 626–635. Springer, Heidelberg (2008)
10. Juršič, M., Mozetič, I., Lavrač., N.: Learning Ripple Down Rules for Efficient Lemmatization. In: Proceedings of the 10th International Multiconference Information Society 2007, vol. A, pp. 206–209 (2007)
11. Koestler, A.: The Act of Creation. The Macmillan Co. (1964)

12. Kötter, T., Berthold, M.R.: From Information Networks to Bisociative Information Networks. In: Berthold, M.R. (ed.) Bisociative Knowledge Discovery. LNCS (LNAI), vol. 7250, pp. 33–50. Springer, Heidelberg (2012)
13. Ohsawa, Y., Benson, N.E., Yachida, M.: KeyGraph: Automatic Indexing by Co occurrence Graph based on Building Construction Metaphor. In: Proceedings of the Advances in Digital Libraries Conference (ADL), pp. 12–18 (1998)
14. Petric, I., Urbancic, T., Cestnik, B., Macedoni Luksic, M.: Literature mining method RaJoLink for uncovering relations between biomedical concepts. Journal of Biomedical Informatics 42(2), 219–227 (2009)
15. Petrič, I., Cestnik, B., Lavrač, N., Urbančič, T.: Outlier Detection in Cross Context Link Discovery for Creative Literature Mining. Comput. J., November 2 (2010)
16. Petrič, I., Cestnik, B., Lavrač, N., Urbančič, T.: Bisociative Knowledge Discovery by Literature Outlier Detection. In: Berthold, M.R. (ed.) Bisociative Knowledge Discovery. LNCS (LNAI), vol. 7250, pp. 313–324. Springer, Heidelberg (2012)
17. Porter, M.F.: An algorithm for suffix stripping. Progr. 14(3), 130–137 (1980)
18. Provost, F., Fawcett, T.: Robust classification for imprecise environments. Machine Learning 42(3), 203–231 (2001)
19. Racunas, S., Griffin, C.: Logical data fusion for biological hypothesis evaluation. In: Proceedings of the 8th International Conference on Information Fusion (2005)
20. Sluban, B., Juršič, M., Cestnik, B., Lavrač, N.: Exploring the Power of Outliers for Cross-domain Literature Mining. In: Berthold, M.R. (ed.) Bisociative Knowledge Discovery. LNCS (LNAI), vol. 7250, pp. 325–337. Springer, Heidelberg (2012)
21. Smalheiser, N.R., Swanson, D.R.: Using ARROWSMITH: a computer assisted approach to formulating and assessing scientific hypotheses. Comput Methods Programs Biomed. 57(3), 149–153 (1998)
22. Smirnov, A., Pashkin, M., Shilov, N., Levashova, T., Krizhanovsky, A.: Intelligent Support for Distributed Operational Decision Making. In: Proceedings of the 9th International Conference on Information Fusion (2006)
23. Srinivasan, P., Libbus, B., Sehgal, A.K.: Mining MEDLINE: Postulating a beneficial role for curcumin longa in retinal diseases. In: Hirschman, L., Pustejovsky, J. (eds.) BioLINK 2004: Linking Biological Literature, Ontologies, and Databases, Boston, Massachusetts, pp. 33–40 (2004)
24. Swanson, D.R.: Migraine and magnesium: Eleven neglected connections. Perspectives in Biology and Medicine 31(4), 526–557 (1988)
25. Swanson, D.R.: Medical literature as a potential source of new knowledge. Bull. Med. Libr. Assoc. 78(1), 29–37 (1990)
26. Swanson, D.R., Smalheiser, N.R., Torvik, V.I.: Ranking Indirect Connections in Literature Based Discovery: The Role of Medical Subject Headings (MeSH). Journal of the American Society for Inf. Science and Technology 57, 1427–1439 (2006)
27. Urbančič, T., Petrič, I., Cestnik, B., Macedoni-Lukšič, M.: Literature Mining: Towards Better Understanding of Autism. In: Bellazzi, R., Abu-Hanna, A., Hunter, J. (eds.) AIME 2007. LNCS (LNAI), vol. 4594, pp. 217–226. Springer, Heidelberg (2007)
28. Weeber, M., Vos, R., Klein, H., de Jong van den Berg, L.T.W.: Using concepts in literature based discovery: Simulating Swanson's Raynaud–fish oil and migraine–magnesium discoveries. J. Am. Soc. Inf. Sci. Tech. 52(7), 548–557 (2001)

Discovery of Novel Term Associations in a Document Collection

Teemu Hynönen, Sébastien Mahler, and Hannu Toivonen

Department of Computer Science and HIIT, University of Helsinki, Finland
firstname.lastname@cs.helsinki.fi

Abstract. We propose a method to mine novel, document-specific associations between terms in a collection of unstructured documents. We believe that documents are often best described by the relationships they establish. This is also evidenced by the popularity of conceptual maps, mind maps, and other similar methodologies to organize and summarize information. Our goal is to discover term relationships that can be used to construct conceptual maps or so called BisoNets.

The model we propose, tpf–idf–tpu, looks for pairs of terms that are associated in an individual document. It considers three aspects, two of which have been generalized from tf–idf to term pairs: term pair frequency (tpf; importance for the document), inverse document frequency (idf; uniqueness in the collection), and term pair uncorrelation (tpu; independence of the terms). The last component is needed to filter out statistically dependent pairs that are not likely to be considered novel or interesting by the user.

We present experimental results on two collections of documents: one extracted from Wikipedia, and one containing text mining articles with manually assigned term associations. The results indicate that the tpf–idf–tpu method can discover novel associations, that they are different from just taking pairs of tf–idf keywords, and that they match better the subjective associations of a reader.

1 Introduction

Documents are routinely characterized by their keywords, and keyword extraction is also a popular topic in text mining. Keywords certainly are useful, but they fail to describe relations between concepts in a document. In this chapter, we propose methods to mine characteristic term associations from unstructured documents in a given collection.

An example application is automatic generation of conceptual maps from news stories: such a map is a graph with terms or concepts as nodes and relations between them as edges. (Different flavors of such representations are known, e.g., as concept maps, mind maps, cognitive maps, and topic maps.) Conceptual maps are a well-known learning tool used to study and organize information, and one of our goals is to facilitate this process by automatic construction of rough conceptual maps.

M.R. Berthold (Ed.): Bisociative Knowledge Discovery, LNAI 7250, pp. 91–103, 2012.

In the context of creative information exploration and bisociative reasoning, such graphical representations are called BisoNets [1]. BisoNets can then be used to explore and discover novel information and unforseen connections between concepts.

As an example application, consider an online service that aggregates news stories from many sources and presents those to the user. Illustrating the novel association as a conceptual map together with suitable associations from the background knowledge provides a good overview of what is new in any particular story, and how it relates to existing information. As an example, consider the mining incident in 2010 in Chile, where 33 miners were trapped in a collapsed mine for more than two months before eventually being rescued via a newly drilled tunnel. In the first news stories, associations such as (*Chile, mine*), (*mine, collapse*) and (*miner, trapped*) were central. However, when more and more stories were written about the incident, these associations became part of the background. As the rescue operation advanced, new information became available about drilling and the tunnel, the rescue vessel to be used in it, the dates of the approaching final rescue operation, and eventually the success of the operation.

We are building such a prototype system, currently harvesting news from 7 online sources and with approximately 30 000 stories indexed so far. As an example, Figure 1 illustrates the essential associations, extracted with methods proposed in this chapter, from a news story published by The Washington Post[1] just before the lifting operation was to start. To highlight the news value of this story, the background associations relating the event to Chile, the capsule, etc. are not shown.

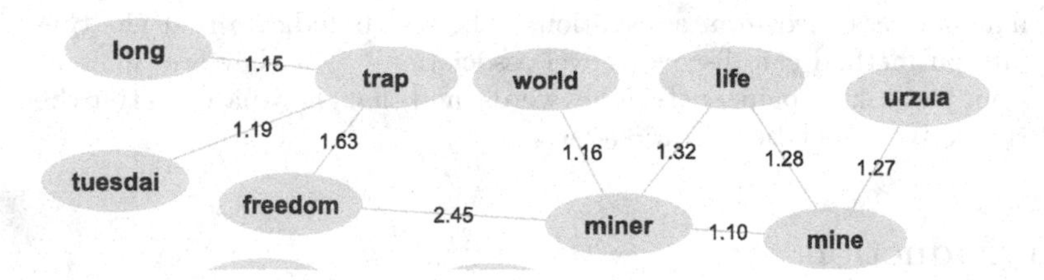

Fig. 1. Conceptual map of novel associations in a Washington Post news story "Chilean miners to begin emerging tonight" (Tue, Oct. 12, 2010). The miners had been trapped for over two months and were now about the be freed in an operation followed all around the world. Urzua is the name of the shift chief in the mine, a spokesman for the miners. Edge labels describe their importance.

Our goal is to extract interesting associations between terms in text document collections, to be presented, for instance, as simple conceptual maps or BisoNets. Roughly speaking, there are two different term association discovery tasks. The more standard one is discovering semantic similarities of terms, e.g.,

[1] `http://www.washingtonpost.com/wp-dyn/content/article/2010/10/12/ AR2010101203510.html?wprss=rss_world`

by their frequent co-occurrences. The other task, on which we focus in this chapter, is finding non-obvious, document-specific associations between terms. Note the strong contrast: in the latter task our aim is to discover novel associations between terms that are usually *not* related.

The remainder of the chapter is organized as follows: We will briefly review related work in Section 2. In Section 3 we propose a new method that finds exceptional relations in the sense that they are independent in the collection and specific to the document. Section 4 contains experimental results on two collections of documents: one extracted from Wikipedia, one containing text mining articles with manually assigned term associations. Section 5 contains concluding remarks and proposes further research on this topic.

2 Related Work

Conceptual maps, concept maps, mind maps, topic maps and many other similar formalisms exist for organizing and representing concepts and their relations as a graph. Many of them have been developed to be used as note taking and learning tools (see, e.g., [2]). Topic maps, on the other hand, are an ISO-standardized representation for interchange of knowledge. Unlike many of these techniques, we do not currently label edges by relation types. This could perhaps be done with information extraction methods (see below) after the associations have been discovered. We are not aware of methods for automatic, domain-independent construction of conceptual maps for documents in a given collection. We next review methods for finding various kinds of relations between terms or concepts.

There is abundant literature on finding statistical relations between terms. Most of the work is focused on discovering *semantically related terms*, such as *car* and *wheel*. Typically these techniques either use lexical databases and ontologies or measure co-occurrences of words, or combine these two. For instance, Hirst and St-Onge [3], as well as Patwardhan and Pedersen [4] measure semantic relatedness using *WordNet* as background knowledge. WordNet is a lexical database that consists of a thesaurus and several types relations between terms. WordNet-based similarity measures use path lengths between terms as the basis of relatedness. The Normalized Google Distance Measure (NGD) [5], in turn, uses Google search engine to measure the semantic relatedness of two terms. NGD has theoretical background in information theory, but in practice the idea is to compute the ratio of web pages where the terms occur independently to the pages where both of the terms occur. Latent Semantic Indexing (LSI) [6] goes beyond direct co-occurrence of terms, and uses singular value decomposition and reduction of matrix dimensions. Co-occurrence measures specifically aimed at *bisociation* are proposed by Segond and Borgelt [7]. They use keywords as the nodes of the BisoNet and focus on selecting appropriate edges between them. For the example application of producing conceptual maps, such semantic relations across documents are needed, and constitute an essential part of the background. The method proposed in this chapter addresses an opposite problem: find associations that are relatively specific to a document.

Our approach shares some mental similarity with *RaJoLink* [8] even though it works in a different setting. Given a collection of articles on some topic, RaJoLink starts by finding rare terms in it. The motivation is that these may be used to generate hypotheses about novel connections to other topics in further steps of the RaJoLink process. RaJoLink's emphasis is, however, on finding indirect relations of topics across documents, not on finding associations within documents.

The goal of *information extraction* is to extract certain structured information from textual documents (see, e.g., [9]). Information extraction methods are also routinely used to discover associations between terms. Examples include news story analysis (who did what, where and when) and automatic extraction of biomedical facts from scientific articles (which proteins interact, which gene contributes to which phenotype, etc.). While information extraction methods are tuned to look for specific types of facts (including relations), our goal is to be able to discover associations between arbitrary terms.

In *topic detection and tracking* the goal is to recognize events in news stories and to relate stories to each other [10]. In this task, information extraction is one of the key technologies. While we use news stories as an example application, our approach is largely complementary to topic detection and tracking: our emphasis is on relations between terms, both within stories (the novel associations looked for with methods introduced here) as well over several stories (semantic associtions in the background).

The technique we propose in this chapter is inspired by the well-known *tf–idf* (*term frequency–inverse document frequency*) keyword extraction method [11,12]. Term frequency $\text{tf}(t,d)$ is the relative frequency of term t within a document d, and it measures how essential the term is for the document. The inverse document frequency $\text{idf}(t)$ of term t measures, in turn, how specific the term is in the document collection. It is defined as the logarithm of the inverse of the relative number of documents that contain the term. Tf–idf for term t in document d is then the product $\text{tf-idf}(t,d) = \text{tf}(t,d) \cdot \text{idf}(t)$. Tf–idf and other methods to extract keywords (e.g., Keygraph [13]) have been highly successful in that task. However, they do not attempt to highlight associations between terms. Our aim is to discover association even if the individual terms are not important.

3 The tpf–idf–tpu Model of Important Term Pair Associations

We now propose and formalize a model for extracting important term associations from unstructured documents in a collection. The starting point is tf–idf [11,12], which we first generalize to pairs of terms. This generalization has, however, a serious shortcoming: term pair frequency and inverse document frequency do not sufficiently outrule possible correlation of the terms. We therefore add a third component, term pair uncorrelation.

We introduce two variants of the model that differ in the way the terms are paired in the documents. We use subscripts "sen" and "doc" to separate these variants where necessary. The *sentence-level variant*, tpf–idf–tpu_{sen}, creates pairs from terms that occur in a same sentence. The *document-level variant*, tpf–idf–tpu_{doc}, pairs every term in the document with every other term in the document.

3.1 Term Pair Frequency (tpf) and Inverse Document Frequency (idf)

Term pair frequency $\text{tpf}_{\text{sen}}(\{t,u\},d)$ is defined as the relative number of sentences s in document d that contain both terms t and u:

$$\text{tpf}_{\text{sen}}(\{t,u\},d) = \frac{|\{s \in d \mid \{t,u\} \subset s\}|}{|\{s \in d\}|}. \tag{1}$$

The inverse document frequency $\text{idf}_{\text{sen}}(t,u)$ of term pair $\{t,u\}$ is the logarithm of the inverse of the relative number of documents in the given collection C that contain both terms in the same sentence:

$$\text{idf}_{\text{sen}}(t,u) = \log \frac{|C|}{|\{d \in C \mid \exists s \in d : \{t,u\} \subset s\}|}. \tag{2}$$

For the document-level variant, there are corresponding definitions of term pair frequency and inverse document frequency:

$$\text{tpf}_{\text{doc}}(\{t,u\},d) = \min(\text{tf}(t,d), \text{tf}(u,d)), \tag{3}$$

where $\text{tf}(t,d)$ is the relative frequency of term t within a document d, and

$$\text{idf}_{\text{doc}}(t,u) = \log \frac{|C|}{|\{d \in C \mid \{t,u\} \subset d\}|}. \tag{4}$$

There is no natural direct measure of term pair frequency in a document. Following a common practice, we use the minimum of the frequencies of the two terms as the frequency of the pair.

3.2 Term Pair Uncorrelation (tpu)

Use of tpf–idf fails to recognize if there is a statistical (and possibly semantic) correlation between the terms. This is because tpf–idf only considers the joint occurrences of them, not if and how they occur without each other.

A pair that scores high on tpf–idf may be uninteresting for a number of reasons, but technically the reason usually is that the occurrence of one term (t) implies an occurrence of the other (u). Different instances of this problem include the following.

1. *Term t hardly ever occurs without term u.* For instance, articles that talk about "information retrieval" almost always mention "document", too.
2. *The two terms t and u occur roughly in the same set of documents.* For instance, "data mining" and "knowledge discovery" are roughly synonyms and obviously tend to occur in the same documents.

3. *Term u occurs in almost all documents.* For instance, "example" has a high document frequency. Paired with any less frequent term t, the tpf and especially idf scores can be high, but the association is trivial.
4. *Term t only occurs in few documents.* For instance, "tpf–idf–tpu" occurs so far only in this chapter. While associations with it are specific to this chapter, they are also trivial in a sense: any other term of this document would make a great pair with "tpf–idf–tpu", since the pair would trivially have an excellent idf score just because "tpf–idf–tpu" is so rare in a document collection.

In cases 1 and 2, the association between t and u is real but not document-specific, and therefore it should be part of the background. Cases 3 and 4 are trivial and therefore not interesting.

To rule all the above-mentioned cases out, we add a third component to the model: term pair uncorrelation, or tpu. We define tpu in terms of the relative amounts $\mathrm{r}(v)$ (where $v = t$ or u) of co-occurrences in the document collection:

$$\mathrm{r}_{\mathrm{sen}}(v) = \frac{|\{d \in C \mid \exists s \in d \text{ s.t. } \{t,u\} \subset s\}|}{|\{d \in C \mid v \in d\}|}. \tag{5}$$

The value of $\mathrm{r}(t)$ is 1 if t and u always co-occur, 0 if they never co-occur, and 0.5 if u co-occurs in half of the documents in which t occurs.

We prefer that both terms occur often independently, i.e., that both $\mathrm{r}(t)$ and $\mathrm{r}(u)$ are small. To measure this, we simply take their maximum. (Alternative measures for tpu include Jaccard index and Tanimoto coefficient. We prefer the measure based on $\max(\mathrm{r}(t), \mathrm{r}(u))$, however, since it more strongly requires that both terms also occur independently.)

In order to have a tpu measure that has larger values for the preferred situations, we define tpu as

$$\mathrm{tpu}(\{t,u\}) = \gamma - \max(\mathrm{r}(t), \mathrm{r}(t)), \tag{6}$$

where γ tunes the relative importance of the tpu component, $\gamma \geq 1$. Smaller values of γ give tpu more weight. An analysis of the effects of γ is outside the scope of this chapter. In our experiments we use $\gamma = 2$ based on some preliminary experiments.

For document-level analysis, we define $\mathrm{r}_{\mathrm{doc}}(v)$ as

$$\mathrm{r}_{\mathrm{doc}}(v) = \frac{|\{d \in C \mid \{t,u\} \subset d\}|}{|\{d \in C \mid v \in d\}|}. \tag{7}$$

Finally, tpf–idf–tpu($\{t,u\}, d$) of term pair $\{t,u\}$ in document d is defined as the product of the three components defined above:

$$\text{tpf–idf–tpu}(\{t,u\}, d) = \mathrm{tpf}(\{t,u\}, d) \cdot \mathrm{idf}(\{t,u\}) \cdot \mathrm{tpu}(\{t,u\}).$$

4 Experiments

In the following subsections we experimentally evaluate the performance of the tpf–idf–tpu model. We contrast the discovered term pairs to keywords obtained using tf–idf, and we also compare the sentence and document-based variants to each other.

Unfortunately, we are not aware of existing data sets with documents and corresponding conceptual maps, so we have to resort to other test methods. We use two different test settings.

In the first setting (used in Sections 4.1 and 4.2) the document collection consists of 425 articles on everyday life (and its subtopics), obtained from the Wikipedia Selection for Schools[2]. We use this document collection to compare the sets of term pairs produced by the different variants.

In the second test setting (Section 4.3), we created a collection of annotated text mining documents. One of the authors of this chapter manually annotated 23 documents with term associations that he considered most descriptive for the topic of each document. The document collection additionally contains another 15 text mining articles, so the total size of the collection is 38 documents. The manually assigned 229 term pairs were considered equally important and thus not ordered nor weighted in any way. Subjective annotation of key terms (or term pairs, in our case) is criticized in the literature, as the background, interests and viewpoint of the annotator affect what he or she considers to be relevant [14]. With this precaution in mind, we believe that such an evaluation can give indications of the performance of the method.

In both settings, the documents were preprocessed by removing stopwords and by stemming with Porter stemmer [15]. In addition, automatic multiword unit extraction was performed with Text-NSP program [16] using log-likelihood measure. Consecutive sequences of two terms, or *bigrams*, that got log-likelihood score of 70 or higher were treated as one term.

The goal of these tests is to give a first evaluation and illustration of the potential of the method. More systematic experiments on different data sets are left for future work.

4.1 Tpf–idf–tpu vs. tf–idf

Let us first address the question if and how different the results of term pair extraction are from single keyword extraction. To study this, we performed the following experiment with the everyday document collection.

First, n best tpf–idf–tpu term pairs were extracted from each document. Then the pair structure was ignored and we simply considered the set of terms in these top pairs. Then, an equal number of top tf–idf terms were extracted from each document. As an evaluation measure, we used the ratio of the number of terms produced by both methods divided by the total number of terms produced by the methods. The ratios were computed for a wide range of values of n, the number of top pairs to be picked in the first phase. For each n the average of the ratios from all documents was computed. The results are shown in Figure 2(a) as a function of n.

---

[2] `http://schools-wikipedia.org/`, downloaded in 2010.

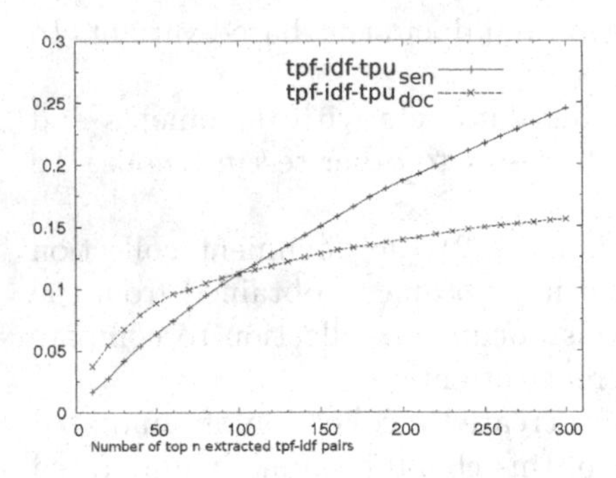

(a) Average ratio of identical terms to all terms in the top n results of tf–idf and the tpf–idf–tpu variants.

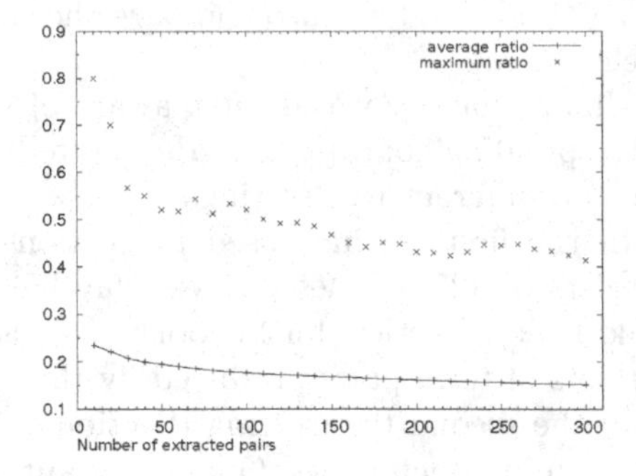

(b) The ratio of identical pairs to all extracted pairs in the top n pairs extracted by the two tpf–idf–tpu variants.

Fig. 2. Overlap of results from tpf–idf–tpu variants and tf–idf

The results of this experiment clearly show that the terms extracted by the tpf–idf–tpu and tf–idf methods differ considerably, even with large numbers of extracted pairs. The tpf–idf–tpu method does not just create pairs of top ranking tf–idf terms, but actually does extract other relations. At ten top pairs, the ratio of identical tpf–idf–tpu_{sen} and tf–idf terms is only about 2% on average and rises to approximately 25% at 300 pairs. The ratio of identical tpf–idf–tpu_{doc} and tf–idf terms in top ten pairs is about 2%, and rises to about 15% at 300 pairs.

4.2 Sentence vs. Document-Level tpf–idf–tpu Methods

We next compare the sentence and document-level tpf–idf–tpu directly to each other. We will consider three related but different aspects: (1) how similar are the term pairs chosen by the methods, (2) how similar are the terms in the pairs chosen by the methods, and (3) are the pairs dominated by a small number of terms.

First, the similarity of tpf–idf–tpu_{sen} and tpf–idf–tpu_{doc} is examined by comparing their top scoring pairs. This is done by extracting top n pairs with each method, and computing the ratio of identical pairs in the top n pairs to the total number of pairs, that is, to $2 \cdot n$. To combine the ratios yielding from different documents, the average, minimum and maximum of the ratios were taken. The results are shown in Figure 2(b) as a function of n, the number of extracted top pairs. The minimum ratio was zero for all n.

The experiment indicates that the top pairs produced by tpf–idf–tpu_{sen} and tpf–idf–tpu_{doc} differ considerably. The average ratio is slightly higher for small numbers of extracted pairs. This indicates that the highest ranking pairs tend to be slightly more similar. At top ten term pairs extracted by tpf–idf–tpu_{sen} and tpf–idf–tpu_{doc}, the average ratio is about 25% and maximum ratio is about 80%. At 300 top pairs the ratio of identical pairs lowers to about 15% and the maximum ratio to about 40%.

Next, the ratio of identical terms in the top pairs produced by tpf–idf–tpu$_{sen}$ and tpf–idf–tpu$_{doc}$ was studied. The motivation for this experiment was to see if the methods generate the pairs from a similar set of terms but pair them in different ways. The experiment was performed by selecting top n pairs for a document by both tpf–idf–tpu$_{sen}$ and tpf–idf–tpu$_{doc}$ methods. Then we again computed the ratio of the number of identical terms in the top n pairs divided by the total number of distinct terms in the pairs. Like in the previous experiment, the average of these ratios from different documents was taken. In addition to the average ratio, the minimum and maximum ratios are considered (Figure 3(a)).

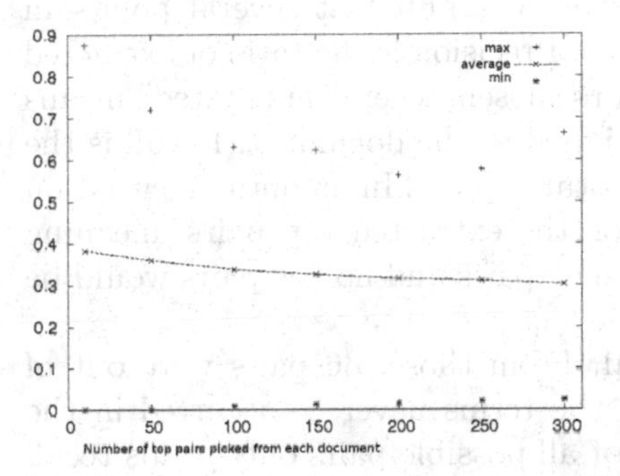

(a) The ratio of identical terms to all terms in the top n pairs extracted by the two tpf–idf–tpu variants.

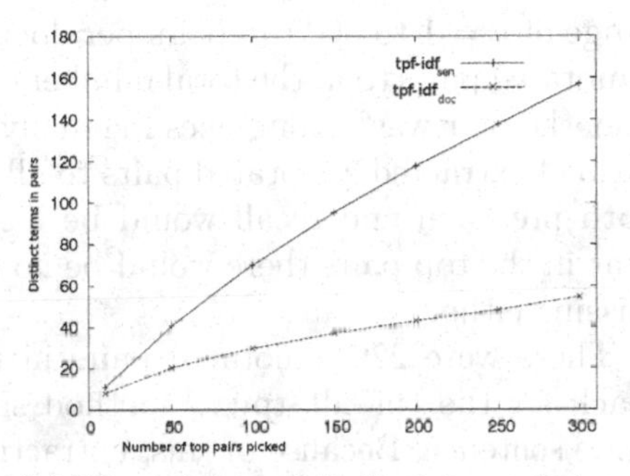

(b) The number of distinct terms in top n pairs extracted either with tpf–idf–tpu$_{sen}$ or with tpf–idf–tpu$_{doc}$.

Fig. 3. Overlap of results from tpf–idf–tpu variants and tf–idf, and internal variability in tpf–idf–tpu results

The ratio of identical terms in the pairs is about 40 percent on average and almost 90 percent at maximum when comparing top ten pairs. The ratios of identical terms in Figure 3(a) are clearly higher than the ratios of identical pairs in Figure 2(b), although on average the ratio is not very large.

Next we consider the number of distinct terms in the pairs produced by tpf–idf–tpu. The goal is to see if the top pairs are dominated by a small set of distinct terms. For this test, the top n pairs were picked from each document and the average number of distinct terms was computed over the documents (Figure 3(b)).

The number of distinct terms is relatively low for both of the methods. Especially pairs produced by tpf–idf–tpu$_{doc}$ are dominated by a small set of terms. For top ten pairs the number of distinct terms is about ten on average for both tpf–idf–tpu$_{sen}$ and tpf–idf–tpu$_{doc}$. At 300 top pairs the number of distinct term rises to about 160 for tpf–idf–tpu$_{sen}$ and to about 60 for tpf–idf–tpu$_{doc}$. In comparison, 25 terms is the minimum number of terms to produce 300 pairs; in tpf–idf–tpu$_{doc}$ there are about 60 terms on average that occur in the 300 top pairs.

It is not clear from these results if a smaller or larger number of distinct terms should lead to a better result. It is possible that the smaller term set used by tpf–idf–tpu$_{doc}$ contains less noise than the larger set extracted by tpf–idf–tpu$_{sen}$. On the other hand, it could also miss relevant terms and term pairs.

4.3 Comparison of tpf–idf–tpu and tf–idf Using Annotated Test Set

We now move to experimental tests with the other document collection, text mining articles, and compare the results of the methods against pairs annotated by hand. As a simple baseline method, we used tf–idf to rank pairs of terms by simply taking the sum of the terms' individual tf–idf scores.

For each method, precision and recall were computed at several points in range of $n = 1$ to 300 top pairs per document. Precision is the ratio of extracted annotated pairs to n, the total number of pairs chosen, where "annotated" means that the pair was among ones manually assigned to the document. Recall is the ratio of extracted annotated pairs to all annotated pairs. In an optimal situation both precision and recall would be high for the extracted top pairs, meaning that in the top pairs there would be no non-key pairs and no key pairs would be missing either.

There were 229 annotated pairs in total. From those, 66 pairs were out of reach for the tpf–idf–tpu$_{sen}$ method since the terms never co-occurred in the same sentence. Because of this, extraction of all possible pairs only yields recall of 0.71 for tpf–idf–tpu$_{sen}$. On the other hand, the number of term pairs per document varied from 3 561 to 55 552 for tpf–idf–tpu$_{sen}$ and from 118 341 to 3 386 503 for tpf–idf–tpu$_{doc}$ and tf–idf–sum.

The results for recall and precision (Figure 4) indicate the following. First, due to the small number of documents, the results for $n = 1$ to 5 are very noisy, and it is difficult to observe systematic differences between any of the three methods. Then, however, for $n = 10$ to 100 extracted pairs, the sentence-based method consistently outperforms the other two, in terms of both precision and recall. The document-based method has a slight systematic edge over the tf–idf-baseline in the mid-range. For $n \geq 100$, the tf–idf-baseline in turn outperforms the document-based method.

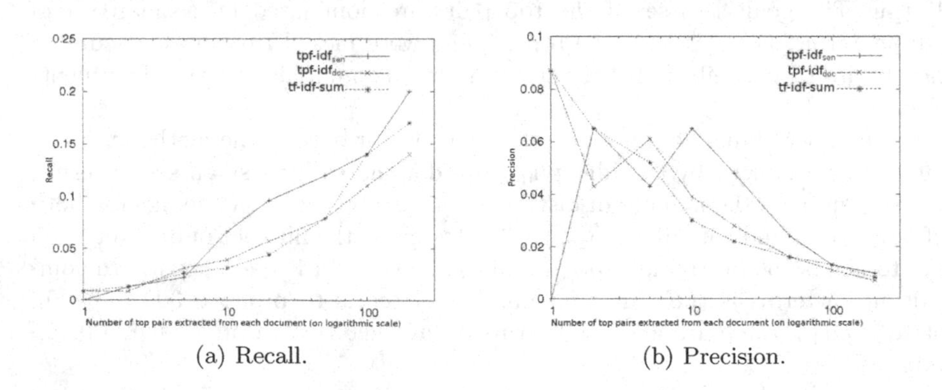

(a) Recall. (b) Precision.

Fig. 4. Recall and precision at different numbers of extracted pairs

The recall and precision values may seem low. Notice, first, that the setup of this experiment differs from the usual precision and recall experiments in document retrieval. In this experiment only the annotated associations are classified relevant; all the other pairs are implicitly classified as irrelevant even though they are not inspected in any way for relevance or novelty. It is thus possible that there are pairs that could be considered relevant for the document even though they were not selected as key pairs in the manual annotation. Second, consider the extreme challenge in the task: on average, 10 pairs were manually extracted from each document, whereas the number of different pairs per document ranges approximately from 3 500 to 3 400 000, depending on the method. In other words, the fraction of manually tagged pairs ranges from 0,0000003 to 0,003. Compared to this scale, the numbers are high.

According to the results, we believe that tpf–idf–tpu_{sen} has great potential to discover important associations between terms. The document-based variant performs less consistently. Since the two variants find largely different pairs, it will be an interesting topic for future research to try to combine their best properties.

5 Conclusion

We have proposed to discover novel associations of terms in unstructured documents, and to use these to summarize the key concepts and relationships of the documents. A term pair has a novel association in a document if the pair is frequent in it (tpf), specific to it (idf), and uncorrelated in the document collection (tpu). The proposed method, tpf–idf–tpu, is a generalization of tf–idf to pairs of terms, with the tpu component added to avoid statistically related pairs of terms.

We proposed two variants of tpf–idf–tpu: the document-level version checks if the terms co-occur within a document, while the sentence-level variant only considers the terms to co-occur if they are in the same sentence. For comparison, we also implemented a simple tf–idf-based method that outputs pairs of keywords.

We experimentally observed that tpf–idf–tpu produces pairs (and terms) significantly different from tf–idf. The sentence and document-based variants also produced results quite different from each other. In a recall/precision analysis with a smaller, manually annotated set of documents, the tpf–idf–tpu_{sen} variant based on sentence-level pairing of terms performed clearly better than the other methods when 10-100 term pairs were extracted per document. For smaller numbers of extracted associations, the results are noisy and inconclusive. Systematic experiments on different data sets are a topic for future work.

We are currently building an experimental online news summary system to try out how an incremental version of tpf–idf–tpu manages to identify and summarize the novelties in news stories and to visualize them as simple conceptual graphs. For this task, semantic associations should also be extracted and visualized as background knowledge.

We plan to apply graph mining and bisociation methods on the conceptual graphs, e.g., to discover more distant relationships between concepts. For such use, it could be useful to keep the tpu score separate from the tpf–idf scores, and allow the graph mining algorithms to consider the strength of the link (tpf · idf) and its unobviousness (tpu) separately.

Acknowledgment. This work has been supported by the Algorithmic Data Analysis (Algodan) Centre of Excellence of the Academy of Finland (Grant 118653) and by the European Commission under the 7th Framework Programme FP7-ICT-2007-C FET-Open, contract no. BISON-211898.

References

1. Kötter, T., Berthold, M.R.: From Information Networks to Bisociative Information Networks. In: Berthold, M.R. (ed.) Bisociative Knowledge Discovery. LNCS (LNAI), vol. 7250, pp. 33–50. Springer, Heidelberg (2012)
2. Novak, J.D., Cañas, A.J.: The theory underlying concept maps and how to construct them. Technical Report IHMC CmapTools 2006-01 Rev 01-2008, Florida Institute for Human and Machine Cognition (2008)
3. Hirst, G., St-Onge, D.: Lexical chains as representations of context for the detection and correction of malapropisms. In: Fellbaum, C. (ed.) WordNet: An Electronic Lexical Database, pp. 305–332. MIT Press (1998)
4. Patwardhan, S., Pedersen, T.: Using WordNet based context vectors to estimate the semantic relatedness of concepts. In: EACL 2006 Workshop Making Sense of Sense - Bringing Computational Linguistics and Psycholinguistics Together, pp. 1–8 (April 2006)
5. Cilibrasi, R.L., Vitányi, P.M.: The Google similarity distance. IEEE Transactions on Knowledge and Data Engineering 19(3), 370–383 (2007)
6. Deerwester, S., Dumais, S.T., Furnas, G.W., Landauer, T.K., Harshman, R.: Indexing by latent semantic analysis. Journal of the American Society for Information Science 41, 391–407 (1990)
7. Segond, M., Borgelt, C.: Selecting the Links in BisoNets Generated from Document Collections. In: Berthold, M.R. (ed.) Bisociative Knowledge Discovery. LNCS, vol. 7250, pp. 56–67. Springer, Heidelberg (2011)
8. Petrič, I., Urbančič, T., Cestnik, B., Macedoni-Lukšič, M.: Literature mining method RaJoLink for uncovering relations between biomedical concepts. Journal of Biomedical Informatics 42(2), 219–227 (2009)
9. Cowie, J.R., Lehnert, W.G.: Information extraction. Communications of ACM, 80–91 (1996)
10. Allan, J., Carbonell, J., Doddington, G., Yamron, J., Yang, Y.: Topic detection and tracking pilot study. In: Proceedings of the DARPA Broadcast News Transcription and Understanding Workshop, pp. 194–218 (1998)
11. Spärck Jones, K.: A statistical interpretation of term specificity and its application in retrieval. Journal of Documentation 28, 11–21 (1972)

12. Salton, G., Wong, A., Yang, C.S.: A vector space model for automatic indexing. Communications of the ACM 18(11), 613–620 (1975)
13. Ohsawa, Y., Benson, N.E., Yachida, M.: Keygraph: Automatic indexing by co-occurrence graph based on building construction metaphor. In: ADL 1998: Proceedings of the Advances in Digital Libraries Conference, vol. 12. IEEE Computer Society, Washington, DC (1998)
14. Luhn, H.P.: The automatic creation of literature abstracts. IBM Journal of Research Development 2(2), 159–165 (1958)
15. Porter, M.F.: An algorithm for suffix stripping. Program 14(3), 130–137 (1980)
16. Banerjee, S., Pedersen, T.: The design, implementation and use of the ngram statistics package. In: Proceedings of the Fourth International Conference on Intelligent Text Processing and Computational Linguistics, pp. 370–381 (2003)

Cover Similarity Based Item Set Mining

Marc Segond and Christian Borgelt

European Centre for Soft Computing,
Calle Gonzalo Gutiérrez Quirós s/n, E-33600 Mieres (Asturias), Spain
{marc.segond,christian.borgelt}@softcomputing.es

Abstract. In standard frequent item set mining one tries to find item sets the support of which exceeds a user-specified threshold (minimum support) in a database of transactions. We, instead, strive to find item sets for which the similarity of the covers of the items (that is, the sets of transactions containing the items) exceeds a user-defined threshold. This approach yields a much better assessment of the association strength of the items, because it takes additional information about their occurrences into account. Starting from the generalized Jaccard index we extend our approach to a total of twelve specific similarity measures and a generalized form. In addition, standard frequent item set mining turns out to be a special case of this flexible framework. We present an efficient mining algorithm that is inspired by the well-known Eclat algorithm and its improvements. By reporting experiments on several benchmark data sets we demonstrate that the runtime penalty incurred by the more complex (but also more informative) item set assessment is bearable and that the approach yields high quality and more useful item sets.

1 Introduction

Frequent item set mining and association rule induction are among the most intensely studied topics in data mining and knowledge discovery in databases. The enormous research efforts devoted to these tasks have led to a variety of sophisticated and efficient algorithms, among the best-known of which are Apriori [1,2], Eclat [38,39] and FP-growth [19,16,17].

Unfortunately, a standard problem in this research area is that the output (that is, the set of reported item sets or association rules) is often huge and can easily exceed the size of the transaction database to mine. As a consequence, the (usually few) interesting item sets and rules drown in a sea of irrelevant ones. One of the reasons for this is that the support measure for item sets and the confidence measure for rules are not very informative, because they do not say that much about the actual strength of association of the items in the set or rule: a set of items may be frequent simply because its elements are frequent and thus their frequent co-occurrence can even be expected by chance. In association rule induction adding an item to the antecedent may be possible without affecting the confidence much, because the association is actually brought about by the other items in the antecedent. Therefore a considerable number of redundant and/or irrelevant item sets and rules is often produced.

M.R. Berthold (Ed.): Bisociative Knowledge Discovery, LNAI 7250, pp. 104–121, 2012.

Approaches to cope with this problem include, for instance, [36,37], which rely on subsequent filtering and statistical tests in order to single out the relevant rules and patterns. In this chapter, however, we pursue a different direction, namely changing the search criterion for item sets, so that fewer irrelevant item sets are produced in the first place. The core idea is to replace the support measure with a more expressive measure that better captures whether the items in a set are associated. To obtain such a measure we draw on the insight that for associated items their covers—that is, the sets of transactions containing them—are more similar than for independent items. Since the Jaccard index is a very natural and straightforward measure for the similarity of sets, this leads us to the definition of a Jaccard item set, which is an item set for which the generalized Jaccard index of the covers of its items exceeds a user-specified threshold. This index has the advantage that it is also anti-monotone, so that the same search and pruning techniques can be employed as in frequent item set mining.

We then extend our approach to a total of twelve specific similarity measures that can be generalized from pairs of sets (or, equivalently, binary vectors). We present a generalized form, from which all of these measures can be obtained by proper parameterization, but which also allows for other options. Finally, it turns out that standard frequent item set mining is a special case of this flexible framework, which, however, also offers several better alternatives.

The rest of this chapter is organized as follows: in Section 2 we briefly review frequent item set mining and a core search procedure and introduce our notation. In Section 3 we present the generalized Jaccard index with the help of which we then define Jaccard item sets. Section 4 reviews the Eclat algorithm, the processing scheme of which we employ in the search for Jaccard item sets. In Section 5 we show how the difference set idea for Eclat can be adapted to efficiently compute the value of the denominator of the generalized Jaccard index, thus completing our JIM algorithm (for Jaccard Item set Mining). In Section 6 we consider a total of twelve specific similarity measures that can be used in place of the Jaccard index, together with a generalized form. In Section 7 we apply our algorithm to standard benchmark data sets and to the 2008/2009 Wikipedia Selection for schools to demonstrate the speed and usefulness of our algorithm. Finally, in Section 8, we draw conclusions from our discussion.

2 Frequent Item Set Mining

Frequent item set mining is a data analysis method that was originally developed for market basket analysis. It aims mainly at finding regularities in the shopping behavior of the customers of supermarkets, mail-order companies, online shops etc. In particular, it tries to identify sets of products (or generally items) that are associated or frequently bought together. Once identified, such sets of associated products may be used to optimize the organization of the offered products on the shelves of a supermarket or the pages of a mail-order catalog or web shop. They can also give hints which products may conveniently be bundled or may be suggested to a new customer, or to a current customer after a purchase.

Formally, the task of frequent item set mining can be described as follows: we are given a set B of *items*, called the *item base*, and a database T of *transactions*. Each item represents a product, and the item base represents the set of all products on offer. The term *item set* refers to any subset of the item base B. Each transaction is an item set and represents a set of products that has been bought by an actual customer. Since two or even more customers may have bought the exact same set of products, the total of all transactions must be represented as a vector, a bag, or a multiset, since in a simple set each transaction could occur at most once.[1] Note that the item base B is usually not given explicitly, but only implicitly as the union of all transactions in the given database.

We write $T = (t_1, \ldots, t_n)$ for a transaction database with n transactions. Thus we are able to distinguishing equal transactions by their position in the database vector (that is, the transaction index is an implicit identifier). In order to conveniently refer to the index set of the transactions, we introduce the abbreviation $\mathbb{N}_n := \{k \in \mathbb{N} \mid k \leq n\} = \{1, \ldots, n\}$. Given an item set $I \subseteq B$ and a transaction database T, the *cover* $K_T(I)$ of I w.r.t. T is defined as $K_T(I) = \{k \in \mathbb{N}_n \mid I \subseteq t_k\}$, that is, as the set of indices of transactions that contain I. The *support* $s_T(I)$ of an item set $I \subseteq B$ is the number of transactions in the database T it is contained in, that is, $s_T(I) = |K_T(I)|$. Given a user-specified *minimum support* $s_{\min} \in \mathbb{N}$, an item set I is called *frequent* in T iff $s_T(I) \geq s_{\min}$. The goal of frequent item set mining is to identify all item sets $I \subseteq B$ that are frequent in a given transaction database T. Note that the task of frequent item set mining may also be defined with a *relative* minimum support, which is the fraction of transactions in T that must contain an item set I in order to make I frequent. This alternative definition is obviously equivalent.

A standard approach to find all frequent item sets w.r.t. a given database T and a minimum support $s_{\min}$, which is adopted by basically all frequent item set mining algorithms (except those of the Apriori family), is a *depth-first search* in the subset lattice of the item base B. Viewed properly, this approach can be seen as a simple *divide-and-conquer* scheme. For some chosen item i, the problem to find all frequent item sets is split into two subproblems: (1) find all frequent item sets containing the item i and (2) find all frequent item sets *not* containing the item i. Each subproblem is then further divided based on another item $j \neq i$: find all frequent item sets containing (1.1) both items i and j, (1.2) item i, but not j, (2.1) item j, but not i, (2.2) neither item i nor j etc.

All subproblems that occur in this divide-and-conquer recursion can be defined by a *conditional transaction database* and a *prefix*. The prefix is a set of items that has to be added to all frequent item sets that are discovered in the conditional database, from which all items in the prefix have been removed. Formally, all subproblems are tuples $S = (T_C, P)$, where T_C is a conditional transaction database and $P \subseteq B$ is a prefix. The initial problem, with which the recursion is started, is $S = (T, \emptyset)$, where T is the given transaction database to mine and the prefix is empty. A subproblem $S_0 = (T_0, P_0)$ is processed as follows: Choose an

[1] Alternatively, each transaction may be enhanced by a unique *transaction identifier*, and these enhanced transactions may then be combined in a simple set.

item $i \in B_0$, where B_0 is the set of items occurring in T_0. This choice is arbitrary, but usually follows some predefined order of the items. A common choice is to process the items in the order of increasing frequency in the transaction database to mine, as this often leads to the shortest search times. If $s_{T_0}(i) \geq s_{\min}$, then report the item set $P_0 \cup \{i\}$ as frequent with the support $s_{T_0}(i)$, and form the subproblem $S_1 = (T_1, P_1)$ with $P_1 = P_0 \cup \{i\}$. The conditional transaction database T_1 comprises all transactions in T_0 that contain the item i, but with the item i removed. This also implies that transactions that contain no other item than i are entirely removed: no empty transactions are ever kept. If T_1 is not empty, process S_1 recursively. In any case (that is, regardless of whether $s_{T_0}(i) \geq s_{\min}$ or not), form the subproblem $S_2 = (T_2, P_2)$, where $P_2 = P_0$ and the conditional transaction database T_2 comprises all transactions in T_0 (including those that do not contain the item i), but again with the item i removed. If T_2 is not empty, process S_2 recursively.

Eclat, FP-growth, and several other frequent item set mining algorithms all follow this basic recursive processing scheme [15,5]. They differ mainly in how they represent the conditional transaction databases. There are basically two fundamental approaches, namely horizontal and vertical representations. In a horizontal representation, the database is stored as a list (or array) of transactions, each of which is a list (or array) of the items contained in it. In a vertical representation, a transaction database is stored by first referring with a list (or array) to the different items. For each item a list of transaction identifiers is stored, which indicate the transactions that contain the item.

However, this distinction is not pure, since there are many algorithms that use a combination of the two forms of representing a database. For example, while Eclat [38,39] uses a purely vertical representation and SaM (Split and Merge) [6] uses a purely horizontal representation, FP-growth [19,16,17] combines in its FP-tree structure a (compressed) horizontal representation (prefix tree of transactions) and a vertical representation (links between the tree branches).[2]

The basic processing scheme outlined above can easily be improved with so-called *perfect extension pruning*, which relies on the following simple idea: given an item set I, an item $i \notin I$ is called a *perfect extension* of I, iff I and $I \cup \{i\}$ have the same support, that is, if i is contained in all transactions containing I. Perfect extensions have the following obvious properties: (1) if the item i is a perfect extension of an item set I, then it is also a perfect extension of any item set $J \supseteq I$ as long as $i \notin J$ and (2) if I is a frequent item set and K is the set of all perfect extensions of I, then all sets $I \cup J$ with $J \in 2^K$ (where 2^K denotes the power set of K) are also frequent and have the same support as I.

These properties can be exploited by collecting in the recursion not only prefix items, but also, in a third element of a subproblem description, perfect extension items. Once identified, perfect extension items are no longer processed in the recursion, but are only used to generate all supersets of the prefix that have the

[2] Note that Apriori, which also uses a purely horizontal representation, is not mentioned here, because it relies on a different processing scheme: it traverses the subset lattice level-wise rather than depth-first.

same support. Depending on the data set, this method, which is also known as *hypercube decomposition* [34,35], can lead to a considerable acceleration of the search. It should be clear that this optimization can, in principle, be applied in all frequent item set mining algorithms.[3]

3 Jaccard Item Sets

As outlined in the introduction, we base our item set mining approach on the similarity of item covers rather than on item set support. In order to measure the similarity of a set of item covers, we start from the Jaccard index [22], which is a well-known statistic for comparing sets. For two arbitrary sets A and B it is defined as

$$J(A, B) = \frac{|A \cap B|}{|A \cup B|}.$$

Obviously, $J(A, B)$ is 1 if the sets coincide (i.e. $A = B$) and 0 if they are disjoint (i.e. $A \cap B = \emptyset$). For overlapping sets its value lies between 0 and 1.

The core idea of using the Jaccard index for item set mining lies in the insight that the covers of (positively) associated items are likely to have a high Jaccard index, while a low Jaccard index rather indicates independent or even negatively associated items. However, since we consider also item sets with more than two items, we need a generalization to more than two sets (here: item covers). In order to achieve this, we define, in a perfectly straightforward manner, the *carrier* $L_T(I)$ of an item set I w.r.t. a transaction database T as

$$L_T(I) = \{k \in \mathbb{N}_n \mid I \cap t_k \neq \emptyset\} = \{k \in \mathbb{N}_n \mid \exists i \in I \colon i \in t_k\} = \bigcup_{i \in I} K_T(\{i\}).$$

The *extent* $r_T(I)$ of an item set I w.r.t. a transaction database T is the size of its carrier, that is, $r_T(I) = |L_T(I)|$. Recall also that, in analogy, the *cover* $K_T(I)$ of an item set I w.r.t. a transaction database T is

$$K_T(I) = \{k \in \mathbb{N}_n \mid I \subseteq t_k\} = \{k \in \mathbb{N}_n \mid \forall i \in I \colon i \in t_k\} = \bigcap_{i \in I} K_T(\{i\})$$

and that the *support* $s_T(I)$ of an item set I is the size of this cover, that is, $s_T(I) = |K_T(I)|$. With these two notions we can simply define the generalized Jaccard index of an item set I w.r.t. a transaction database T as its support divided by its extent, that is, as

$$J_T(I) = \frac{s_T(I)}{r_T(I)} = \frac{|K_T(I)|}{|L_T(I)|} = \frac{|\bigcap_{i \in I} K_T(\{i\})|}{|\bigcup_{i \in I} K_T(\{i\})|}.$$

[3] Note that perfect extension pruning is *not* the same as restricting the output to *closed* frequent item sets [26], even though a closed item set can be defined as an item set that does not possess a perfect extension. The reason is that the search, in order to avoid redundant work, usually does not consider all possible extensions. Hence there may be perfect extensions which are not detected in the search.

Clearly, this is a very natural and straightforward generalization of the Jaccard index. Since for an arbitrary item $a \in B$ it is obviously $K_T(I \cup \{a\}) \subseteq K_T(I)$ and equally obviously $L_T(I \cup \{a\}) \supseteq L_T(I)$, we have $s_T(I \cup \{a\}) \leq s_T(I)$ and $r_T(I \cup \{a\}) \geq r_T(I)$. From these two relations it follows

$$J_T(I \cup \{a\}) \leq J_T(I).$$

Therefore the generalized Jaccard index w.r.t. a transaction database T over an item base B is an anti-monotone function on the partially ordered set $(2^B, \subseteq)$.

Given a user-specified minimum Jaccard value $J_{\min}$, an item set I is called *Jaccard-frequent* if $J_T(I) \geq J_{\min}$. The goal of Jaccard item set mining is to identify all item sets that are Jaccard-frequent in a given transaction database T. Since the generalized Jaccard index is anti-monotone, this task can be addressed with the same basic scheme as the task of frequent item set mining. The only problem to be solved is to find an efficient scheme for computing the extent $r_T(I)$.

4 The Eclat Algorithm

Since we will draw on the scheme of the well-known Eclat algorithm for mining Jaccard item sets, we briefly review some of its core ideas in this section. As already mentioned, Eclat [38] uses a purely vertical representation of conditional transaction databases. That is, it uses lists of transaction indices, which represent the cover of an item or an item set. It then exploits the obvious relation

$$K_T(I_1 \cup I_2) = K_T(I_1) \cap K_T(I_2),$$

which can easily be verified by inserting the definition of a cover. In particular, Eclat exploits the special case

$$K_T(I \cup \{a, b\}) = K_T(I \cup \{a\}) \cap K_T(I \cup \{b\}),$$

which allows to extend an item set by an item. This is used in the recursive divide-and-conquer scheme described above by intersecting the list of transaction indices associated with the split item with the lists of transaction indices of all items that have not yet been considered in the recursion. In this case the set I in the formula above is the prefix P of the conditional transaction database.

An alternative to the intersection approach, which is particularly useful for mining dense transaction databases[4], relies on so-called *difference sets* (or *diffsets* for short) [39]. The diffset $D_T(a \mid I)$ of an item a w.r.t. an item set I and a transaction database T is defined as

$$D_T(a \mid I) = K_T(I) - K_T(I \cup \{a\}).$$

[4] A transaction database is called *dense* if the average fraction of all items that occur per transaction is relatively high. Formally, we may define the density of a transaction database T as $\delta(T) = \frac{1}{n \cdot |B|} \sum_{k=1}^{n} |t_k|$, which is equivalent to the fraction of ones in a binary matrix representation of the transaction database T.

That is, a diffset $D_T(a \mid I)$ lists the indices of all transactions that contain I, but not a. Since obviously

$$s_T(I \cup \{a\}) = s_T(I) - |D_T(a \mid I)|,$$

diffsets are equally effective for finding frequent item sets, provided one can derive a formula that allows to compute diffsets with a larger conditional item set I without going through covers (using the above definition of a diffset). However, this is easily achieved, because the following equality holds [39]:

$$D_T(b \mid I \cup \{a\}) = D_T(b \mid I) - D_T(a \mid I).$$

This formula allows to formulate the search entirely with the help of diffsets. It may be started either with the complements of the covers of the items, which are the diffsets for an empty condition, or by forming the differences of the covers of individual items to obtain the diffsets for condition sets with only a single item.

5 The JIM Algorithm (Jaccard Item Set Mining)

The diffset approach as it was reviewed in the previous section can easily be transferred in order to find an efficient scheme for computing the *carrier* and thus the *extent* of item sets. To this end we define the *extra set* $E_T(a \mid I)$ as

$$E_T(a \mid I) = K_T(\{a\}) - \bigcup_{i \in I} K_T(\{i\}) = \{k \in \mathbb{N}_n \mid a \in t_k \wedge \forall i \in I\colon i \notin t_k\}.$$

That is, $E_T(a \mid I)$ is the set of indices of all transactions that contain a, but no item in I. Thus it identifies the extra transaction indices that have to be added to the carrier if item a is added to the item set I. For extra sets we have

$$E_T(a \mid I \cup \{b\}) = E_T(a \mid I) - E_T(b \mid I),$$

which corresponds to the analogous formula for diffsets reviewed above. This relation is easily verified as follows:

$$\begin{aligned}
&E_T(a \mid I) - E_T(b \mid I) \\
&= \{k \in \mathbb{N}_n \mid a \in t_k \wedge \forall i \in I\colon i \notin t_k\} - \{k \in \mathbb{N}_n \mid b \in t_k \wedge \forall i \in I\colon i \notin t_k\} \\
&= \{k \in \mathbb{N}_n \mid a \in t_k \wedge \forall i \in I\colon i \notin t_k \wedge \neg(b \in t_k \wedge \forall i \in I\colon i \notin t_k)\} \\
&= \{k \in \mathbb{N}_n \mid a \in t_k \wedge \forall i \in I\colon i \notin t_k \wedge (b \notin t_k \vee \exists i \in I\colon i \in t_k)\} \\
&= \{k \in \mathbb{N}_n \mid (a \in t_k \wedge \forall i \in I\colon i \notin t_k \wedge b \notin t_k) \\
&\qquad\qquad \vee \underbrace{(a \in t_k \wedge \forall i \in I\colon i \notin t_k \wedge \exists i \in I\colon i \in t_k)}_{=\text{false}}\} \\
&= \{k \in \mathbb{N}_n \mid a \in t_k \wedge \forall i \in I\colon i \notin t_k \wedge b \notin t_k\} \\
&= \{k \in \mathbb{N}_n \mid a \in t_k \wedge \forall i \in I \cup \{b\}\colon i \notin t_k\} \\
&= E_T(a \mid I \cup \{b\})
\end{aligned}$$

In order to see how extra sets can be used to compute the extent of item sets, let $I = \{i_1, \ldots, i_m\}$, with some arbitrary, but fixed order of the items that is indicated by the index. This will be the order in which the items are used as split items in the recursive divide-and-conquer scheme. It is

$$\begin{aligned} L_T(I) &= \bigcup_{k=1}^{m} K_T(\{i_k\}) \;=\; \bigcup_{k=1}^{m} \left(K_T(\{i_k\}) - \bigcup_{l=1}^{k-1} K_T(\{i_l\})\right) \\ &= \bigcup_{k=1}^{m} E(i_k \mid \{i_1, \ldots, i_{k-1}\}), \end{aligned}$$

and since the terms of the last union are clearly all disjoint, we have immediately

$$r_T(I) = \sum_{k=1}^{m} |E(i_k \mid \{i_1, \ldots, i_{k-1}\})| = r_T(I - \{i_m\}) + |E(i_m \mid I - \{i_m\})|.$$

Thus we have a simple recursive scheme to compute the extent of an item set from its parent in the search tree (as defined by the divide-and-conquer scheme).

The search algorithm for Jaccard item sets can now easily be implemented as follows: we start be creating a vertical representation of the given transaction database. The only difference to the Eclat algorithm is that we have not only one, but two transaction lists per item i: one represents $K_T(\{i\})$ as in standard Eclat, and the other represents $E_T(i \mid \emptyset)$, which happens to be equal to $K_T(\{i\})$. That is, for the initial transaction database the two lists are identical. However, this will obviously not be maintained in the recursive processing. In the recursion the first list for the split item is intersected with the first list of all other items to form the lists representing the covers of the corresponding pairs. The second list of the split item is subtracted from the second list of all other items, thus yielding the extra sets of transactions for these items given the split item. From the sizes of the resulting lists the support and the extent of the enlarged item sets and thus their generalized Jaccard index can easily be computed.

Note that the support computation may, as in the Eclat algorithm, also be based on diffsets. Likewise, an analogous scheme can be derived for the extent computation. In addition, Jaccard item set mining can also exploit perfect extension pruning. The only difference is that an item a is now called a perfect extension of an item set I w.r.t. a transaction database T only if $s_T(I \cup \{a\}) = s_T(I)$ and $r_T(I \cup \{a\}) = r_T(I)$, while standard frequent item set mining only requires the first equality. Such perfect extensions are handled exactly in the same way: they are not employed as split items, but collected in a third element of the subproblem description, and are used only to generate all supersets of an item set that share the same generalized Jaccard index.

6 Other Similarity Measures

Up to now we focused on the generalized Jaccard index to measure the similarity of sets (item covers). However, there is a large number of other similarity measures for sets (or, equivalently, for binary vectors, because a set may be represented by its indicator vector w.r.t. some base set). Recent extensive overviews of such measures for the pairwise case include [7] and [8].

Table 1. Quantities in terms of which the considered similarity measures are specified, together with their behavior as functions on the partially ordered set $(2^B, \subseteq)$

quantity	requirement on transaction	behavior
n_T	none (independent of the set I)	constant
$s_T(I) = \lvert K_T(I)\rvert = \lvert\bigcap_{i\in I} K_T(\{i\})\rvert$	contains all items	anti-monotone
$r_T(I) = \lvert L_T(I)\rvert = \lvert\bigcup_{i\in I} K_T(\{i\})\rvert$	contains at least one item	monotone
$q_T(I) = r_T(I) - s_T(I)$	contains some, but not all items	monotone
$z_T(I) = n_T \quad - r_T(I)$	contains no item	anti-monotone

By relying on the same scheme that we used to generalize the Jaccard index to more than two sets, a large number of such set similarity or binary vector similarity measures can be generalized beyond pairwise comparisons as follows: with the JIM algorithm we presented in the preceding section, we can easily compute the five quantities listed in Table 1. These quantities count the number of transactions that satisfy different requirements w.r.t. a given item set I (see the second column of Table 1). With these quantities a wide range of similarity measures for sets or binary vectors can be generalized.

Exceptions are measures for comparing two sets X and Y that refer explicitly to the number $|X - Y|$ of elements that are contained in the set X, but not in the set Y, and distinguish this number from the number $|Y - X|$ of elements that are contained in the set Y, but not in the set X. This distinction is difficult to generalize beyond the pairwise case, because the number of possible containment patterns of an element to the members of a family of sets grows exponentially with the number of the sets (here: covers, and thus: items). As a generalization would have to consider all of these containment patterns separately, it becomes quickly infeasible. Note, however, that an occurrence of the sum $|X-Y|+|Y-X|$ does not pose a problem, because this sum corresponds to the value $q_T(I)$.

By collecting from [8] similarity measures that can be specified in terms of the quantities listed in Table 1, we compiled Table 2. Note that the index T and the argument I are omitted to make the formulas more easily readable. Note also that the Gower & Legendre measure $S_{\mathrm{G}} = \frac{s+z}{s+q/2+z}$ [18] listed in [8] is exactly the same as the second Sokal & Sneath measure (it is just written differently, with a factor of 2 canceled from both numerator and denominator). Furthermore, note that the Hamann measure $S_{\mathrm{H}} = \frac{x+z-s}{n} = \frac{n-2s}{n}$ [20] listed in [8] is equivalent to the Sokal & Michener measure S_{M}, because $S_{\mathrm{H}} + 1 = 2S_{\mathrm{M}}$, and hence omitted. Likewise, the second Baroni-Urbani & Buser measure $S_{\mathrm{U}} = \frac{\sqrt{xz}+x-q}{\sqrt{xz}+o}$ [4] listed in [8] is equivalent to the one given in Table 2, because $S_{\mathrm{U}} + 1 = 2S_{\mathrm{B}}$. Finally, note that all of the measures listed in Table 2 have range $[0, 1]$ except S_{K} (Kulczynski) and S_{O} (Sokal & Sneath 3), which have range $[0, \infty)$.

Table 2 is split into two parts depending on whether the numerator of a measure refers only to the support s or to both the support s and the number z of transactions that do not contain any of the items in the considered set I. The former are referred to as based on the inner product, because in the pairwise case

Table 2. Considered similarity measures for sets/binary vectors

Measures derived from inner product:

Russel & Rao [28]	$S_R = \frac{s}{n} = \frac{s}{r+z}$
Kulczynski [25]	$S_K = \frac{s}{q} = \frac{s}{r-s}$
Jaccard [22] Tanimoto [33]	$S_J = \frac{s}{s+q} = \frac{s}{r}$
Dice [10] Sørensen [32] Czekanowski [9]	$S_D = \frac{2s}{2s+q} = \frac{2s}{r+s}$
Sokal & Sneath 1 [31,29]	$S_S = \frac{s}{s+2q} = \frac{s}{r+q}$

Measures derived from Hamming distance:

Sokal & Michener Hamming [30,21]	$S_M = \frac{s+z}{n} = \frac{n-q}{n}$
Faith [12]	$S_F = \frac{2s+z}{2n} = \frac{s+\frac{1}{2}z}{n}$
AZZOO [7] $\sigma \in [0,1]$	$S_Z = \frac{s+\sigma z}{n}$
Rogers & Tanimoto [27]	$S_T = \frac{s+z}{n+q} = \frac{n-q}{n+q}$
Sokal & Sneath 2 [31,29]	$S_N = \frac{2(s+z)}{n+s+z} = \frac{n-q}{n-\frac{1}{2}q}$
Sokal & Sneath 3 [31,29]	$S_O = \frac{s+z}{q} = \frac{n-q}{q}$
Baroni-Urbani & Buser [4]	$S_B = \frac{\sqrt{sz}+s}{\sqrt{sz}+r}$

s is the value of the inner (or scalar) product of the binary vectors that are compared. The latter measures (that is, those with both s and z in the numerator) are referred to as based on the Hamming distance, because in the pairwise case q is the Hamming distance of the two vectors and $n - q = s + z$ their Hamming similarity. The decision whether for a given application the term z should be considered in the numerator of a similarity measure or not is difficult. Discussions of this issue for the pairwise case can be found in [29] and [11].

Note that the Russel & Rao measure is simply normalized support, demonstrating that our framework comprises standard frequent item set mining as a special case. The Sokal & Michener measure is simply the normalized Hamming similarity. The Dice/Sørensen/Czekanowski measure may be defined without the factor 2 in the numerator, changing the range to $[0, 0.5]$. The Faith measure is equivalent to the AZZOO measure (Alter Zero Zero One One) for $\sigma = 0.5$ and the Sokal & Michener/Hamming measure results for $\sigma = 1$. AZZOO is meant to introduce flexibility in how much weight should be placed on z, the number of transactions which lack all items in I (zero zero), relative to s (one one).

All measures listed in Table 2 are anti-monotone on the partially ordered set $(2^B, \subseteq)$, where B is the underlying item base. This is obvious if in at least one of the formulas given for a measure the numerator is (a multiple of) a constant or anti-monotone quantity or a (weighted) sum of such quantities, and the numerator is (a multiple of) a constant or monotone quantity or a (weighted) sum of such quantities (see Table 1). This is the case for all but S_D, S_N and S_B.

That S_D is anti-monotone can be seen by considering its reciprocal value

$$S_D^{-1} = \frac{2s+q}{2s} = 1 + \frac{q}{2s}.$$

Since q is monotone and s is anti-monotone, S_D^{-1} is clearly monotone and thus S_D is anti-monotone. Applying the same approach to S_B, we arrive at

$$S_\mathrm{B}^{-1} = \frac{\sqrt{sz}+r}{\sqrt{sz}+s} = \frac{\sqrt{sz}+s+q}{\sqrt{sz}+s} = 1 + \frac{q}{\sqrt{sz}+s}.$$

Since q is monotone and both s and $\sqrt{sz}$ are anti-monotone, S_B^{-1} is clearly monotone and thus S_B is anti-monotone. Finally, S_N can be written as

$$S_\mathrm{N} = \frac{2n-2q}{2n-q} = 1 - \frac{q}{2n-q} = 1 - \frac{q}{n+s+z}.$$

Since q is monotone, the numerator is monotone, and since n is constant and s and z are anti-monotone, the denominator is anti-monotone. Hence the fraction is monotone and since it is subtracted from 1, S_N is anti-monotone.

Note that all measures in Table 2 can be expressed as

$$S = \frac{c_0 s + c_1 z + c_2 n + c_3\sqrt{sz}}{c_4 s + c_5 z + c_6 n + c_7\sqrt{sz}} \tag{1}$$

by specifying appropriate coefficients $c_0, \ldots, c_7$. For example, we obtain S_J for $c_0 = c_6 = 1$, $c_5 = -1$ and $c_1 = c_2 = c_3 = c_4 = c_7 = 0$, since $S_\mathrm{J} = \frac{s}{r} = \frac{s}{n-z}$. Similarly, we obtain S_O for $c_0 = c_1 = c_6 = 1$, $c_4 = c_5 = -1$ and $c_2 = c_3 = c_7 = 0$, since $S_\mathrm{O} = \frac{s+z}{q} = \frac{s+z}{n-s-z}$. This general form allows for a flexible specification of various similarity measures. Note, however, that not all selections of coefficients lead to an anti-monotone measure and hence one has to carefully check this property before using a measure that differs from the pre-specified ones.

7 Experiments

We implemented the described item set mining approach as a C program that was derived from an Eclat implementation by adding the second transaction identifier list for computing the extent of item sets. All similarity measures listed in Table 2 are included as well as the general form (1). This implementation has been made publicly available under the GNU Lesser (Library) Public License.[5]

In a first set of experiments we applied the program to five standard benchmark data sets, which exhibit different characteristics, especially different densities, and compared it to a standard Eclat search. The data sets we used are: BMS-Webview-1 (a web click stream from a leg-care company that no longer exists, which has been used in the KDD cup 2000 [23,40]), T10I4D100K (an artificial data set generated with IBM's data generator [41]), census (a data set derived from an extract of the US census bureau data of 1994, which was preprocessed by discretizing numeric attributes), chess (a data set listing chess end game positions for king vs. king and rook), and mushroom (a data set describing poisonous and edible mushrooms by different attributes). The first two data

[5] See http://www.borgelt.net/jim.html

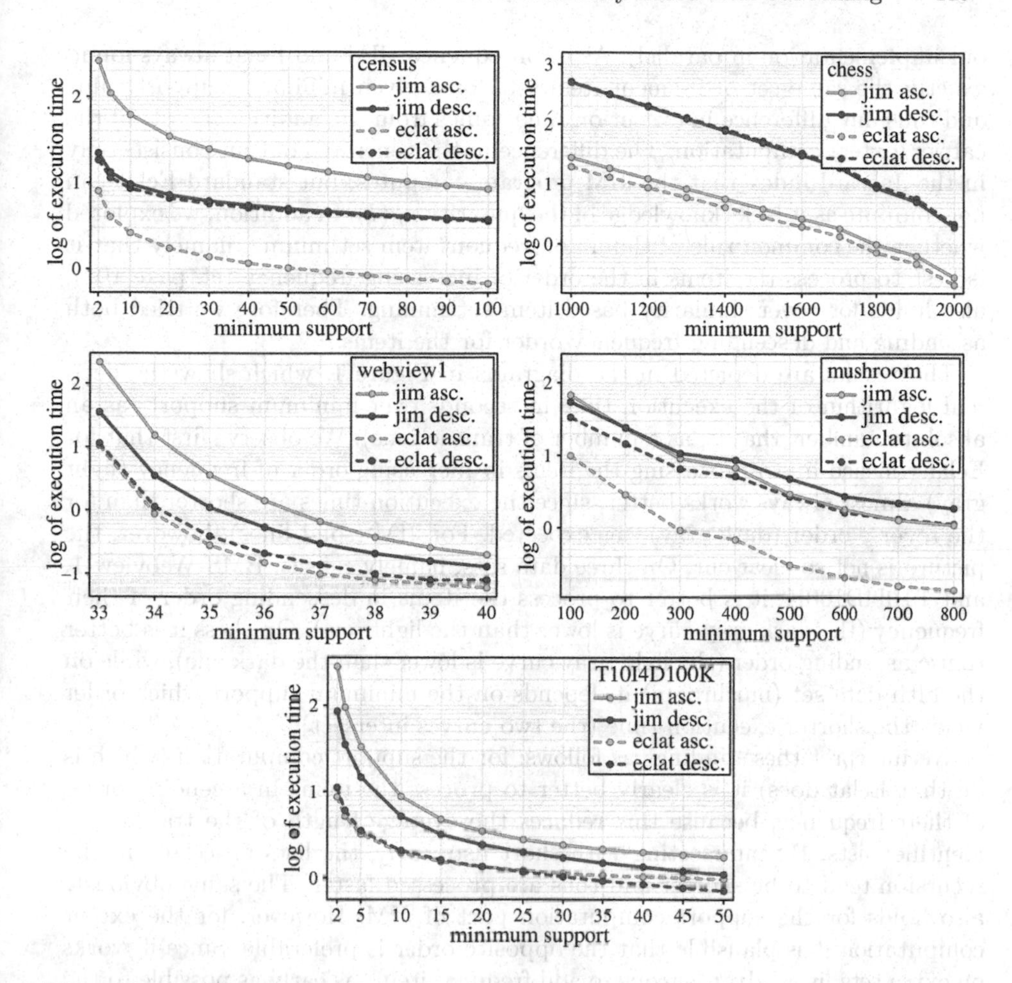

Fig. 1. Logarithms of execution times, measured in seconds, over absolute minimum support for Jaccard item set mining compared to standard Eclat frequent item set mining. Items were processed in ascending or descending order w.r.t. their frequency. Jaccard item set mining was executed with $J_{\min} = 0$, thus ensuring that exactly the same item sets are found.

sets are available in the FIMI repository [14], the last three in the UCI machine learning repository [3]. The discretization of the numeric attributes in the census data set was done with a shell/gawk script that can be found on the web page given in Footnote 5 (previous page). For the experiments we used an Intel Core 2 Quad Q9650 (3GHz) machine with 8 GB main memory running Ubuntu Linux 10.04 (64 bit) and gcc version 4.4.3.

The goal of these experiments was to determine how much the computation of the carrier/extent of an item set affected the execution time. Therefore we ran the JIM algorithm without any threshold for the similarity measure (we used the Jaccard index, i.e. $J_{\min} = 0$, but any other measure gives basically the same results), using only a minimum support threshold (which is supported by our implementation in parallel). As a consequence, JIM and Eclat always found exactly the same set of frequent item sets for a given minimum support value and thus any difference in execution time comes from the additional costs of the carrier/extent computation. The difference in the generated output consists only in the Jaccard index that the JIM program computes, but standard Eclat can not compute as it lacks knowledge of the quantity $r_T(I)$. In addition, we explored whether the common rule of thumb of frequent item set mining, namely that it is best to process the items in the order of increasing frequency (cf. page 107), also holds for cover similarity based item set mining. Therefore we tried both ascending and descending frequency order for the items.

The results are depicted in the diagrams in Figure 1, which show the decimal logarithm of the execution time in seconds over minimum support (as an absolute number, that is, as a number of transactions). We observe first that for Eclat (dashed lines) processing the items in increasing order of frequency (light gray) almost always works better, since the execution times are shorter than for the reverse order (dark gray)—as expected. For JIM (solid lines), however, the picture is not so clear cut. On three data sets, namely census, BMS-Webview-1, and T10I4D100K, it is better to process the items in descending order of their frequency (the dark gray curve is lower than the light one). On chess it is better to use ascending order (the light gray curve is lower than the dark one), while on the fifth data set (mushroom) it depends on the minimum support which order yields the shorter execution time (the two curves intersect).

We interpret these findings as follows: for the support computation (which is all that Eclat does) it is clearly better to process the items in ascending order of their frequency, because this reduces the average length of the transaction identifier lists. By intersecting with short lists early, the lists processed in the recursion tend to be shorter and thus are processed faster. The same obviously also holds for the support computation part of JIM. However, for the extent computation it is plausible that the opposite order is preferable. Since it works on extra sets, it is advantageous to add frequent items as early as possible to the carrier, because this increases the size of the already covered carrier and thus reduces the average length of the extra lists that are processed in the recursion. Therefore, since there are different preferences, it depends on the data set which operation governs the complexity and thus which item order is better.

From Figure 1 we conjecture that dense data sets (high fraction of ones in a bit matrix representation), like chess and mushroom, favor ascending order, while sparse data sets, like census, BMS-Webview-1 and T10I4D100K, favor descending order. This is plausible, because in dense data sets the intersection lists tend to be long, so it is important to reduce them. In sparse data sets, however, the extra lists tend to be long, so here it is more important to focus on them. The mushroom data set behaves more like a dense data set for lower minimum support and more like a sparse data set for higher minimum support.

Naturally, the execution times of JIM are always greater than those of the corresponding Eclat runs (with the same order of the items), but the execution times are still bearable. This shows that even if one does *not* use a similarity measure to *prune* the search, this additional information can be computed fairly

Table 3. Jaccard item sets found in the 2008/2009 Wikipedia Selection for schools

item set	s_T	J_T
Reptiles, Insects	12	1.0000
phylum, chordata, animalia	34	0.7391
planta, magnoliopsida, magnoliophyta	14	0.6667
wind, damag, storm, hurrican, landfal	23	0.1608
tournament, doubl, tenni, slam, Grand Slam	10	0.1370
dinosaur, cretac, superord, sauropsida, dinosauria	10	0.1149
decai, alpha, fusion, target, excit, dubna	12	0.1121
conserv, binomi, phylum, concern, animalia, chordata	14	0.1053

efficiently. However, it should be kept in mind that the idea of the approach is to set a threshold for the similarity measure, which can effectively prune the search, so that the actual execution times found in applications are much lower. In our own practice we basically always achieved execution times that were lower than for the Eclat algorithm (but, of course, with a different output).

In another experiment we used an extract from the 2008/2009 Wikipedia Selection for schools[6], which consisted of 4861 web pages. Each of these web pages was taken as a transaction and processed with standard text processing methods (like name detection, stemming, stop word removal etc.) to extract a total of 59330 terms/keywords. The terms occurring on a web page are the items occurring in the corresponding transaction. The resulting data file was then mined for Jaccard item sets with thresholds of $J_{\min} = 0.1$ and $s_{\min} = 10$. Some examples of term associations found in this way are listed in Table 3.

Clearly, there are several term sets with surprisingly high Jaccard indices and thus strongly associated terms. For example, "Reptiles" and "Insects" always appear together (on a total of 12 web pages) and never alone (as their Jaccard index is 1, so their covers are identical). A closer inspection revealed, however, that this is an artifact of the name detection, which extracts these terms from the Wikipedia category title "Insects, Reptiles and Fish" (but somehow treats "Fish" not as a name, but as a normal word). All other item sets contain normal terms, though (only "Grand Slam" is another name), and are not artifacts of the text processing step. The second item set captures several biology pages, which describe different vertebrates, all of which belong to the phylum "chordata" and the kingdom "animalia". The third set indicates that this selection contains a surprisingly high number of pages referring to magnolias. The remaining item sets show that term sets with five or even six terms can exhibit a quite high Jaccard index, even though they have a fairly low support (only 10–20 transactions, which corresponds to 0.2–0.4% of the 4861 transactions/web pages).

An impression of the filtering power can be obtained by comparing the size of the output to standard frequent item set mining: for $s_{\min} = 10$ there are 83130 frequent item sets and 19394 closed item sets with at least two items. A threshold of $J_{\min} = 0.1$ for the generalized Jaccard index reduces the output

[6] See `http://schools-wikipedia.org/`

Table 4. Some Jaccard item sets that were found in BMS-Webview-1

item set	s_T	J_T
35201, 35205, 35193, 35189, 35197, 35209	37	0.1034
18767, 18751, 18755, 18763, 18743, 18747, 18759	33	0.1467
18543, 18567, 18751, 18539, 18763, 18743, ... 18747, 18571, 18759	27	0.1089
18543, 18567, 18751, 18539, 18763, 18743, ... 18747, 18571, 18759, 18767	27	0.0951

to 5116 (frequent) item sets. From manual inspection, we gathered the impression that the Jaccard item sets contained more meaningful sets and that the Jaccard index was a valuable additional piece of information. It has to be conceded, though, that whether item sets are more "meaningful" or "interesting" is difficult to assess in a convincing fashion. Such an assessment would require an objective measure, which is not available (and if it were available, it could be used directly for the mining). What can be said, though, is that the support and the generalized Jaccard index assess item sets in very different ways, since for the 5116 item sets mentioned above, the correlation coefficient of the support and the generalized Jaccard index is merely 0.18. That is, neither does a high support imply a high generalized Jaccard index nor vice versa.

As an additional example, Table 4 lists Jaccard item sets that were found in BMS-Webview-1. Despite their low support (25–40 transactions, which corresponds to 0.04%–0.07% of the 59602 transactions), they could quickly and effectively be identified with Jaccard item set mining. This result is particularly impressive, because standard frequent item set mining without a restriction to e.g. closed item sets is not possible in reasonable time on BMS-Webview-1 for a minimum support less than about 32 transactions. Restricting the output to closed item sets makes mining feasible and yields 110427 item sets for $s_{\min} = 32$. Jaccard item set mining with thresholds of $s_{\min} = 32$ and $J_{\min} = 0.1$ (but without a restriction to closed item sets) reduces the output to 982 item sets. Again item sets with fairly many items and surprisingly high generalized Jaccard index are found. As for the 2008/2009 Wikipedia Selection for schools the correlation coefficient of the support and the generalized Jaccard index is very low, in this case actually even slightly negative, namely −0.02.

An example of a Jaccard item set from the census data set is

{loss=none, gain=none, country=United-States, race=White,
workclass=Private, sex=Male, age=middle-aged,
marital_status=Married-civ-spouse, relationship=Husband},

that is, an item set with 9 items with a support of 5245 (10.7%) and a Jaccard index of 0.1074. Again it is surprising to see how large an item set can possess a high generalized Jaccard index. Although this set would be discovered with standard frequent item set mining as well, the generalized Jaccard index provides a relevant additional assessment and thus distinguishes it from other item sets.

As a final remark we would like to point out that the usefulness of our method is indirectly supported by a successful application of the Jaccard item set mining approach for (missing) concept detection (see [24] as well as the chapter by Kötter and Berthold in this book, which describes the application).

8 Conclusions

In this chapter we introduced the notion of a Jaccard item set as an item set for which the generalized Jaccard index of the covers of its items exceeds a user-specified threshold. In addition, we extended this basic idea to a total of twelve similarity measures for sets or binary vectors, all of which can be generalized in the same way and can be shown to be anti-monotone. By exploiting an idea that is similar to the difference set approach for the well-known Eclat algorithm, we derived an efficient search scheme that is based on forming intersections and differences of sets of transaction indices in order to compute the quantities that are needed to compute the similarity measures. Since it contains standard frequent item set mining as a special case, mining item sets based on cover similarity yields a flexible and versatile framework. Furthermore, the similarity measures provide highly useful additional assessments of found item sets and thus help us to select the interesting ones. By running experiments on standard benchmark data sets we showed that mining item sets based on cover similarity can be done fairly efficiently, and by evaluating the results obtained with a threshold for the cover similarity measure we demonstrated that the output is considerably reduced, while expressive and meaningful item sets are preserved.

References

1. Agrawal, R., Srikant, R.: Fast Algorithms for Mining Association Rules. In: Proc. 20th Int. Conf. on Very Large Databases, VLDB 1994, Santiago de Chile, pp. 487–499. Morgan Kaufmann, San Mateo (1994)
2. Agrawal, R., Mannila, H., Srikant, R., Toivonen, H., Verkamo, A.: Fast Discovery of Association Rules. In: [13], pp. 307–328
3. Asuncion, A., Newman, D.J.: UCI Machine Learning Repository. School of Information and Computer Science, University of California at Irvine, CA, USA (2007), `http://www.ics.uci.edu/~mlearn/MLRepository.html`
4. Baroni-Urbani, C., Buser, M.W.: Similarity of Binary Data. Systematic Zoology 25(3), 251–259 (1976)
5. Bayardo, R., Goethals, B., Zaki, M.J. (eds.): Proc. Workshop Frequent Item Set Mining Implementations, FIMI 2004, Brighton, UK, Aachen, Germany. CEUR Workshop Proceedings, vol. 126 (2004), `http://www.ceur-ws.org/Vol-126/`

6. Borgelt, C., Wang, X.: SaM: A Split and Merge Algorithm for Fuzzy Frequent Item Set Mining. In: Proc.13th Int. Fuzzy Systems Association World Congress and 6th Conf. of the European Society for Fuzzy Logic and Technology, IFSA/EUSFLAT 2009. IFSA/EUSFLAT Organization Committee, Lisbon (2009)
7. Cha, S.-H., Tappert, C.C., Yoon, S.: Enhancing Binary Feature Vector Similarity Measures. J. Pattern Recognition Research 1, 63–77 (2006)
8. Choi, S.-S., Cha, S.-H., Tappert, C.C.: A Survey of Binary Similarity and Distance Measures. Journal of Systemics, Cybernetics and Informatics 8(1), 43–48 (2010); Int. Inst. of Informatics and Systemics, Caracas, Venezuela (2010)
9. Czekanowski, J.: Zarys metod statystycznych w zastosowaniu do antropologii (An Outline of Statistical Methods Applied in Anthropology). Towarzystwo Naukowe Warszawskie, Warsaw, Poland (1913)
10. Dice, L.R.: Measures of the Amount of Ecologic Association between Species. Ecology 26, 297–302 (1945)
11. Dunn, G., Everitt, B.S.: An Introduction to Mathematical Taxonomy. Cambridge University Press, Cambirdge (1982)
12. Faith, D.P.: Asymmetric Binary Similarity Measures. Oecologia 57(3), 287–290 (1983)
13. Fayyad, U.M., Piatetsky-Shapiro, G., Smyth, P., Uthurusamy, R. (eds.): Advances in Knowledge Discovery and Data Mining. AAAI Press / MIT Press, Cambridge (1996)
14. Goethals, B. (ed.): Frequent Item Set Mining Dataset Repository. University of Helsinki, Finland (2004), http://fimi.cs.helsinki.fi/data/
15. Goethals, B., Zaki, M.J. (eds.): Proc. Workshop Frequent Item Set Mining Implementations, FIMI 2003, Melbourne, FL, USA. CEUR Workshop Proceedings 90, Aachen, Germany (2003), http://www.ceur-ws.org/Vol-90/
16. Grahne, G., Zhu, J.: Efficiently Using Prefix-trees in Mining Frequent Itemsets. In: Proc. Workshop Frequent Item Set Mining Implementations, FIMI, Melbourne, FL [15] (2003)
17. Grahne, G., Zhu, J.: Reducing the Main Memory Consumptions of FPmax* and FPclose. In: Proc. Workshop Frequent Item Set Mining Implementations, FIMI, Brighton, UK [5] (2004)
18. Gower, J.C., Legendre, P.: Metric and Euclidean Properties of Dissimilarity Coefficients. Journal of Classification 3, 5–48 (1986)
19. Han, J., Pei, H., Yin, Y.: Mining Frequent Patterns without Candidate Generation. In: Proc. Conf. on the Management of Data, SIGMOD 2000, Dallas, TX, pp. 1–12. ACM Press, New York (2000)
20. Hamann, V.: Merkmalbestand und Verwandtschaftsbeziehungen der Farinosae. Ein Beitrag zum System der Monokotyledonen. Willdenowia 2, 639–768 (1961)
21. Hamming, R.V.: Error Detecting and Error Correcting Codes. Bell Systems Tech. Journal 29, 147–160 (1950)
22. Jaccard, P.: Étude comparative de la distribution florale dans une portion des Alpes et des Jura. Bulletin de la Société Vaudoise des Sciences Naturelles 37, 547–579 (1901)
23. Kohavi, R., Bradley, C.E., Frasca, B., Mason, L., Zheng, Z.: KDD-Cup 2000 Organizers' Report: Peeling the Onion. SIGKDD Exploration 2(2), 86–93 (2000)
24. Kötter, T., Berthold, M.R.: Concept Detection. In: Proc. 8th Conf. on Computing and Philosophy, ECAP 2010. University of Munich, Munich (2010)
25. Kulczynski, S.: Classe des Sciences Mathématiques et Naturelles. Bulletin Int. de l'Acadamie Polonaise des Sciences et des Lettres Série B (Sciences Naturelles) (Supp. II), 57–203 (1927)

26. Pasquier, N., Bastide, Y., Taouil, R., Lakhal, L.: Discovering Frequent Closed Itemsets for Association Rules. In: Beeri, C., Bruneman, P. (eds.) ICDT 1999. LNCS, vol. 1540, pp. 398–416. Springer, Heidelberg (1998)
27. Rogers, D.J., Tanimoto, T.T.: A Computer Program for Classifying Plants. Science 132, 1115–1118 (1960)
28. Russel, P.F., Rao, T.R.: On Habitat and Association of Species of Anopheline Larvae in South-eastern Madras. J. Malaria Institute 3, 153–178 (1940)
29. Sneath, P.H.A., Sokal, R.R.: Numerical Taxonomy. Freeman Books, San Francisco (1973)
30. Sokal, R.R., Michener, C.D.: A Statistical Method for Evaluating Systematic Relationships. University of Kansas Scientific Bulletin 38, 1409–1438 (1958)
31. Sokal, R.R., Sneath, P.H.A.: Principles of Numerical Taxonomy. Freeman Books, San Francisco (1963)
32. Sørensen, T.: A Method of Establishing Groups of Equal Amplitude in Plant Sociology based on Similarity of Species and its Application to Analyses of the Vegetation on Danish Commons. Biologiske Skrifter / Kongelige Danske Videnskabernes Selskab 5(4), 1–34 (1948)
33. Tanimoto, T.T.: IBM Internal Report, November 17 (1957)
34. Uno, T., Kiyomi, M., Arimura, H.: LCM ver. 2: Efficient Mining Algorithms for Frequent/Closed/Maximal Itemsets. Proc. Workshop Frequent Item Set Mining Implementations, FIMI 2004, Brighton, UK. CEUR Workshop Proceedings 126, Aachen, Germany (2004)
35. Uno, T., Kiyomi, M., Arimura, H.: LCM ver. 3: Collaboration of Array, Bitmap and Prefix Tree for Frequent Itemset Mining. Proc. 1st Open Source Data Mining on Frequent Pattern Mining Implementations, OSDM 2005, Chicago, IL, pp. 77–86. ACM Press, New York (2005)
36. Webb, G.I., Zhang, S.: k-Optimal-Rule-Discovery. Data Mining and Knowledge Discovery 10(1), 39–79 (2005)
37. Webb, G.I.: Discovering Significant Patterns. Machine Learning 68(1), 1–33 (2007)
38. Zaki, M.J., Parthasarathy, S., Ogihara, M., Li, W.: New Algorithms for Fast Discovery of Association Rules. In: Proc. 3rd ACM SIGKDD Int. Conf. on Knowledge Discovery and Data Mining, KDD 1997, Newport Beach, CA, pp. 283–296. AAAI Press, Menlo Park (1997)
39. Zaki, M.J., Gouda, K.: Fast Vertical Mining Using Diffsets. In: Proc. 9th ACM SIGKDD Int. Conf. on Knowledge Discovery and Data Mining, KDD 2003, Washington, DC, pp. 326–335. ACM Press, New York (2003)
40. Zheng, Z., Kohavi, R., Mason, L.: Real World Performance of Association Rule Algorithms. In: Proc. 7th ACM SIGKDD Int. Conf. on Knowledge Discovery and Data Mining, KDD 2001, San Francisco, CA, ACM Press, New York (2001)
41. Synthetic Data Generation Code for Associations and Sequential Patterns. Intelligent Information Systems, IBM Almaden Research Center, `http://www.almaden.ibm.com/software/quest/Resources/index.shtml`

Patterns and Logic for Reasoning with Networks

Angelika Kimmig[1], Esther Galbrun[2], Hannu Toivonen[2], and Luc De Raedt[1]

[1] Departement Computerwetenschappen, K.U. Leuven
Celestijnenlaan 200A - bus 2402, B-3001 Heverlee, Belgium
{angelika.kimmig,luc.deraedt}@cs.kuleuven.be
[2] Department of Computer Science and
Helsinki Institute for Information Technology HIIT
P.O. Box 68, FI-00014 University of Helsinki, Finland
{esther.galbrun,hannu.toivonen}@cs.helsinki.fi

Abstract. `Biomine` and ProbLog are two frameworks to implement bisociative information networks (BisoNets). They combine structured data representations with probabilities expressing uncertainty. While `Biomine` is based on graphs, ProbLog's core language is that of the logic programming language Prolog. This chapter provides an overview of important concepts, terminology, and reasoning tasks addressed in the two systems. It does so in an informal way, focusing on intuition rather than on mathematical definitions. It aims at bridging the gap between network representations and logical ones.

1 Introduction

Nowadays, large, heterogeneous collections of uncertain data exist in many domains, calling for reasoning tools that support such data. Networks and logical theories are two common representations used in this context. In the setting of bisociative knowledge discovery, such networks are called BisoNets [1]. The `Biomine` project has constructed a large network (or BisoNet) of biological knowledge and provided several reasoning mechanisms to explore this network [2]. ProbLog [3], on the other hand, provides a logic-based representation language and corresponding inference methods that have been used in the context of the same network. Both `Biomine` and ProbLog allow one to associate probabilities to network edges and thereby to reason about uncertainty. For ProbLog, this idea has recently also been extended to other types of labels, such as for instance costs, connection strengths, or revenues [4]. In this chapter, we highlight the common underlying ideas of these two frameworks, focusing on illustrative examples rather than formal detail. We provide an overview of network-related inference techniques from a logical perspective. These techniques can potentially be used to support bisociative reasoning and knowledge discovery. The aim is to bridge the gap between the two views and to point out similarities and opportunities for cross-fertilization.

The chapter is organized as follows: We first introduce the `Biomine` and ProbLog frameworks and their underlying concepts in Section 2. Section 3 then

M.R. Berthold (Ed.): Bisociative Knowledge Discovery, LNAI 7250, pp. 122–143, 2012.

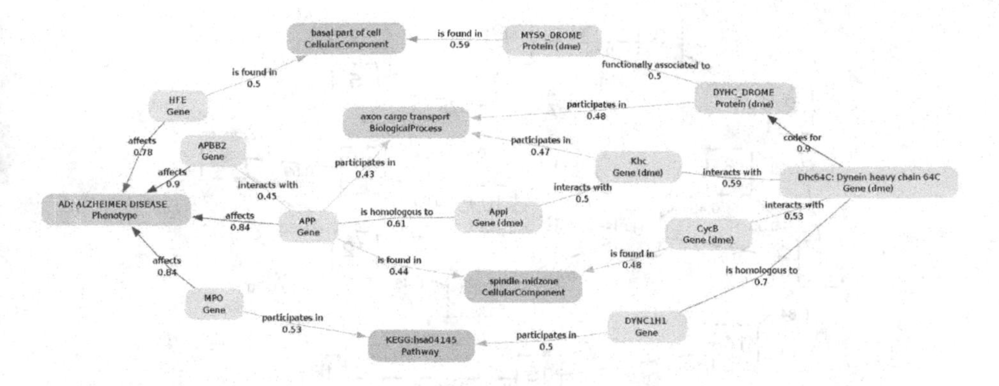

Fig. 1. An example of a subgraph extracted from Biomine

gives an overview of various inference and reasoning tasks, focusing on the structural aspect, before Section 4 discusses their extension towards the use of probabilities and other types of labels.

2 The Biomine and ProbLog Frameworks

The Biomine project has contributed a large network of biological entities and relationships between them, represented as typed nodes and edges, respectively [2]. The Biomine network is probabilistic; to each edge is associated a value that represents the probability that the link between the entities exists. A subnetwork extracted from this database is shown in Figure 1. Inspired on the Biomine network, ProbLog [3] extends the logic programming language Prolog with independent random variables in the form of probabilistic facts, corresponding to Biomine's probabilistic edges. In the remainder of this section, we will introduce the basic terminology used in the context of these frameworks for reasoning about networks.

2.1 Using Graphs: Biomine

Figure 2 gives a simplified representation of the Biomine subnetwork of Figure 1. We will use this representation for illustration throughout the chapter. Nodes have numbers as identifiers. There are five node types (tn1 to tn5). The number of edge types has been reduced to three (te1, te2 and te3) and their directions have been removed. We use colors and border styles to represent the node types, and line styles to represent the edge types; see Figure 5 for the exact mapping.

In general, nodes and edges could have several types simultaneously. Also, edges can be directed and there may exist multiple edges between a given pair of nodes. For ease of explanation, we will only consider the simpler case where

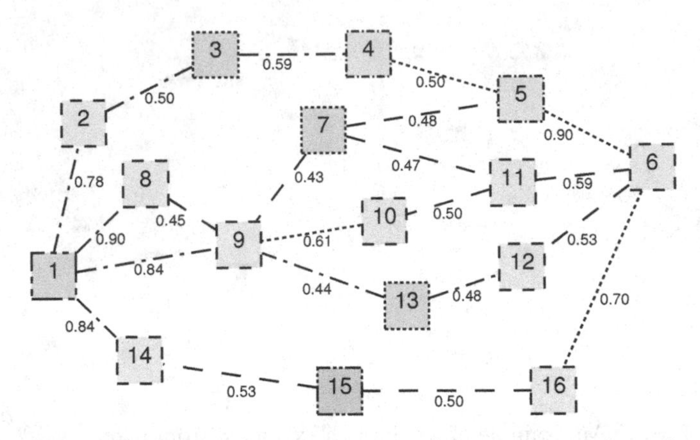

Fig. 2. A simplified representation of the `Biomine` subgraph

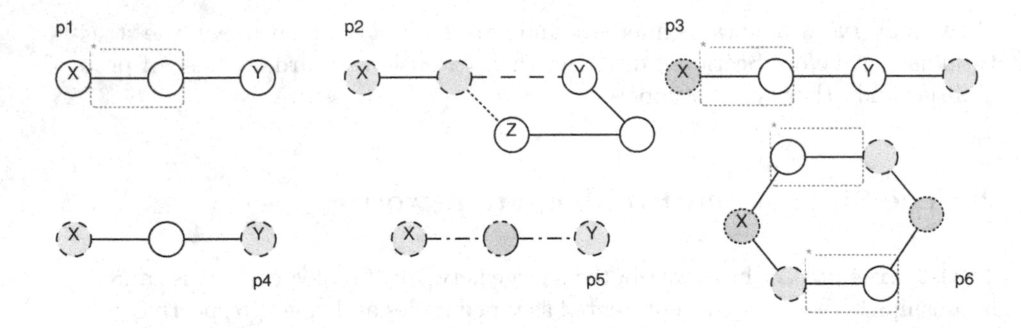

Fig. 3. Examples of graph patterns

edges and nodes have a single type and there is at most one edge between any pair of nodes.[1]

A *graph pattern* is an expression over node and edge types. It is an abstract graph that defines a subgraph by means of a set of constraints over the connection structure and edge and node types. Six example patterns, `p1` to `p6`, are presented in Figure 3.

The pattern nodes are represented using circles to distinguish them from network nodes, which are represented as squares. Pattern nodes and edges are either required to be of a given type, or can be of arbitrary type. The latter is denoted using white nodes and solid edges. *Query nodes* are labeled with capital letters, these are the main points of interest when querying the network. As in regular expressions, the star denotes unlimited repetitions of substructures.

For instance, pattern `p1` corresponds to a path of length at least one between the query nodes `X` and `Y`, using arbitrary node and edge labels, whereas `p5` specifies the exact number of edges and all edge and node types.

[1] Allowing multiple edges between the same pair of nodes can be done by introducing explicit edge identifiers, both in the network and, where needed, also in the patterns.

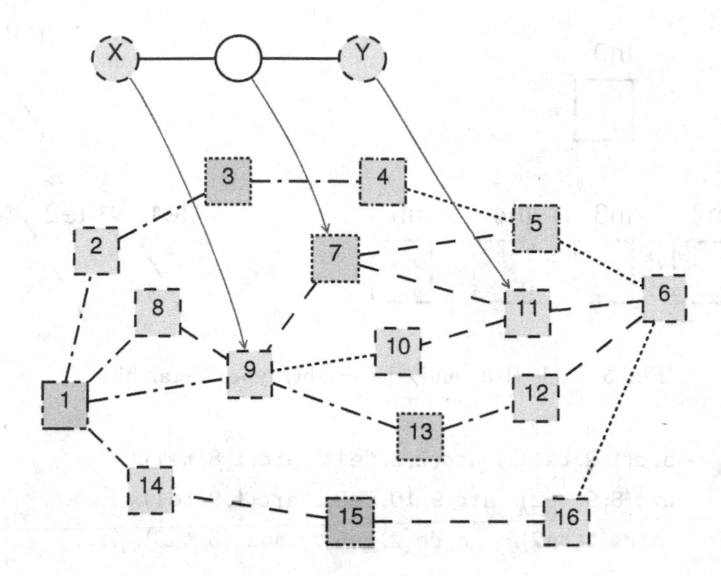

Fig. 4. Example of instantiation of pattern p4

A *substitution* assigns network nodes to nodes in a pattern. An *instantiation* maps a pattern onto the network using subgraph isomorphism. Thus, it is a substitution of all nodes in the pattern in such a way that a corresponding edge mapping exists as well. An *answer substitution* is a restriction of an instantiation to the query nodes.

An example instantiation of pattern p4 is shown in Figure 4. There might be several possible instantiations of a pattern with the same answer substitution. For instance, `p4{X/9, Y/11}` can be instantiated in two ways, either by mapping the middle node to 7, as in the illustration, or by mapping it to 10.

While we here consider a flat type system, where types are either given or completely undefined, it is also possible to use type hierarchies. When instantiating patterns, a node (respectively an edge) can then be mapped to a node (edge) of same type or one of its descendant types. The hierarchies used in our example are shown in Figure 5, where the undefined type is the root node of the hierarchy.

2.2 Using Logic: ProbLog

As ProbLog is based on the logic programming language Prolog, we first illustrate the key concepts of Prolog by means of an example; for a more detailed introduction, we refer to [5]. We defer discussion of the probabilistic aspects of ProbLog to Section 4.

In Prolog, the network of Figure 2 (ignoring the probability labels) can be represented as a set of *facts*:

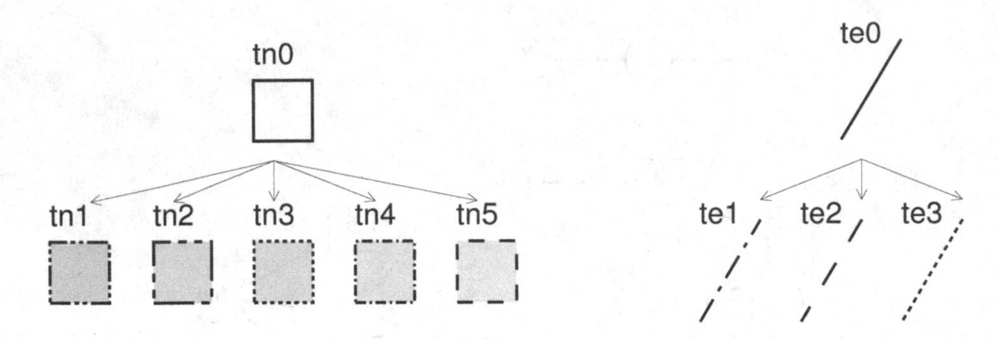

Fig. 5. Node (left) and edge (right) types hierarchies

$$\begin{array}{lll} \texttt{arc(1,2,te1).} & \texttt{arc(2,3,te1).} & \texttt{arc(1,8,te1).} \\ \texttt{arc(8,9,te2).} & \texttt{arc(9,10,te3).} & \texttt{arc(1,9,te1).} \ \ldots \\ \texttt{node(1,tn2).} & \texttt{node(2,tn5).} & \texttt{node(3,tn3).} \ \ldots \end{array} \quad (1)$$

Here, `arc(1,2,te1)` states that there is a directed edge from node `1` to node `2` of type `te1`; `node(1,tn1)` specifies that node `1` is of type `tn1`, and so forth.[2] `arc/3` is a *predicate* of *arity* 3, that is, with 3 *arguments.* To obtain undirected edges, a Prolog program would define an additional predicate `edge/3` as follows:

$$\texttt{edge(X,Y,T)} \ :- \ \texttt{arc(X,Y,T).} \quad (2)$$
$$\texttt{edge(X,Y,T)} \ :- \ \texttt{arc(Y,X,T).} \quad (3)$$

Here, uppercase letters indicate *logical variables* that can be instantiated to constants such as `1` or `tn3`. The definition of `edge/3` above consists of two *clauses* or rules. The first clause states that `edge(X,Y,T)` is true for some nodes `X` and `Y` and node type `T` if `arc(X,Y,T)` is true. The second clause gives an alternative precondition for the same conclusion. Together, they provide a disjunctive definition, that is, `edge(X,Y,T)` is true if at least one of the rules is true.

For instance, `edge(2,1,te1)` is true due to the second clause and the fact `arc(1,2,te1)`, where we use the substitution {`X/2, Y/1, T/te1`} to map rule variables to constants. `edge(2,1,te1)` is said to *follow* from or to be *entailed* by the Prolog program.

More formally, Prolog answers a given query by trying to *prove* the query using the facts and clauses in the program. The answer will be `yes` (possibly together with a substitution for the query variables, which are considered to be existentially quantified), if the query follows from the program (for that substitution). Query `?- edge(2,1,te1)` results in the answer `yes` due to clause (3) and fact `arc(1,2,te1)`. For `?- edge(2,1,te2)`, the answer is `no`, as Prolog terminates without finding a corresponding fact to complete the proof. For `?- edge(A,B,C)`,

[2] Alternatively, one could also use facts such as `te1(1,2)` and `tn1(1)`.

Prolog will return the substitution {A/1, B/2, C/te1}, and will allow the user to keep asking for alternative answers, such as {A/2, B/3, C/te1}, {A/1, B/8, C/te1}, and so forth, until no more substitutions can be generated from the program.

For convenience, we also define edges of arbitrary type:

$$\texttt{edge(X,Y)} \; :- \; \texttt{edge(X,Y,T)}. \tag{4}$$

Alternatively, one could encode the type hierarchy in Figure 5:

$$\begin{aligned} \texttt{edge(X,Y,te0)} &\; :- \; \texttt{edge(X,Y,te1)}. \\ \texttt{edge(X,Y,te0)} &\; :- \; \texttt{edge(X,Y,te2)}. \\ \texttt{edge(X,Y,te0)} &\; :- \; \texttt{edge(X,Y,te3)}. \end{aligned}$$

Prolog also allows for more complex predicate definitions, such as a path between two nodes:

$$\begin{aligned} \texttt{path(X,Y)} &\; :- \; \texttt{edge(X,Y)}. \\ \texttt{path(X,Y)} &\; :- \; \texttt{edge(X,Z),path(Z,Y)}. \end{aligned} \tag{5}$$

The set of facts in a Prolog program are also called the *database*, and the set of clauses the *background knowledge*.

To simplify notation and to closely follow the network view, in the remainder of this chapter we assume that different logical variables are mapped onto different constants; this could be enforced in Prolog by adding atoms of the form $\texttt{X} \neq \texttt{Y}$ to predicate definitions.

So far, we have focused on encoding information about a specific network. However, Prolog allows one to encode both data and algorithms within the same logical language, and thus makes it easy to implement predicates that reason about the program itself, for instance, by simulating proofs of a query in order to generate additional information. As we will see in Section 3, this provides a powerful means to cast reasoning tasks in terms of queries; we refer to [5] for a detailed discussion.

In the logical setting, a pattern corresponds to a *predicate*. As in the graph setting, its definition imposes constraints on the types and connection structure. For example, the predicate path(X,Y) defined above directly corresponds to pattern p1 in Figure 3. The *query variables* X and Y correspond to the query nodes in graph patterns. Query variables are mapped to constants using *answer substitutions* as in the network setting. Building on the definitions above, the full set of patterns in Figure 3 can be encoded as follows:

$$\texttt{p1(X,Y)} \; :- \; \texttt{path(X,Y)}. \tag{6}$$

$$\begin{aligned} \texttt{p2(X,Y,Z)} \; :- \; & \texttt{node(X,tn5),edge(X,A),node(A,tn5),edge(A,Y,te2),} \\ & \texttt{edge(A,Z,te3),edge(Y,B),edge(Z,B)}. \end{aligned} \tag{7}$$

$$\texttt{p3(X,Y)} \; :- \; \texttt{node(X,tn3),path(X,Y),edge(Y,A),node(A,tn5)}. \tag{8}$$

$$\texttt{p4(X,Y)} \; :- \; \texttt{node(X,tn5),edge(X,A),edge(A,Y),node(Y,tn5)}. \tag{9}$$

$$\begin{aligned}
\texttt{p5(X,Y)} \; :- \;& \texttt{node(X,tn5),edge(X,A,te1),node(A,tn2),} \\
& \texttt{edge(A,Y,te1),node(Y,tn5).} \qquad (10) \\
\texttt{p6(X)} \; :- \;& \texttt{node(X,tn3),edge(X,A),node(A,tn5),path(A,B),} \\
& \texttt{node(B,tn3),edge(B,C),node(C,tn5),path(C,X).} \qquad (11)
\end{aligned}$$

Furthermore, using logic also allows one to easily express additional constraints on patterns. For instance, pattern p7 states that node X has a neighboring node whose type is not tn5, while pattern p8 states that it has two outgoing edges of the same type:

$$\begin{aligned}
\texttt{p7(X)} \; :- \;& \texttt{edge(X,Y),not(node(Y,tn5)).} \\
\texttt{p8(X)} \; :- \;& \texttt{edge(X,Y,T),edge(X,Z,T).}
\end{aligned}$$

For ease of presentation, we assume that patterns are always defined by a single clause. Note that this does not preclude disjunctive patterns, as can be seen for p1 in (6) above.

A substitution θ is an *answer substitution* for a pattern p if the query $p\theta$ follows from the Prolog program. For instance, {X/9, Y/11} is an answer substitution for pattern p4(X,Y), as p4(9, 11) follows from our example program.

An *explanation* for a pattern is a minimal set S of database facts such that the pattern follows from S and the background knowledge. For instance, {node(9,tn5), arc(9,7,te2), arc(7,11,te2), node(11,tn5)} is an explanation for p4(9, 11).

2.3 Summary

Table 1 summarizes the key terms introduced in this section.

Table 1. Correspondence of the different terminologies

logic view	graphical view
background knowledge	set of patterns
set of facts, database	graph
predicate	pattern
query variables	query nodes
explanation	instantiation

3 Inference and Reasoning Techniques

This section provides an overview of a broad range of reasoning techniques. We start with the classical tasks of deduction and abduction, that are both concerned with matching given patterns against the graph or database. Next, we discuss various settings for induction, that is, for inferring patterns under different conditions. We then in turn consider techniques that combine pattern creation and pattern matching, that identify nodes in the graph, and that modify the database or the background knowledge in a number of different settings. Throughout the discussion, we assume a Prolog program encoding the graph and possible background knowledge as discussed in Section 2.2. This allows us to view the different reasoning tasks as queries asked to a Prolog system.

3.1 Deduction: Reasoning about Node Tuples

The question answered by deduction is whether there exists an instantiation of a pattern in a graph, or, equivalently, whether the pattern follows from the Prolog program encoding the graph. It thus directly corresponds to answering Prolog queries as discussed in Section 2.2.

In our example, given the ground query `?- p2(8,7,10)`, deduction will produce an affirmative answer, as there is an instantiation of the pattern using the real nodes `9` and `11`. Similarly, `p4(8,10)` is true but `p2(10,7,8)`, `p2(9,6,11)` and `p4(14,10)` are false.

To summarize, *given a Prolog program, a pattern* `p` *and a substitution* θ *that grounds* `p`*, deduction corresponds to answering the query* `?- p`θ *from the program.*

The decision problem of deduction as described here forms the basis for many other reasoning tasks on the level of node tuples; we discuss some examples next.

Answer Enumeration. For non-ground patterns, the enumeration problem associated to deduction corresponds to finding *all* answer substitutions for the pattern. Alternatively, one can ask for *some* answer substitution chosen from the set of all possible ones. For instance, one possible answer substitution for `?- p2(X,Y,Z)` would be `{X/8, Y/7, Z/10}`, whereas `?- p4(15,Y)` does not produce an answer substitution, as there is no proof of this query.

Thus, *given a Prolog program and a pattern* `p`*, the answer substitution and enumeration problem of deduction correspond to finding one or all answer substitutions for the query* `?- p` *from the program, respectively.*

Representative Nodes. A binary pattern `p(X,Y)` can be used to find a set of *representative nodes*, that is, nodes `r` that, when substituted for `X`, lead to a set of patterns `p(r,Y)` such that all other nodes appear in an answer substitution for at least one such pattern. For instance, using `p(X,Y) :- edge(X,Z),edge(Z,Y)`, one set of representative nodes is $\{1, 4, 16\}$. Note that here, some nodes are associated to several representative nodes, in this example node `7` is associated to both the representative nodes `1` and `4`. A harder variant of the problem would be to require that there is exactly one such representative for each node.

In a nutshell, *given a constant k, a Prolog program encoding a network with nodes N, and a pattern* `p(X,Y)`*, the task of finding representative nodes is to find a subset $S \subseteq N$ of size k such that for each node $y \notin S$ there is a node $s \in S$ for which the query* `?- p(s,y)` *is answered affirmatively.*

Spread of Influence. Recursive patterns such as `path(X,Y)` can be used to measure distances from a given node in a network to all other nodes in terms of the minimal number of edges needed to reach the other node. This principle can be regarded as the basis of techniques measuring spread of influence, and can be used to enumerate nodes by increasing distance.

For instance, using pattern `path(1,X)`, thus measuring the distance from node 1, the closest set of nodes is $\{2, 8, 9, 14\}$, the next one $\{3, 7, 10, 13, 15\}$, and so forth.

In Prolog, this could easily be realized by extending the path predicate with a third argument that counts the number of edges traversed:

$$\begin{aligned}\mathtt{path(X,Y,1)} &\ :- \ \mathtt{edge(X,Y)}.\\ \mathtt{path(X,Y,L)} &\ :- \ \mathtt{edge(X,Z),path(Z,Y,L),L\ is\ N+1}.\end{aligned}$$

One would then ask a sequence of queries `?- path(1,X,i)` with $\mathtt{i} = 1, \ldots, n$ up to a maximum length n, though some extra book-keeping would be required to filter out nodes that have been returned as an answer on previous levels already.

Thus, *given a Prolog program, a maximum distance n, and a recursive pattern* `p(x,Y,D)` *with source node* `x`*, spread of inference corresponds to answer enumeration for the sequence of queries* `?- p(x,Y,i)` *for* $\mathtt{i} = 1, \ldots, n$.

3.2 Abduction: Reasoning about Subgraphs

The task of abduction is closely related to that of generating an explanation for a query as discussed in Section 2.2. In terms of graphs, it directly corresponds to finding a minimal instantiation of a pattern. In the logical setting, abduction is not restricted to database predicates, but can use all predicates marked *abducible.* In the context of patterns and networks, one could simply assume all predicates used in pattern definitions to be abducible and implement a predicate `abduce`; see [5] for a general definition of this predicate and more details.

For instance, when calling the query `?- abduce(p6(7),E)` (cf. Equation (11)) and assumming that all predicates are abducible, the answer `E` would be the conjunction of `node(7,tn3)`, `edge(7,9)`, `node(9,tn5)`, `path(9,13)`, `node(13, tn3)`, `edge(13,12)`, `node(12,tn5)`, and `path(12,7)`.

To summarize, *given a Prolog program and a pattern* `p`*, abduction corresponds to answering the query* `?- abduce(p,E)`.

Again, one can also consider the corresponding enumeration problem, where the task is to find all explanations or instantiations.

3.3 Induction: Finding Patterns

Frequent Patterns. The usual frequent subgraph mining problem corresponds to the problem of finding all patterns from a given pattern language with more than a chosen number of instantiations. The pattern language will specify both allowed structures of patterns and which nodes in patterns can be query nodes.

For instance, for a frequency threshold of 3, all patterns in Figure 3 would be frequent.

Similarly to `abduce/2` above, one could implement a Prolog predicate

```
frequent(P,T) :- pattern(P), count(P,N), N >= T.
```

that returns patterns with a frequency greater or equal to a user-defined threshold `T`. It relies on suitable definitions of `pattern/1` (defining elements of the pattern language) and `count/2` (counting instantiations of a given pattern). Then, finding frequent patterns for a given frequency threshold `t` corresponds to answering the query `?- frequent(P,t)`. While this simple approach illustrates the basic idea, an efficient implementation would clearly be more involved.

To summarize, *given a frequency threshold* `t` *and a Prolog program including definitions of a pattern language and a counting function, finding frequent patterns corresponds to answering the query* `?- frequent(P,t)`.

Concept Learning. The aim of concept learning is to construct a definition of a new predicate that covers all positive examples, but none of the negative ones. In our context, examples are node tuples, but for convenience we represent them as ground instances of the pattern to be found. Again, this could be realized in Prolog based on a suitable definition of a predicate

```
concept(C) :- hypothesis(C),
              findall(P, (pos(P), not(covers(C,P))), []),
              findall(N, (neg(N), covers(C,N)), []).
```

Here, `hypothesis/1` enumerates possible concepts, `covers/2` checks whether the concept covers an example, and `pos/1` and `neg/1` define examples. The Prolog builtin `findall/3` is used here to verify that there is no positive example that is not covered by the concept, and no negative one that is covered. In general, its third argument is a list of all instantiations of the variable in the first argument for which the query in the second argument holds, and `[]` denotes the empty list.

For instance, assume we are given examples `pos(q(3,1))`, `pos(q(7,15))` and `neg(q(7,4))`. Then, querying `?- concept(C)` could return `C = (q(X,Y) :- p3(X,Y))` as a possible solution.

Thus, *given a Prolog program including definitions of a hypothesis language and positive and negative examples, concept learning corresponds to answering the query* `?- concept(C)`.

Generalisation. Comparing patterns based on a *generality* relation provides a means to choose between alternative solutions. Given a Prolog program including two patterns

$$p_a :- a_1, \ldots, a_n.$$
$$p_b :- b_1, \ldots, b_m.$$

p_a is *more general* than p_b if the query `?- pa` follows from the program that is obtained by adding the facts b_1 to b_m to the original program, where each variable is replaced by a new constant symbol. For instance, `p4(X,Y)` (Eq. (9)) is more general than `p5(X,Y)` (Eq. (10)), as `?- p4(x,y)` can be proven from our example program extended with the facts `node(x,tn5)`, `edge(x,a,te1)`, `node(a,tn2)`, `edge(a,y,te1)`, and `node(y,tn5)`.

From the perspective of graphs, generality can again be seen as a form of subgraph isomorphism, this time between patterns where the nodes and edges of the more general pattern are mapped to those of the more specific one of same type or children type. Notice that in the literature on logical and relational learning there are multiple notions of generality that can be employed [6].

The notion of generality can also be used to find a maximally specific common generalisation of two given patterns, that is, a pattern that is more general than each of the input patterns, but for which there is no more specific pattern that also fulfills this criterion. A corresponding Prolog predicate `generalize/3`, queried as `?- generalize(p2(X,Y),p4(X,Y),G)` would provide the answer `C = (node(X,tn5), edge(X,A), edge(A,Y))`.

Thus, *given a Prolog program and two patterns* p_a *and* p_b*, the task of generalization corresponds to answering the query* `?- generalize(pa,pb,P)`.

Clustering. Patterns can also be used to *cluster* node tuples: all node tuples that satisfy a given pattern fall into the same cluster. The task of clustering a given set of node tuples then corresponds to that of finding k patterns that cluster the node tuples into k disjoint (or possibly overlapping) subsets based on characteristics of their local connection.

A very simple set of clustering patterns in our example would be the set containing `node(X,tni)` for $i = 1, \ldots, 5$ that would simply cluster single nodes by their types.

In a nutshell, *given a Prolog program, a constant* k *and a set of node tuples* T*, clustering is the task of finding a set* P *of* k *patterns such that for each* $t \in T$*, there is exactly one pattern* $p \in P$ *for which* t *is an answer substitution for* p*.*

3.4 Combining Induction and Deduction

As deduction matches patterns against the database, while induction constructs new patterns, the two approaches can be naturally combined to find both patterns and corresponding substitutions simultaneously.

Analogy. Node tuples can be considered analogous if they are answer substitutions for the same pattern. Given a substitution, the problem of finding analogous tuples can be defined as finding a pattern for which this substitution is an answer substitution along with all other answer substitutions for it. The more specific the pattern, the stronger the analogy.

For example, the pairs of nodes `(2,8)`, `(12,16)`, `(14,9)` and `(9,11)` are analogous with respect to pattern `p4`, i.e., in the sense that they are all pairs of nodes of type `tn5` separated by an intermediate node. However, pattern `p5` defines a stronger analogy that only relates the pairs `(2,8)` and `(14,9)`. In the case of an asymmetric pattern, it can be interesting to consider the sets of real nodes assigned to one particular query node in the pattern, in other words, nodes that take the same role in the analogy.

That is, *given a Prolog program and a substitution θ, the task of reasoning by analogy is to find a pattern* `p` *for which θ is an answer substitution as well as the set S of all answer substitutions for* `p`.

Synonyms. Two structurally distinct patterns are *synonyms* of one another if they have the same answer substitutions. Synonyms are also known as *redescriptions* or *syntactic variants.* Finding synonym patterns and their answer substitutions can be one way of finding sets of objects of special interest. Furthermore, given two networks with node and edge types from different domains, finding synonyms can help to establish mappings between these domains.

For instance, the following two patterns are synonyms (albeit only covering a single node due to the simplicity of the example graph):

```
s1(X) :- edge(X,Y,te1),edge(X,Z,te1),edge(Y,Z).
s2(X) :- node(X,tn2).
```

Thus, *given a Prolog program, finding synonyms means finding a set of patterns P such that each pattern in P has the same set of answer substitutions.*

3.5 Modifying the Knowledge Base

We now turn to a set of techniques that modify the graph, database, or background knowledge. The key difference to the techniques discussed so far is that we now allow for loosing information.

Graph Simplification. The goal of graph simplification is to remove *redundant* edges from a graph. Here, redundancy is defined with respect to paths: an edge is considered redundant if all pairs of nodes connected in the original graph are also connected in the graph after removing the edge. In the purely structural case, graph simplification thus corresponds to finding a spanning tree; we will come back to the use of additional quality measures in Section 4.6.

That is, *given a Prolog program including a set E of facts representing edges and a predicate* `path/2`, *graph simplification finds a minimal set $S \subseteq E$ such that the set of answer substitutions for* `path(X,Y)` *remains the same when reducing E to S in the program.*

Subgraph Extraction. The aim of subgraph extraction is to find a subgraph of a given maximal size while retaining as much information as possible with respect to a given set of examples, that is, answer substitutions for a pattern.

For instance, given examples `path(3,7)` and `path(3,13)` and an upper limit of 5 edges, our network could be compressed to contain `edge(2,3)`, `edge(1,2)`, `edge(1,9)`, `edge(9,7)` and `edge(9,13)` only.

Thus, *given a Prolog program including a set E of facts representing edges, a constant k and a set T of answer substitutions for pattern* `p`, *subgraph extraction finds a set $S \subseteq E$ of size at most k such that all $\theta \in T$ are also answer substitutions for* `p` *when reducing E to S in the program.*

Abstraction. The task of abstraction is to rewrite the database using new predicates that abstract away some of the information present in the initial database. While techniques such as graph simplification and subgraph extraction also loose information, abstraction differs in that it replaces database predicates by a new predicate, obtained by computing answers for the pattern defining the new predicate. For instance, one could replace the predicate `arc/3`, that is, the directed, typed edges, using

$$\texttt{p(X, Y)} \;:-\; \texttt{node(X, T), node(Y, T), T} \neq \texttt{tn5, path(X, Y).}$$

that is, edges that correspond to paths between pairs of nodes of the same type (different from `tn5`) in the original network. This would result in the new database

$$\begin{array}{llll} \texttt{p(3,7).} & \texttt{p(3,13).} & \texttt{p(7,13).} & \texttt{p(4,5).} \\ \texttt{p(7,3).} & \texttt{p(13,3).} & \texttt{p(13,7).} & \texttt{p(5,4).} \end{array}$$

Abstractions can be created using any technique that identifies patterns and thus predicate definitions. Instead of adding the definitions of these predicates to the database, it computes all groundings of the new predicate, adds these to the database, and deletes the old facts.

Thus, *given a Prolog program, a database predicate* `d` *and a pattern* `p`, *abstraction adds* $\texttt{p}\theta$ *for all answer substitutions θ for* `p` *to the program and deletes the definition of* `d`.

Predicate Invention. The key idea of predicate invention is to introduce new patterns that can be used to represent the background knowledge more compactly. For instance, the DUCE system [7] measures compactness using the minimum description length principle. As an example, consider the following set of rules:

$$\begin{array}{lll} \texttt{q1(Z)} & :- & \texttt{edge(Z, Y), edge(Y, X), edge(X, W), edge(W, V), node(V, tn1).} \\ \texttt{q2(Z)} & :- & \texttt{edge(Z, Y), edge(Y, X), edge(X, W), edge(W, V), node(V, tn2).} \\ \texttt{q3(Z)} & :- & \texttt{edge(Z, Y), edge(Y, X), edge(X, W), edge(W, V), node(V, tn3).} \end{array}$$

Inventing a predicate dist4(X,Y) allows one to rewrite these definitions more compactly as

```
dist4(Z,V) :- edge(Z,Y),edge(Y,X),edge(X,W),edge(W,V).
     q1(Z) :- dist4(Z,V),node(V,tn1).
     q2(Z) :- dist4(Z,V),node(V,tn2).
     q3(Z) :- dist4(Z,V),node(V,tn3).
```

While this transformation has preserved the meaning of the original fragment, this need not be the case in general. Similar principles can also be used to compress graphs by replacing instantiations of a pattern by new nodes [8].

In general, *given a Prolog program, the task of predicate invention is to introduce new pattern definitions which are then used to rewrite the program more compactly.*

3.6 Summary

Table 2 summarizes the different reasoning techniques presented in previous sections. It recapitulates the information provided to and the problem solved by each of them.

4 Using Probabilistic or Algebraic Labels

So far, we have restricted our discussion to crisply defined networks and logical theories. However, in both Biomine and ProbLog, the information provided is uncertain. This uncertainty is expressed by attaching probabilities to edges or facts, and can be exploited in various ways for reasoning. Furthermore, ProbLog has recently been generalized to aProbLog [4], where probabilities can be replaced by other types of labels, such as costs or distances. In this section, we first briefly review the probabilistic model underlying Biomine and ProbLog, and then illustrate how the techniques from Section 3 can benefit from the probabilistic setting. While some of these techniques have already been realized in Biomine, ProbLog, or other probabilistic frameworks, for others, the details of such a transfer are still open. Finally, we touch upon the perspectives opened by aProbLog.

4.1 The Probabilistic Model of Biomine and ProbLog

In the probabilistic graph model underlying Biomine, a value is associated to each edge, indicating the probability that the relationship exists. In Biomine, these values are obtained as the product of three factors, indicating the *reliability*, the *relevance*, and the *rarity* (or specificity) of the information, cf. [2], but they can be obtained in a different way as well. Existences of the edges are considered independent from each other. This actually defines a probability distribution over

Table 2. Summary of the different reasoning methods

Sec.	Methods	Information	Problem
3.1	Deduction	pattern, substitution	is it an answer substitution?
3.1	Answer Enumeration	pattern	list all answer substitutions
3.1	Representative Nodes	pattern, integer k	find k representative nodes
3.1	Spread of Influence	recursive pattern, node	enumerate nodes by distance
3.2	Abduction	pattern	find an/all instantiation(s)
3.3	Frequent Patterns	frequency threshold	list all frequent patterns
3.3	Concept Learning	pos./neg. examples	find a discriminative pattern
3.3	Generalisation	two patterns	find a generalized pattern
3.3	Clustering	substitutions, integer k	find k clustering patterns
3.4	Analogy	substitution	find a pattern and answer substitutions
3.4	Synonyms		find a set of patterns with same answer substitutions
3.5	Graph Simplification		maximally reduce graph keeping answer substitutions for path
3.5	Subgraph Extraction	examples, integer k	reduce graph to size $\leq k$ respecting examples
3.5	Abstraction	database predicate, pattern	replace predicate definition by pattern instances
3.5	Predicate invention		reduce program size via new predicates

possible subnetworks, i.e., deterministic instances of the probabilistic network. Each subnetwork E_i has probability

$$P(E_i) = \prod_{x \in E \setminus E_i} (1 - p_x) \prod_{x \in E_i} p_x \tag{12}$$

where E is the set of edges in the probabilistic network, E_i is the set of edges realised in the deterministic instance and p_x the existence probability of edge x. For instance, the network in Figure 6 has probability (starting with the edges involving node 1) $0.78 \cdot (1 - 0.9) \cdot 0.84 \cdot 0.84 \cdot \ldots = 1.237e - 06$.

In ProbLog, probabilities are associated to ground facts instead of edges, and again, these facts are considered to correspond to independent random variables. The directed edges of (1) are now represented as follows:

```
0.78 :: arc(1,2,te1).   0.50 :: arc(2,3,te1).   0.90 :: arc(1,8,te1).
0.45 :: arc(8,9,te2).   0.61 :: arc(9,10,te3).  0.84 :: arc(1,9,te1).
```

In analogy to Equation (12), ProbLog thus defines a probability distribution over instances E_i of a probabilistic database with facts E.

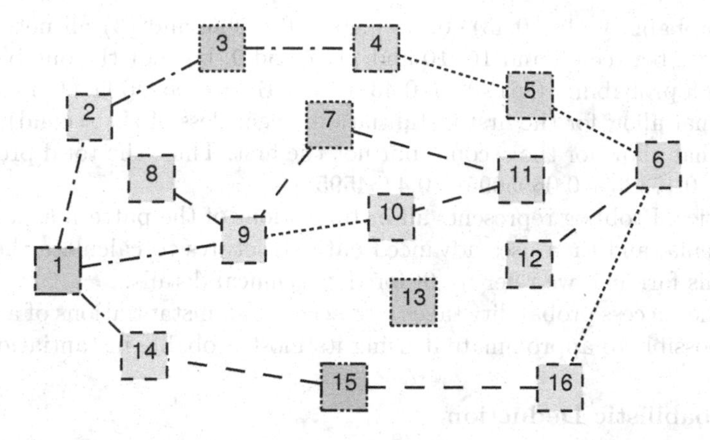

Fig. 6. A network sampled from the probabilistic graph in Figure 2

One can now ask for the probability of a specific pattern instantiation, which corresponds to the probability that this subgraph is present in a randomly sampled network. Due to the independence assumption, this probability is obtained by simply multiplying the probabilities of the instance's edges. Put differently, it corresponds to the sum of probabilities of all subnetworks of the probabilistic network that contain the instance. For example, the probability of the instantiation of pattern `p4(9,11)` presented in Figure 4 is $0.43 \cdot 0.47 = 0.2021$. The probability of its instantiation using node `10` as the middle node instead is $0.61 \cdot 0.50 = 0.305$.

The same principle of summing over all relevant subnetworks is also used to define the probability of a pattern q, called *success probability* in ProbLog:

$$P_s(q) = \sum_{E_i \subseteq E : q \text{ follows from } E_i} P(E_i). \tag{13}$$

Clearly, directly following this definition to calculate probabilities is infeasible in any network of realistic size. However, several alternative approaches have been developed, either based on sampling large numbers of networks or on enumerating pattern instantiations instead of full subnetworks. The latter approach, followed by ProbLog, requires to address the *disjoint-sum-problem*, that is, the fact that more than one instantiation of the same pattern can exist in the same subnetwork. It is therefore not possible to simply sum the probabilities of all instantiations, as this would count such subnetworks multiple times. Consider again the two instantiations of pattern `p4(9,11)` above. There are many subnetworks that allow for both instantiations (including the one in Figure 6), and we thus cannot simply sum these probabilities. Instead, we could split the relevant set of subnetworks into three disjoint parts based on the edges occurring in the instantiations: (1) all networks including the edges between `9` and `7` and between `7` and `11`, with probability 0.2021, (2) all networks that do not contain the edge between `7` and `9`, but the edges between `9` and `10` and between `10` and

11, with probability $(1-0.43)\cdot 0.61\cdot 0.50 = 0.17385$, and (3) all networks including edges between 9 and 10, 10 and 11, 7 and 9, but not the one between 7 and 11, with probability $0.61\cdot 0.50\cdot 0.43\cdot(1-0.47) = 0.0695095$. (1) includes all networks that allow for the first instantiation (regardless of the second), (2) and (3) those that allow for the second, but not the first. Thus, the total probability is $0.2021 + 0.17385 + 0.0695095 = 0.4454595$.

In practice, ProbLog represents all instantiations of the pattern as a propositional formula, and then uses advanced data structures to calculate the probability of this formula; we refer to [9] for the technical details.

While the success probability takes into account all instantiations of a pattern, it is also possible to approximate it using its most probable instantiation only.

4.2 Probabilistic Deduction

While deduction in the classical sense is concerned with deciding whether a substitution is an answer substitution for a given pattern in the network, in a probabilistic setting, it asks for the *probability* that this is the case, and thus solves Equation (13).

Answer Enumeration. When considering the set of all answer substitutions for a pattern, probabilities provide a natural means of ranking these. For instance, each answer substitution for `p5(X,Y)` corresponds to a single instantiation, that is, two edges linking node `1` to two of its neighbors. The most likely answer substitutions (omitting symmetric cases for brevity) thus are {`X/8,Y/9`} and {`X/8,Y/14`}, each with probability $0.9\cdot 0.84 = 0.756$, followed by {`X/9,Y/14`} (probability 0.7056), {`X/2,Y/8`} (0.702), and finally {`X/2,Y/9`} and {`X/2,Y/14`} (0.6552 each).

Non-redundant Set of Representatives. Finding a non-redundant set of representatives as proposed in [10] consists in solving the representative nodes problem (cf. Section 3.1) in a probabilistic setting.

Using the `path` predicate, the aim is to find a set X of k representative nodes such that the probability that {`Y/y`} is an answer substitution for `path(x,Y)` for some `x` in X is maximum for each original node `y`. More formally, the set of representative nodes is defined as

$$argmax_{X \subset N, |X|=k} \sum_{y \in N} max_{x \in X} P_{path(x,y)}$$

where $P_{path(x,y)}$ is the probability of the best instance of `path(x,y)`, namely the most probable path between nodes `x` and `y`.

Spread of Influence. Instead of using the number of recursive steps or the size of the instantiation as a measure for the distance, in a probabilistic context, spread of inference can use the probability that a substitution is an answer substitution for a pattern. It would thus prefer more distant nodes (in terms of path length) if their probability of being connected to the source node is higher.

4.3 Probabilistic Abduction and Top-k Instantiations

In the presence of probabilities, one might not be interested in finding an explanation for a query but rather in finding the most probable one.

In that setting an interesting alternative to the enumeration problem of abduction is the task of finding the k most probable instantiations of a given pattern.

Note that identifying the k most probable instantiations of a pattern might return rather uninteresting results if they are all about the same node tuple. In order to obtain a more diverse set of answers one might look for the k tuples with most probable instantiations instead (corresponding to deductive answer enumeration approximating probabilities by those of the most likely instantiations), or even require the tuples to not overlap.

4.4 Patterns and Probabilities

When looking for patterns, probabilities can again provide a natural way to select between various alternative solutions.

Pattern Mining. Probabilistic local pattern mining in ProbLog [11] extends pattern mining in multi-relational databases to the probabilistic setting. Instead of a counting function, it uses a scoring function based on the probabilities of candidate patterns on given node tuples. It thus basically replaces the 0/1-membership function of frequent pattern mining with a gradual one based on probabilities. Probabilities of individual instances are combined using sum (resulting in a kind of probabilistic frequency) or product (resulting in a kind of likelihood function).

Concept Learning. Concept learning in the context of ProbLog has been studied in [12], where the relational rule learner FOIL is lifted to work with probabilistic data and examples.

Generalisation. In a probabilistic setting, generalisation can be used in different contexts and ways. For instance, probabilistic explanation based learning in ProbLog [13] generalizes an explanation of an example query in terms of database predicates by replacing constants by variables, thus obtaining a new pattern definition. Stochastic logic programs, a probabilistic logic language inspired on probabilistic grammars, can be learned from examples in the form of proofs by generalizing pairs of clauses extracted from these examples [14]. In the latter case, probabilities are associated to clauses, and need to be adapted during generalisation as well.

Clustering. In the context of clustering, probabilities can express the degree to which an example belongs to a cluster. One would then no longer require that node tuples are assigned to single clusters.

4.5 Combining Induction and Deduction

Analogy. While the generality of a pattern provides a first means to assess the strength of an analogy, the probabilities of the different groundings additionally provide a means to rank all node tuples that are analogous with respect to a certain pattern. For instance, while `(2,8)`, `(12,16)`, `(14,9)` and `(9,11)` are all analogous with respect to pattern `p4`, the probabilities are much higher for `(2,8)` and `(14,9)` than for the other two pairs. In the context of ProbLog, both local pattern mining [11] and probabilistic explanation based learning [13] have been used for reasoning by analogy.

Synonyms. In the context of finding synonyms, probabilities allow for choosing a subset of candidate synonyms based on the probabilities of the corresponding answer substitutions, and to thus restrict a possibly large set of synonyms to a set that is more suitable for manual inspection.

4.6 Modifying the Probabilistic Knowledge Base

Simplification of a Probabilistic Graph. The problem introduced by Toivonen *et al* [15] consists in simplifying probabilistic networks while maintaining the connectivity. It refines the task of graph simplification as defined in Section 3.5 by using the probabilities as an additional quality measure.

The aim is to find a minimal database by dropping edges while keeping the probability of the `path` predicate for each pair of nodes constant. With the probability of `path(x,y)` for a pair of nodes `x` and `y` defined as the probability of the best instantiation, this corresponds to maintaining the best paths between all pairs of nodes.

This definition might be too strict, as it might not allow for significant reductions of database size. In a later work [16], the condition is relaxed to maintaining the overall best path quality as close to the original as possible.

Subgraph Extraction. Various approaches to extract subgraphs with strong connections among given nodes have been developed in the context of `Biomine` and ProbLog [17,18,19,20]. These works all aim at maintaining high probabilities for connections between selected nodes. In `Biomine`, connections are typically defined as paths between pairs of nodes from a given set, while ProbLog theory compression [18] provides them as positive examples in the form of ground patterns whose definitions are included in the background knowledge. The latter also takes into account corresponding negative examples by using a score that encourages high probabilities for positive and low probabilities for negative examples.

Abstraction. In a probabilistic database or network, abstraction would need to take into account the probability labels as well. However, simply labeling the new facts with their probabilities as deduced from the old program may introduce hidden dependencies between facts that might be undesirable.

Predicate Invention. When applying predicate invention to a probabilistic database, the probabilities provide a means to measure the information loss and balance it against the compactness of the representation obtained. While the underlying probability distributions could be maintained for transformations that maintain the meaning of the program, how to adapt probabilities for transformations that generalize the program is an open question.

4.7 Beyond Probabilities

While probability labels provide one way of defining a quality measure on different subnetworks or databases, in certain situations, it can be more convenient to use different types of labels, such as for instance costs, capacities, or numbers of co-occurrences. For instance, in the context of a transportation network where edges are labeled with travel times, prices, or the number of available seats, one could be interested in shortest or cheapest routes, or in routes allowing for the largest group of passengers traveling together, or even in some criterion balancing these requirements. In a co-authorship graph where edges are labeled with the number of joint papers, one could be interested in patterns suggesting strong collaboration networks.

aProbLog [4] generalizes ProbLog to labels from arbitrary commutative semirings, that is, sets of labels together with two binary operators with certain characteristics.[3] Multiplication is used to define labels of subsets of the database sets (as done for the semiring of probabilities in Equation (12)), while addition is used to define labels of queries in terms of these (as done in Equation (13)). In the case of probabilities, negative literals are naturally labeled with $1-p$, where p is the label of the database facts; in the general case considered in aProbLog, these labels need to be given explicitly. By replacing summation with maximization, one obtains another probabilistic semiring that can be used to obtain most likely database instances. The examples given above can be formalized in this framework.

Inference in aProbLog generalizes that in ProbLog, and the framework thus allows one to explore the tasks discussed in this chapter in the context of different types of labels on basic relations without the need to redefine the underlying machinery.

5 Conclusions

We have given an overview of network inference tasks from the perspective of the `Biomine` and ProbLog frameworks. These tasks provide information at the node, subgraph, or pattern level, and they differ in the types of input they assume in addition to the basic graph, such as training examples or background knowledge.

[3] More formally, a commutative semiring is a tuple $(\mathcal{A}, \oplus, \otimes, e^{\oplus}, e^{\otimes})$ where *addition* $\oplus$ and *multiplication* $\otimes$ are associative and commutative binary operations over the set $\mathcal{A}$, $\otimes$ distributes over $\oplus$, $e^{\oplus} \in \mathcal{A}$ is the neutral element with respect to $\oplus$, $e^{\otimes} \in \mathcal{A}$ that of $\otimes$, and for all $a \in \mathcal{A}$, $e^{\oplus} \otimes a = a \otimes e^{\oplus} = e^{\oplus}$.

They all have been or can be extended to exploit the probabilistic information present in both frameworks, or other types of labels as supported in aProbLog, a recent generalization of ProbLog to algebraic labels.

Acknowledgments. A. Kimmig is supported by the Research Foundation Flanders (FWO Vlaanderen). This work is partially supported by the GOA project 2008/08 Probabilistic Logic Learning and by the European Commission under the 7th Framework Programme, contract no. BISON-211898. This work has been supported by the Algorithmic Data Analysis (Algodan) Centre of Excellence of the Academy of Finland (Grant 118653).

References

1. Kötter, T., Berthold, M.R.: From Information Networks to Bisociative Information Networks. In: Berthold, M.R. (ed.) Bisociative Knowledge Discovery. LNCS (LNAI), vol. 7250, pp. 33–50. Springer, Heidelberg (2012)
2. Sevon, P., Eronen, L., Hintsanen, P., Kulovesi, K., Toivonen, H.: Link Discovery in Graphs Derived from Biological Databases. In: Leser, U., Naumann, F., Eckman, B. (eds.) DILS 2006. LNCS (LNBI), vol. 4075, pp. 35–49. Springer, Heidelberg (2006)
3. De Raedt, L., Kimmig, A., Toivonen, H.: ProbLog: A probabilistic Prolog and its application in link discovery. In: Veloso, M.M. (ed.) Proceedings of the 20th International Joint Conference on Artificial Intelligence (IJCAI 2007), pp. 2462–2467 (2007)
4. Kimmig, A., Van den Broeck, G., De Raedt, L.: An algebraic Prolog for reasoning about possible worlds. In: Burgard, W., Roth, D. (eds.) Proceedings of the 25th AAAI Conference on Artificial Intelligence (AAAI 2011), pp. 209–214. AAAI Press (2011)
5. Flach, P.: Simply logical - Intelligent Reasoning by Example. John Wiley (1994), `http://www.cs.bris.ac.uk/~flach/SimplyLogical.html`
6. De Raedt, L.: Logical and Relational Learning. Springer (2008)
7. Muggleton, S.: Duce, an oracle-based approach to constructive induction. In: McDermott, J.P. (ed.) Proceedings of the 10th International Joint Conference on Artificial Intelligence (IJCAI 1987), pp. 287–292 (1987)
8. Cook, D.J., Holder, L.B.: Substructure discovery using minimum description length and background knowledge. J. Artif. Intell. Res. (JAIR) 1, 231–255 (1994)
9. Kimmig, A., Demoen, B., De Raedt, L., Santos Costa, V., Rocha, R.: On the implementation of the probabilistic logic programming language ProbLog. Theory and Practice of Logic Programming (TPLP) 11(2-3), 235–262 (2011)
10. Langohr, L., Toivonen, H.: Finding representative nodes in probabilistic graphs. In: Proceedings of the Workshop on Explorative Analytics of Information Networks at ECML PKDD, WEAIN 2009 (2009)

11. Kimmig, A., De Raedt, L.: Local query mining in a probabilistic Prolog. In: Boutilier, C. (ed.) Proceedings of the 21st International Joint Conference on Artificial Intelligence (IJCAI 2009), pp. 1095–1100 (2009)
12. De Raedt, L., Thon, I.: Probabilistic Rule Learning. In: Frasconi, P., Lisi, F.A. (eds.) ILP 2010. LNCS, vol. 6489, pp. 47–58. Springer, Heidelberg (2011)
13. Kimmig, A., De Raedt, L., Toivonen, H.: Probabilistic Explanation Based Learning. In: Kok, J.N., Koronacki, J., Lopez de Mantaras, R., Matwin, S., Mladenič, D., Skowron, A. (eds.) ECML 2007. LNCS (LNAI), vol. 4701, pp. 176–187. Springer, Heidelberg (2007)
14. De Raedt, L., Kersting, K., Torge, S.: Towards learning stochastic logic programs from proof-banks. In: Veloso, M.M., Kambhampati, S. (eds.) Proceedings of the 20th National Conference on Artificial Intelligence (AAAI 2005), pp. 752–757. AAAI Press/The MIT Press (2005)
15. Toivonen, H., Mahler, S., Zhou, F.: A Framework for Path-Oriented Network Simplification. In: Cohen, P.R., Adams, N.M., Berthold, M.R. (eds.) IDA 2010. LNCS, vol. 6065, pp. 220–231. Springer, Heidelberg (2010)
16. Zhou, F., Mahler, S., Toivonen, H.: Network simplification with minimal loss of connectivity. In: Webb, G.I., Liu, B., Zhang, C., Gunopulos, D., Wu, X. (eds.): Proceedings of the 10th IEEE International Conference on Data Mining (ICDM 2010), pp. 659–668 (2010)
17. Hintsanen, P.: The Most Reliable Subgraph Problem. In: Kok, J.N., Koronacki, J., Lopez de Mantaras, R., Matwin, S., Mladenič, D., Skowron, A. (eds.) PKDD 2007. LNCS (LNAI), vol. 4702, pp. 471–478. Springer, Heidelberg (2007)
18. De Raedt, L., Kersting, K., Kimmig, A., Revoredo, K., Toivonen, H.: Compressing probabilistic Prolog programs. Machine Learning 70(2-3), 151–168 (2008)
19. Hintsanen, P., Toivonen, H.: Finding reliable subgraphs from large probabilistic graphs. Data Mining and Knowledge Discovery 17(1), 3–23 (2008)
20. Kasari, M., Toivonen, H., Hintsanen, P.: Fast Discovery of Reliable k-terminal Subgraphs. In: Zaki, M.J., Yu, J.X., Ravindran, B., Pudi, V. (eds.) PAKDD 2010, Part II. LNCS, vol. 6119, pp. 168–177. Springer, Heidelberg (2010)

Network Analysis: Overview

Hannu Toivonen

Department of Computer Science and HIIT
P.O. Box 68, FI-00014 University of Helsinki, Finland
firstname.lastname@cs.helsinki.fi

Heterogeneous information networks or BisoNets, as they are called in the context of bisociative knowledge discovery, are a flexible and popular form of representing data in numerous fields. Additionally, such networks can be created or derived from other types of information using, e.g., the methods given in Part II of this volume.

This part of the book describes various network algorithms for the exploration and analysis of BisoNets. Their general goal is to support and partially even automate the process of bisociation. More specific goals are to allow navigation of BisoNets by indirect and predicted relationships and by analogy, to produce explanations for discovered relationships, and to help abstract and summarise BisoNets for more effective visualisation.

Contributions

In the first chapter of this part, Dries *et al.* [1] propose BiQL, a novel query language for BisoNets. It is motivated by the observation that graph and network databases have specific needs for query tools, but the tools are much less developed than for relational data. For instance, a statistic such as the shortest path between two given nodes cannot be computed by a relational database. BiQL allows for querying and analyzing databases, especially probabilistic graphs, by using such aggregates and ranking.

The next three chapters address the problem of simplifying a large BisoNet and providing a smaller version instead, both to aid visual exploration and to ease the use of computationally more demanding methods. The first of these chapters, by Zhou *et al.* [2], is an overview of existing approaches to this problem.

The next two chapters then propose novel methods for two specific network abstraction tasks. Zhou *et al.* [3] provide methods for so called network simplification. There, the goal is to remove least important edges, i.e., those that have least effect on the quality of connections between any nodes. In this approach, nodes are left intact. In the chapter on network compression, in turn, Toivonen *et al.* [4] obtain a smaller network by merging nodes that have similar neighbours (or roles) in the network. Such a graph can also be uncompressed to obtain an approximate copy of the original graph. Both of these abstraction methods are designed specifically for BisoNets, paying attention to edge weights and maintaining strengths of (indirect) relations between nodes.

Langohr and Toivonen [5] then introduce a method to identify representative nodes in BisoNets, also motivated by the need to produce different simple views

M.R. Berthold (Ed.): Bisociative Knowledge Discovery, LNAI 7250, pp. 144–146, 2012.

to large networks. They define a probabilistic similarity measure for nodes, and then apply clustering methods to find groups of nodes. Finally, a representative (the medoid) is output from each cluster, to obtain a sample of nodes that is representative for the whole network.

Kötter and Berthold [6] propose a new approach to extract existing concepts, or to detect missing ones, from a BisoNet by means of concept graph detection. Extracted concepts can then be used to create a higher level representation of the data, while discovered missing concepts might lead to new insights by connecting seemingly unrelated information units.

The final two chapters propose two different approaches to discover similarities or associations — or bisociations — in BisoNets. Thiel and Berthold [7] propose a novel way to find non-trivial structural similarities between nodes in a BisoNet. The basic idea is to compare the neighborhoods of the given nodes, also indirect neighbors. The clue of the method is to do this by comparing the patterns of activation spreading from each of the given nodes.

Finally, Nagel *et al.* [8] address the problem of finding domain bridging associations between otherwise weakly connected domains. They propose a method based purely on structural properties of the connections between entities. It first identifies domains and then assesses interestingness of connections between these domains.

Conclusions

The chapters in this part of the book cover a wide range of methods for bisociation network analysis. Many of the methods are directed to making large BisoNets easier to handle and grasp. Also, a multitude of methods were developed to measure relationships or similarities between entities in BisoNets, and to discover interesting relations or concepts.

Automated discovery of actual, useful bisociations seems to be a very difficult problem. This observation is also supported by the experimental work and applications that are described in Part V of this book. Instead, it is more useful to offer the user tools and mechanisms that help her explore the data, and that facilitate her bisociative processes. Part IV below will continue with even stronger focus on interactive exploration methods for BisoNets.

The applications and evaluations in Part V indicate that the overall bisociative methodology, including network analysis methods as its key components, has potential for helping users make genuine discoveries. At the time of writing, network analysis methods and tools descibed in this part have already been adopted for regular use by end users.

References

1. Dries, A., Nijssen, S., Raedt, L.D.: BiQL: A Query Language for Analyzing Networks. In: Berthold, M.R. (ed.) Bisociative Knowledge Discovery. LNCS (LNAI), vol. 7250, pp. 147–165. Springer, Heidelberg (2012)
2. Zhou, F., Mahler, S., Toivonen, H.: Review of BisoNet Abstraction Techniques. In: Berthold, M.R. (ed.) Bisociative Knowledge Discovery. LNCS (LNAI), pp. 166–178. Springer, Heidelberg (2012)
3. Zhou, F., Mahler, S., Toivonen, H.: Simplification of Networks by Edge Pruning. In: Berthold, M.R. (ed.) Bisociative Knowledge Discovery. LNCS (LNAI), vol. 7250, pp. 179–198. Springer, Heidelberg (2012)
4. Toivonen, H., Zhou, F., Hartikainen, A., Hinkka, A.: Network Compression by Node and Edge Mergers. In: Berthold, M.R. (ed.) Bisociative Knowledge Discovery. LNCS (LNAI), vol. 7250, pp. 199–217. Springer, Heidelberg (2012)
5. Langohr, L., Toivonen, H.: Finding Representative Nodes in Probabilistic Graphs. In: Berthold, M.R. (ed.) Bisociative Knowledge Discovery. LNCS (LNAI), vol. 7250, pp. 218–229. Springer, Heidelberg (2012)
6. Kötter, T., Berthold, M.R.: (Missing) Concept Discovery in Heterogeneous Information Networks. In: Berthold, M.R. (ed.) Bisociative Knowledge Discovery. LNCS (LNAI), vol. 7250, pp. 230–245. Springer, Heidelberg (2012)
7. Thiel, K., Berthold, M.R.: Node Similarities from Spreading Activation. In: Berthold, M.R. (ed.) Bisociative Knowledge Discovery. LNCS (LNAI), vol. 7250, pp. 246–262. Springer, Heidelberg (2012)
8. Nagel, U., Thiel, K., Kötter, T., Piatek, D., Berthold, M.R.: Towards Discovery of Subgraph Bisociations. In: Berthold, M.R. (ed.) Bisociative Knowledge Discovery. LNCS (LNAI), vol. 7250, pp. 263–284. Springer, Heidelberg (2012)

BiQL: A Query Language for Analyzing Information Networks

Anton Dries[1,2], Siegfried Nijssen[1], and Luc De Raedt[1]

[1] Katholieke Universiteit Leuven, Belgium
[2] Universitat Pompeu Fabra, Barcelona, Spain

Abstract. One of the key steps in data analysis is the exploration of data. For traditional relational data, this process is facilitated by relational database management systems and the aggregates and rankings they can compute. However, for the exploration of graph data, relational databases may not be most practical and scalable. Many tasks related to exploration of information networks involve computation and analysis of connections (e.g. paths) between concepts. Traditional relational databases offer no specific support for performing such tasks. For instance, a statistic such as the shortest path between two given nodes cannot be computed by a relational database. Surprisingly, tools for querying graph and network databases are much less well developed than for relational data, and only recently an increasing number of studies are devoted to graph or network databases. Our position is that the development of such graph databases is important both to make basic graph mining easier and to prepare data for more complex types of analysis.

In this chapter, we present the BiQL data model for representing and manipulating information networks. The BiQL data model consists of two parts: a data model describing objects, link, domains and networks, and a query language describing basic network manipulations. The main focus here lies on data preparation and data analysis, and less on data mining or knowledge discovery tasks directly.

1 Introduction

Information networks are a popular way of representing information. In its most basic form, such a network can be seen as a set of objects, interconnected by links. Because of this link structure, these networks are capable of representing complex information using a simple data model. Information networks can be found in a wide variety of domains, for example, as social networks, bibliographical networks, and biological networks such as gene-protein interaction networks and pathways. Although all these examples seem very different, their analysis requires many similar operations. For example, determining the influence of a publication in a citation network is similar to finding the role of a gene in a biological pathway, finding the well-connected users in a social network corresponds to finding the important traffic hubs in a road network, and network analysis algorithms such as PageRank can be applied to different types of networks such

M.R. Berthold (Ed.): Bisociative Knowledge Discovery, LNAI 7250, pp. 147–165, 2012.

as the world wide web and social networks. Because of this common structure it seems natural to look for a common infrastructure to deal with these networks.

Currently, different graph databases are available (e.g. DEX [31] and Neo4j [32]). However, many of these systems focus mainly on low-level aspects such as data structures and algorithms, instead of higher level concepts such as providing a simple data model and query language. In this article, we take a different approach and we focus on developing a data model for information networks that is suitable for network analysis and data mining. This data model, called BiQL (or **Bi**son **Q**uery **L**anguage), aims at providing a powerful set of operations for manipulating a wide variety of heterogeneous networks. Within the knowledge discovery process, BiQL mainly focusses on preprocessing, transformation, analysis, and, to a lesser extent, data mining.

In this chapter, we give a general overview of the BiQL system. For a more in-depth discussion on the query language and its underlying operations we refer the reader to [19, chapter 6].

2 Motivating Example

Consider the bibliographic network shown in Figure 1. This network contains authors, publications, keywords, citations, authorship and keyword relations.

Such a network can be used and analyzed in many ways. For example, one could be interested in doing co-authorship analysis. In that case the 'publication' nodes are considered to be edges between 'authors' and the network can be represented as shown in Figure 2. The co-author relationship can be expressed using regular edges (Figure 2a) or using hyperedges (Figure 2b).

Alternatively, one may be interested in analyzing publications for each domain separately by splitting up the network into a set of networks, one for each keyword, as can be seen in Figure 3.

Many more cases can be imagined, for example, citation analysis between publications, authors, or even keyword domains. In order to be able to perform all these tasks, we need a data representation and query language that are capable of representing, manipulating and transforming information networks. Moreover, we also want to analyse such networks, that is, calculate aggregate measures, apply ranking functions, and store the results back in the network for future querying. In general, we can identify a number of key tasks that a network management system should support:

1. *Introduce new relationships in the network*, for example, create a 'co-author' relationship between authors that have published a paper together, or create a citation relation between authors based on the citation relation between publications.
2. *Find connections between objects*, for example, find co-citations between authors, that is, author A cites author B and author B cites author A (possibly indirectly).

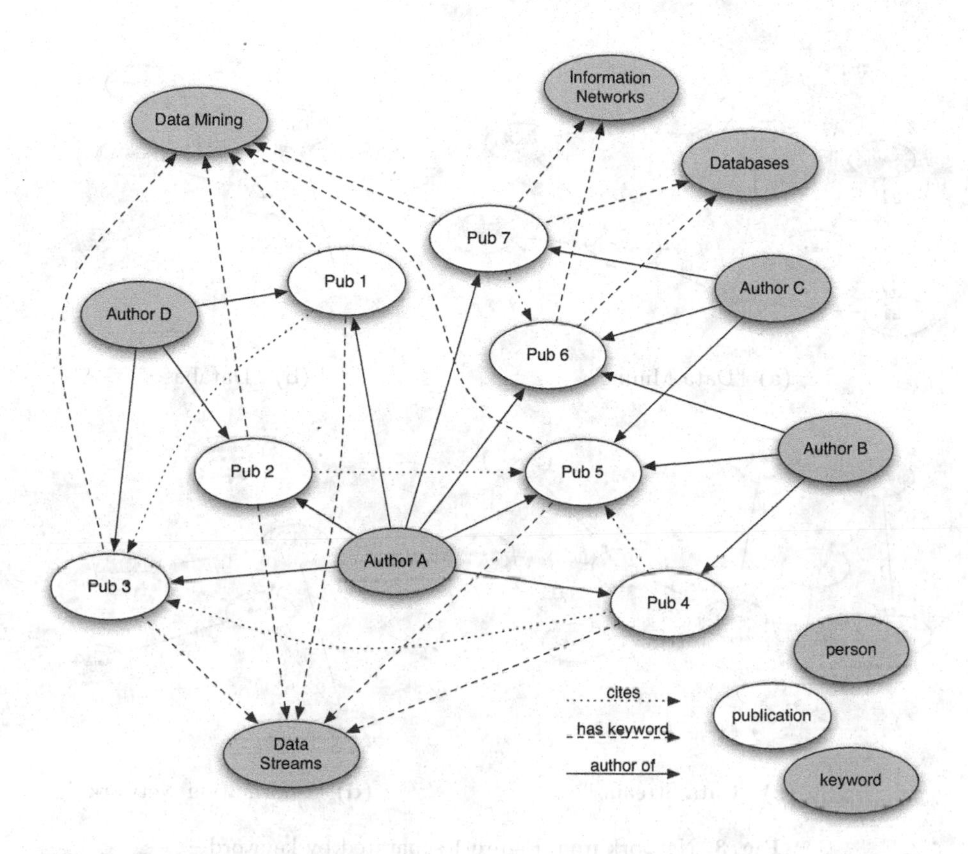

Fig. 1. Bibliographic network containing the entities 'authors', 'publications' and 'keywords', and the 'author of', 'has keyword', and 'cites' relationships

3. *Find the transitive closure of a relation*, for example, find the influence graph of a publication based on citations, or the co-author neighbourhood of an author.
4. *Rank results*, for example, find the authors with the most co-authors, or with the largest co-author network.

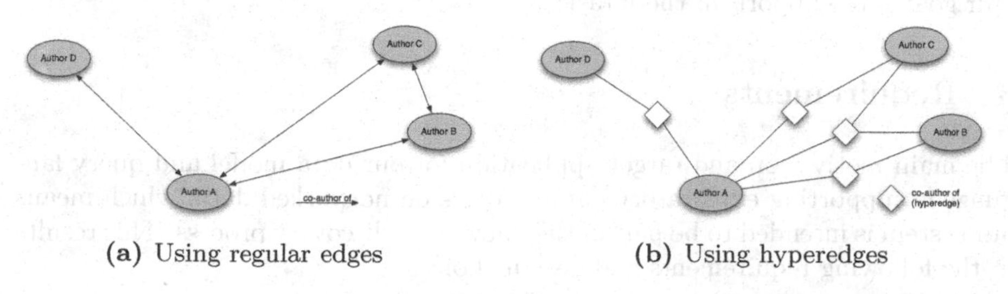

(a) Using regular edges (b) Using hyperedges

Fig. 2. Network from Figure 1 transformed for co-authorship analysis

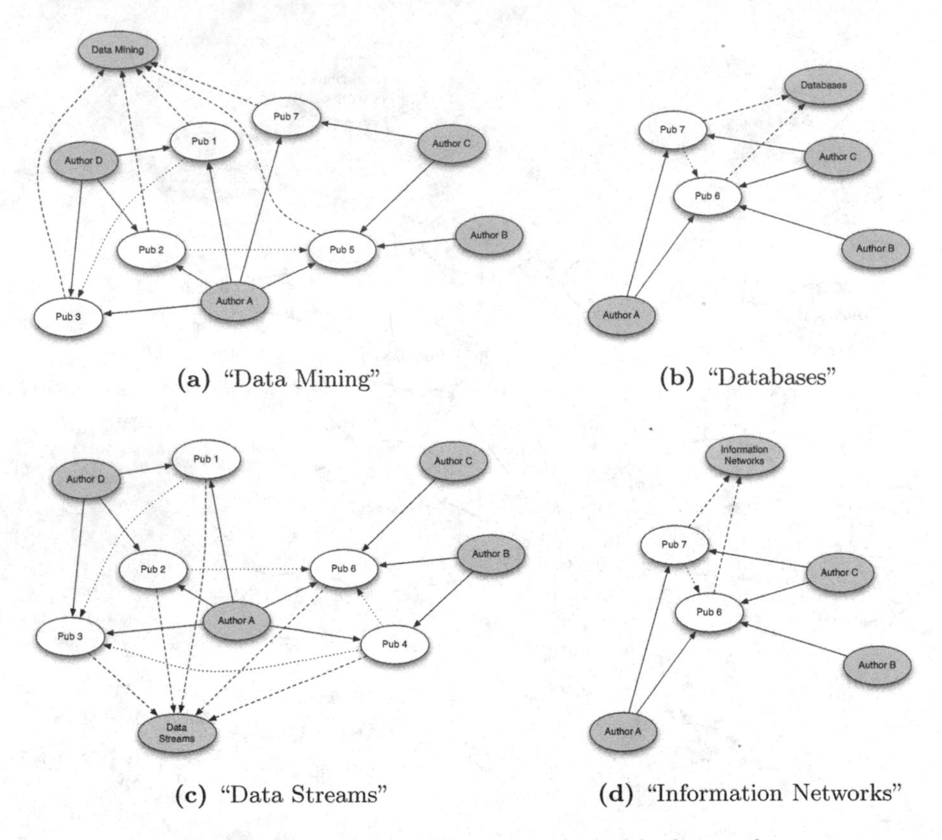

(a) "Data Mining"

(b) "Databases"

(c) "Data Streams"

(d) "Information Networks"

Fig. 3. Network from Figure 1 separated by keyword

5. *Calculate network analysis metrics*, for example, centrality of an author in the co-author network.
6. *Introduce weights and probabilities*, and use them in probabilistic queries.
7. *Discover bisociations or other non-obvious connections*, for example, by comparing different distance measures.
8. *Apply external algorithms on the network*, for example, for finding quasi-cliques [42].

Our goal is to support all these tasks.

3 Requirements

The main motivation and target application for our data model and query language is supporting exploratory data analysis on networked data, which means our system is intended to be part of the knowledge discovery process. This results in the following requirements and design choices.

Small is beautiful. The data model should consist of a small number of concepts and primitives. As a consequence, we do not wish to introduce special language constructs to deal with complicated types of networks (directed, undirected, labeled, hypergraphs, etc.) or sets of graphs.

Uniform representation of nodes and edges. The most immediate consequence of the former choice is that we wish edges and nodes to be represented in a uniform way. We will do this by representing both edges and nodes as objects that are linked together by links that have no specific semantics. This also allows one to generate different views on a network. For instance, in a bibliographic database, we may have objects such as papers, authors and citations. In one context one could analyze the co-author relationship, in which case the authors are viewed as nodes and the papers as edges, while in another context, one could be more interested in citation-analysis, in which case the papers are the nodes and the citations the edges.

Closure property. The result of any operation or query can be used as the starting point for further queries and operations. The information created by a query combined with the original database can therefore be queried again.

SQL-based. There are many possible languages that could be taken as starting point, such as SQL, relational algebra or Datalog. We aimed for a data model on which multiple equivalent ways to represent queries can be envisioned. The queries that we propose on this model are expressed in an SQL-like notation here, as this notation is more familiar to many users of databases, and is the prime example of a declarative query language.

Aggregates. To support a basic analysis of graphs, we need to be able to calculate statistics such as

- the degree of nodes;
- the number of nodes reachable from a certain node (connected component size);
- the length of a shortest path between two nodes;
- the length of the longest shortest path from one node to all other nodes (closeness centrality);
- the sum or product of weights on edges on paths.

These statistics are not only useful when obtaining an initial insight in data. It is also important that these statistics can be attached to the newly created graph (representing another context). For instance, in simple random walk models the probability of going from one node to another node may be determined by the degrees of the nodes involved. These probabilities can be seen as attributes of the edges; ideally, a database query would be sufficient to put these probabilities in a graph. The closure property entails that we can also run queries on the attributes generated in this way. One such type of query could be a *probabilistic* query, which calculates new probabilities from probabilities present in the network.

Ranking. Once an aggregate is computed, it can be desirable to rank results on aggregate values; for instance, one may not be interested in the centrality of all nodes, but only in the nodes that are most central. A database system should support such ranking queries and ideally be optimized to answer them more efficiently than by post-processing a sorted list of all results.

In the following two sections, we translate these requirements into a specification for a data representation and a data manipulation language.

4 Data Representation

An important choice for any data management system is the representation of the data it operates on. For example, in Codd's relational database model [16], data is represented as sets of tuples. The challenge is to find a data model that is capable of storing any kind of information network, and that fulfils the requirements described in the previous section.

In its most basic form, an information network is a collection of objects with links between them. It is therefore natural to use objects and links as the basic building blocks for a network representation. However, as we have seen in the examples of the previous section, it is not always clear which concepts to consider as objects, and which as links. For example, is a publication an object, or a link between (co-)authors? Usually, the answer to this question depends on the application at hand. However, BiQL is intended as an application-independent data management system. This means that the data should be modelled in the most general way, and the term "object" should be taken as broad as possible. Intuitively, we define it as *any entity that has meaning in reality*, or, less abstractly, as any entity that can have additional properties or roles assigned to it. Following this guideline, we only allow features on objects. That is, links are modelled as nothing more than ordered pairs of object identifiers, and they only express that two objects are connected.

Hence, the main choice that we have made is, in a sense, that also edges are represented as objects. An edge object is linked to the nodes it connects. Even though this may not seem intuitive, or could seem a bloated representation, the advantages of this choice outweigh the disadvantages because:

- by treating both edges and nodes as objects, we obtain simplicity and uniformity in dealing with attributes;
- it is straightforward to treat (hyper)edges as nodes (or vice versa);
- it is straightforward to link two edges, for instance, when one wishes to express a similarity relationship between two edges.

In this way, the data representation fulfills the requirements of simplicity, uniformity between nodes and edges, and flexibility.

However, not all objects in the network have the same meaning or role. In the bibliographic network, we had objects that represented authors, publications, citations, etc. In our data model, we use *domains* to indicate these categories of objects. Such a domain is a named set of objects. Objects can belong to any number of domains, for example, an 'author' in the bibliographic network can also be a 'person', or an 'employee', at the same time.

Apart from domain membership, each object can have an arbitrary set of features described by a list of name-value pairs.

5 Basic Data Manipulation

Now that we have a basic understanding of how the data is organized in the database, we can focus on manipulating this information. In this section, we give a general overview of BiQL's query language. For a more in-depth discussion on the query language and its underlying operations we refer the reader to [19, chapter 6].

The primary goal of BiQL is to manipulate a network by querying, analyzing, and modifying its objects and links. The main operations offered by the query language are

- adding an existing object to a new domain,
- adding links and attributes to an existing object,
- creating new objects (with links and attributes) and adding them to a new domain.

Each of these tasks can be specified as an CREATE/UPDATE query of the following form.

```
CREATE/UPDATE "domain name" <"variables"> {"object properties"}
FROM "selection from domains"
WHERE "predicate on attributes of objects"
LIMIT "k" ON "sorting criteria"
```

For example, the query

```
UPDATE Pubs2010<p>
FROM Publ p
WHERE p.year = 2010
```

creates a new domain Pubs2010 that contains all articles published in 2010. The UPDATE keyword indicates that existing objects are used instead of newly created ones. This means that all existing features for the objects are preserved (unless they are overwritten by an object property definition in the query).

In general, a query in BiQL consists of the following statements:

The FROM statement defines the structural component of the query and introduces variables that can be used in the other statements.

The WHERE statement defines constraints on these variables based on the features of the objects.

The CREATE/UPDATE statement describes the output of the query, that is, how objects should be created or updated based on the retrieved information, and where they should be stored.

The LIMIT statement allows for ranking the results of a query and returning only the top k results.

FROM statement. The primary function of the FROM statement is to define a graph pattern that must be matched in the network. Within this pattern, variables are defined that can be used in the other parts of the query. In a sense, this statement has the same role as SQL's FROM statement, that is, determining which sources of information to use, and how these sources are related. In BiQL, the FROM statement consists of a list of path expressions, where each path expression consists of an alternating sequence of object definitions and link expressions indicating how the objects are connected. For example, a co-authorship relation in the publication network can be expressed as the following sequence of objects and links.

```
Author a -> AuthorOf -> Publ p <- AuthorOf <- Author b
```

Every object is described by a domain it belongs to (e.g. `Author`), and, optionally, a variable name (e.g. `a`). The arrows between the objects indicate the direction of the links between them. A path expression by itself can only express a sequence. However, the FROM statement can contain multiple path expressions that can be connected by references. For example, if we are interested in co-authorship within certain topics, we can include the domain 'Keyword' in the graph pattern by using the path expression

```
#p -> HasKeyword -> Keyword k
```

where `#p` is a reference to the variable p in the previous expression. This pattern is shown in Figure 4. Variable references can also be used to point to variables defined in the same path expression, for example, for expressing cycles.

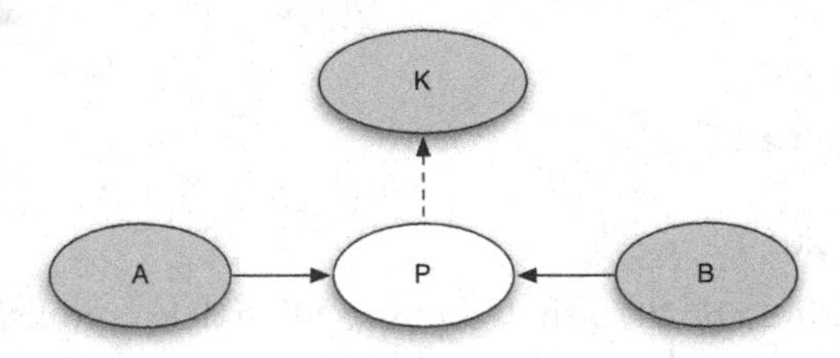

Fig. 4. Example of a graph pattern

Many problems in network analysis are based on finding paths of arbitrary length. To express such paths, we use regular expression operators. For example, the path expression

```
Node (-> Edge -> Node)* -> Edge -> Node
```

defines a path as an alternating sequence of nodes and edges of arbitrary length. To specify constraints on this path we can use *list variables.* These variables capture a sequence of objects instead of a single object. For example, we can use this to restrict the length of a path as shown in the following example.

```
FROM Node (-> Edge [e] -> Node)* -> Edge [e] -> Node
WHERE length(e) <= 4
```

The list aggregate `length` counts the number of objects assigned to the variable `e`.

WHERE statement. The `FROM` statement generates a set of tuples corresponding to the possible assignments of objects in the network to the variables defined in the path expressions. The `WHERE` statement of the query can impose further constraints on this set of tuples based on the features of the resulting objects. For example, given the path expressions above, one can express the constraints

```
WHERE p.year = 2009 AND k.keyword = 'Data Mining'
```

to find only publications from 2009 in the field of data mining. This statement is equivalent to the `WHERE/HAVING` statements of SQL.

CREATE/UPDATE statement. The previous operations produce a set of tuples. However, the final result of the query should fit into BiQL's data representation model. This means that this set of tuples should be transformed into a set of objects, links and domains. This transformation is defined in the `CREATE/UPDATE` statement, which is written as

```
CREATE/UPDATE DomainName<Var1,Var2,...> {<object properties>}.
```

A key part of this statement is the *partition operation* `<Var1,Var2,...>`, which splits the set of tuples and creates a separate partition for each distinct combination of the variables *Var1*, *Var2*,.... The final results of the query will contain a separate object for each of these partitions. The features and links of this object are described by the object properties. After construction, the set of objects is stored in the domain with the given name. This partition operation is comparable to the `GROUP BY` statement in SQL.

For example, if we want to define the co-author relationship, we can use the following query.

```
CREATE CoAuthor<a,b> { a ->, -> b, strength: count<p> }
FROM Author a -> AuthorOf -> Publ p <- AuthorOf <- Author b
WHERE a != b
```

This query creates a new object for each pair of authors a and b that have published at least one article together, that is, for whom the path expression can be mapped onto part of the network. The object properties specify that this new object is linked to both authors and that it contains an attribute `strength` indicating the number of articles the authors have co-written. The created objects are added to the new domain `CoAuthor`.

The partition operator is also used in the calculation of aggregate functions, for example, the function `count<p>` counts the number of distinct p, that is, the number of partitions `<p>` creates. Other aggregates include `sum<...>(expr)`, `min<...>(expr)`, `max<...>(expr)`, etc.

LIMIT statement. Apart from feature-based selection, BiQL also supports rank-based selection through the LIMIT statement.

```
LIMIT k BY criteria
```

For example, we can select the three strongest co-authorship relationships using the statement

```
LIMIT 3 BY count<p> DESC
```

where DESC indicates a descending sort order. This statement is a *global* limit statement, that is, it is used to reduce the number of objects returned by the query. The operation of this statement is comparable to the ORDER BY statement in SQL, combined with a statement for selecting the top-k results (e.g. FETCH FIRST in SQL.2008 [25]).

In this section, we provided a limited overview of the features present in the BiQL query language. For an extensive description of this query language and its operational model, we refer the reader to [19, chapter 6].

6 Illustrative Examples

In Section 2, we introduced a list of key tasks that we want to support in BiQL. We now revisit this list to evaluate BiQL's capabilities. Unless stated otherwise, each of these queries can be evaluated using the prototype implementation on the ILPnet2 publication database. This database is structurally similar to the network shown in Figure 1.

1. Introduce new relationships in the network. Throughout this chapter, we have repeatedly used the co-author relationship as an example of a new relationship. Here, we express this relationship as a connection between authors that have more than one publication in common.

```
CREATE CoAuthor<a,b> { a->, b<-, strength: count<p> }
FROM Author a -> AuthorOf -> Publ p <- AuthorOf <- Author b
WHERE count<p> > 1 AND a != b
```

Another example introduces the 'InArea' relation, which expresses whether an author has published a paper within a certain research area (indicated by a 'Keyword').

```
CREATE InArea<a,k> { a->, k<-, weight: count<pk>/count<p> }
FROM Author a -> AuthorOf -> Publ pk -> HasKeyword -> Keyword k,
    #a -> AuthorOf ->Publ p
```

The attribute 'weight' indicates the fraction of the author's publications that contain this keyword.

2. Find connections between objects. Using the 'InArea' and 'CoAuthor' relations, we can express how far an author is removed from any given research area.

```
CREATE RelatedToArea<a,k>{ a->, k<-, distance: min<b>(length(b))}
FROM Author a (-> CoAuthor -> Author [b])* -> InArea -> Keyword k
```

The expression `min<b>(length(b)` computes the length of the shortest path (i.e. the number of intermediate authors) from a specific author to a specific keyword.

3. Find the transitive closure of a relation. The previous query already used the transitive closure of the 'CoAuthor' relation to find a relationship between authors and research areas. We can also use such a relationship to determine the size of the neighborhood of an author.

```
UPDATE <a> { a->, b<-, networksize: count<b> }
FROM Author a (-> CoAuthor [co] -> Author)*
                                    -> CoAuthor [co] -> Author b
WHERE length(co) < 4
```

4. Rank results. Often we are interested in finding the top-k results according to some criteria. For example, we might be interested in the top 3 authors with most co-authors.

```
SELECT¹ <a>
FROM Author a -> CoAuthor co
LIMIT 3 BY count<co> DESC
```

We can also find the authors with the largest network of co-authors up to a certain distance.

```
SELECT <a> { network_size: count<b> }
FROM Author a -> CoAuthor [co] ->
          (Author -> CoAuthor [co] ->)* -> Author b
WHERE length(co) < 4
LIMIT 3 BY count<b> DESC
```

5. Calculate network analysis metrics. Another interesting task is calculating network analysis metrics such as centrality measures. Perhaps the simplest centrality measure is degree centrality which calculates, for a given node v, the fraction of all nodes that v is connected to. In BiQL, this measure can be calculated, for all authors simultaneously, using the following query.

```
UPDATE <a> { Cdegree: count<b>/(count<n> - 1)}
FROM Author a -- CoAuthor -- Author b, Author n
```

[1] In our prototype implementation, a `SELECT` query can be used to output a list of results without causing changes to the database.

Another common centrality measure is closeness centrality, which involves determining the length of the shortest path to all other nodes in the network. First let us define the notion of shortest path between two authors using the co-authorship relation.

```
CREATE ShortestPath<a,b>{ a->, b<-, len: min<co>(length(co))}
FROM Author a -> CoAuthor [co] ->
         (Author -> CoAuthor [co] ->)* -> Author b
WHERE a != b
```

This query creates for each pair of (connected) authors a and b an object with as attribute the length of a shortest path between them. Using these new objects, we can easily calculate the closeness centrality as follows.

```
UPDATE <a>{ Cclose: 1/sum<b>(min2<sp>(sp.len))}
FROM Author a -> ShortestPath sp -> Author b
```

Another type of centrality measure is the betweenness centrality. This measure expresses the importance of a node based on its occurrence on the shortest paths in the network. In BiQL this measure can be expressed using the following two queries. The first query computes the length of the shortest path between each pair of authors and calculates how many paths of this length there are.[3]

```
CREATE ShortestPathCount<a,b> { a ->, b <-,
              count: count<co>, length: min<co>(length(co)) }
FROM Author a -> CoAuthor [co] ->
                         (Author -> CoAuthor [co] ->)* -> Author b
LIMIT 1 KEYS ON length(co) ASC
```

The second query uses this information to calculate the betweenness centrality of a node v as a fraction of shortest paths in the network that contain v.

```
UPDATE <v> { Cb: sum<s,t>((sv.count*vt.count)/st.count) }
FROM Author s -> ShortestPathCount sv -> Author v ->
                             ShortestPathCount vt -> Author t,
      #s -> ShortestPathCount st -> #t
WHERE st.length = sv.length + vt.length
  AND s != t AND s != v AND t != v
```

This query uses the calculation approach for betweenness centrality described in [3, section 3].

---

[2] Given the definition of ShortestPath we expect the variable sp to be uniquely identified when a and b are fixed (i.e. there is only one shortest path length between two given nodes). However, BiQL currently does not support such type of constraint reasoning across queries. This is why we need the additional aggregation min<sp> even though there is only one value for sp.len in this context.

[3] For clarity, we omitted the extra aggregation operations on the variables sv, vt and st as described in the previous footnote.

6. Introduce weights and probabilities. Another important aspect of BiQL is its ability to deal with probabilistic networks. To illustrate this, we first need to introduce probabilities in our network. For this we assume that the information in the network is very unreliable by stating that for each publication in the network there is only 10% probability that it actually exists. Under this assumption we can attach a probability to each co-author connection using the following query.

```
UPDATE <co>{ prob: 1-(0.9^co.strength) }
FROM CoAuthor co
```

We can now calculate for each pair of authors the probability that they are connected using the probabilistic aggregate `problog_connect`.

```
CREATE ProbConnect<a,b>{a->, b<-, prob: problog_connect(co.prob)}
FROM Author a -> CoAuthor [co] ->
        (Author -> CoAuthor [co] ->)* -> Author b
WHERE a != b
```

The `problog_connect` aggregate uses ProbLog's [18] approach to calculate the connection probability between each pair of nodes in the network.

7. Discover bisociations. We can use this domain in combination with the shortest path to find authors that are very likely connected, but that are relatively far apart in the co-author network.

```
SELECT <a,b,pc,sp>{nameA: a.name, nameB: b.name,
                              prob: pc.prob, dist: sp.length }
FROM Author a -> ProbConnect pc -> Author b,
     #a -> ShortestPath sp -> #b
WHERE sp.length > 2
LIMIT 3 BY pc.prob DESC
```

Another example of bisociative discovery consists of finding bridging nodes between different domains. In Example 5 we described betweenness centrality. If we modify the second part of that query we can express the interdomain betweenness centrality as the occurrence of a node on the shortest paths between *concepts in different domains.*

```
UPDATE <v> { Cb: sum<s,t>((sv.count*vt.count)/st.count) }
FROM DomainA s -> ShortestPathCount sv -> DomainC v ->
                        ShortestPathCount vt -> DomainB t,
     #s -> ShortestPathCount st -> #t
WHERE st.length = sv.length + vt.length
  AND s != t AND s != v AND t != v
```

8. Apply external algorithms on the network. In the final task of section 2, we want to apply external algorithms on the networks in BiQL. Unfortunately, there are still many open questions on how this integration should work in practice.

However, instead of providing integration of external tools within BiQL, we have integrated BiQL in the data analysis integration platform KNIME [4]. Through this platform, networks can be passed from BiQL to external algorithms and back, allowing BiQL to be used as part of a broader knowledge discovery process.

7 Related Work

7.1 Knowledge Discovery

Graph mining. Graph mining aims at extending the field of pattern mining towards graphs. Most graph mining techniques work in the transactional setting, that is, on data consisting of sets of graphs. As in item set mining, the focus lies on finding subgraphs that, for example, occur frequently in such a set [33,41,24]. However, many other interestingness measures have been translated towards graph patterns (e.g. correlated patterns [8,15]), and new graph-specific measures have been introduced (e.g. for finding quasi-cliques [42]). Several techniques have been developed that target subsets of graph representations, such as sequences or trees [39]. Recently, there has been increasing interest in applying graph mining techniques to the network setting, that is, to a single graph [10,9,26].

Network analysis. Network analysis is concerned with analyzing the properties of networks, by use of graph theoretical concepts such as node degrees and paths [6]. The primary tool in network analysis are measures such as centrality [36], and specialized algorithms for calculating them efficiently have been developed (e.g. [5]). This field has also gained a lot of interest in domains outside computer science, for example, in social sciences (social network analysis) [40].

Another part of network analysis focusses on the spread of information in a network. This can be used to determine the importance of, for example, web pages on the World Wide Web [7], or to analyze the transmission of infectious diseases [38].

7.2 Databases

General-purpose database systems. The best-known general purpose database systems are based on Codd's relational data model [16]. Many of these database systems (e.g. Oracle Database, Microsoft SQL Server, MySQL, PostgresQL) use (a variant of) ISO SQL [25] as the query language of choice. Datalog [13] is an alternative query language that is based on first order logic. Syntactically, it is a subset of Prolog restricted as to make efficient query answering possible.

A more recent development is that of object-oriented database systems and query languages such as OQL (Object Query Language) [12]. These systems use objects instead of tuples, and they allow for nested objects. Part of the OQL standard focusses on a tight integration with object-oriented languages such as Java and C++. However, due to the overall complexity of object databases, there are few systems that fully support the OQL standard.

Recently, there is a increasing interest in so-called NoSQL databases. These database systems focus on applications that require extremely large databases. Such databases typically use non-relational representations specialized for specific applications, such as Google's BigTable [14] or Facebook's Cassandra [11]. Current graph databases such as Dex [31] and Neo4J [32] also fall under this category, and arguably BiQL does as well.

Graph query languages. A number of query languages for graph databases have been proposed, many of which have been described in a recent survey [2]. However, none of these languages was designed for supporting the knowledge discovery process and each language satisfies at most a few of the requirements mentioned in Section 3. For instance, GraphDB [20] and GOQL [37] are based on an object-oriented approach, with provisions for specific types of objects for use in networks such as nodes, edges and paths. This corresponds to a more structured data model that does not uniformly represent nodes and edges. In addition, these languages target other applications: GraphDB has a strong focus on representing spatially embedded networks such as highway systems or power lines, while GOQL [37], which extends the Object Query Language (OQL), is meant for querying and traversing paths in small multimedia presentation graphs. Both languages devote a lot of attention to querying and manipulating paths: for example, GraphDB supports regular expressions and path rewriting operations.

GraphQL [22] provides a query language that is based on formal languages for strings. It provides an easy, yet powerful way of specifying graph patterns based on graph structure and node and edge attributes. In this model graphs are the basic unit and graph specific optimizations for graph structure queries are proposed. The main objective of this language is to be general and to work well on both large sets of small graphs as well as small sets of large graphs. However, extending existing graphs is not possible in this language; flexible contexts are not supported.

PQL [27] is an SQL-based query language focussed on dealing with querying biological pathway data. It is mainly focussed on finding paths in these graphs and it provides a special path expression syntax to this end. The expressivity of this language is, however, limited and it has no support for complex graph operations.

GOOD [21] was one of the first systems that used graphs as its underlying representation. Its main focus was on the development of a database system that could be used in a graphical interface. To this end it defines a graphical transformation language, which provides limited support for graph pattern queries. This system forms the basis of a large group of other graph-oriented object data models such as Gram [1] and GDM [23].

Hypernode [29] uses a representation based on hypernodes, which make it possible to embed graphs as nodes in other graphs. This recursive nature makes them very well suited for representing arbitrarily complex objects, for example as underlying structure of an object database. However, the data model is significantly different from a traditional network structure, which makes it less suitable for modeling information networks as encountered in data mining.

A similar, but slightly less powerful representation based on hypergraphs is used in GROOVY [30]. This system is primarily intended as an object-oriented data model using hypergraphs as its formal model. It has no support for graph specific queries and operations.

More recently, approaches based on XML and RDF are being developed, such as SPARQL [34]. They use a semi-structured data model to query graph networks in heterogenous web environments; support for creating new nodes and flexible contexts is not provided.

While most of the systems discussed here use a graph-based data model and are capable of representing complex forms of information, none of them uses a uniform representation of edges and nodes (and its resulting flexible contexts), nor supports advanced aggregates.

Graph databases. Whereas the previous studies propose declarative query languages, recently several storage systems have been proposed that do not provide a declarative query language. Notable examples here are Neo4J [32] and DEX [31], which provide Java interfaces to graphs persistently stored on disk. For Neo4J an alternative programming language called Gremlin is under development [35].

Graph libraries. Finally, in some communities, Java or C++ libraries are used for manipulating graphs in the memory of the computer (as opposed to the above graph databases which support typical database concepts such as transactions). Examples are SNAP [28] and igraph [17].

8 Conclusions

In this article, we gave an introduction to BiQL, a novel system for representing, querying and analyzing information networks. The key features of this system are:

- It uses a *simple, yet powerful representation model.* Using only objects (with attributes), links (as pairs of objects), and domains (as named sets of objects), it is capable of representing a wide variety of network types, such as labelled graphs, directed hypergraphs, and even sets of graphs.
- Its query language is *declarative.* This means that the queries only describe what the results should be, but not how they should be obtained. This makes the language more accessible to the average user.
- Its query language uses a powerful mechanism for *expressing graph patterns based on regular expressions.* This makes it possible to, for example, express paths of arbitrary length.
- Its query language allows for the use of *nested aggregates* with a syntax that closely resembles mathematical notation. These aggregates allow the user to perform all kinds of analysis tasks, such as calculating distances and centrality measures.

– Its query language provides a powerful mechanism for *object creation*, which makes it possible to return structured output from a query. However, the result of a query always produces a new network that can be queried again.
– The system itself is developed *from a knowledge discovery perspective*. It focusses on providing specific support for knowledge discovery operations such as network analysis, ranking, and tool integration.

In this chapter, we focussed on defining a data model and the syntax and semantics of the corresponding query language. In future work, the main challenge is to develop a query optimization model that would form the basis of a scalable implementation of the BiQL system.

References

1. Amann, B., Scholl, M.: Gram: a graph data model and query language. In: Proceedings of the ACM Conference on Hypertext, pp. 201–211. ACM (1993)
2. Angles, R., Gutierrez, C.: Survey of graph database models. ACM Computing Surveys 40(1), 1–39 (2008)
3. Batagelj, V.: Semirings for social network analysis. Journal of Mathematical Sociology 19(1), 53–68 (1994)
4. Berthold, M.R., Cebron, N., Dill, F., Gabriel, T.R., Kötter, T., Meinl, T., Ohl, P., Sieb, C., Thiel, K., Wiswedel, B.: Knime: The Konstanz information miner. In: Data Analysis, Machine Learning and Applications, pp. 319–326 (2008)
5. Brandes, U.: A faster algorithm for betweenness centrality. The Journal of Mathematical Sociology 25(2), 163–177 (2001)
6. Brandes, U., Erlebach, T. (eds.): Network Analysis. LNCS, vol. 3418. Springer, Heidelberg (2005)
7. Brin, S., Page, L.: The anatomy of a large-scale hypertextual web search engine. Computer Networks 30(1-7), 107–117 (1998)
8. Bringmann, B.: Mining Patterns in Structured Data. PhD thesis, Katholieke Universiteit Leuven (2009)
9. Bringmann, B., Nijssen, S.: What Is Frequent in a Single Graph? In: Washio, T., Suzuki, E., Ting, K.M., Inokuchi, A. (eds.) PAKDD 2008. LNCS (LNAI), vol. 5012, pp. 858–863. Springer, Heidelberg (2008)
10. Calders, T., Ramon, J., Van Dyck, D.: Anti-monotonic overlap-graph support measures. In: Proceedings of the 8th International Conference on Data Mining, pp. 73–82. IEEE (2009)
11. Cassandra. The Apache Cassandra project, `http://cassandra.apache.org`
12. Cattell, R.G.G., Barry, D.K. (eds.): The Object Data Standard: ODMG 3.0. Morgan Kaufmann Publishers (2000)
13. Ceri, S., Gottlob, G., Tanca, L.: What you always wanted to know about DataLog (and never dared to ask). IEEE Transactions on Knowledge and Data Engineering 1(1), 146–166 (1989)

14. Chang, F., Dean, J., Ghemawat, S., Hsieh, W.C., Wallach, D.A., Burrows, M., Chandra, T., Fikes, A., Gruber, R.E.: Bigtable: A distributed storage system for structured data. In: Seventh Symposium on Operating System Design and Implementation (2006)
15. Cheng, H., Yan, X., Han, J., Hsu, C.-W.: Discriminative frequent pattern analysis for effective classification. In: Proceedings of the 23rd International Conference on Data Engineering, pp. 716–725. IEEE (2007)
16. Codd, E.F.: A relational model of data for large shared data banks. Communications of the ACM 13(6), 377–387 (1970)
17. Csárdi, G., Nepusz, T.: The igraph library, http://igraph.sourceforge.net/
18. De Raedt, L., Kimmig, A., Toivonen, H.: ProbLog: A probabilistic prolog and its application in link discovery. In: Veloso, M.M. (ed.) Proceedings of the 20th International Joint Conference on Artificial Intelligence, pp. 2462–2467 (2007)
19. Dries, A.: Data streams and information networks: a knowledge discovery perspective. PhD thesis, Katholieke Universiteit Leuven (2010)
20. Güting, R.H.: GraphDB: Modeling and querying graphs in databases. In: Bocca, J.B., Jarke, M., Zaniolo, C. (eds.) Proceedings of the 20th International Conference on Very Large Data Bases, pp. 297–308. Morgan Kaufmann (1994)
21. Gyssens, M., Paredaens, J., Van den Bussche, J., van Gucht, D.: A graph-oriented object database model. IEEE Transactions on Knowledge and Data Engineering 6(4), 572–586 (1994)
22. He, H., Singh, A.K.: Graphs-at-a-time: query language and access methods for graph databases. In: Wang, J.T.-L. (ed.) Proceedings ACM SIGMOD International Conference on Management of Data, pp. 405–418. ACM (2008)
23. Hidders, J.: Typing Graph-Manipulation Operations. In: Calvanese, D., Lenzerini, M., Motwani, R. (eds.) ICDT 2003. LNCS, vol. 2572, pp. 391–406. Springer, Heidelberg (2002)
24. Inokuchi, A., Washio, T., Motoda, H.: Complete mining of frequent patterns from graphs: Mining graph data. Machine Learning 50(3), 321–354 (2003)
25. International Organization for Standardization. SQL Language. ISO/IEC 9075(1-4,9-11,13,14):2008 (2008)
26. Kuramochi, M., Karypis, G.: Finding frequent patterns in a large sparse graph. Data Mining and Knowledge Discovery 11(3), 243–271 (2005)
27. Leser, U.: A query language for biological networks. Bioinformatics 21(2), 33–39 (2005)
28. Leskovec, J.: The SNAP library, http://snap.stanford.edu/snap/
29. Levene, M., Poulovassilis, A.: The hypernode model and its associated query language. In: Proceedings of the Fifth Jerusalem Conference on Information Technology, pp. 520–530. IEEE Computer Society Press (1990)
30. Levene, M., Poulovassilis, A.: An object-oriented data model formalised through hypergraphs. Data and Knowledge Engineering 6(3), 205–224 (1991)
31. Martínez-Bazan, N., Muntés-Mulero, V., Gómez-Villamor, S., Nin, J., Sánchez-Martínez, M., Larriba-Pey, J.: Dex: high-performance exploration on large graphs for information retrieval. In: Silva, M.J., Laender, A.H.F., Baeza-Yates, R.A., McGuinness, D.L., Olstad, B., Olsen, Ø.H., Falcão, A.O. (eds.) Proceedings of the Sixteenth ACM Conference on Information and Knowledge Management, pp. 573–582. ACM (2007)
32. Neo Technology. The Neo4J project, http://neo4j.org
33. Nijssen, S.: Mining Structured Data. PhD thesis, Universiteit Leiden (2006)
34. Prud'hommeaux, E., Seaborne, A.: SPARQL query language for RDF (2008), http://www.w3.org/TR/rdf-sparql-query/

35. Rodriguez, M.A.: Gremlin, `http://wiki.github.com/tinkerpop/gremlin/`
36. Sabidussi, G.: The centrality index of a graph. Psychometrika 31, 581–603 (1966)
37. Sheng, L., Ozsoyoglu, Z.M., Ozsoyogly, G.: A graph query language and its query processing. In: Proceedings of the 15th International Conference on Data Engineering, pp. 572–581. IEEE Computer Society (1999)
38. Van Segbroeck, S., Santos, F.C., Pacheco, J.M.: Adaptive contact networks change effective disease infectiousness and dynamics. PLoS Computational Biology 6(8), 1–10 (2010)
39. Washio, T., Kok, J.N., De Raedt, L. (eds.): Advances in Mining Graphs, Trees and Sequences. Frontiers in Artificial Intelligence and Applications, vol. 124. IOS Press (2005)
40. Wasserman, S., Faust, K.: Social Network Analysis: Methods and Applications. Cambridge University Press (1994)
41. Yan, X., Han, J.: gspan: Graph-based substructure pattern mining. In: Proceedings of the 2002 IEEE Conference on Data Mining, p. 721. IEEE Computer Society (2002)
42. Zeng, Z., Wang, J., Zhou, L., Karypis, G.: Coherent closed quasi-clique discovery from large dense graph databases. In: Eliassi-Rad, T., Ungar, L.H., Craven, M., Gunopulos, D. (eds.) Proceedings of the Twelfth ACM SIGKDD International Conference on Knowledge Discovery and Data Mining, pp. 797–802. ACM (2006)

Review of BisoNet Abstraction Techniques*

Fang Zhou, Sébastien Mahler, and Hannu Toivonen

Department of Computer Science and
Helsinki Institute for Information Technology HIIT,
P.O. Box 68, FI-00014 University of Helsinki, Finland
{fang.zhou,sebastien.mahler,hannu.toivonen}@cs.helsinki.fi

Abstract. BisoNets represent relations of information items as networks. The goal of BisoNet abstraction is to transform a large BisoNet into a smaller one which is simpler and easier to use, although some information may be lost in the abstraction process. An abstracted BisoNet can help users to see the structure of a large BisoNet, or understand connections between distant nodes, or discover hidden knowledge. In this paper we review different approaches and techniques to abstract a large BisoNet. We classify the approaches into two groups: preference-free methods and preference-dependent methods.

1 Introduction

Bisociative information networks (BisoNets) [2] are a representation for many kinds of relational data. The BisoNet model is a labeled and weighted graph $G = (V, E)$. For instance, in a BisoNet describing biological information, elements of the vertex set V are biological entities, such as genes, proteins, articles, or biological processes. Connections between vertexes are represented by edges E, which have types such as "codes for", "interacts with", or "is homologous to", and have weights to show how strong they are.

BisoNets are often large. One example is Biomine[1]. It currently consists of about 1 million vertices and 10 million edges, so that it is difficult for users to directly visualize and explore it. One solution is to present to a user an abstract view of a BisoNet. We call this *BisoNet abstraction*.

The goal of BisoNet abstraction is to transform a large BisoNet into one that is simpler and therefore easier to use, even though some information is lost in the abstraction process. An abstracted view can help users see the structure of a large BisoNet, or understand connections between distant nodes, or even discover new knowledge difficult to see in a large BisoNet. This chapter is a literature review of applicable approaches to BisoNet abstraction.

An abstracted BisoNet can be obtained through different approaches. For example, a BisoNet can be simplified by removing irrelevant nodes or edges.

* This chapter is a modified version of article "Review of Network Abstraction Techniques" in Workshop on Explorative Analytics of Information Networks, Sep 2009, Bled, Slovenia [1].

[1] http://biomine.cs.helsinki.fi/

M.R. Berthold (Ed.): Bisociative Knowledge Discovery, LNAI 7250, pp. 166–178, 2012.

Another example is that a BisoNet can be divided into several components, or some parts of a BisoNet can be replaced by general structures. Furthermore, user preference can be considered during abstraction. For instance, a user can specify which parts of a BisoNet should retain more details.

Structure of the review. Although this chapter reviews potential techniques with the goal to abstract large BisoNets, the techniques present here are also applicable to general networks. In the rest of this chapter, we therefore use the general term "network" instead of "BisoNet". We first review methods which do not take user preference into account in Section 2, and then review methods in which a user can specify preference in Section 3. We conclude in Section 4.

2 Preference-Free Methods

In this section, we discuss network abstraction methods where the user has no control over how specific parts of the graph are handled (but there may be numerous other parameters for the user to set).

2.1 Relative Neighborhood Graph

The Relative Neighborhood Graph (RNG) [3, 4] only contains edges whose two endpoints are relatively close: by definition, nodes a and b are connected by an edge if and only if there is no third node c which is closer to both endpoints a and b than a and b are to each other. RNG has originally been defined for points, but it can also be used to prune edges between nodes a and b that do have a shared close neighbor c. The relative neighborhood graph then is a superset of the Minimum Spanning Tree (MST) and a subset of Delaunay Triangulation (DT). According to Toussaint [3], RNG can in most cases capture a perceptually more significant subgraph than MST and DT.

2.2 Node Centrality

The field of social network analysis has produced several methods to measure the importance or centrality of nodes [5–8]. Typical definitions of node importance are the following.

1. Degree centrality simply means that nodes with more edges are more central.
2. Betweenness centrality [9–11] measures how influential a node is in connecting pairs of nodes. A node's betweenness is the number of times the node appears on the paths between all other nodes. It can be computed for shortest paths or for all paths [12]. Computation of a node's betweenness involves all paths between all pairs of nodes of a graph. This leads to high computational costs for large networks.

3. Closeness centrality [13] is defined as the sum of graph-theoretic distances from a given node to all others in the network. The distance can be defined as mean geodesic distance, or as the reciprocal of the sum of geodesic distances. Computation of a node's closeness also involves all paths between all pairs of nodes, leading to a high complexity.
4. Feedback centrality of a vertex is defined recursively by the centrality of its adjacent vertices.
5. Eigenvector centrality has also been proposed [14].

Node centrality measures focus on selecting important nodes, not on selecting a subgraph (of a very small number of separate components). Obviously, centrality measures can be used to identify least important nodes to be pruned. For large input networks and small output networks, however, the result of such straightforward pruning would often consist of individual, unconnected nodes, not an abstract network in the intended sense.

Methods in the following subsections (2.3 and 2.4) are similar in this sense: they help to rank nodes individually based on their importance, but do not as such produce (connected) subgraphs.

2.3 PageRank and HITS

In Web graph analysis, PageRank algorithm [15, 16] is proposed to find the most important web pages according to the web's link structure. The process can be understood as the probability of a random walk on a directed graph; the quality of each page depends on the number and quality of all pages that link to it. It emphasizes highly linked pages and their links. A closely related link analysis method is HITS (Hyperlink-Induced Topic Search) [17, 18], which also aims to discover web pages of importance. Unlike PageRank, it has two values for each page, and is processed on a small subset of pages, not the whole web. Haveliwala [19] discusses the relative benefits of PageRank and HITS.

In their basic forms, both PageRank and HITS value a node just according to the graph topology. An open question is to add edge weights to them.

2.4 Birnbaum's Component Importance

Birnbaum importance [20] is defined on (Bernoulli) random graphs where edge weights are probabilities of the existence of the edge. The Birnbaum importance of an edge depends directly on the overall effect of the existence of the edge. An edge whose removal has a large effect on the probability of other nodes to be connected, has a high importance. The importance of a node can be defined in terms of the total importance of its edges. This concept has been extended for two edges by Hong and Lei [21].

2.5 Graph Partitioning

Inside a network, there often are clusters of nodes (called communities in social networks), within which connections are stronger, while connections between clusters are weaker and less frequent. In such a situation, a useful abstraction is to divide the network into clusters and present each one of them separately to the user.

A prevalent class of approaches to dividing a network into small parts is based on graph partitioning [22, 23]. The basic goal is to divide the nodes into subsets of roughly equal size and minimize the sum of weights of edges crossing different subsets. This problem is NP-complete. However, many algorithms have been proposed to find a reasonably good partition.

Popular graph partitioning techniques include spectral bisection methods [24, 25] and geometric methods [26, 27]. While they are quite elegant, they have some downsides. Spectral bisection in its standard form is computationally expensive for very large networks. The geometric methods in turn require coordinates of vertices of the graph.

Another approach is multilevel graph partitioning [28, 29]. It first collapses sets of nodes and edges to obtain a smaller graph and partitions the small graph, and then refines the partitioning while projecting the smaller graph back to the original graph. The multilevel method combines a global view with local optimization to reduce cut sizes.

An issue with many of these partitioning methods is that they only bisect networks [30]. Good results are not guaranteed by repeating bisections when more than two subgroups are needed. For example, if the graph essentially has three subgroups, there is no guarantee that these three subgroups can be discovered by finding the best division into two and then dividing one of them again.

Other methods take a rough partitioning as input. A classical representative is Kernighan-Lin (K-L) algorithm [31]. It iteratively looks for a subset of vertices, from each part of the given graph, so that swapping them will lead to a partition with smaller edge-cut. It does not create partitions but rather improves them. The first (very!) rough partitioning can be obtained by randomly partitioning the set of nodes. A weakness of the The K-L method is that it only has a local view of the problem. Various modifications of K-L algorithm have been proposed [32, 33], one of them dealing with an arbitrary number of parts [32].

2.6 Hierarchical Clustering

Another popular technique to divide networks is hierarchical clustering [34]. It computes similarities (or distances) between nodes, for which typical choices include Euclidean distance and Pearson correlation (of neighborhood vectors), as well as the count of edge-independent or vertex-independent paths between nodes.

Hierarchical clustering is well-known for its incremental approach. Algorithms for hierarchical clustering fall into agglomerative or divisive class. In an agglomerative process, each vertex is initially taken as an individual group, then the closest pair of groups is iteratively merged until a single group is constructed or some qualification is met. Newman [35] indicates that agglomerative processes frequently fail to detect correct subgroups, and it has tendency to find only the cores of clusters. The divisive process iteratively removes edges between the least similar vertices, thus it is totally the opposite of an agglomerative method. Obviously, other clustering methods can be applied on nodes (or edges) as well to partition a graph.

2.7 Edge Betweenness

One approach to find a partitioning is through removing edges. This is similar to the divisive hierarchical clustering, and is based on the principle that the edges which connect communities usually have high betweenness [36]. Girvan and Newman define edge betweenness as the number of paths that run along that given edge [35]. It can be calculated using shortest-path betweenness, random-walk betweenness and current-flow betweenness. The authors first use edge centrality indices to find community boundaries. They then remove high betweenness edges in a divisive process, which eventually leads to a division of the original network into separate parts. This method has a high computational cost: in order to compute each edge's betweenness, one should consider all paths in which it appears. Many authors have already proposed different approaches to speed up that algorithm [37, 38].

2.8 Frequent Subgraphs

A frequent subgraph may be considered as a general pattern whose instances can be replaced by a label of that pattern (i.e., a single node or edge representing the pattern). Motivation for this is two-fold. Technically, this operation can be seen as compression. On the other hand, frequent patterns possibly reflect some semantic structures of the domain and therefore are useful candidates for replacement.

Two early methods for frequent subgraph mining use frequent probabilistic rules [39] and compression of the database [40]. Some early approaches use greedy, incomplete schemes [41, 42]. Many of the frequent subgraph mining methods are based on the Apriori algorithm [43], for instance AGM [44] and FSG [45, 46]. However, such methods usually suffer from complicated and costly candidate generation, and high computation time of subgraph isomorphism [47]. To circumvent these problems, gSpan [47] explores depth-first search in frequent subgraph mining. CloseGraph [48] in turn mines closed frequent graphs, which reduces the size of output without losing any information. The Spin method [49] only looks for maximal connected frequent subgraphs.

Most of the methods mentioned above consider a database of graphs as input, not a single large graph. More recently, several methods have been proposed to find frequent subgraphs also in a single input graph [50–53].

3 Preference-Dependent Methods

In this section, we discuss abstraction methods in which a user can explicitly indicate which parts or aspects are more important, according to his interests. Such network abstraction methods are useful when providing more flexible ways to explore a BisoNet.

3.1 Relevant Subgraph Extraction

Given two or more nodes, the idea here is to extract the most relevant subnetwork (of a limited size) with respect to connecting the given nodes as strongly as possible. This subnetwork is then in some sense maximally relevant to the given nodes. There are several alternatives for defining the objective function, i.e., the quality of the extracted subnetwork.

An early approached proposed by Grötschel *et al.* [54] bases the definition on the count of edge-disjoint or vertex-disjoint paths from the source to the sink. A similar principle has later been applied to multi-relational graphs [55], where a pair of entities could be linked by a myriad of relatively short chains of relationships.

The problem in its general form was later formulated as the connection subgraph problem by Faloutsos *et al.* [56]. The authors also proposed a method based on electricity analogies, aiming at maximizing electrical currents in a network of resistors. However, Tong and Faloutsos later point out the weaknesses of using delivered current criterion as a goodness of connection [57]: it only deals with query node pair, and is sensible to the order of the query nodes. Thus, they propose method to extract a subgraph with strong connections to any arbitrary number of nodes.

For random graphs, work from reliability research suggests network reliability as suitable measure [58]. This is defined as the probability that query nodes are connected, given that edges fail randomly according to their probabilities. This approach was then formulated more exactly and algorithms were proposed by Hintsanen and Toivonen [59]. Hintsanen and Toivonen restrict the set of terminals to a pair, and propose two incremental algorithms for the problem.

A logical counterpart of this work, in the field of probabilistic logic learning, is based on ProbLog [60]. In a ProbLog program, each Prolog clause is labeled with a probability. The ProbLog program can then be used to compute the success probabilities of queries. In the theory compression setting for ProbLog [61], the goal is to extract a subprogram of limited size that maximizes the success probability of given queries. The authors use subgraph extraction as the application example.

3.2 Detecting Interesting Nodes or Paths

Some techniques aim to detect interesting paths and nodes, with respect to given nodes. Lin and Chapulsky [62] focus on determining novel, previously unknown paths and nodes from a labeled graph. Based on computing frequencies of similar paths in the data, they use rarity as a measure to find interesting paths or nodes with respect to the given nodes.

An alternative would be to use node centrality to measure the relative importance. White and Smyth [63] define and compute the importance of nodes in a graph relative to one or more given root nodes. They have also pointed out advantages and disadvantages of such measurement based on shortest paths, k-short paths and k-short node-disjoint paths.

3.3 Personalized PageRank

On the basis of PageRank, Personlized PageRank (PPR) is proposed to personalize ranking of web pages. It assigns importance according to the query or user preferences. Early work in this area includes Jeh and Widom [64] and Haveliwala [19]. Later, Fogaras *et al.* [65] have proposed improved methods for the problem.

An issue for network abstraction with these approaches is that they can identity relevant individual nodes, but not a relevant subgraph.

3.4 Exact Subgraph Search

Some substructures may represent obvious or general knowledge, which may moreover occur frequently. Complementary to the approach of Subsection 2.8 where such patterns are identified automatically, here we consider user-input patterns or replacement rules. We first introduce methods that find all exact specified subgraphs.

Finding all exact instances of a graph structure reduces to the subgraph isomorphism problem, which is NP-complete. Isomorphisms are mappings of node and edge labels that preserve the connections in the subgraph.

Ullmann [66] has proposed a well-known algorithm to number the isomorphisms with a refinement procedure that overcomes brute-force tree-search enumeration. Cordella *et al.* [67] include more selective feasibility rules to prune the state search space of their VF algorithm.

A faster algorithm, GraphGrep [68], builds an index of a database of graphs, then uses filtering and exact matching to find isomorphisms. The database is indexed with paths, which are easier to manipulate than trees or graphs. As an alternative, GIndex [69] relies on frequent substructures to index a graph database.

3.5 Similarity Subgraph Search

A more flexible search is to find graphs that are similar but not necessarily identical to the query. Two kinds of similarity search seem interesting in the context of network abstraction. The first one is the K-Nearest-Neighbors (K-NN) query that reports the K substructures which are the most similar to the user's input; the other is the range query which returns subgraphs within a specific dissimilarity range to user's input.

These definitions of the problem imply computation of a similarity measure between two subgraphs. The edit distance between two graphs has been used for that purpose [70]: it generally refers to the cost of transforming one object into the other. For graphs, the transformations are the insertion and removal of vertices and edges, and the changing of attributes on vertices and edges. As graphs have mappings, the edit distance between graphs is the minimum distance over all mappings.

Tian *et al.* [71] propose a distance model containing three components: one measures the structural differences, a second component is the penalty associated with matching two nodes with different labels, and the third component measures the penalty for the gap nodes, nodes in the query that cannot be mapped to any nodes in the target graph.

Another family of similarity measures is based on the maximum common subgraph of two graphs [72]. Fernandez and Valiente [73] propose a graph distance metric based on both maximum common subgraph and minimum common supergraph. The maximum percentage of edges in common has also been used as a similarity measure [74].

Processing pairwise comparisons is very expensive in term of computational time. Grafil [74] and PIS [75] are both based on GIndex [69], indexing the database by frequent substructures.

The concept of graph closure [70] represents the union of graphs, by recording the union of edge labels and vertex labels, given a mapping.

The derived algorithm, Closure-tree, organizes graphs in a hierarchy where each node summarizes its descendants by a graph closure: efficiency of similarity query may improve, and that may avoid some disadvantages of path-based and frequent substructure methods.

The authors of SAGA (Substructure Index-based Approximate Graph Alignment) [71] propose the FragmentIndex technique, which indexes small and frequent substructures. It is efficient for small graph queries, however, processing large graph queries is much more expensive. TALE (Tool for Approximate Subgraph Matching of Large Queries Efficiently) [76] is another approximate subgraph matching system. The authors propose to use NH-Index (Neighborhood Index) to index and capture the local graph structure of each node. An alternative approach uses structured graph decomposition to index a graph database [77].

4 Conclusion

There is a large literature on methods suitable for BisoNet abstraction. We reviewed some of the most important approaches, classified by whether they allow user focus or not. Even though we did not cover the literature exhaustively, we can propose areas for further research based on the gaps and issues observed in the review.

First, we noticed that different node ranking measures (Sections 2.2–2.4) are useful for picking out important nodes, as evidenced by search engines, but the result is just that – a set of nodes. How to better use those ideas to find a connected, relevant subBisoNet is an open question.

Second, although there are lots of methods for partitioning a BisoNet (Section 2.5–2.7), the computational complexity usually is prohibitive for large BisoNets, such as Biomine, with millions of nodes and edges. Obviously, partitioning would be a valuable tool for BisoNet abstraction there.

Third, we observed that some more classical graph problems have been researched much more intensively for graph databases consisting of a number of graphs, rather than for a single large graph. This holds especially for frequent subgraphs (Section 2.8) and subgraph search (Section 3.5).

Finally, a practical exploration system needs an integrated approach to abstraction, using several of the techniques reviewed here to complement each other in producing a simple and useful abstract BisoNet.

Acknowledgements. This work has been supported by the Algorithmic Data Analysis (Algodan) Centre of Excellence of the Academy of Finland and by the European Commission under the 7th Framework Programme FP7-ICT-2007-C FET-Open, contract no. BISON-211898.

References

1. Zhou, F., Mahler, S., Toivonen, H.: Review of Network Abstraction Techniques. In: Workshop on Explorative Analytics of Information Networks at ECML PKDD 2009, pp. 50–63 (2009)
2. Dubitzky, W., Kötter, T., Schmidt, O., Berthold, M.R.: Towards Creative Information Exploration Based on Koestler's Concept of Bisociation. In: Berthold, M.R. (ed.) Bisociative Knowledge Discovery. LNCS (LNAI), pp. 11–32. Springer, Heidelberg (2012)
3. Toussaint, G.T.: The Relative Neighbourhood Graph of a Finite Planar Set. Pattern Recogn. 12(4), 261–268 (1980)
4. Jaromczyk, J., Toussaint, G.: Relative Neighborhood Graphs and Their Relatives. Proc. IEEE 80(9), 1502–1517 (1992)
5. Freeman, L.C.: Centrality in social networks: Conceptual clarification. Soc. Networks 1(3), 215–239 (1979)
6. Stephenson, K.Z.M.: Rethinking centrality: Methods and examples. Soc. Networks 11(1), 1–37 (1989)
7. Wasserman, S., Faust, K.: Social Network Analysis: Methods and Applications. Cambridge University Press, Cambridge (1994)
8. Katz, L.: A new status index derived from sociometric analysis. Psychometrika 18(1), 39–43 (1953)
9. Everett, M., Borgatti, S.P.: Ego network betweenness. Soc. Networks 27(1), 31–38 (2005)
10. Brandes, U.: A Faster Algorithm for Betweenness Centrality. J. Math. Sociol. 25(2), 163–177 (2001)
11. Freeman, L.C.: A Set of Measures of Centrality Based on Betweenness. Sociometry 40, 35–41 (1977)
12. Friedkin, N.E.: Theoretical Foundations for Centrality Measures. Am. J. Sociol. 96(6), 1478–1504 (1991)
13. Gert, S.: The centrality index of a graph. Psychometrika 31(4), 581–603 (1966)

14. Bonacich, P.: Factoring and weighting approaches to status scores and clique identification. J. Math. Sociol. 2(1), 113–120 (1972)
15. Lawrence, P., Sergey, B., Rajeev, M., Terry, W.: The PageRank Citation Ranking: Bringing Order to the Web. Technical report, Stanford Digital Library Technologies Project (1998)
16. Brin, S., Page, L.: The anatomy of a large-scale hypertextual Web search engine. Comput. Netw. ISDN Syst. 30, 107–117 (1998)
17. Kleinberg, J.M.: Authoritative sources in a hyperlinked environment. J. ACM 46(5), 604–632 (1999)
18. Li, L., Shang, Y., Zhang, W.: Improvement of HITS-based algorithms on web documents. In: WWW 2002: Proc. 11th International Conf. on World Wide Web, pp. 527–535. ACM, New York (2002)
19. Haveliwala, T.H.: Topic-Sensitive PageRank. In: WWW 2002: Proc. 11th International Conf. World Wide Web, pp. 517–526. ACM, New York (2002)
20. Birnbaum, Z.W.: On the importance of different components in a multicomponent system. In: Multivariate Analysis - II, pp. 581–592. Academic Press, New York (1969)
21. Hong, J., Lie, C.: Joint reliability-importance of two edges in an undirected network. IEEE Trans. Reliab. 42, 17–23, 33 (1993)
22. Fjällström, P.O.: Algorithms for graph partitioning: A Survey. Linköping Electronic Atricles in Computer and Information Science. Linköping University Electronic Press, Linköping (1998)
23. Elsner, U.: Graph Partitioning - A Survey. Technical Report SFB393/97-27, Technische Universität Chemnitz (1997)
24. Pothen, A., Simon, H.D., Liou, K.P.: Partitioning Sparse Matrices with Eigenvectors of Graphs. SIAM J. Matrix Anal. Appl. 11(3), 430–452 (1990)
25. Hendrickson, B., Leland, R.: An improved spectral graph partitioning algorithm for mapping parallel computations. SIAM J. Sci. Comput. 16(2), 452–469 (1995)
26. Miller, G.L., Teng, S.H., Thurston, W., Vavasis, S.A.: Geometric Separators for Finite-Element Meshes. SIAM J. Sci. Comput. 19(2), 364–386 (1998)
27. Berger, M.J., Bokhari, S.H.: A Partitioning Strategy for Nonuniform Problems on Multiprocessors. IEEE Trans. Comput. 36(5), 570–580 (1987)
28. Karypis, G., Kumar, V.: A Fast and High Quality Multilevel Scheme for Partitioning Irregular Graphs. SIAM J. Sci. Comput. 20, 359–392 (1998)
29. Hendrickson, B., Leland, R.: A Multi-Level Algorithm For Partitioning Graphs. In: Proc. 1995 ACM/IEEE Conf. Supercomputing (CDROM). ACM, New York (1995)
30. Newman, M.E.J.: Detecting community structure in networks. Eur. Phy. J. B - Condensed Matter and Complex Systems 38(2), 321–330 (2004)
31. Kernighan, B.W., Lin, S.: An Efficient Heuristic Procedure for Partitioning Graphs. Bell Sys. Tech. J. 49(1), 291–307 (1970)
32. Fiduccia, C.M., Mattheyses, R.M.: A Linear-Time Heuristic for Improving Network Partitions. In: DAC 1982: P. 19th Conf. Des. Autom., pp. 175–181. ACM, New York (1982)
33. Diekmann, R., Monien, B., Preis, R.: Using Helpful Sets to Improve Graph Bisections. In: Interconnection Networks and Mapping and Scheduling Parallel Computations, pp. 57–73. American Mathematical Society, USA (1995)
34. Scott, J.: Social Network Analysis: A Handbook. SAGE Publications, UK (2000)
35. Newman, M.E.J., Girvan, M.: Finding and evaluating community structure in networks. Phys. Rev. E 69, 026113 (2004)

36. Girvan, M., Newman, M.E.J.: Community structure in social and biological networks. Proc. Natl. Acad. Sci. USA 99(12), 7821–7826 (2002)
37. Radicchi, F., Castellano, C., Cecconi, F., Loreto, V., Parisi, D.: Defining and identifying communities in networks. Proc. Natl. Acad. Sci. USA 101, 2658–2663 (2004)
38. Wu, F., Huberman, B.: Finding Communities in Linear Time: A Physics Approach. Eur. Phys. J. B - Condensed Matter and Complex Systems 38, 331–338 (2004)
39. Dehaspe, L., Toivonen, H., King, R.D.: Finding Frequent Substructures in Chemical Compounds. In: Agrawal, R., Stolorz, P., Piatetsky-Shapiro, G. (eds.) 4th International Conf. Knowl. Disc. Data Min., USA, pp. 30–36. AAAI Press (1998)
40. Holder, L.B., Cook, D.J., Djoko, S.: Substructure Discovery in the SUBDUE System. In: Proc. AAAI Workshop Knowl. Disc. Databases, pp. 169–180. AAAI, Menlo Park (1994)
41. Cook, D.J., Holder, L.B.: Substructure Discovery Using Minimum Description Length and Background Knowledge. J. Artif. Intell. Res. 1, 231–255 (1994)
42. Yoshida, K., Motoda, H.: CLIP: Concept Learning from Inference Patterns. Artif. Intell. 75(1), 63–92 (1995)
43. Agrawal, R., Srikant, R.: Fast Algorithms for Mining Association Rules. In: Bocca, J.B., Jarke, M., Zaniolo, C. (eds.) Proc. 20th International Conf. Very Large Data Bases, VLDB 1994, pp. 487–499. Morgan Kaufmann, San Francisco (1994)
44. Inokuchi, A., Washio, T., Motoda, H.: An Apriori-Based Algorithm for Mining Frequent Substructures from Graph Data. In: Zighed, D.A., Komorowski, J., Żytkow, J.M. (eds.) PKDD 2000. LNCS (LNAI), vol. 1910, pp. 13–23. Springer, Heidelberg (2000)
45. Kuramochi, M., Karypis, G.: Frequent Subgraph Discovery. In: Proc. 2001 IEEE International Conf. Data Min., ICDM 2001, pp. 313–320. IEEE Computer Society, Washington, DC (2001)
46. Kuramochi, M., Karypis, G.: An Efficient Algorithm for Discovering Frequent Subgraphs. IEEE Trans. on Knowl. and Data Eng. 16(9), 1038–1051 (2004)
47. Yan, X., Han, J.: gSpan: Graph-Based Substructure Pattern Mining. In: Proceedings of the 2002 IEEE International Conf. Data Min., pp. 721–724. IEEE Computer Society, Washington, DC (2002)
48. Yan, X., Han, J.: CloseGraph: Mining Closed Frequent Graph Patterns. In: KDD 2003: Proc. 9th ACM SIGKDD International Conf. Knowl. Disc. Data Min., pp. 286–295. ACM, New York (2003)
49. Huan, J., Wang, W., Prins, J., Yang, J.: SPIN: Mining Maximal Frequent Subgraphs from Graph Databases. In: KDD 2004: Proc. 10th ACM SIGKDD International Conf. Knowl. Disc Data Min., pp. 581–586. ACM, New York (2004)
50. Bringmann, B., Nijssen, S.: What Is Frequent in a Single Graph? In: Washio, T., Suzuki, E., Ting, K.M., Inokuchi, A. (eds.) PAKDD 2008. LNCS (LNAI), vol. 5012, pp. 858–863. Springer, Heidelberg (2008)
51. Fiedler, M., Borgelt, C.: Subgraph Support in a Single Large Graph. In: ICDMW 2007: Proc. 7th IEEE International Conf. Data Min. Workshops, pp. 399–404. IEEE Computer Society, Washington, DC (2007)
52. Fiedler, M., Borgelt, C.: Support Computation for Mining Frequent Subgraphs in a Single Graph. In: Proc. 5th Int. Workshop on Mining and Learning with Graphs, MLG 2007, Florence, Italy, pp. 25–30 (2007)
53. Kuramochi, M., Karypis, G.: Finding Frequent Patterns in a Large Sparse Graph. Data Min. Knowl. Disc. 11(3), 243–271 (2005)
54. Grötschel, M., Monma, C.L., Stoer, M.: Design of survivable networks. In: Handbooks in Operations Research and Management Science, vol. 7, pp. 617–672 (1995)

55. Ramakrishnan, C., Milnor, W.H., Perry, M., Sheth, A.P.: Discovering Informative Connection Subgraphs in Multi-relational Graphs. SIGKDD Explor. Newsl. 7(2), 56–63 (2005)
56. Faloutsos, C., McCurley, K.S., Tomkins, A.: Fast Discovery of Connection Subgraphs. In: KDD 2004: Proc. 10th ACM SIGKDD International Conf. Knowl. Disc. Data Min., pp. 118–127. ACM, New York (2004)
57. Tong, H., Faloutsos, C.: Center-Piece Subgraphs: Problem Definition and Fast Solutions. In: KDD 2006: Proc. 12th ACM SIGKDD International Conf. Knowl. Disc. Data Min., pp. 404–413. ACM, New York (2006)
58. Sevon, P., Eronen, L., Hintsanen, P., Kulovesi, K., Toivonen, H.: Link Discovery in Graphs Derived from Biological Databases. In: Leser, U., Naumann, F., Eckman, B. (eds.) DILS 2006. LNCS (LNBI), vol. 4075, pp. 35–49. Springer, Heidelberg (2006)
59. Hintsanen, P., Toivonen, H.: Finding reliable subgraphs from large probabilistic graphs. Data Min. Knowl. Discov. 17, 3–23 (2008)
60. Raedt, L.D., Kimmig, A., Toivonen, H.: ProbLog: A Probabilistic Prolog and its Application in Link Discovery. In: Proc. 20th International Joint Conf. Artif. Intel., pp. 2468–2473. AAAI Press, Menlo Park (2007)
61. Raedt, L., Kersting, K., Kimmig, A., Revoredo, K., Toivonen, H.: Compressing probabilistic Prolog programs. Mach. Learn. 70(2-3), 151–168 (2008)
62. Lin, S., Chalupsky, H.: Unsupervised Link Discovery in Multi-relational Data via Rarity Analysis. In: ICDM 2003: Proc. 3rd IEEE International Conf. Data Min., p. 171. IEEE Computer Society, Washington, DC (2003)
63. White, S., Smyth, P.: Algorithms for Estimating Relative Importance in Networks. In: KDD 2003: Proc. 9th ACM SIGKDD International Conf. Knowl. Disc. Data Min., pp. 266–275. ACM, New York (2003)
64. Jeh, G., Widom, J.: Scaling Personalized Web Search. In: WWW 2003: Proc. 12th International Conf. World Wide Web, pp. 271–279. ACM, New York (2003)
65. Forgaras, D., Rácz, B., Csalogány, K., Sarlós, T.: Towards Scaling Fully Personalized PageRank: Algorithms, Lower Bounds and Experiments. Internet Mathematics 2(3), 335–358 (2005)
66. Ullmann, J.R.: An Algorithm for Subgraph Isomorphism. J. ACM 23(1), 31–42 (1976)
67. Cordella, L.P., Foggia, P., Sansone, C., Vento, M.: A (Sub)Graph Isomorphism Algorithm for Matching Large Graphs. IEEE Trans. Pattern Anal. 26(10), 1367–1372 (2004)
68. Shasha, D., Wang, J.T.L., Giugno, R.: Algorithmics and Applications of Tree and Graph Searching. In: PODS 2002: Proc. 21st ACM SIGMOD-SIGACT-SIGART Symposium on Principles of Database Systems, pp. 39–52. ACM, New York (2002)
69. Yan, X., Yu, P.S., Han, J.: Graph Indexing: A Frequent Structure-based Approach. In: SIGMOD 2004: Proc. 2004 ACM SIGMOD International Conf. Management of Data, pp. 335–346. ACM, New York (2004)
70. He, H., Singh, A.K.: Closure-Tree: An Index Structure for Graph Queries. In: ICDE 2006: Proc. 22nd International Conf. Data Eng., p. 38. IEEE Computer Society, Los Alamitos (2006)
71. Tian, Y., Mceachin, R.C., Santos, C., States, D.J., Patel, J.M.: SAGA: a subgraph matching tool for biological graphs. Bioinformatics 23(2), 232–239 (2007)
72. Bunke, H., Shearer, K.: A graph distance metric based on the maximal common subgraph. Pattern Recogn. Lett. 19(3-4), 255–259 (1998)

73. Fernández, M.-L., Valiente, G.: A graph distance metric combining maximum common subgraph and minimum common supergraph. Pattern Recogn. Lett. 22(6-7), 753–758 (2001)
74. Yan, X., Yu, P.S., Han, J.: Substructure Similarity Search in Graph Databases. In: SIGMOD 2005: Proc. 2005 ACM SIGMOD International Conf. Management of Data, pp. 766–777. ACM, New York (2005)
75. Yan, X., Zhu, F., Han, J., Yu, P.S.: Searching Substructures with Superimposed Distance. In: ICDE 2006: Proc. 22nd International Conf. Data Eng. IEEE Computer Society, Washington, DC (2006)
76. Tian, Y., Patel, J.M.: TALE: A Tool for Approximate Large Graph Matching. In: Proc. 2008 IEEE 24th International Conf. Data Eng., pp. 963–972. IEEE Computer Society, Los Alamitos (2008)
77. Williams, D., Huan, J., Wang, W.: Graph Database Indexing Using Structured Graph Decomposition. In: Proc. 2007 IEEE 23rd International Conf. Data Eng, pp. 976–985. IEEE Computer Society Press, Los Alamitos (2007)

Simplification of Networks by Edge Pruning*

Fang Zhou, Sébastien Mahler, and Hannu Toivonen

Department of Computer Science and HIIT, University of Helsinki, Finland
firstname.lastname@cs.helsinki.fi

Abstract. We propose a novel problem to simplify weighted graphs by pruning least important edges from them. Simplified graphs can be used to improve visualization of a network, to extract its main structure, or as a pre-processing step for other data mining algorithms.

We define a graph connectivity function based on the best paths between all pairs of nodes. Given the number of edges to be pruned, the problem is then to select a subset of edges that best maintains the overall graph connectivity. Our model is applicable to a wide range of settings, including probabilistic graphs, flow graphs and distance graphs, since the path quality function that is used to find best paths can be defined by the user. We analyze the problem, and give lower bounds for the effect of individual edge removal in the case where the path quality function has a natural recursive property. We then propose a range of algorithms and report on experimental results on real networks derived from public biological databases.

The results show that a large fraction of edges can be removed quite fast and with minimal effect on the overall graph connectivity. A rough semantic analysis of the removed edges indicates that few important edges were removed, and that the proposed approach could be a valuable tool in aiding users to view or explore weighted graphs.

1 Introduction

Graphs are frequently used to represent information. Some examples are social networks, biological networks, the World Wide Web, and so called BisoNets, used for creative information exploration [2]. Nodes usually represent objects, and edges may have weights to indicate the strength of the associations between objects. Graphs with a few dozens of nodes and edges may already be difficult to visualize and understand. Therefore, techniques to simplify graphs are needed. An overview of such techniques is provided in reference [3].

In this chapter, we propose a generic framework and methods for simplification of weighted graphs by pruning edges while keeping the graph maximally connected. In addition to visualization of graphs, such techniques could have applications in various network design or optimization tasks, e.g., in data communications or traffic.

* This chapter is a modified version of article "Network Simplification with Minimal Loss of Connectivity" in the 10th IEEE International Conference on Data Mining (ICDM), 2010 [1].

M.R. Berthold (Ed.): Bisociative Knowledge Discovery, LNAI 7250, pp. 179–198, 2012.

The framework is built on two assumptions: the connectivity between nodes is measured using the best path between them, and the connectivity of the whole graph is measured by the average connectivity over all pairs of nodes. We significantly extend and generalize our previous work [4]. The previous work prunes edges while keeping the full original connectivity of the graph, whereas here we propose to relax this constraint and allow removing edges which result in loss of connectivity. The intention is that the user can flexibly choose a suitable trade-off between simplicity and connectivity of the resulting network. The problem then is to simplify the network structure while minimizing the loss of connectivity.

We analyze the problem in this chapter, and propose four methods for the task. The methods can be applied to various types of weighted graphs, where the weights can represent, e.g., distances or probabilities. Depending on the application, different definitions of the connectivity are possible, such as the shortest path or the maximum probability.

The remainder of this article is organized as follows. We first formalize the problem of lossy network simplification in Section 2, and then analyze the problem in Section 3. We present a range of algorithms to simplify a graph in Section 4, and present experimental results in Section 5. We briefly review related work in Section 6, and finally draw some conclusions in Section 7.

2 Lossy Network Simplification

Our goal is to simplify a given weighted graph by removing some edges while still keeping a high level of connectivity. In this section we define notations and concepts, and also give some example instances of the framework.

2.1 Definitions

Let $G = (V, E)$ be a weighted graph. We assume in the rest of the chapter that G is undirected. An *edge* $e \in E$ is a pair $e = \{u, v\}$ of nodes $u, v \in V$. Each edge has a *weight* $w(e) \in \mathbb{R}$. A *path* P is a set of edges $P = \{\{u_1, u_2\}, \{u_2, u_3\}, \ldots, \{u_{k-1}, u_k\}\} \subset E$. We use the notation $u_1 \overset{P}{\rightsquigarrow} u_k$ to say that P is a path between u_1 and u_k, or equivalently, to say that u_1 and u_k are the endvertices of P. A path P can be regarded as the concatenation of several sub-paths, i.e., $P = P_1 \cup \ldots \cup P_n$, where each P_i is a path.

We parameterize our problem and methods with a *path quality function* $q(P) \rightarrow \mathbb{R}^+$. The form of the path quality function depends on the type of graph and the application at hand. For example, in a probabilistic or random graph, it can be the probability that a path exists. Without loss of generality, we assume that the value of any path quality function is positive, and that a larger value of q indicates better quality.

Given two nodes u and v in a weighted graph, they might be linked by a direct edge or a path, or none in a disconnected graph. A simple way to quantify how strongly they are connected is to examine the quality of the best path between them [4]. Thus, the *connectivity between two nodes* u *and* v *in the set* E *of edges* is defined as

$$C(u, v; E) = \begin{cases} \max_{P \subset E: u \overset{P}{\rightsquigarrow} v} q(P) & \text{if such } P \text{ exists} \\ -\infty & \text{otherwise.} \end{cases} \quad (1)$$

A natural measure for the *connectivity of a graph* is then the average connectivity over all pairs of nodes,

$$C(V, E) = \frac{2}{|V|(|V| - 1)} \sum_{u,v \in V, u \neq v} C(u, v; E), \quad (2)$$

where $|V|$ is the number of nodes in the graph. Without loss of generality, in the rest of the chapter we assume the graph is connected, so $C(V, E) > 0$. (If the graph is not connected, we simplify each connected component separately, so the assumption holds again.)

Suppose a set of edges $E_R \subset E$ is removed from the graph. The connectivity of the resulting graph is $C(V, E \setminus E_R)$, and the *ratio of connectivity kept* after removing E_R is

$$rk(V, E, E_R) = \frac{C(V, E \setminus E_R)}{C(V, E)}. \quad (3)$$

Clearly, connectivity can not increase when removing edges. $rk = 1$ means the removal of edges does not affect the graph's connectivity. $0 < rk < 1$ implies that the removal of edges causes some loss of connectivity, while $rk = -\infty$ implies the graph has been cut into two or more components.

Our goal is to remove a fixed number of edges while minimizing the loss of connectivity. From the definitions in Equations (1)–(3) it follows that cutting the input graph drops the ratio to $-\infty$. In this chapter, we thus want to keep the simplified graph connected (and leave simplification methods that may cut the graph for future work). Under the constraint of not cutting the input graph, possible numbers of edges remaining in the simplified graph range from $|V| - 1$ to $|E|$. This follows from the observation that a maximally pruned graph is a spanning tree, which has $|V| - 1$ edges. Thus numbers of removable edges range from 0 to $|E| - (|V| - 1)$.

In order to allow users to specify different simplification scales, we introduce a parameter γ, with values in the range from 0 to 1, to indicate the strength of pruning. Value 0 indicates no pruning, while value 1 implies that the result should be a spanning tree. Thus, the number of edges to be removed by an algorithm is $|E_R| = \lceil \gamma(|E| - (|V| - 1)) \rceil$. Based on notations and concepts defined above, we can now present the problem formally.

Given a weighted graph $G = (V, E)$, a path quality function q, and a parameter γ, the *lossy network simplification* task is to produce a simplified graph $H = (V, F)$, where $F \subset E$ and $|E \setminus F| = \lceil \gamma(|E| - (|V| - 1)) \rceil$, such that $rk(V, E, E \setminus F)$ is maximized. In other words, the task is to prune the specified amount of edges while keeping a maximal ratio of connectivity.

2.2 Example Instances of the Framework

Consider a random (or uncertain) graph where edge weight $w(e)$ gives the probability that edge e exists. A natural quality of a path P is then its probability, i.e., the probability that all of its edges co-exist: $q(P) = \Pi_{\{u,v\}\in P} w(\{u,v\})$. Intuitively, the best path is the one which has the highest probability.

If edge weights represent lengths of edges, then the shortest path is often considered as the best path between two given nodes. Since in this case smaller values (smaller distances) indicate higher quality of paths, one can either reverse the definitions where necessary, or simply define the path quality as the inverse of the length, i.e., $q(P) = 1/\text{length}(P)$.

A flow graph is a directed graph where each edge has a capacity $w(e)$ to transport a flow. The capacity $c(P)$ of a path is limited by the weakest edge along that path: $c(P) = \min_{\{u,v\}\in P} w(\{u,v\}) = q(P)$. The best path is one that has the maximal flow capacity. If the flow graph is undirected, the graph can be simplified without any loss of quality to a spanning tree that maximizes the smallest edge weight in the tree.

3 Analysis of the Problem

In this section, we investigate some properties of the problem of lossy network simplification. We first note that the ratio of connectivity kept $rk(V, E, E_R)$ is multiplicative with respect to successive removals of sets of edges. Based on this we then derive two increasingly fast and approximate ways of bounding $rk(V, E, E_R)$: These bounds will be used by algorithms we give in Section 4.

3.1 Multiplicativity of Ratio of Connectivity Kept

Let E_R be any set of edges to be removed. Consider an arbitrary partition of E_R into two sets E_R^1 and E_R^2, such that $E_R = E_R^1 \cup E_R^2$ and $E_R^1 \cap E_R^2 = \emptyset$. Using Equation (3), we can rewrite the ratio of connectivity kept by E_R as

$$\begin{aligned} rk(V, E, E_R^1 \cup E_R^2) & \\ &= \frac{C(V,E\setminus(E_R^1\cup E_R^2))}{C(V,E)} \\ &= \frac{C(V,E\setminus E_R^1)}{C(V,E)} \cdot \frac{C(V,E\setminus E_R^1\setminus E_R^2)}{C(V,E\setminus E_R^1)} \\ &= rk(V,E,E_R^1) \cdot rk(V, E \setminus E_R^1, E_R^2). \end{aligned}$$

In other words, the ratio of connectivity kept $rk(\cdot)$ is multiplicative with respect to successive removals of sets of edges.

An immediate consequence is that the ratio of connectivity kept after removing set E_R of edges can also be represented as the product of ratios of connectivity kept for each edge, in any permutation:

$$rk(V, E, E_R) = \Pi_{i=1}^{|E_R|} rk(V, E \setminus E_{i-1}, e_i),$$

where e_i the ith edge in the chosen permutation and $E_i = \{e_1, \ldots, e_i\}$ is the set of i first edges of E_R.

Note that the ratio of connectivity kept is *not* multiplicative for the ratios $rk(V, E, \{e_i\})$ of connectivity kept with respect to the original set E of edges. It is therefore not straightforward to select an edge set whose removal keeps the maximal $rk(V, E, E_R)$ value among all possible results.

The multiplicativity directly suggests, however, to greedily select the edge maximizing $rk(V, E \setminus E_{i-1}, e_i)$ at each step. The multiplicativity property tells that the exact ratio of connectivity kept will be known throughout the process, even if it is not guaranteed to be optimal. We will use this approach in the brute force algorithm that we give in Section 4. Two other algorithms will use the greedy search too, but in a more refined form that uses results from the next subsections.

3.2 A Bound on the Ratio of Connectivity Kept

Recall that the connectivity of a graph is the average connectivity among all pairs of nodes. In principle, the removal of an edge may cause the connectivity between any arbitrary pair of nodes to decrease. We now derive a lower bound for the connectivity kept, based on the effect of edge removal only on the endpoints of the edge itself.

Many path quality functions are recursive in the sense that sub-paths of a best path are also best paths between their own endpoints. (This is similar to the property known as *optimal substructure* in dynamic programming.) Additionally, a natural property for many quality functions q is that the effect of a local change is at most as big for the whole path P as it is for the modified segment $R \subset P$.

Formally, let $P = \arg\max_{P \subset E: u \rightsquigarrow v} q(P)$ be a best path (between any pair of nodes u and v), let $m \in P$ be a node on the path, let $R \subset P$ be a subpath (segment) of P and S a path (not in P) with the same endvertices as R. Function q is a *local recursive path quality function* if

$$q(P) = q(\arg\max_{P_1 \subset E: u \rightsquigarrow m} q(P_1) \cup \arg\max_{P_2 \subset E: m \rightsquigarrow v} q(P_2))$$

and

$$\frac{q(P \setminus R \cup S)}{q(P)} \geq \frac{q(S)}{q(R)}.$$

Examples of local recursive quality functions include the (inverse of the) length of a path (when edge weights are distances), the probability of a path (when edge weights are probabilities), and minimum edge weight on a path (when edge weights are flow capacities). A negative example is average edge weight.

The local recursive property allows to infer that over all pairs of nodes, the biggest effect of removing a particular edge will be seen on the connectivity of the edge's own endvertices. In other words, the ratio of connectivity kept for any pair of nodes is at least as high as the ratio kept for the edge's endvertices.

To formalize this bound, we denote by $\kappa(E,e)$ the ratio of connectivity kept between the endvertices of an edge $e = \{u,v\}$ after removing it from the set E of edges:

$$\kappa(E,e) = \begin{cases} -\infty & \text{if } C(u,v;E\setminus\{e\}) = -\infty; \\ \frac{C(u,v;E\setminus\{e\})}{q(\{e\})} & \text{if } C(u,v;E\setminus\{e\}) < q(\{e\}); \\ 1 & \text{if } C(u,v;E\setminus\{e\}) \geq q(\{e\}). \end{cases} \tag{4}$$

The first two cases directly reflect the definition of ratio of connectivity kept (Equation 3) when edge e is the only path (case one) or the best path (case two) between its endpoint. The third case applies when $\{e\}$ is not the best path between between its endpoints. Then, its absence will not cause any loss of connectivity between u and v, and $\kappa(E,e) = 1$.

Theorem 1. *Let $G = (V,E)$ be a graph and $e \in E$ an edge, and let q be a local recursive path quality function. The ratio of connectivity kept if e is removed is lower bounded by $rk(V,E,e) \geq \kappa(E,e)$.*

Sketch of a proof. The proof is based on showing that the bound holds for the ratio of connectivity kept for any pair of nodes. (1) Case one: $\kappa(E,e) = -\infty$ clearly is a lower bound for any ratio of connectivity kept. (2) Case two: Consider any pair of nodes u and v. In the worst case the best path between them contains e and, further, the best alternative path between u and v is the one obtained by replacing e by the best path between the endvertices of e. Since q is local recursive, even in this case at least fraction $\kappa(E,e)$ of connectivity is kept between u and v. (3) Case three: edge e has no effect on the connectivity of its own endvertices, nor on the connectivity of any other nodes.

Theorem 1 gives us a fast way to bound the effect of removing an edge and suggests a greedy method to the lossy network simplification problem by removing an edge with the largest κ. Obviously, only based on $\kappa(E,e) < 1$, we can not infer the exact effect of removing edge e, nor the relative difference between removing two alternative edges. However, computing κ is much faster than computing rk, since only the best path between the edge's endvertices needs to be examined, not all-pairs best paths.

3.3 A Further Bound on the Ratio of Connectivity Kept

Previously, we suggested two ways to compute or approximate the best alternative path for an edge [4]. The *global best path search* finds the best path with unlimited length and thus gives the exact $C(u,v;E\setminus\{e\})$ and κ values. However, searching the best path globally takes time. A faster alternative, called *triangle search*, is to find the best path of length two, denoted by $S_2(e)$. That is, let $S_2(e) = \{\{u,w\}\{w,v\}\} \subset E$, $e \notin S_2(e)$, be a path between the endvertices u, v such that $q(S_2(e))$ is maximized. Obviously, path $S_2(e)$ may not be the best path between the edge's endvertices, and therefore $q(S_2(e))$ is a lower bound for the quality of the best path between the endvertices of e.

To sum up the results from this section, we have two increasingly loose lower bounds for the ratio of connectivity kept for local recursive functions. The first one is based on only looking at the best alternative path for an edge. The second one is a further lower bound for the quality of this alternative path. Denoting by $S_2(e)$ the best path of length two as defined above, we have

$$rk(V, E, e) \geq \kappa(E, e) \geq \min(\frac{q(S_2(e))}{q(\{e\})}, 1).$$

In the next section, we will give algorithms that use these lower bounds to complete the simplification task with different trade-offs between connectivity kept and time complexity.

4 Algorithms

We next present four algorithms to simplify a given graph by pruning a fixed number of edges while aiming to keep a high connectivity. All algorithms take as input a weighted graph G, a path function q and a ratio γ. They prune $n = \gamma(|E| - (|V| - 1))$ edges. The first algorithm is a naive approach, simply pruning a fraction of the weakest edges by sorting edges according to the edge weight. The second one is a computationally demanding brute-force approach, which greedily removes an edge with the highest rk value in each iteration. The third and fourth algorithms are compromises between these extremes, aimed at a better trade-off between quality and efficiency. The third one iteratively prunes the edge which has the largest κ value through global search. The fourth algorithm prunes edges with the combination of triangle search and global search.

4.1 Naive Approach

Among the four algorithms that we present, the simplest approach is the naive approach (NA), outlined in Algorithm 1. It first sorts edges by their weights in an ascending order (Line 1). Then, it iteratively checks the edge from the top of the sorted list (Line 7), and prunes the one whose removal will not lead to disconnected components (Line 8). The algorithm stops when the number of edges removed reaches n, derived from G and γ.

The computational cost of sorting edges is $O\big(|E| \log |E|\big)$ (Line 1). On Line 7, we use Dijkstra's algorithm with a complexity of $O\big((|E| + |V|) \log |V|\big)$ to check whether there exists a path between the edge's endvertices. So, the total computational complexity of the naive approach is $O\big(|E| \log |E| + n(|E| + |V|) \log |V|\big)$.

4.2 Brute Force Approach

The brute force approach (BF), outlined in Algorithm 2, prunes edges in a greedy fashion. In each iteration, it picks the edge whose removal best keeps the connectivities, i.e., has the largest rk value. It first calculates the $rk(V, F, e)$ value

Algorithm 1. NA algorithm

Input: A weighted graph $G = (V, E)$, q and γ
Output: Subgraph $H \subset G$
1: Sort edges E by weights in an ascending order.
2: $F \leftarrow E$
3: $n \leftarrow \gamma(|E| - (|V| - 1))$
4: { Iteratively prune the weakest edge which does not cut the graph }
5: $i \leftarrow 1, j \leftarrow 1$ { j is an index to the sorted list of edges }
6: **while** $i \leq n$ **do**
7:    **if** $C(u, v; F \setminus \{e_j\})$ is not $-\infty$ **then**
8:       $F \leftarrow F \setminus \{e_j\}$
9:       $i \leftarrow i + 1$
10:    $j \leftarrow j + 1$
11: Return $H = (V, F)$

for every edge e (Line 10), and then stores the information of the edge whose $rk(V, F, e)$ value is the highest at the moment (Line 11), and finally prunes the one which has the highest rk value among all existing edges (Line 16). As an optimization, set M is used to store edges that are known to cut the remaining graph (Lines 9 and 15), and the algorithm only computes $rk(V, F, e)$ for the edges which are not in M (Line 8).

When computing $rk(V, F, e)$ for an edge (Line 10), all-pairs best paths need to be computed with a cost of $O\big(|V|(|E| + |V|) \log |V|\big)$. (This dominates the connectivity check on Line 9.) Inside the loop, $rk(V, F, e)$ is computed for all edges in each of n iterations, so the total time complexity is $O\big(n|E||V|(|E| + |V|) \log |V|\big)$.

Algorithm 2. BF algorithm

Input: A weighted graph $G = (V, E)$, q and γ
Output: Subgraph $H \subset G$
1: $F \leftarrow E$
2: $n \leftarrow \gamma(|E| - (|V| - 1))$
3: { Iteratively prune the edge with the highest rk value. }
4: $M \leftarrow \emptyset$ { edges whose removal is known to cut the graph. }
5: **for** $r = 1$ to n **do**
6:    $rk\_largest \leftarrow -\infty$
7:    $e\_largest \leftarrow null$
8:    **for** $e = \{u, v\}$ in F and $e \notin M$ **do**
9:       **if** graph $(V, F \setminus \{e\})$ is connected **then**
10:          compute $rk(V, F, e) = \frac{C(V, F \setminus \{e\})}{C(V, F)}$
11:          **if** $rk(V, F, e) > rk\_largest$ **then**
12:             $rk\_largest \leftarrow rk(V, F, e)$
13:             $e\_largest \leftarrow e$
14:       **else**
15:          $M \leftarrow M + e$
16:    $F \leftarrow F \setminus \{e\_largest\}$
17: Return $H = (V, F)$

4.3 Path Simplification

The outline of the path simplification approach (PS) is in Algorithm 3. The main difference to the brute force approach is that PS calculates κ instead of $rk(V, F, e)$ for each edge.

The method finds, for each edge, the best possible alternative path S globally (Line 9). It then prunes in each loop the edge with the largest lower bound κ of connectivity kept. As an efficient shortcut, as soon as we find an edge whose κ is equal to 1, we remove it immediately. Again, list M is used to store information of those edges whose removal cuts the graph.

Algorithm 3. PS algorithm

Input: A weighted graph $G = (V, E)$, q and γ
Output: Subgraph $H \subset G$

```
1: F ← E
2: n ← γ(|E| − (|V| − 1))
3: {Iteratively prune the edge with the largest κ value.}
4: M ← ∅
5: for r = 1 to n do
6:    κ_largest ← −∞
7:    e_largest ← null
8:    for e = {u, v} in F and e ∉ M do
9:       Find path S such that q(S) = C(u, v; F \ {e})
10:      if q(S) ≥ q({e}) then
11:         κ ← 1
12:         F ← F \ {e}
13:         break
14:      else if 0 < q(S) < q({e}) then
15:         κ ← q(S) / q({e})
16:      else
17:         κ ← −∞
18:         M ← M + e
19:      if κ > κ_largest then
20:         κ_largest ← κ
21:         e_largest ← e
22:   F ← F \ {e_largest}
23: Return H = (V, F)
```

The complexity of the innermost loop is dominated by finding the best path between the edge's endvertices (Line 9), which has time complexity $O((|E| + |V|)\log|V|)$. This is done n times for $O(|E|)$ edges, so the total time complexity is $O(n|E|(|E| + |V|)\log|V|)$. While still quadratic in the number of edges, this is a significant improvement over the brute force method.

4.4 Combinational Approach

The fourth and final algorithm we propose is the combinational approach (CB), outlined in Algorithm 4. The difference to the path simplification (PS) method

Algorithm 4. CB algorithm

Input: A weighted graph $G = (V, E)$, q and γ
Output: Subgraph $H \subset G$

```
 1: F ← E
 2: n ← γ(|E| − (|V| − 1))
 3: { Iteratively prune the edge with the largest κ using triangle search }
 4: r ← 1
 5: find ← true
 6: while r ≤ n and find = true do
 7:    κ_largest ← −∞
 8:    e_largest ← null
 9:    for e = {u, v} in F do
10:       Find path S2 = {{u, w}{w, v}} ⊂ F \ {e} that maximizes q(S2)
11:       if q(S2) ≥ q({e}) then
12:          κ ← 1
13:          F ← F \ {e}
14:          r ← r + 1
15:          break
16:       else if 0 < q(S2) < q({e}) then
17:          κ ← q(S2)/q({e})
18:       else
19:          κ ← −∞
20:       if κ > κ_largest then
21:          κ_largest ← κ
22:          e_largest ← e
23:    if κ_largest > 0 then
24:       F ← F \ {e_largest}
25:       r ← r + 1
26:    else
27:       find ← false
28: if r < n then
29:    apply the path simplification (PS) method in Algorithm 3 to prune n − r edges
30: Return H = (V, F)
```

above is that the best path search is reduced to triangle search (Line 10). However, triangle search is not always able to identify a sufficient number of edges to be removed, depending on the number and quality of triangles in the graph. Therefore the combinational approach invokes the PS method to remove additional edges if needed (Line 29).

The computational complexity of triangle search for a single edge is $O(|V|)$ (Line 10). Thus, if we only apply triangle search, the total cost is $O(n|E||V|)$. However, if additional edges need to be removed, the worst case computational complexity equals the complexity of the path simplification method (PS).

5 Experiments

To assess the problem and methods proposed in this chapter, we carried out experiments on real graphs derived from public biological databases. With the experiments, we want to evaluate the trade-off between the size of the result and the loss of connectivity, compare the performances of the proposed algorithms, study the scalability of the methods, and assess what the removed edges are like semantically in the biological graphs.

5.1 Experimental Setup

We have adopted the data and test settings from Toivonen et al. [4]. The data source is the Biomine database [5] which integrates information from twelve major biomedical databases. Nodes are biological entities such as genes, proteins, and biological processes. Edges correspond to known or predicted relations between entities. Each edge weight is between 0 and 1, and is interpreted as the probability that the relation exists. The path quality function is the probability of the path, i.e., the product of weights of the edges in the path. This function is local recursive.

For most of the tests, we use 30 different graphs extracted from Biomine. The number of nodes in each of them is around 500, and the number of the edges ranges from around 600 to 900. The graphs contain some parallel edges that can be trivially pruned. For more details, see reference [4]. For scalability tests, we use a series of graphs with up to 2000 nodes, extracted from the same Biomine database.

The algorithms are coded in Java. All tests were run on standard PCs with x86_64 architecture with Intel Core 2 Duo 3.16GHz, running Linux.

5.2 Results

Trade-Off between Size of the Result and Connectivity Kept. By construction, our methods work on a connected graph and keep it connected. As described in Section 2, maximally simplified graphs are then spanning trees, with $|V|-1$ edges. The number of edges removed is algorithm independent: they all remove fraction γ of the $|E|-(|V|-1)$ edges that can be removed. The distribution of the number of edges to be removed in our test graphs, relative to the total number of edges, are shown as a function of γ in Figure 1. These graphs are relatively sparse, and approximately at most 35% of edges can be removed without cutting the graph.

In this chapter, we extend a previous simplification task [4] from lossless to lossy simplification (with respect to the connectivity of the graph). In other words, in the previous proposal the ratio of connectivity kept must always stay at 1. We now look at how many more edges and with how little loss of connectivity our new methods can prune. We use the path simplification method as a representative here (and will shortly compare the proposed methods).

In Figure 2, we plot the ratio of connectivity kept by the four methods of Toivonen et al. [4] for two different graphs, randomly selected from our 30 graphs. Four different types of points are positioned horizontally according to n, the number of edges pruned by the previous methods. The x-axis shows the number of edges pruned in terms of γ, computed as $\gamma = n/(|E|-(|V|-1))$. Results of the

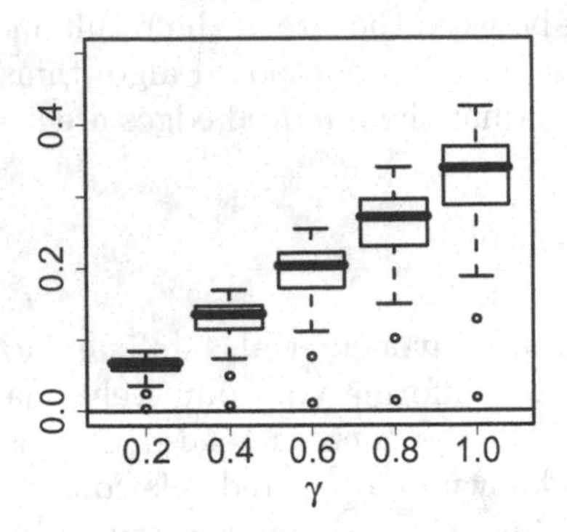

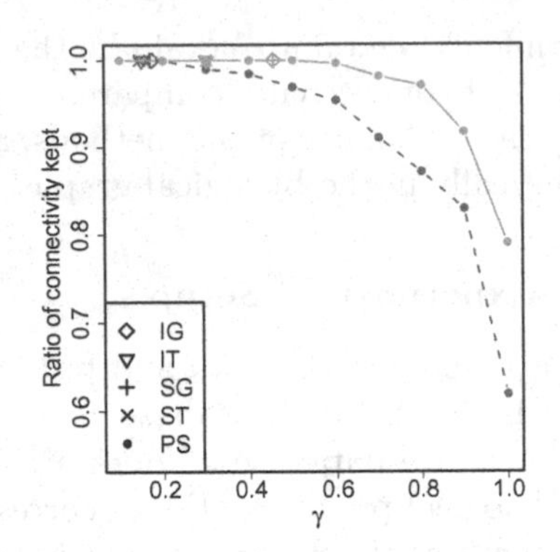

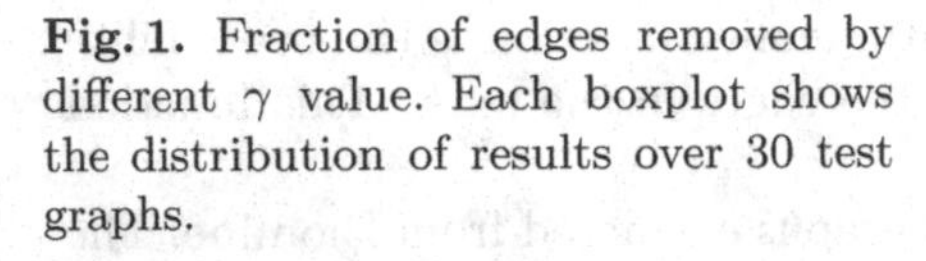

Fig. 1. Fraction of edges removed by different γ value. Each boxplot shows the distribution of results over 30 test graphs.

Fig. 2. Ratio of connectivity kept by the four methods of Toivonen et al. [4] and by the path simplification method for two graphs (green and red). IG=Iterative Global, IT=Iterative Triangle, SG=Static Global, ST=Static Triangle.

path simplification method proposed in this chapter are shown as lines. Among the four previous methods, the Iterative Global (IG) method prunes the maximal number of edges. Significantly more edges can be pruned, with larger values of γ, while keeping a very high ratio of connectivity. This indicates that the task of lossy network simplification is useful: significant pruning can be achieved with little loss of connectivity.

Comparison of Algorithms. Let us next compare the algorithms proposed in this chapter. Each of them prunes edges in a somewhat different way, resulting in different ratios of connectivity kept. These ratios with respect to different γ are shown in Figure 3. For $\gamma = 1$ (Figure 3(e)), where the result of all methods is a spanning tree, we also plot the results of a standard maximum spanning tree method [6] that maximizes the sum of edge weights.

Among all methods, the brute force approach expectedly always keeps the highest ratio of graph connectivity. When γ is between 0.2 and 0.6, the brute force method can actually keep the original connectivity, and even when $\gamma = 1$ it still keeps around 93% connectivity.

Overall, the four proposed methods perform largely as expected. The second best method is path simplification, followed by the combinational approach. They both keep high connectivities for a wide range of values for γ, still approximately 90% with $\gamma = 0.8$. The naive approach is clearly inferior, but it also produces useful results for smaller values of γ.

An interesting observation can be made from Figure 3(e) where $\gamma = 1$. The maximum spanning tree has similar ratios of connectivity kept with all methods except the brute force method, which can produce significantly better results.

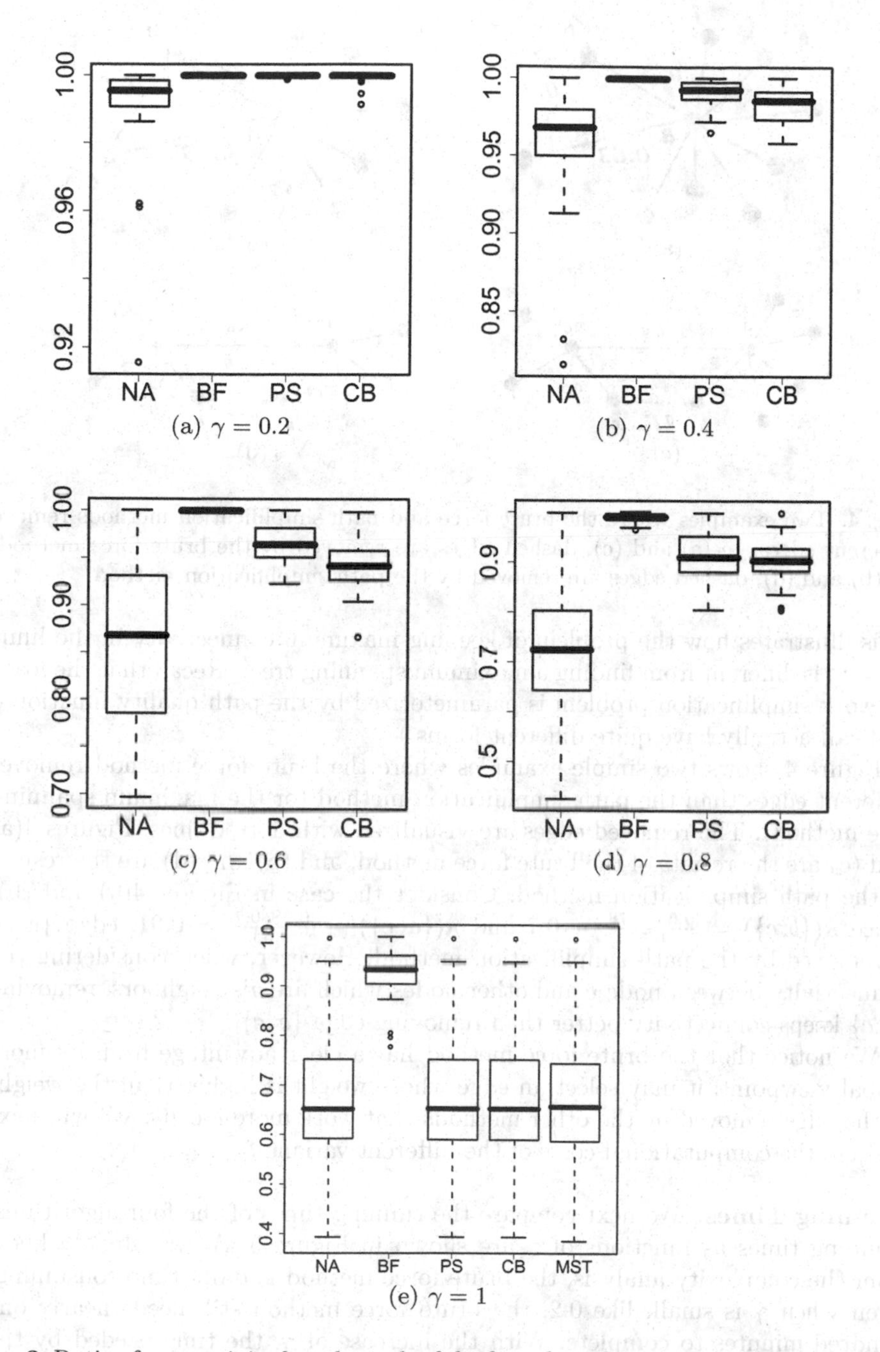

Fig. 3. Ratio of connectivity kept by each of the four algorithmic variants. Each boxplot shows the distribution of results over 30 test graphs. NA = Naive approach, BF = Brute Force, PS = Path Simplification, CB = Combinational approach, MST = Maximum Spanning Tree.

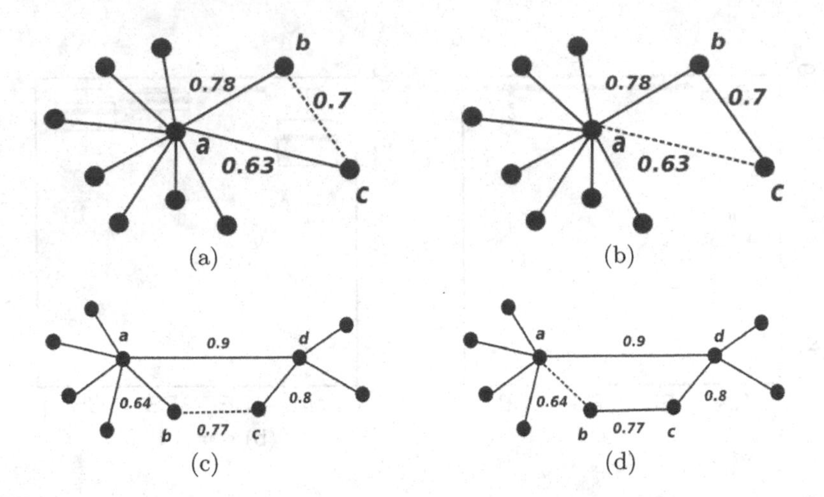

Fig. 4. Two examples where the brute force and path simplification methods remove different edges. In (a) and (c), dashed edges are removed by the brute force method. In (b) and (d), dashed edges are removed by the path simplification method.

This illustrates how the problem of keeping maximum connectivity in the limit ($\gamma = 1$) is different from finding a maximum spanning tree. (Recall that the lossy network simplification problem is parameterized by the path quality function q and can actually have quite different forms.)

Figure 4 shows two simple examples where the brute force method removes different edges than the path simplification method (or the maximum spanning tree method). The removed edges are visualized with dotted lines; Figures 4(a) and (c) are the results of the brute force method, and (b) and (d) are the results of the path simplification method. Consider the case in Figures 4(a) and (b). Since $\kappa(\{b,c\}) = \frac{0.63*0.78}{0.7} = 0.7$ and $\kappa(\{a,c\}) = \frac{0.78*0.7}{0.63} = 0.91$, edge $\{a,c\}$ is removed by the path simplification method. However, when considering the connectivity between node c and other nodes which are a's neighbors, removing $\{b,c\}$ keeps connectivity better than removing edge $\{a,c\}$.

We notice that the brute force method has a clear advantage from its more global viewpoint: it may select an edge whose weight is higher than the weight of the edge removed by the other methods that work more locally. We will next address the computational costs of the different variants.

Running Times. We next compare the running times of the four algorithms. Running times as functions of γ are shown in Figure 5. As we already know from the complexity analysis, the brute force method is quite time consuming. Even when γ is small, like 0.2, the brute force method still needs nearly one hundred minutes to complete. With the increase of γ, the time needed by the brute force increases from 100 to more than 400 minutes, while the other three methods only need a few seconds to complete. The second slowest method is the path simplification, which running time increases linearly with γ from 5 to 50 seconds. The naive approach always needs less than 1 second to complete.

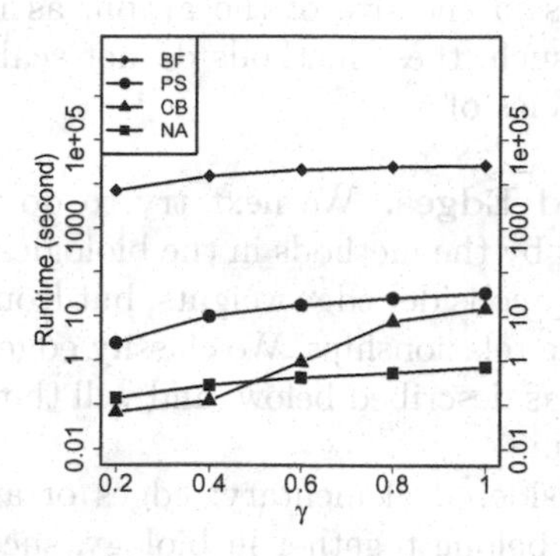

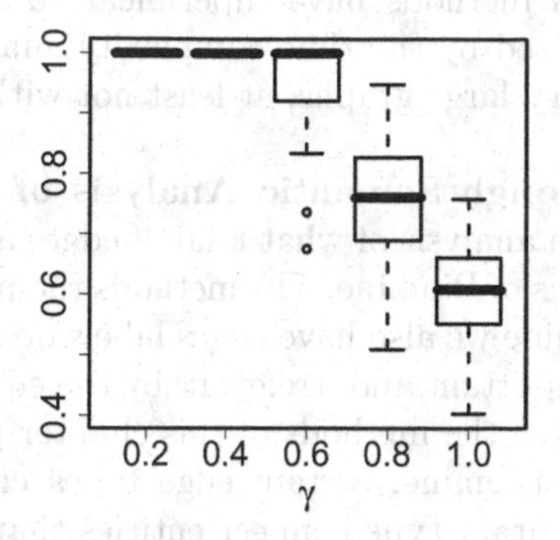

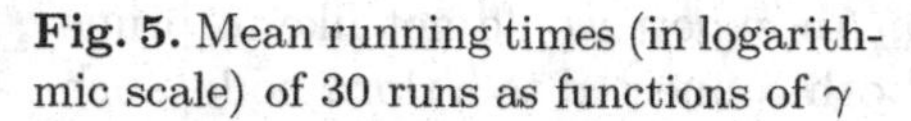

Fig. 5. Mean running times (in logarithmic scale) of 30 runs as functions of γ

Fig. 6. Fraction of edges removed by triangle search in the combinational method

The combinational approach is the fastest one when γ is very small, but it comes close to the time the path simplification method needs when γ is larger. The reason for this behavior is that the combinational approach removes varying shares of edges using the computationally more intensive global search: Figure 6 shows that, with small values of γ, all or most edges are removed with the efficient triangle search. When γ increases, the fraction of edges removed by global search correspondingly increases.

In order to evaluate the scalability of the methods, we ran experiments with a series of graphs with up to 2000 nodes. The node degree is around 2.5. The running times as functions of graph size are shown in Figures 7 (with $\gamma = 0.4$) and 8 (with $\gamma = 0.8$).

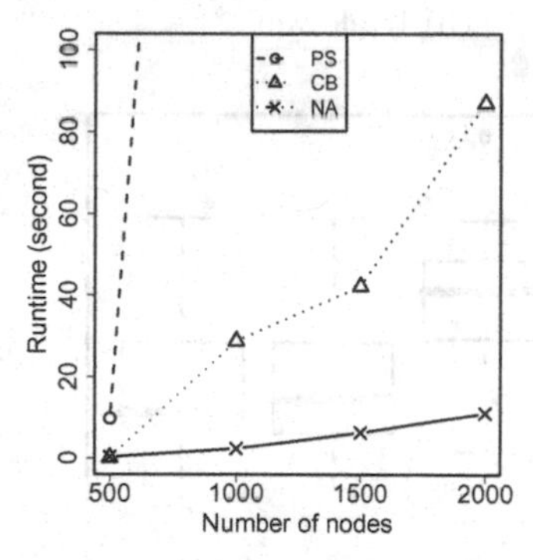

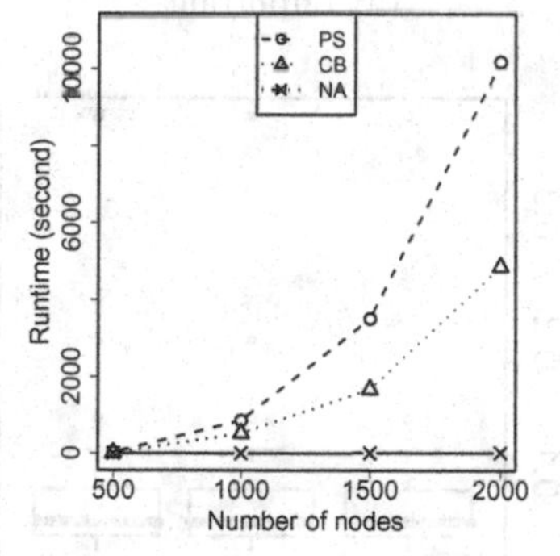

Fig. 7. Running times as functions of graph size (number of nodes) with $\gamma = 0.4$. The running time of the brute force method for a graph of 500 nodes is 15 000 seconds.

Fig. 8. Running times as functions of graph size (number of nodes) with $\gamma = 0.8$. The running time of the brute force method for a graph of 500 nodes is 25 000 seconds.

All methods have superlinear running times in the size of the graph, as is expected by the time complexity analysis. As such, these methods do not scale to very large graphs, at least not with large values of γ.

A Rough Semantic Analysis of Removed Edges. We next try to do a rough analysis of what kind of edges are pruned by the methods in the biological graphs of Biomine. The methods themselves only consider edge weights, but from Biomine we also have edges labels describing the relationships. We classify edges to important and irrelevant by the edge labels, as described below, and will then see how the methods of this chapter prune them.

In Biomine, certain edge types can be considered elementary: edges of an elementary type connect entities that strongly belong together in biology, such as a protein and the gene that codes for it. An expert would not like to prune these links. On the other hand, if they are both connected to a third node, such as a biological function, then one of these edges could be considered redundant. Since the connection between the protein and gene is so essential, any connections to either one could be automatically considered to hold also for the other one.

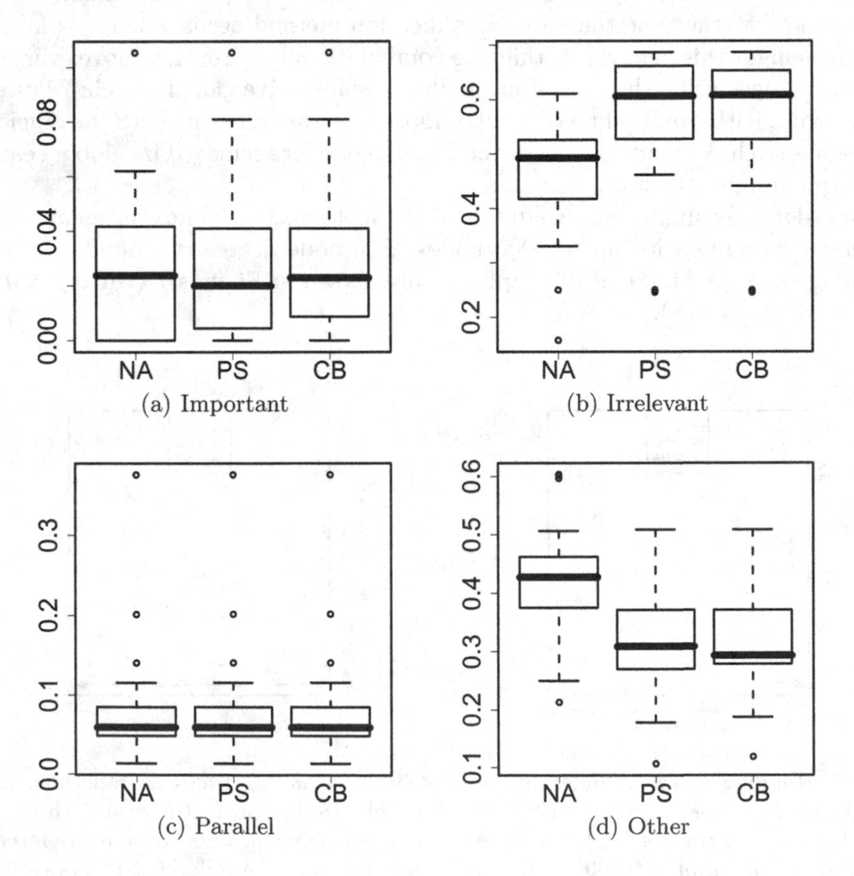

Fig. 9. Shares of different semantic categories among all removed edges with $\gamma = 0.8$

An explicit representation of such an edge would be considered "semantically irrelevant".

Following the previous setting [4], we considered the edge types *codes for*, *is homologous to*, *subsumes*, and *has synonym* as "important." Then, we computed the number of those edges that are "semantically irrelevant." Additionally, we marked the edges which have the same endvertices as "Parallel" edges. For the sake of completeness, we also counted the number of "other edges" that are neither important nor semantically irrelevant, nor parallel edges.

The semantic categories of the edges removed with $\gamma = 0.8$ are shown in Figure 9. Among edges removed by the naive approach, 3% are important, 45% are irrelevant, 8% are parallel and around 44% are other edges. The results of the path simplification and the combinational approach are quite similar: with edges removed by them there are around 2% important edges, 60% irrelevant edges, around 8% parallel edges and 30% other edges. (We do not analyze the semantic types of edges removed by the brute force method due to its time complexity.)

We notice that the path simplification and the combinational approach remove more irrelevant edges than the naive approach does. The reason is that these irrelevant edges may have a high weight, but they also have high κ value, in most cases, $\kappa = 1$.

The results indicate that the path simplification and the combinational approaches could considerably complement and extend expert-based or semantic methods, while not violating their principles.

6 Related Work

Network simplification has been addressed in several variants and under different names. Simplification of flow networks [7, 8] has focused on the detection of vertices and edges that have no impact on source-to-sink flow in the graph. Network scaling algorithms produce so-called Pathfinder networks [9–11] by pruning edges for which there is a better path of at most q edges, where q is a parameter. Relative Neighborhood Graphs [12] only connect relatively close pairs of nodes. They are usually constructed from a distance matrix, but can also be used to simplify a graph: indeed, relative neighborhood graphs use the triangle test only.

The approach most closely related to ours is path-oriented simplification [4], which removes edges that do not affect the quality of best paths between any pair of nodes. An extreme simplification that still keeps the graph connected, can be obtained by Minimum Spanning Tree (MST) [6, 13] algorithms. Our approach differs from all these methods in an important aspect: we measure and allow loss of network quality, and let the user choose a suitable trade-off.

There are numerous measures for edge importance. These can be used to rank and prune edges with varying results. Representative examples include edge betweenness [14], which is measured as the number of paths that run along the edge, and Birnbaum's component importance [15], defined as the probability that the edge is critical to maintain a connected graph.

The goal of extracting a subgraph graph is similar to the problem of reliable subgraph or connection subgraph extraction [16–18]). Their problem is, however, related to a set of (query) nodes, while our problem is independent of query nodes. They also prune least useful nodes, while we only prune edges.

We have reviewed related work more extensively in [3].

7 Conclusion

We have addressed the problem of network simplification given that the loss of connectivity should be minimized. We have introduced and formalized the task of selecting an edge set whose removal keeps the maximal ratio of the connectivity. Our framework is applicable to many different types of networks and path qualities. We have demonstrated the effect on random (or uncertain) graphs from a real-world application.

Based on our definition of ratio of connectivity kept, we have proposed a naive approach and a brute force method. Moreover, we have shown that the property of local recursive path quality functions allows to design a simpler solution: when considering the removal of one edge, the ratio of connectivity kept between the edge's endvertices can be used to bound the ratio for all pairs of nodes. Based on this observation, we have proposed two other efficient algorithms: the path simplification method and the combinational approach.

We have conducted experiments with 30 real biological networks to illustrate the behavior of the four methods. The results show that the naive approach is in most cases the fastest one, but it induces a large loss of connectivity. The brute force approach is very slow in selecting the best set of edges. The path simplification and the combinational approach were able to select a good set in few seconds for graphs with some hundreds of nodes. A rough semantic analysis of the simplification indicates that, in our experimental setting, both the path simplification and the combinational approach have removed very few important edges, and a relatively high number of irrelevant edges. We suggest those two approaches can well complement a semantic-based simplification.

Future work includes development of more scalable algorithms for the task of lossy network simplification. The problem and algorithms we proposed here are objective techniques: they do not take into account any user-specific emphasis on any region of the network. Future work may be to design query-based simplification techniques that would take user's interests into account when simplifying a network. It would also be interesting to combine different network abstraction techniques with network simplification, such as a graph compression method to aggregate nodes and edges [19].

Acknowledgment. We would like to thank Lauri Eronen for his help. This work has been supported by the Algorithmic Data Analysis (Algodan) Centre of Excellence of the Academy of Finland (Grant 118653) and by the European Commission under the 7th Framework Programme FP7-ICT-2007-C FET-Open, contract no. BISON-211898.

References

1. Zhou, F., Mahler, S., Toivonen, H.: Network Simplification with Minimal Loss of Connectivity. In: The 10th IEEE International Conference on Data Mining (ICDM), Sydney, Australia, pp. 659–668 (2010)
2. Dubitzky, W., Kötter, T., Schmidt, O., Berthold, M.R.: Towards Creative Information Exploration Based on Koestler's Concept of Bisociation. In: Berthold, M.R. (ed.) Bisociative Knowledge Discovery. LNCS (LNAI), vol. 7250, pp. 11–32. Springer, Heidelberg (2012)
3. Zhou, F., Mahler, S., Toivonen, H.: Review of BisoNet Abstraction Techniques. In: Berthold, M.R. (ed.) Bisociative Knowledge Discovery. LNCS (LNAI), vol. 7250, pp. 166–178. Springer, Heidelberg (2012)
4. Toivonen, H., Mahler, S., Zhou, F.: A Framework for Path-Oriented Network Simplification. In: Cohen, P.R., Adams, N.M., Berthold, M.R. (eds.) IDA 2010. LNCS, vol. 6065, pp. 220–231. Springer, Heidelberg (2010)
5. Sevon, P., Eronen, L., Hintsanen, P., Kulovesi, K., Toivonen, H.: Link Discovery in Graphs Derived from Biological Databases. In: Leser, U., Naumann, F., Eckman, B. (eds.) DILS 2006. LNCS (LNBI), vol. 4075, pp. 35–49. Springer, Heidelberg (2006)
6. Kruskal Jr., J.: On the shortest spanning subtree of a graph and the traveling salesman problem. Proceedings of the American Mathematical society 7(1), 48–50 (1956)
7. Biedl, T.C., Brejová, B., Vinař, T.: Simplifying Flow Networks. In: Nielsen, M., Rovan, B. (eds.) MFCS 2000. LNCS, vol. 1893, pp. 192–201. Springer, Heidelberg (2000)
8. Misiołek, E., Chen, D.Z.: Efficient Algorithms for Simplifying Flow Networks. In: Wang, L. (ed.) COCOON 2005. LNCS, vol. 3595, pp. 737–746. Springer, Heidelberg (2005)
9. Schvaneveldt, R., Durso, F., Dearholt, D.: Network structures in proximity data. In: The Psychology of Learning and Motivation: Advances in Research and Theory, vol. 24, pp. 249–284. Academic Press, New York (1989)
10. Quirin, A., Cordon, O., Santamaria, J., Vargas-Quesada, B., Moya-Anegon, F.: A new variant of the Pathfinder algorithm to generate large visual science maps in cubic time. Information Processing and Management 44, 1611–1623 (2008)
11. Hauguel, S., Zhai, C., Han, J.: Parallel PathFinder Algorithms for Mining Structures from Graphs. In: Proceedings of the 2009 Ninth IEEE International Conference on Data Mining, ICDM 2009, pp. 812–817. IEEE Computer Society, Washington, DC (2009)
12. Toussaint, G.T.: The Relative Neighbourhood Graph of a Finite Planar Set. Pattern Recognition 12(4), 261–268 (1980)
13. Osipov, V., Sanders, P., Singler, J.: The Filter-Kruskal Minimum Spanning Tree Algorithm. In: ALENEX, pp. 52–61. SIAM (2009)
14. Girvan, M., Newman, M.E.J.: Community structure in social and biological networks. Proc. Natl. Acad. Sci. U S A 99(12), 7821–7826 (2002)

15. Birnbaum, Z.W.: On the importance of different components in a multicomponent system. In: Multivariate Analysis - II, pp. 581–592 (1969)
16. Grötschel, M., Monma, C.L., Stoer, M.: Design of Survivable Networks. In: Handbooks in Operations Research and Management Science, vol. 7, pp. 617–672 (1995)
17. Faloutsos, C., McCurley, K.S., Tomkins, A.: Fast Discovery of Connection Subgraphs. In: KDD 2004: Proceedings of the Tenth ACM SIGKDD International Conference on Knowledge Discovery and Data Mining, pp. 118–127. ACM, New York (2004)
18. Hintsanen, P., Toivonen, H.: Finding reliable subgraphs from large probabilistic graphs. Data Min. Knowl. Discov. 17, 3–23 (2008)
19. Toivonen, H., Zhou, F., Hartikainen, A., Hinkka, A.: Compression of Weighted Graphs. In: The 17th ACM SIGKDD Conference on Knowledge Discovery and Data Mining (KDD), San Diego, CA, USA (2011)

Network Compression by Node and Edge Mergers*

Hannu Toivonen, Fang Zhou, Aleksi Hartikainen, and Atte Hinkka

Department of Computer Science and HIIT, University of Helsinki, Finland
firstname.lastname@cs.helsinki.fi

Abstract. We give methods to compress weighted graphs (i.e., networks or BisoNets) into smaller ones. The motivation is that large networks of social, biological, or other relations can be complex to handle and visualize. Using the given methods, nodes and edges of a give graph are grouped to supernodes and superedges, respectively. The interpretation (i.e. decompression) of a compressed graph is that a pair of original nodes is connected by an edge if their supernodes are connected by one, and that the weight of an edge equals the weight of the superedge. The compression problem then consists of choosing supernodes, superedges, and superedge weights so that the approximation error is minimized while the amount of compression is maximized.

In this chapter, we describe this task as the 'simple weighted graph compression problem'. We also discuss a much wider class of tasks under the name of 'generalized weighted graph compression problem'. The generalized task extends the optimization to preserve longer-range connectivities between nodes, not just individual edge weights. We study the properties of these problems and outline a range of algorithms to solve them, with different trade-offs between complexity and quality of the result. We evaluate the problems and algorithms experimentally on real networks. The results indicate that weighted graphs can be compressed efficiently with relatively little compression error.

1 Introduction

Graphs and networks are used in numerous applications to describe relationships between entities, such as social relations between persons, links between web pages, flow of traffic, or interactions between proteins. We are also interested in conceptual networks called BisoNets which allow creative information exploration and support bisociative reasoning [2]. In many applications, including most BisoNets, relationships have weights that are central to any use or analysis of graphs: how frequently do two persons communicate or how much do they influence each other's opinions; how much web traffic flows from one page

* This chapter is a modified version of article "Compression of Weighted Graphs" in the 17th ACM SIGKDD Conference on Knowledge Discovery and Data Mining, 2011 [1].

M.R. Berthold (Ed.): Bisociative Knowledge Discovery, LNAI 7250, pp. 199–217, 2012.

to another or how many cars drive from one crossing to another; or how strongly does one protein regulate another one?

We describe models and methods for the compression of BisoNets (weighted graphs) into smaller ones that contain approximately the same information. In this process, also known as graph simplification in the context of unweighted graphs [3, 4], nodes are grouped to supernodes, and edges are grouped into superedges between supernodes. A superedge then represents all possible edges between two nodes, one from each of the conneceted supernodes.

This problem is different from graph clustering or partitioning where the aim is to find groups of strongly related nodes. In graph compression, nodes are grouped based on the similarity of their relationships to other nodes, not by their (direct) mutual relations.

As a small example, consider the co-authorship social network in Figure 1a. It contains an excerpt from the DBLP Computer Science Bibliography[1], a subgraph containing *Jiawei Han* and *Philip S. Yu* and a dozen related authors. Nodes in this graph represent authors and edges represent co-authorships. Edges are weighted by the number of co-authored articles.

Compressing this graph by about 30% gives a simpler graph that highlights some of the inherent structure or roles in the original graph (Figure 1b). For instance, *Ke Wang* and *Jianyong Wang* have identical sets of co-authors (in this excerpt from DBLP) and have been grouped together. This is also an example of a group that would not be found by traditional graph clustering methods, since the two nodes grouped together are not directly connected. *Daxin Jiang* and *Aidong Zhang* have been grouped, but additionally the self-edge of their supernode indicates that they have also authored papers together.

Groups that could not be obtained by the existing compression algorithms of [3, 4] can be observed among the six authors that (in this excerpt) only connect to *Jiawei Han* and *Philip S. Yu*. Instead of being all grouped together as structurally equivalent nodes, we have three groups that have different weight profiles. *Charu C. Aggarwal* is a group by himself, very strongly connected with *Philip S. Yu*. A second group includes *Jiong Yang*, *Wei Fan*, and *Xifeng Yan*, who are roughly equally strongly connected to both *Jiawei Han* and *Philip S. Yu*. The third group, *Hong Cheng* and *Xiaoxin Yin*, are more strongly connected to *Jiawei Han*. Such groups are not found with methods for unweighted graphs [3, 4].

In what we define as the *simple weighted graph compression problem*, the approximation error of the compressed graph with respect to original edge weights is minimized by assigning each superedge the mean weight of all edges it represents. For many applications on weighted graphs it is, however, important to preserve relationships between faraway nodes, too, not just individual edge weights. Motivated by this, we also introduce the *generalized weighted graph compression problem* where the goal is to produce a compressed graph that maintains connectivities across the graph: the best path between any two nodes should be approximately equally good in the compressed graph as it is in the original graph, but the path does not have to be the same.

[1] `http://dblp.uni-trier.de/`

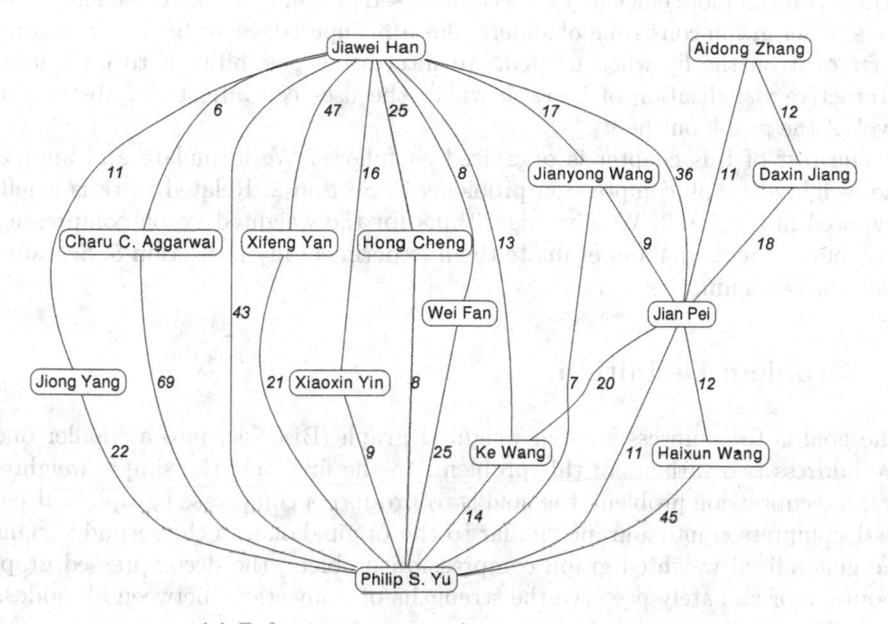

(a) Before compression (14 nodes, 26 edges)

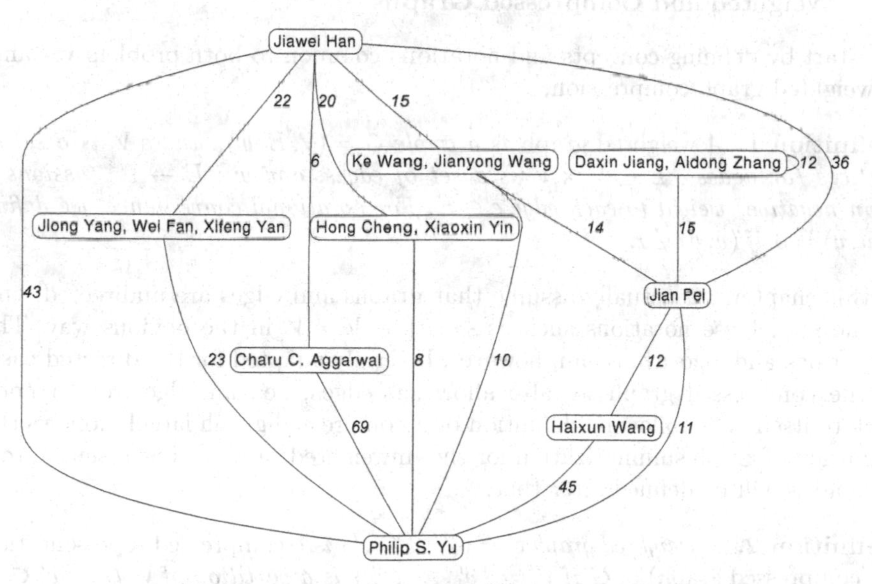

(b) After some compression (9 nodes, 16 edges)

Fig. 1. A neighborhood graph of *Jiawei Han* in the DBLP bibliography

Compressed weighted graphs can be utilized in a number of ways. Graph algorithms can run more efficiently on a compressed graph, either by considering just the smaller graph consisting of supernodes and superedges, or by decompressing parts of it on the fly when needed. An interesting possibility is to provide an interactive visualization of a graph where the user can adjust the abstraction level of the graph on the fly.

The rest of this chapter is organized as follows. We formulate and analyze the weighted graph compression problems in Section 2. Related work is briefly reviewed in Section 3. We give algorithms for the weighted graph compression problems in Section 4 and evaluate them experimentally in Section 5. Section 6 contains concluding remarks.

2 Problem Definition

The goal is to compress a given weighted graph (BisoNet) into a smaller one. We address two variants of this problem. In the first one, the simple weighted graph compression problem, the goal is to produce a compressed graph that can be decompressed into a graph similar to the original one. In the second variant, the generalized weighted graph compression problem, the decompressed graph should approximately preserve the strengths of connections between all nodes.

2.1 Weighted and Compressed Graphs

We start by defining concepts and notations common to both problem variants of weighted graph compression.

Definition 1. *A* weighted graph *is a triple $G = (V, E, w)$, where V is a set of vertices (or nodes), $E \subset V \times V$ is a set of edges, and $w : E \to \mathbb{R}^+$ assigns a (non-negative) weight to each edge $e \in E$. For notational convenience, we define $w(u, v) = 0$ if $(u, v) \notin E$.*

In this chapter, we actually assume that graphs and edges are undirected, and in the sequel use notations such as $\{u, v\} \in V \times V$ in the obvious way. The definitions and algorithms can, however, be easily adapted for the directed case. In the compressed graph we also allow self-edges, i.e., an edge from a node back to itself. The following definition of a compressed graph largely follows the definition of graph summarization for the unweighted case [3]. The essential role of weights will be defined after that.

Definition 2. *A weighted graph $S = (V', E', w')$ is a* compressed representation (or compressed graph) *of G if $V' = \{v'_1, \ldots, v'_n\}$ is a partition of V (i.e., $v'_i \subset V$ for all i, $\cup_i v'_i = V$, and $v'_i \cap v'_j = \emptyset$ for all $i \neq j$). The nodes $v' \in V'$ are also called* supernodes, *and edges $e' \in E'$ are also called* superedges.

We use the notation $com : V \to V'$ to map original nodes to the corresponding supernodes: $com(u) = v'$ if and only if $u \in v' \in V'$.

The idea is that a supernode represents all original nodes within it, and that a single superedge represents all possible edges between the corresponding original nodes, whether they exist in G or not. Apparently, this may cause structural errors of two types: a superedge may represent edges that do not exist in the original graph, or edges in the original graph are not represented by any superedge. (Our algorithms in Section 4 only commit the first kind of errors, i.e., they will not miss any edges, but may introduce new ones.) In addition, edge weights may have changed in compression. We will next formalize these issues using the concepts of decompressed graphs and graph dissimilarity.

Definition 3. *Given G and S as above, the* decompressed graph $dec(S)$ of S *is a weighted graph $dec(S) = (V, E'', w'')$ such that $E'' = \{\{u,v\} \in V \times V \mid \{com(u), com(v)\} \in E'\}$ and $w''(\{u,v\}) = w'(\{com(u), com(v)\})$. (By the definition of compressed representation, $V = \cup_{i=1}^{n} V_i'$.)*

In other words, a decompressed graph has the original set of nodes V, and there is an edge between two nodes exactly when there is a superedge between the corresponding supernodes. The weight of an edge equals the weight of the corresponding superedge.

Definition 4. *Given G and S as above, the* compression ratio *of S (with respect to the original graph G) is defined as $cr(S) = \frac{|E'|}{|E|}$.*

The compression ratio measures how much smaller the compressed graph is. The number of supernodes vs. original nodes is not included in the definition since nodes are actually not compressed, in the sense that their identities are preserved in the supernodes and hence no space is saved. They are also always completely recovered in decompression.

2.2 Simple Weighted Graph Compression

Compression ratio does not consider the amount of errors introduced in edges and their weights. This issue is addressed by a measure of dissimilarity between graphs. We first present a simple distance measure that leads to the simple weighted graph compression problem.

Definition 5. *The* simple distance *between two graphs $G_a = (V, E_a, w_a)$ and $G_b = (V, E_b, w_b)$, with an identical set of nodes V, is*

$$dist_1(G_a, G_b) = \sqrt{\sum_{\{u,v\} \in V \times V} (w_a(\{u,v\}) - w_b(\{u,v\}))^2}. \tag{1}$$

This distance measure has an interpretation as the Euclidean distance between G_a and G_b in a space where each pair of nodes $\{u,v\} \in V \times V$ has its own dimension. Given the distance definition, the dissimilarity between a graph G and its compressed representation S can then be defined simply as $dist_1(G, dec(S))$.

Edges introduced by the compression/decompression process are considered in this measure, since edge weight $w(\{u,v\}) = 0$ if edge $\{u,v\}$ is not in G.

The distance can be seen as the cost of compression, whereas the compression ratio represents the savings. Our goal is to produce a compressed graph which optimizes the balance between these two. In particular, we will consider the following form of the problem.

Definition 6. *Given a weighted graph G and a compression ratio cr, $0 < cr < 1$, the* simple weighted graph compression problem *is to produce a compressed representation S of G with $cr(S) \leq cr$ such that $dist_1(G, dec(S))$ is minimized.*

Other forms can be just as useful. One obvious choice would be to give a maximum distance as parameter, and then seek for a minimum compression ratio. In either case, the problem is complex, as the search space consists of all partitions of V. However, the compression ratio is non-increasing and graph distance non-decreasing when nodes are merged to supernodes, and this observation can be used to devise heuristic algorithms for the problem, as we do in Section 4.

2.3 Generalized Weighted Graph Compression

We next generalize the weighted graph compression problem. In many applications, it is not the individual edge weights but the overall connectivity between nodes that matters, and we propose a model that takes this into account. The model is based on measuring the best paths between nodes, and trying to preserve these qualities. We start with some preliminary definitions and notations.

Definition 7. *Given a graph $G = (V, E, w)$, a* path *P is a set of edges $P = \{\{u_1, u_2\}, \{u_2, u_3\}, \ldots, \{u_{k-1}, u_k\}\} \subset E$. We use the notation $u_1 \overset{P}{\rightsquigarrow} u_k$ to say that P is a path between u_1 and u_k, and that u_1 and u_k are the endnodes of P.*

The definition of how good a path is and which is the best one depends on the kind of graph and the application. For the sake of generality, we parameterize our formulation by a path quality function q. For example, in a flow graph where edge weights are capacities of edges, path quality q can be defined as the maximum flow through the path (i.e., as the minimum edge weight on the path). In a probabilistic or uncertain graph where edge weights are probabilities that the edge exists, q often is defined as the probability that the path exists (i.e., as the product of the edge weights). Without loss of generality, we assume that the value of any path quality function is positive, that a larger value of q indicates better quality, and that q is monotone in the sense that a path with a cycle is never better than the same path without the cycle. We also parameterize the generalized definition by a maximum path length λ. The goal of generalized weighted graph compression will be to preserve all pairwise connectivities of length at most λ.

Definition 8. *Given a weighted graph $G = (V, E, w)$, a path quality function q, and a positive integer λ, the* λ-connection *between a pair of nodes u and v is defined as*

$$Q_\lambda(u,v;G) = \begin{cases} \max_{P \subset E: u \overset{P}{\leadsto} v, |P| \leq \lambda} q(P) & \textit{if such } P \textit{ exists} \\ 0 & \textit{otherwise,} \end{cases}$$

i.e., as the quality of the best path, of length at most λ*, between* u *and* v*. If* G *is obvious in the context, we simply write* $Q_\lambda(u,v)$*.*

Definition 9. *Let* G_a *and* G_b *be weighted graphs with an identical set* V *of nodes, and let* λ *be a positive integer and* q *a path quality function as defined above. The* generalized distance between G_a and G_b (with respect to λ and q) *is*

$$dist_\lambda(G_a, G_b) = \sqrt{\sum_{\{u,v\} \in V \times V} (Q_\lambda(u,v;G_a) - Q_\lambda(u,v;G_b))^2}. \tag{2}$$

Definition 10. *Given a weighted graph* G *and a compression ratio* cr*,* $0 < cr < 1$*, the* generalized weighted graph compression problem *is to produce a compressed representation* S *of* G *with* $cr(S) \leq cr$ *such that* $dist_\lambda(G, dec(S))$ *is minimized.*

The simple weighted graph compression problem defined earlier is an instance of this generalized problem with $\lambda = 1$ and $q(\{e\}) = w(e)$. In this chapter, we will only consider the two extreme cases with $\lambda = 1$ and $\lambda = \infty$. For notational convenience, we often write $dist(\cdot)$ instead of $dist_\lambda(\cdot)$ if the value of λ is not significant.

2.4 Optimal Superedge Weights and Mergers

Given a compressed graph structure, it is easy to set the weights of superedges to optimize the simple distance measure $dist_1(\cdot)$. Each pair $\{u,v\} \in V \times V$ of original nodes is represented by exactly one pair $\{u',v'\} \in V' \times V'$ of supernodes, including the cases $u = v$ and $u' = v'$. In order to minimize Equation 1, given the supernodes V', we need to minimize for each pair $\{u',v'\}$ of supernodes the sum $\sum_{\{u,v\} \in u' \times v'} (w(\{u,v\}) - w'(\{u'v'\}))^2$. This sum is minimized when the superedge weight is the mean of the original edge weights (including "zero-weight edges" for those pairs of nodes that are not connected by an edge):

$$w'(\{u',v'\}) = \frac{\sum_{\{u,v\} \in u' \times v'} w(\{u,v\})}{|u'|\,|v'|}, \tag{3}$$

where $|x|$ is the number of original nodes in supernode x.

The compression algorithms that we propose below work in an incremental, often greedy fashion, merging two supernodes at a time into a new supernode (following the ideas of references [3, 4]). The merge operation that these algorithms use is specified in Algorithm 1. It takes a graph and two of its nodes as parameters, and it returns a graph where the given nodes are merged into one and the edge weights of the new supernode are set according to Equation 3. Line 6 of the merge operation sets the weight of the self-edge for the supernode.

When $\lambda = 1$, function $W(x, y)$ returns the sum of weights of all original edges between x and y using their mean weight $Q_1(\{x, y\}; S)$. The weight of the self-edge is then zero and the edge non-existent if neither u nor v has a self-edge and if there is no edge between u and v.

Algorithm 1. merge(u, v, S)

Input: Nodes u and v, and a compressed graph $S = (V, E, w)$ s.t. $u, v \in V$
Output: A compressed graph S' obtained by merging u and v in S
1: $S' \leftarrow S$ {i.e., $(V', E', w') \leftarrow (V, E, w)$}
2: $z \leftarrow \{u \cup v\}$
3: $V' \leftarrow V' \setminus \{u, v\} \cup \{z\}$
4: **for all** $x \in V$ s.t. $u \neq x \neq v$, and $\{u, x\}$ or $\{v, x\} \in E$ **do**
5: $\quad w'(\{z, x\}) = \frac{|u| Q_\lambda(\{u,x\};S) + |v| Q_\lambda(\{v,x\};S)}{|u|+|v|}$
6: $w'(\{z, z\}) = \frac{W(u,u)+W(v,v)+W(u,v)}{|z|(|z|-1)/2}$
7: **return** S'

8: **function** $W(x, y)$:
9: **if** $x \neq y$ **then**
10: $\quad$ **return** $Q_\lambda(\{x, y\}; S)|x||y|$
11: **else**
12: $\quad$ **return** $Q_\lambda(\{x, x\}; S)|x|(|x| - 1)/2$

Setting superedge weights optimally is much more complicated for the generalized distance (Equation 2) when $\lambda > 1$: edge weights contribute to best paths and therefore distances up to λ hops away, so the distance cannot be optimized in general by setting each superedge weight independently. We use the merge operation of Algorithm 1 as an efficient, approximate solution also in these cases, and leave better solutions for future work.

2.5 Bounds for Distances between Graphs

Our compression algorithms produce the compressed graph S by a sequence of merge operations, i.e., as a sequence $S_0 = G, S_1, \ldots, S_n = S$ of increasingly compressed graphs. Since the distance function $dist(\cdot)$ is a metric and satisfies the triangle inequality (recall its interpretation as Euclidian distance), the distance of the final compressed graph S from the original graph G can be upper-bounded by

$$dist(G, dec(S)) \leq \sum_{i=1}^{n} dist(dec(S_{i-1}), dec(S_i)).$$

An upper bound for the distance between two graphs can be obtained by considering only the biggest distance over all pairs of nodes. Let G_a and G_b be weighted graphs with an identical set V of nodes, and denote the maximum distance for any pair of nodes by

$$d_{\max}(G_1, G_2) = \max_{\{u,v\} \in V \times V} |(Q_\lambda(u, v; G_a) - Q_\lambda(u, v; G_b))|.$$

We now have the following bound:

$$dist(G_1, G_2) \leq \sqrt{\sum_{\{u,v\} \in V \times V} d_{\max}(G_1, G_2)^2} \propto d_{\max}(G_1, G_2). \tag{4}$$

This result can be used by compression algorithms to bound the effects of potential merge operations.

2.6 A Bound on Distances between Nodes

We now derive an upper bound for $d_{\max}(S, S')$ for the case where $S' =$ merge(u, v, S) for some nodes u and v in V (cf. Algorithm 1). Let $d_{\max}(u, v; S)$ be the maximum difference of weights between any two edges merged together as the result of merging u and v:

$$\begin{aligned} d_{\max}(u, v; S) = & \\ \max\{ & \max_{x:\{u,x\} \text{ or } \{v,x\} \in E} (|Q_\lambda(u, x; S) - Q_\lambda(v, x; S)|), \\ & |Q_\lambda(u, u; S) - Q_\lambda(v, v; S)|, \\ & |Q_\lambda(u, u; S) - Q_\lambda(u, v; S)|, \\ & |Q_\lambda(v, v; S) - Q_\lambda(u, v; S)| \ \}. \end{aligned} \tag{5}$$

The first element is the maximum over all edges to neighboring nodes x, and the rest are the differences between edges that are merged into the self-edge.

For $\lambda = 1$ it is fairly obvious that we have the bound

$$d_{\max}(S, \text{merge}(u, v, S)) \leq d_{\max}(u, v; S), \tag{6}$$

since all effects of merge operations are completely local to the edges adjacent to the merged nodes. The situation is more complicated for $\lambda = \infty$ since a merger can also affect arbitrary edges. Luckily, many natural path quality functions q have the property that a change in the weight of an edge (from $w(e)$ to $w'(e)$) changes the quality of the whole path (from $q(P)$ to $q'(P)$) at most as much as it changes the edge itself:

$$\frac{|q(P) - q'(P)|}{q(P)} \leq \frac{|w(e) - w'(e)|}{w(e)}.$$

Path quality functions q that have this property include the sum of edge weights (e.g., path length), product of edge weights (e.g., path probability), minimum edge weight (e.g., maximum flow), maximum edge weight, and average edge weight. Based on this property, we can infer that the biggest distance after merging u and v will be seen on the edges connecting u, v and their neighbors, i.e., that the bound of Equation 6 holds also for $\lambda = \infty$ for many usual path quality functions.

Based on Equations 4 and 6, we have a fast way to bound the effect of merging any two nodes. We will use this bound in some of our algorithms.

3 Related Work

Graph compression as presented in this chapter is based on merging nodes that have similar relationships to other entities i.e., that are structurally most equivalent — a classic concept in social network analysis [5]. Structural equivalence and many other types of relations between (super)nodes have been considered in social networks under block modeling (see, e.g., [6]), where the goal is both to identify supernodes and to choose among the different possible types of connections between them. Our approach (as well as that of references [3, 4], see below) uses only two types: "null" (no edges) and "complete" (all pairs are connected), as these seem to be best suited for compression.

Graph compression has recently attracted new interest. The work most closely related to ours is by Navlakha et al. [3] and Tian et al. [4], who independently proposed to construct graph summaries of unweighed graphs by grouping nodes and edges to supernodes and superedges. We generalize these approaches in two important and related directions: to weighted graphs, and to long-range, indirect (weighted) connections between nodes.

Both above-mentioned papers also address issues we do not consider here. Navlakha et al. [3] propose a representation which has two parts: one is a graph summary (in our terminology, an unweighted compressed graph), the other one is a set of edge corrections to fix the errors introduced by mergers of nodes and edges to superedges. Tian et al. [4] consider labeled graphs with categorical node and edge attributes, and the goal is to find relatively homogeneous supernodes and superedges. This approach has been generalized by Zhang et al. [7] to numerical node attributes which are then automatically categorized. They also addressed interactive drill-down and roll-up operations on graphs. Tian et al. used both top-down (divisive) and bottom-up (agglomerative) algorithms, and concluded that top-down methods are more practical in their problem [4], whereas Navlakha et al. had the opposite experience [3]. This difference is likely due to different use of node and edge labels. The methods we propose work bottom-up since we have no categorical attributes to guide a divisive approach like Tian et al. had.

Unweighted graph compression techniques have been used to simplify graph storage and manipulation. For example, Chen et al. [8] successfully applied a graph compression method to reduce the number of embeddings when searching frequent subgraphs in a large graph. Navlakha et al. [9] revealed biological modules with the help of compressed graphs. Furthermore, Chen et al. [10] incorporated the compressed graph notion with a generic topological OLAP framework to realize online graph analysis.

There are many related but subtly different problems. Graph partitioning methods (e.g. [11, 12]) aim to find groups of nodes that are more strongly connected to each other than to nodes in other groups. Extraction of a subgraph, whether based on a user query (e.g. [13, 14]) or not (e.g., [15–17]) produces a smaller graph by just throwing out edges and nodes. Web graph compression algorithms aim to produce as compact a representation of a graph as possible, in different formats (e.g., [18, 19]). For more related work, we refer to the good overviews given in references [3, 4]. A wider review of network abstraction techniques is available in [20].

4 Algorithms

We next propose a series of algorithms for the weighted graph compression problem. All of the proposed algorithms work more or less in a greedy fashion, merging two (super)nodes and their edges at a time until the specified compression rate is achieved. All these algorithms have the following input and output:

Input: weighted graph $G = (V, E, w)$, compression ratio cr $(0 < cr < 1)$, path quality function q, and maximum path length $\lambda \in \mathbb{N}$.
Output: compressed weighted graph $S = (V', E', w')$ with $cr(S) \leq cr$, such that $dist(G, dec(S))$ is minimized.

Brute-force greedy algorithm. The brute-force greedy method (Algorithm 2) computes the effects of all possible pairwise mergers (Line 4) and then performs the best merger (Line 5), and repeats this until the requested compression rate is achieved. The algorithm generalizes the greedy algorithm of Navlakha et al. [3] to distance functions $dist_\lambda(\cdot)$ that take the maximum path length λ and the path quality function q as parameters.

Algorithm 2. Brute-force greedy search

1: $S \leftarrow G$ {i.e., $(V', E', w') \leftarrow (V, E, w)$}
2: **while** $cr(S) > cr$ **do**
3: **for all** pairs $\{u, v\} \in V' \times V'$ **do** {(*)}
4: $d_{\{u,v\}} \leftarrow dist(G, dec(merge(u, v, S)))$
5: $S \leftarrow merge(\arg\min_{\{u,v\}} d_{\{u,v\}}, S)$
6: **return** S
(*) 2-hop optimization can be used, see text.

The worst-case time complexity for simple weighted graph compression is $O(|V|^4)$, and for generalized compression $O(|V|^3|E| \log |V|)$. We omit the details for brevity.

2-hop optimization. The brute-force method, as well as all other methods we present here, can be improved by 2-hop optimization. Instead of arbitrary pairs of nodes, the 2-hop optimized version only considers u and v for a potential merger if they are exactly two hops from each other. Since 2-hop neighbors have a shared neighbor that can be linked to the merged supernode with a single superedge, some compression may result. The 2-hop optimization is safe in the sense that any merger by Algorithm 1 that compresses the graph involves 2-hop neighbors.

The time saving by 2-hop optimization can be significant: for the brute-force method, for instance, there are approximately $O(deg\,|E|)$ feasible node pairs with the optimization, where *deg* is the average degree, instead of the $O(|V|^2)$ pairs in the unoptimized algorithm.

For the randomized methods below, a straight-forward implementation of 2-hop optimization by random walk has a nice property. Assume that one node has been chosen, then find a random pair for it by taking two consequtive random hops starting from the first node. Now 2-hop neighbors with many shared neighbors are more likely to get picked, since there are several 2-hop paths to them, and a merger between nodes with many shared neighbors will lead to better compression. A uniform selection among all 2-hop neighbors does not have this property.

Thresholded algorithm. We next propose a more practical algorithmic alternative, the thresholded method (Algorithm 3). It iterates over all pairs of nodes and merges all pairs (u, v) such that $d_{\max}(u, v; S) \leq T_i$ (Lines 5–6). The threshold value T_i is increased iteratively in a heuristic manner whenever no mergers can be done with the current threshold (Lines 2 and 4).

Algorithm 3. Thresholded algorithm

1: **for all** $0 \leq i \leq K$ **do**
2: $T_i \leftarrow 2^{-K+i}$
3: $S \leftarrow G$ {i.e., $(V', E', w') \leftarrow (V, E, w)$}
4: **for all** $i = 0, \ldots, K$ **do**
5: **while** there exists a pair $\{u, v\} \in V' \times V'$ such that $d_{\max}(u, v; S) \leq T_i$ **do** {(*)}
6: $S \leftarrow merge(u, v, S)$
7: **if** $cr(S) \leq cr$ **then**
8: **return** S
(*) 2-hop optimization can be used, see text.

Different schemes for setting the thresholds would give different results and time complexity. The heuristic we have used has $K = 20$ exponentially growing steps and aims to produce relatively high-quality results faster than the brute-force method. Increasing the threshold in larger steps would give a faster method, but eventually a random compression (cf. Algorithm 5 below). We will give better informed, faster methods below.

The time complexity is $O(|V|^4)$ for the simple and $O(|V|^4 + |V|^2|E| \log |V|)$ for the generalized problem. These are upper bounds for highly improbable worst cases, and in practice the algorithm is much faster. See experiments in Section 5 for details on real world performance.

Randomized semi-greedy algorithm. The next algorithm is half random, half greedy (Algorithm 4). In each iteration, it first picks a node v at random (Line 3). Then it chooses node u so that the merge of u and v is optimal with respect to $d_{\max}(u, v; S)$ (Line 6). This algorithm, with 2-hop optimization, is a generalized version of the randomized algorithm of Navlakha et al. [3].

The worst-case time complexity of the algorithm is $O(|V|^3)$ for the simple and $O(|V|^2|E| \log |V|)$ for the generalized problem.

Random pairwise compression. Finally, we present a naive, random method which simply merges pairs of nodes at random without any aim to produce

Algorithm 4. Randomized semi-greedy algorithm

1: $S \leftarrow G$ {i.e., $(V', E', w') \leftarrow (V, E, w)$}
2: **while** $cr(S) > cr$ **do**
3: randomly choose $v \in V'$
4: **for all** nodes $u \in V'$ **do** {(*)}
5: $d_u \leftarrow d_{\max}(v, u; S)$
6: $S \leftarrow merge(\arg\min_u d_u, v, S)$
7: **return** S
(*) 2-hop optimization can be used, see text.

Algorithm 5. Random pairwise compression

1: $S \leftarrow G$ {i.e., $(V', E', w') \leftarrow (V, E, w)$}
2: **while** $cr(S) > cr$ **do**
3: randomly choose $\{u, v\} \in V' \times V'$ {(*)}
4: $S \leftarrow merge(u, v, S)$
5: **return** S
(*) 2-hop optimization can be used, see text.

a good compression (Algorithm 5). The uninformed random method provides a baseline for the quality of other methods that make informed decisions about mergers.

The time complexity of the random algorithm is $O(|V|^2)$ for the simple and $O(|V||E| \log |V|)$ for the generalized problem. The random algorithm is essentially the fastest possible compression algorithm that uses pairwise mergers. It therefore provides a baseline (lower bound) for runtime comparisons.

Interactive compression. Thanks to the simple agglomerative structure of the methods, all of them lend themselves to interactive visualization of graphs where the abstraction level can be adjusted dynamically. This simply requires that the merge operations save the hierarchical composition of the supernodes produced. A drill-down operation then corresponds to backtracking merge operations, and a roll-up operation corresponds to mergers.

5 Experiments

We next present experimental results on the weighted graph compression problem using algorithms introduced in the previous section and real data sets. With these experiments we aim to address the following questions. (1) How well can weighted graphs be compressed: what is the trade-off between compression (lower number of edges) and distance to the original graph? (2) How do the different algorithms fare in this task: how good are the results they produce? (3) What are the running times of the algorithms? And, finally: (4) How does compression affect the use of the graph in clustering?

5.1 Experimental Setup

We extracted test graphs from the biological Biomine database[2] and from a co-authorship graph compiled from the DBLP computer science bibliography. Edge weights are in $[0, 1]$, and the path quality function is the product of weights of the edges in the path. Below we briefly describe how the datasets were obtained.

A set of 30 connection graphs, each consisting of around 1000 nodes and 2411 to 3802 edges (median 2987 edges; average node degree 2.94) was used in most of the tests. These graphs were obtained as connection graphs between three sets of related genes (different gene sets for each of the 30 replicates) so that they contain some non-trivial structure. We mostly report mean results over all 30 graphs.

A set of 30 smaller graphs was used for tests with the time-consuming brute-force method. These graphs have 50 nodes each and 76 to 132 edges (median 117 edges; average node degree 2.16).

Two series of increasingly larger graphs were used to compare the scalability of the methods. The sizes in one series range from 1000 to 5000 nodes and from 2000 to 17000 edges, and in the other series from 10 000 to 200 000 nodes and about 12 000 to 400 000 edges.

The algorithms were implemented in Java, and all the experiments were run on a standard PC with 4 GB of main memory and an Intel Core 2 Duo 3.16 GHz processor.

5.2 Results

Compressibility of weighted graphs. Figures 2a and 2b give the distance between the compressed and original graphs as a function of the compression ratio. For better interpretability, the distance is represented as the root mean square error (RMSE) over all pairs of nodes. Overall, the distances are small. Compression to half of the original size can be achieved with errors of 0.03 ($\lambda = 1$) or 0.06 ($\lambda = \infty$) per node pair. Especially for $\lambda = \infty$ graphs compress very nicely.

Comparison of algorithms. Figure 2c complements the comparison with results for the smaller graphs, and now including the brute-force method ($\lambda = 1$). The brute-force method clearly produces the best results (but is very slow as we will see shortly). Note also how small graphs are relatively harder to compress and the distances are larger than for the standard set of larger graphs.

The thresholded method is almost as good for compression ratios 0.8-0.9 but the gap grows a bit for smaller compression ratios. The semi-greedy version, on the other hand, is not as good with the larger compression ratios, but has a relatively good performance with smaller compression ratios. The random method is consistently the worst. A few early bad mergers already raise the distance for high compression ratios. Experiments on larger graphs could not be run with the brute force methods.

---

[2] `http://biomine.cs.helsinki.fi`

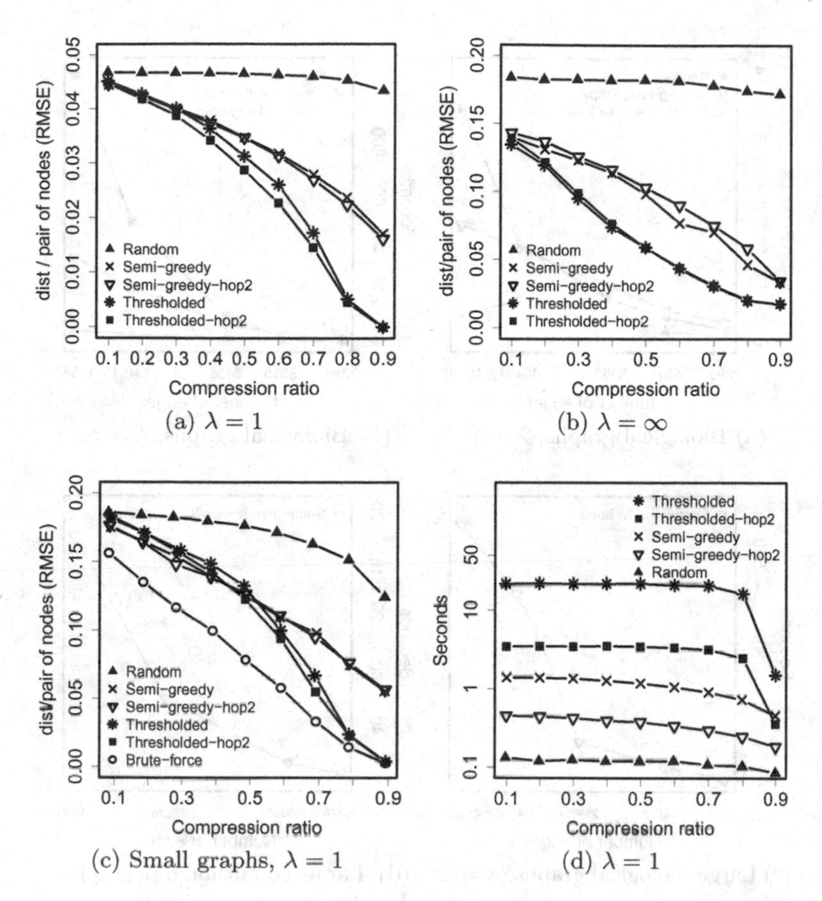

Fig. 2. (a)-(c): Distance between the compressed and the original graph as a function of compression ratio. (d): Running times of algorithms.

Efficiency of algorithms. Mean running times of the algorithms (except brute-force, see below) over the 30 standard graphs are shown in Figure 2d. The differences in the running times are big between the methods, more than two orders of magnitude between the extremes.

The 2-hop-optimized versions are an order of magnitude faster than the un-optimized versions while the results were equally good (cf. Figure 2c). 2-hop optimization thus very clearly pays off.

The brute-force method is very slow compared to the other methods (results not shown). Its running times for the small graphs were 1.5–5 seconds with $\lambda = 1$ where all other methods always finished within 0.4 seconds. With $\lambda = \infty$, the brute-force method spent 20–80 seconds whereas all other methods used less than 0.5 second.

Running times with $\lambda = \infty$ are larger than with $\lambda = 1$ by an order of magnitude, for the semi-greedy versions by two orders of magnitude (not shown).

We evaluated the effect of graph size on running times of the three fastest algorithm, using the series of increasingly large graphs and a fixed compression ratio 0.8

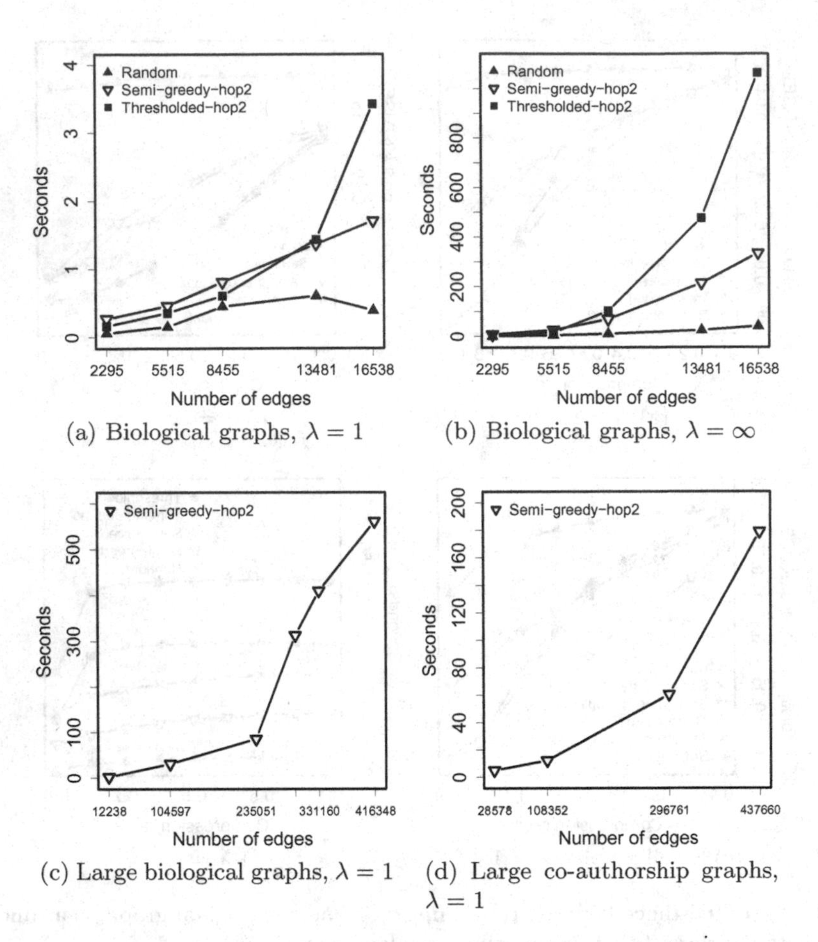

(a) Biological graphs, $\lambda = 1$

(b) Biological graphs, $\lambda = \infty$

(c) Large biological graphs, $\lambda = 1$

(d) Large co-authorship graphs, $\lambda = 1$

Fig. 3. Running times of weighted graph compression algorithms on graphs of various sizes from different sources

(Figures 3a and 3b). For $\lambda = 1$ the random method should be linear (and deviations are likely due to random effects). The thresholded method seems in practice approximately quadratic as is to be expected: for any value of λ, it will iterate over all pairs of nodes. The semi-greedy algorithm has a much more graceful behavior, even if slightly superlinear. Relative results are similar for $\lambda = \infty$.

Additional scalability experiments were run with larger graphs from both biological and co-authorship domains, using the semi-greedy algorithm with 2-hop optimization, compression ratio $cr = 0.8$, and $\lambda = 1$ (Figures 3c and 3d). The algorithm compressed graphs of upto 400000 edges in less than 10 minutes (biology) or in less than 3 minutes (co-authorship). The biological graphs contain nodes with high degrees, and this makes the compression algorithms slower.

Effect of compression on node clustering results. We next study how errors introduced by weighted graph compression affect methods that work on graphs. As a case study, we consider node clustering and measure the difference of clusters in the original graph vs. clusters in (the decompressed version of) the compressed graph.

We applied the k-medoids clustering algorithm on the 30 standard graphs. We set $k = 3$, corresponding to the three gene groups used to obtain the graphs. The proximity between two nodes was computed as the product of weights (probabilities) of edges on the best path. We measure the difference between clusterings by the Rand index (more exactly, by 1 minus Rand index). In other words, we measure the fraction of node pairs that are clustered inconsistently in the clusterings, i.e., assigned to the same cluster in one graph and to different clusters in the other graph.

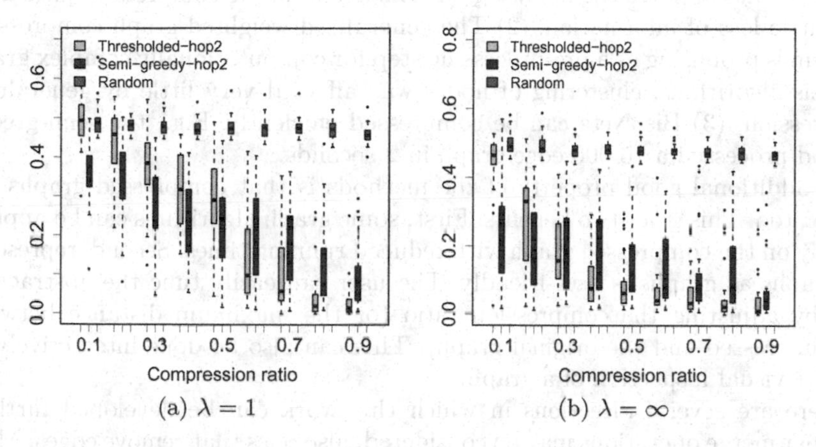

Fig. 4. Effect of compression on node clustering. Y-axis is the fraction of node pairs clustered inconsistently in the original and the compressed graph.

According to the results, the thresholded and semi-greedy compression methods can compress a weighted graph with little effect on node clustering (Figure 4). The effect is small especially when $\lambda = \infty$, where the inconsistency ratio is less than 0.1 (the thresholded method) or 0.3 (the semi-greedy method) for a wide range of compression ratios. The effects of the thresholded and semi-greedy versions are larger for $\lambda = 1$, especially when the compression ratio cr becomes smaller. This is because a clustering based solely on immediate neighborhoods is more sensitive to individual edge weights, whereas $Q_\infty(\cdot)$ can find a new best path elsewhere if an edge on the current best path is strongly changed.

Surprisingly, the semi-greedy method performs best in this comparison with compression ratio $cr \leq 0.2$. With $\lambda = \infty$ even an aggressive compression introduced relatively little changes to node clustering. In the other extreme, clusters found in randomly compressed graphs are quite—but not completely—different from the clusters found in the original graph. Close to 50% of pairs are clustered inconsistently, whereas a random clustering of three equally sized clusters would have about 2/3 inconsistency.

6 Conclusions

We have discussed the problem of compressing BisoNets or, in more general, weighted graphs. We derived bounds for it and gave algorithms and experimental results on real datasets. We presented two forms of the problem: a simple one, where the compressed graph should preserve edge weights, and a generalized one, where the compressed graph should preserve strengths of connections of up to λ hops. The generalized form may be valuable especially for graph analysis algorithms that rely more on strengths of connections than individual edge weights.

The results indicate the following. (1) BisoNets can be compressed quite a lot with little loss of information. (2) The generalized weighted graph compression problem is promising as a pre-processing step for computationally complex graph analysis algorithms: clustering of nodes was affected very little by generalized compression. (3) BisoNets can be compressed efficiently. E.g., the semi-greedy method processed a 16 000 edge graph in 2 seconds.

An additional good property of the methods is that compressed graphs are graphs, too. This gives two benefits. First, some graph algorithms can be applied directly on the compressed graph with reduced running times. Second, representing graphs as graphs is user-friendly. The user can easily tune the abstraction level by adjusting the compression ratio (or the maximum distance between the compressed and the original graph). This can also be done interactively to support visual inspection of a graph.

There are several directions in which this work can be developed further. Different merge operations may be considered, also ones that remove edges. More efficient algorithms can be developed for even better scalability to large graphs. It could be useful to modify the methods to guarantee a bounded edge-wise or node pair-wise error, or to also accommodate categorical labels.

Acknowledgements. This work has been supported by the Algorithmic Data Analysis (Algodan) Centre of Excellence of the Academy of Finland (Grant 118653) and by the European Commission under the 7th Framework Programme FP7-ICT-2007-C FET-Open, contract no. BISON-211898.

References

1. Toivonen, H., Zhou, F., Hartikainen, A., Hinkka, A.: Compression of weighted graphs. In: The 17th ACM SIGKDD Conference on Knowledge Discovery and Data Mining (KDD), San Diego, CA, USA (2011)
2. Kötter, T., Berthold, M.R.: From Information Networks to Bisociative Information Networks. In: Berthold, M.R. (ed.) Bisociative Knowledge Discovery. LNCS (LNAI), vol. 7250, pp. 33–50. Springer, Heidelberg (2012)

3. Navlakha, S., Rastogi, R., Shrivastava, N.: Graph summarization with bounded error. In: SIGMOD 2008: Proceedings of the 2008 ACM SIGMOD International Conference on Management of Data, pp. 419–432. ACM, New York (2008)
4. Tian, Y., Hankins, R., Patel, J.: Efficient aggregation for graph summarization. In: SIGMOD 2008: Proceedings of the 2008 ACM SIGMOD International Conference on Management of Data, pp. 567–580. ACM, New York (2008)
5. Lorrain, F., White, H.C.: Structural equivalence of individuals in social networks. Journal of Mathematical Sociology 1, 49–80 (1971)
6. Borgatti, S.P., Everett, M.G.: Regular blockmodels of multiway, multimode matrices. Social Networks 14, 91–120 (1992)
7. Zhang, N., Tian, Y., Patel, J.: Discovery-driven graph summarization. In: 2010 IEEE 26th International Conference on Data Engineering (ICDE), pp. 880–891. IEEE (2010)
8. Chen, C., Lin, C., Fredrikson, M., Christodorescu, M., Yan, X., Han, J.: Mining graph patterns efficiently via randomized summaries. In: 2009 Int. Conf. on Very Large Data Bases, Lyon, France, pp. 742–753. VLDB Endowment (August 2009)
9. Navlakha, S., Schatz, M., Kingsford, C.: Revealing biological modules via graph summarization. Presented at the RECOMB Systems Biology Satellite Conference; J. Comp. Bio. 16, 253–264 (2009)
10. Chen, C., Yan, X., Zhu, F., Han, J., Yu, P.: Graph OLAP: Towards online analytical processing on graphs. In: ICDM 2008: Proceedings of the 2008 Eighth IEEE International Conference on Data Mining, pp. 103–112. IEEE Computer Society, Washington, DC (2008)
11. Fjällström, P.O.: Algorithms for graph partitioning: A Survey. Linköping Electronic Atricles in Computer and Information Science, vol. 3 (1998)
12. Elsner, U.: Graph partitioning - a survey. Technical Report SFB393/97-27, Technische Universität Chemnitz (1997)
13. Faloutsos, C., McCurley, K.S., Tomkins, A.: Fast discovery of connection subgraphs. In: KDD 2004: Proceedings of the Tenth ACM SIGKDD International Conference on Knowledge Discovery and Data Mining, pp. 118–127. ACM, New York (2004)
14. Hintsanen, P., Toivonen, H.: Finding reliable subgraphs from large probabilistic graphs. Data Mining and Knowledge Discovery 17, 3–23 (2008)
15. Toussaint, G.T.: The relative neighbourhood graph of a finite planar set. Pattern Recognition 12(4), 261–268 (1980)
16. Hauguel, S., Zhai, C.X., Han, J.: Parallel PathFinder algorithms for mining structures from graphs. In: 2009 Ninth IEEE International Conference on Data Mining, pp. 812–817. IEEE (2009)
17. Toivonen, H., Mahler, S., Zhou, F.: A Framework for Path-Oriented Network Simplification. In: Cohen, P.R., Adams, N.M., Berthold, M.R. (eds.) IDA 2010. LNCS, vol. 6065, pp. 220–231. Springer, Heidelberg (2010)
18. Adler, M., Mitzenmacher, M.: Towards compressing web graphs. In: Data Compression Conference, pp. 203–212 (2001)
19. Boldi, P., Vigna, S.: The webgraph framework I: compression techniques. In: WWW 2004: Proceedings of the 13th International Conference on World Wide Web, pp. 595–602. ACM, New York (2004)
20. Zhou, F., Mahler, S., Toivonen, H.: Review of BisoNet Abstraction Techniques. In: Berthold, M.R. (ed.) Bisociative Knowledge Discovery. LNCS (LNAI), pp. 166–178. Springer, Heidelberg (2012)

Finding Representative Nodes in Probabilistic Graphs

Laura Langohr and Hannu Toivonen

Department of Computer Science and
Helsinki Institute for Information Technology HIIT,
University of Helsinki, Finland
{laura.langohr,hannu.toivonen}@cs.helsinki.fi

Abstract. We introduce the problem of identifying representative nodes in probabilistic graphs, motivated by the need to produce different simple views to large BisoNets. We define a probabilistic similarity measure for nodes, and then apply clustering methods to find groups of nodes. Finally, a representative is output from each cluster. We report on experiments with real biomedical data, using both the k-medoids and hierarchical clustering methods in the clustering step. The results suggest that the clustering based approaches are capable of finding a representative set of nodes.

1 Introduction

Bisociative information networks (BisoNets) allow integration and analytical use of information from various sources [1]. However, information contained in large BisoNets is difficult to view and handle by users. The problem is obvious for BisoNets of hundreds of nodes, but the problems start already with dozens of nodes.

In this chapter, we propose identification of a few representative nodes as one approach to help users make sense of large BisoNets. As an example scenario of the approach, consider link discovery. Given a large number of predicted links, it would be useful to present only a small number of representative ones to the user. Or, representatives could be used to abstract a large set of nodes, e.g., all nodes fulfilling some user-specified criteria of relevance, into a smaller but representative sample.

Our motivation for this problem comes from genetics, where current high-throughput techniques allow simultaneous analysis of very large sets of genes or proteins. Often, these wet lab techniques identify numerous genes (or proteins, or other biological components) as potentially interesting, e.g., by the statistical significance of their expression, or association with a phenotype (e.g., disease). Finding representative genes among the potentially interesting ones would be useful in several ways. First, it can be used to remove redundancy, when several genes are closely related and showing all of them adds no value. Second, representatives might be helpful in identifying complementary or alternative components in biological mechanisms.

M.R. Berthold (Ed.): Bisociative Knowledge Discovery, LNAI 7250, pp. 218–229, 2012.

The BisoNet in our application is Biomine [2], an integrated network database currently consisting of about 1 million biological concepts and about 10 million links between them. Concepts include genes, proteins, biological processes, cellular components, molecular functions, phenotypes, articles, etc.; weighted links mostly describe their known relationships. The data originates from well known public databases such as Entrez[1], GO[2], and OMIM[3].

The problem thus is to identify few representative nodes among a set of them, in a given weighted network. The solutions proposed in this chapter are based on defining a probabilistic similarity measure for nodes, then using clustering to group nodes, and finally selecting a representative from each cluster.

In this framework, two design decisions need to be made: how to measure similarities or distances of nodes in a probabilistic network (Section 3), and which clustering method to use on the nodes (Section 4). Experimental results with real datasets are reported in Section 5, and we conclude in Section 6 with some notes about the results and future work.

2 Related Work

Representatives are used to reduce the number of objects in different applications. In an opposite direction to our work, clustering can be approximated by finding representative objects, clustering them, and assigning the remaining objects to the clusters of their representatives. Yan et al. [3] use k-means or RP trees to find representative points, Kaufman and Rousseeuw [4] k-medoids, and Ester et al. [5] the most central object of a data page.

Representatives are also used to reduce the number of datapoints in large databases, i.e., to eliminate irrelevant and redundant examples in databases to be tested by data mining algorithms. Riquelme et al. [6] use ordered projections to find representative patterns, Rozsypal and Kubat [7] genetic algorithms, and Pan et al. [8] measure the representativeness of a set with mutual information and relative entropy.

DeLucia and Obraczaka [9] as well as Liang et al. [10] use representative receivers to limit receiver feedback. Only representatives provide feedback and suppress feedback from the other group members. Representatives are found by utilizing positive and negative acknowledgments in such a way that each congested subtree is represented by one representative.

The cluster approximation and example reduction methods use clustering algorithms to find representatives, but are not applied on graphs. The feedback limitation methods again use graph structures, but not clustering to find representatives. Other applications like viral marketing [11], center-piece subgraphs [12], or PageRank [13] search for special node(s) in graphs, but not for representative nodes. The authors are not aware of approaches to find representatives by clustering nodes and utilizing the graph structure.

---

[1] www.ncbi.nlm.nih.gov/Entrez/

[2] www.geneontology.org/

[3] www.ncbi.nlm.nih.gov/omim/

3 Similarities in Probabilistic Graphs

Probabilistic graphs offer a simple yet powerful framework for modeling relationships in weighted networks. A probabilistic graph is simply a weighted graph $G = (V, E)$ where the weight associated with an edge $e \in E$ is probability $p(e)$ (or can be transformed to a probability). The interpretation is that edge e exists with probability $p(e)$, and conversely e does not exist, or is not true, with probability $1 - p(e)$. Edges are assumed mutually independent.

The probabilistic interpretation of edge weights $p(e)$ gives natural measures for indirect relationships between nodes. In this chapter we call these similarity measures, as is conventional in the context of clustering.

Probability of a Path. Given a path P consisting of edges $e_1, \ldots, e_k$, the probability $p(P)$ of the path is the product $p(e_1) \cdot \ldots \cdot p(e_k)$. This corresponds to the probability that the path exists, i.e., that all of its edges exist.

Probability of the Best Path. Given two nodes $u, v \in V$, a measure of their connectedness or similarity is the probability of the best path connecting them:

$$s(u,v) = \max_{P \text{ is a path from } u \text{ to } v} p(P).$$

Obviously, this is not necessarily the path with the least number of edges. This similarity function $s(\cdot)$ is our choice for finding representatives.

Network Reliability. Given two nodes s and t, an alternative measure of their connectivity is the probability that there exists at least one path (not necessarily the best one) between s and t. This measure is known as the (two-terminal) network reliability (see, e.g., [14]). A classical application of reliability is in communication networks, where each communication link (edge) may fail with some probability. The reliability then gives the probability that s and t can reach each other in the network.

Network reliability is potentially a more powerful measure of connectedness than the probability of the best path, since reliability uses more information — not only the best path. The reliability measure considers alternative paths between s and t as independent evidence for their connectivity, and in effect rewards for such parallelism, while penalizing long paths. The reliability is always at least as high as the probability of the best path, but can also be considerably higher.

However, computing the two-terminal network reliability has been shown to be NP-hard [15]. Fortunately, the probability can be estimated, for instance, by using a straightforward Monte Carlo approach: generate a large number of realizations of the random graph and count the relative frequency of graphs where a path from s to t exists. For very large graphs, we would first extract a smaller neighborhood of s and t, and perform the computation there. These techniques are described in more detail, e.g., by Sevon et al. [2] and Hintsanen and Toivonen [16]. Due to the complexity of computing the network reliability, we stick to the simpler definition of similarity $s(\cdot)$ as the probability of the best path.

4 Clustering and Representatives in Graphs

Our approach to finding representatives in networks is to cluster the given nodes, using the similarity measure defined above, and then select one representative from each cluster (Algorithm 1). The aim is to have representatives that are similar to the nodes they represent (i.e., to other members of the cluster), and also to have diverse representatives (from different clusters). In clustering, we experiment with two methods: k-medoids and hierarchical clustering. Both are well-known and widely used methods which can be applied to our problem of finding representatives; k-medoids is an obvious choice, since it directly produces representatives.

Algorithm 1. Find representative nodes

Input: Set S of nodes, graph G, number k of representatives
Output: k representative nodes from S
1: Find k clusters of nodes in S using similarities $s(\cdot)$ in graph G
2: For each of the k clusters, output its most central node (the node with the maximum similarity to other nodes in the cluster)

---

k-medoids. k-medoids is similar to the better known k-means method, but better suited for clustering nodes in a graph. Given k, the number of clusters to be constructed, the k-medoids method iteratively chooses cluster centers (medoids) and assigns all nodes to the cluster identified by the nearest medoid. The difference to the k-means clustering method is that instead of using the mean value of the objects within a cluster as cluster center, k-medoids uses the best object as a cluster center. This is a practical necessity when working with graphs, since there is no well defined mean for a set of nodes. The k-medoids method also immediately gives the representatives. The method is described in more detail, e.g., by Han and Kamber [17] and Kaufman and Rousseeuw [4].

For very large graphs, a straight forward implementation of k-medoids is not necessarily the most efficient. In our applications we use the Biomine database and tools to facilitate faster clustering. Given a set S of nodes, i.e., biological entities, to be clustered, and k, the number of clusters to be constructed, the method proceeds as follows. First, the Biomine system is queried for a graph G of at most 1000 nodes cross-connecting nodes in S as strongly as possible. The pairwise similarities between nodes are then calculated as the best path probabilities in G.

The Biomine system uses a heuristic to obtain G, details are omitted here. As the Biomine network consists of a million nodes, querying it for a graph exceeds by far the computational complexity of running k-medoids on the extracted graph. For brevity, we here omit discussion of the computational complexities of k-medoids and other approaches.

To start the actual clustering, k nodes from S are chosen randomly as initial medoids. Each remaining node in S is then clustered to the most similar medoid. If the pairwise similarity between a node and all medoids equals zero, the node will be considered an outlier and is not assigned to any medoid in this iteration. Then, a new medoid is calculated for each cluster. The node that has a maximal product of similarities between each other node in the cluster and itself is chosen as the new medoid. The last two steps are then repeated until the clustering converges or the maximum number of iterations is reached.

Example. As an example, k-medoids was run with $k = 3$ and a set of nine genes. The genes belong to three known groups, each group of three genes being associated to the same phenotype. The three OMIM phenotypes used in the example are a pigmentation phenotype (MIM:227220), lactase persistence (MIM:223100), and Alzheimer disease (MIM: 104300).

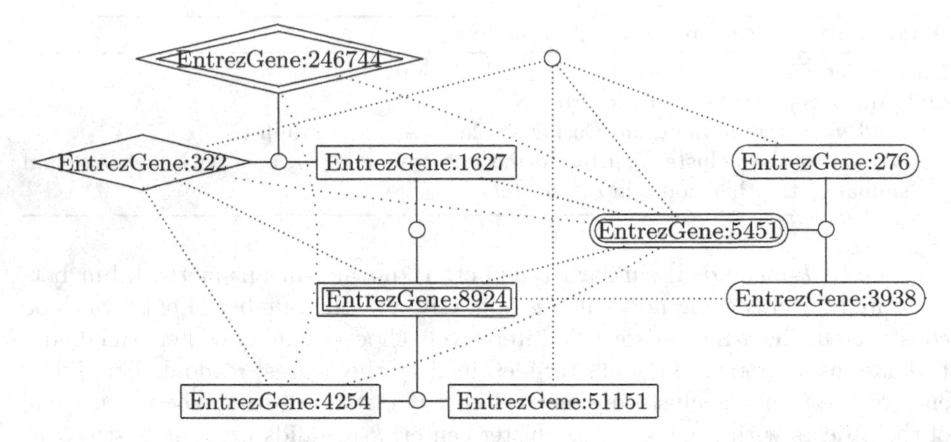

Fig. 1. Clusters (diamonds, boxes, ellipses) and representatives (double borders) of nine given nodes, and some connecting nodes (circles) on best paths between them. Lines represent edges between two nodes, dotted lines represent best paths with several nodes.

The algorithm converged in this case after two iterations. The result of the example run is shown in Figure 1. Looking at the quality of clustering, only one gene (EntrezGene:1627) was assigned to another cluster than it should with respect to the OMIM phenotypes. Apart from this gene, the clustering produced the expected partitioning: each gene was assigned to a cluster close to its corresponding phenotype. The three representatives (medoids) are genes assigned to different phenotypes. Hence, the medoids can be considered representative for the nine genes.

Hierarchical Clustering. As an alternative clustering method we use hierarchical clustering. Hierarchical clustering is a greedy clustering method that iteratively merges pairs of clusters. (Again, see, e.g., Han and Kamber [17] or Kaufman and Rousseeuw [4] for more information.) A possible problem with the k-medoids approach is that it may discover star-shaped clusters, where cluster members are connected mainly through the medoid. To give more weight on cluster coherence, we use the hierarchical clustering method with average linkage, as follows.

In the practical implementation, we again start by querying the Biomine system for a graph G of at most 1000 nodes connecting the given nodes S, and compute similarities of nodes in S as the probabilities of the best paths connecting them in G.

The hierarchical clustering proceeds in the standard, iterative manner, starting with having each node in a cluster of its own. In each iteration, those two clusters are merged that give the best merged cluster as a result, measured by the average similarity of nodes in the merged cluster. The clustering is finished when exactly k clusters remain.

After the clusters have been identified, we find the medoid in each cluster (as in the k-medoids method) and output them as representatives.

Random Selection of Representatives. For experimental evaluation, we also consider a method that selects representatives randomly. We again query the Biomine system for a graph G of at most 1000 nodes connecting the given nodes S, and compute similarities of nodes in S as the probabilities of the best paths connecting them in G.

We randomly select k medoids and cluster the remaining nodes of S to the most similar medoid. If the pairwise similarity between a node and all medoids equals zero, the node will be considered an outlier, as in k-medoids.

5 Experiments

Our goal in this section is to evaluate how successful the method is in finding representative nodes.

5.1 Test Setting

Test Data. We used data published by Köhler et al. [18], who defined 110 disease-gene families based on the OMIM database. The families contain three to 41 genes each; each family is related to one disease. Köhler et al. originally used the families in their experiments on candidate gene prioritization. Given a list of candidate genes they used a protein-protein interaction network to score the given genes by distance to all genes that are known to be related to a particular disease. Then they set up a ranking of the candidate genes based on their scores. Although their aim was different from ours, and the network they used was only a protein interaction network, the data sets also give a real test case for our problem.

Test Setting. In each test run, k gene families were randomly chosen as the nodes to find k representatives for. We performed 100 test runs for $k = 3$ and $k = 10$ of all three variants (k-medoids, hierarchical, random) of the method, and report averages over the 100 runs. As k-medoids is sensitive to the randomly selected first medoids, we applied k-medoids five times in each run and selected the best result. We applied the random selection of representatives 20 times in each run and used average values of the measures in order to compensate the random variation.

Measures of Representativeness. We use two measures of representativeness of the selected nodes. The first one is based on the similarity of nodes to their representatives, the second one on how well the k (known) families of nodes are covered by the k representatives.

The first measure is directly related to the objective of the k-medoids method. The idea is that each node is represented by its nearest representative, and we simply measure the average similarity of objects to their closest representative (ASR):

$$ASR = \frac{1}{|S| - K} \sum_{x \in S, x \neq m(x)} s(x, m(x))$$

where S is the set of given vertices, K is the number of clusters, $m(x)$ is the medoid most similar to x, and $s(x, m(x))$ denotes the similarity (probability of best path) between node x and medoid $m(x)$.

The second measure takes advantage of the known families of genes in our test setting. The rationale here is that a representation is better if it covers more families, i.e., contains a representative in more families. For this purpose, we calculate the fraction of non-represented classes (NRC):

$$NRC = \frac{1}{K} |\{k \mid \nexists j : m_j \in H_k,\ j = 1..K\}|,$$

where K is the number of classes and clusters (equal in our current test setting), m_j is the medoid of the jth cluster, and H_k is the kth original class.

For the k-medoids variant, we also report the number of outliers. Recall that the method outputs as outliers those nodes that are not connected (in the extracted subnetwork) to any medoid.

As additional characteristics of the methods we measure how good the underlying clusterings are. Again, we have two measures, one for the compactness of clusters, and one based on the known classification.

The first additional measure is the average compactness of clusters (ACC), where the compactness of a given cluster is defined as the minimum similarity of two objects in the cluster. The average is computed over clusters having at least two members:

$$ACC = \frac{1}{k'} \sum_{k=1}^{K} \min_{x,y \in C_k} s(x, y), \text{ where } k' = |\{k \mid |C_k| > 1, k = 1..K\}|,$$

i.e., k' is the number on non-trivial clusters. This measure is sensitive to outliers, and thus may favor the k-medoids variant.

The second additional measure compares the clustering to the known classes and measures their difference. We first identify the class best represented by each cluster, and then calculate how many objects were "wrongly assigned" (WAO):

$$WAO = \frac{1}{|S|} \sum_{k=1}^{K} \min_{k'=1..K} |C_k \backslash H_{k'}|.$$

Rand index could have been used here just as well.

5.2 Results

In terms of average similarity of nodes to their representative (ASR), the k-medoids method slightly but clearly outperforms the hierarchical method (Figure 2, left panels). The hierarchical method, in turn, is clearly superior to the random selection of representatives (Figure 2, right panels). For the k-medoids variant and $k = 3$, average similarities in the 100 test runs range from 0.3 to 0.8, and the total average is 0.51. For $k = 10$ the average is 0.55 and range is 0.4 to 0.8. For the hierarchical variant and $k = 3$, the average is 0.48 and range is 0.1 to 0.8. For $k = 10$ the average is 0.51 and range is 0.3 to 0.7. For the random variant and $k = 3$, average is 0.36 and range is 0.2 to 0.7. For $k = 10$ average is 0.43 and range is 0.3 to 0.6. These differences are no big surprise, since the k-medoids method more directly aims to maximize this measure than the hierarchical method, which however performs better than random choice of

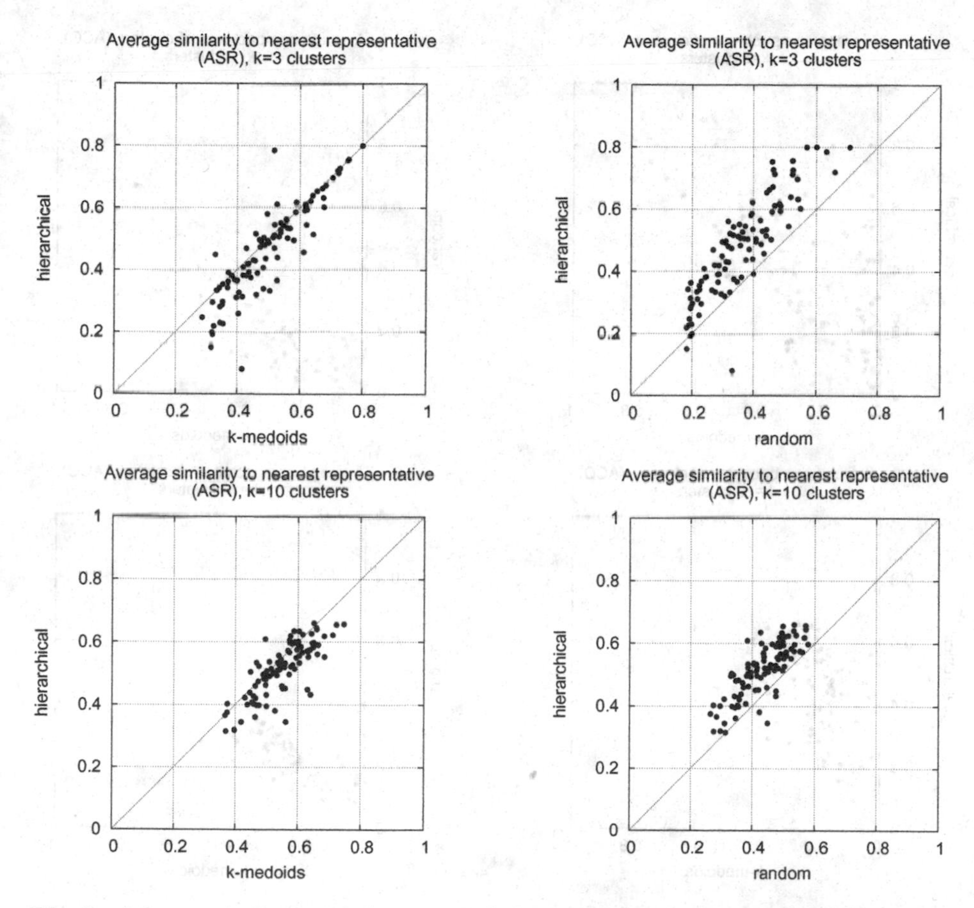

Fig. 2. Average similarity of objects to their nearest representative (ASR). In each panel 100 runs are visualized. Each point represents one run, thereby comparing ASR values of two variants (see x- and y-axis).

Table 1. Fraction of non-represented classes (NRC)

	k=3	k=10
k-medoids	14 %	29 %
hierarchical	16 %	21 %
random	34 %	39 %

representatives. Further, the k-medoids method may output some nodes as outliers. The average fraction of outliers in the experiments was 1.9 % for $k = 3$ and 4.5 % for $k = 10$.

The fraction of non-represented classes is a more neutral measure of performance since neither variant directly maximizes this. The results indicate that the k-medoids variant is slightly better with respect to this measure for $k = 3$

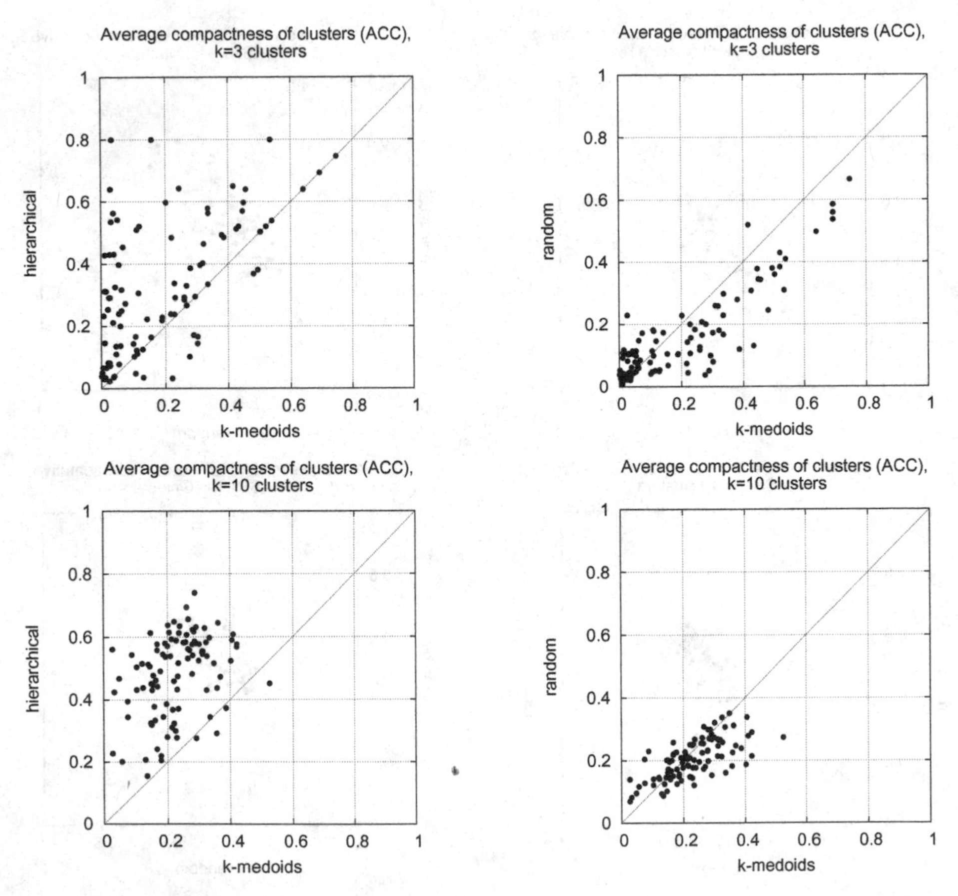

Fig. 3. Average compactness of nontrivial clusters (ACC). In each panel 100 runs are visualized. Each point represents one run, thereby comparing ACC values of two variants (see x- and y-axis).

(Table 1), but for $k = 10$ the hierarchical variant is clearly superior. Both methods clearly outperform the random selection of representatives.

To gain a better understanding of the performance of the methods, we look at the quality of clusterings produced. It is not surprising that clusters produced by the hierarchical method are on average more compact than those produced by the k-medoids method (Figure 3), as the hierarchical method more directly optimizes this measure. It is however somewhat surprising that k-medoids performs only slightly better than the random variant. The average compactness (minimum similarity within a cluster) is 0.20 ($k = 3$) and 0.23 ($k = 10$) for k-medoids, 0.33 ($k = 3$) and 0.48 ($k = 10$) for the hierarchical variant, and 0.16 ($k = 3$) and 0.21 ($k = 10$) for the random variant, with considerable spread and variance in all results.

In terms of wrongly assigned objects, the hierarchical variant clearly outperforms k-medoids (Table 2). The k-medoids variant outperforms the random selection of representatives, but for $k = 10$ only by a small difference.

Table 2. Wrongly assigned objects (WAO)

	k=3	k=10
k-medoids	18 %	44 %
hierarchical	15 %	25 %
random	27 %	46 %

6 Conclusions

We have described the problem of finding representative nodes in large probabilistic graphs. We based our definition of node similarity on a simple probabilistic interpretation of edge weights. We then gave a clustering-based method for identifying representatives, with two variants: one based on the k-medoids methods, one on the hierarchical clustering approach.

We performed a series of 100 experiments on real biomedical data, using published gene families [18] and the integrated Biomine network [2]. We measured the success of finding representatives with two measures: the similarity of nodes to their representatives, and the fraction of classes represented by the output.

In our experimental comparison, the k-medoids based variant and the hierarchical method are promising approaches. A look at the quality of the clusterings indicates that the success of the methods in identifying the underlying clusters depends on the measure used, and may also depend on the number of clusters to be constructed. According to the results, the hierarchical method is more robust, especially when looking for more than just a couple of representatives.

More work is needed to understand the reasons for the differences of the two approaches. Further, the problem of finding representative nodes needs to be validated in real applications. Based on the simple methods introduced here, and the initial experimental results, the clustering approach seems to be capable of reliably identifying a high quality set of representatives.

Acknowledgements. We would like to thank Lauri Eronen for his help with the test data and hierarchical clustering.

This work has been supported by the Algorithmic Data Analysis (Algodan) Centre of Excellence of the Academy of Finland and by the European Commission under the 7th Framework Programme FP7-ICT-2007-C FET-Open, contract no. BISON-211898.

References

1. Kötter, T., Berthold, M.R.: From Information Networks to Bisociative Information Networks. In: Berthold, M.R. (ed.) Bisociative Knowledge Discovery. LNCS (LNAI), vol. 7250, pp. 33–50. Springer, Heidelberg (2012)
2. Sevon, P., Eronen, L., Hintsanen, P., Kulovesi, K., Toivonen, H.: Link Discovery in Graphs Derived from Biological Databases. In: Leser, U., Naumann, F., Eckman, B. (eds.) DILS 2006. LNCS (LNBI), vol. 4075, pp. 35–49. Springer, Heidelberg (2006)
3. Yan, D., Huang, L., Jordan, M.I.: Fast approximate spectral clustering. In: 15th ACM SIGKDD International Conference on Knowledge Discovery and Data Mining (KDD 2009), pp. 907–916. ACM, New York (2009)
4. Kaufman, L., Rousseeuw, P.: Finding Groups in Data: An Introduction to Cluster Analysis. John Wiley Inc., New York (1990)
5. Ester, M., Kriegel, H.P., Xu, X.: Knowledge discovery in large spatial databases: focusing techniques for efficient class identification. In: 4th International Symposium on Advances in Spatial Databases (SDD 1995), pp. 67–82. Springer, London (1995)
6. Riquelme, J.C., Aguilar-Ruiz, J.S., Toro, M.: Finding representative patterns with ordered projections. Pattern Recognition 36(4), 1009–1018 (2003)
7. Rozsypal, A., Kubat, M.: Selecting representative examples and attributes by a genetic algorithm. Intelligent Data Analysis 7(4), 291–304 (2003)
8. Pan, F., Wang, W., Tung, A.K.H., Yang, J.: Finding representative set from massive data. In: The 5th IEEE International Conference on Data Mining (ICDM 2005), pp. 338–345. IEEE Computer Society, Washington, DC (2005)
9. DeLucia, D., Obraczka, K.: Multicast feedback suppression using representatives. In: 16th Annual Joint Conference of the IEEE Computer and Communications Societies. Driving the Information Revolution (INFOCOM 1997), pp. 463–470. IEEE Computer Society, Washington, DC (1997)
10. Liang, C., Hock, N.C., Liren, Z.: Selection of representatives for feedback suppression in reliable multicast protocols. Electronics Letters 37(1), 23–25 (2001)
11. Domingos, P., Richardson, M.: Mining the network value of customers. In: 7th ACM SIGKDD International Conference on Knowledge Discovery and Data Mining (KDD 2001), pp. 57–66. ACM, New York (2001)
12. Tong, H., Faloutsos, C.: Center-piece subgraphs: problem definition and fast solutions. In: 12th ACM SIGKDD International Conference on Knowledge Discovery and Data Mining (KDD 2006), pp. 404–413. ACM, New York (2006)

13. Page, L., Brin, S., Motwani, R., Winograd, T.: The PageRank citation ranking: bringing order to the web. Technical report, Stanford Digital Library Technologies Project (1999)
14. Colbourn, C.J.: The Combinatiorics of Network Reliability. Oxford University Press (1987)
15. Valiant, L.: The complexity of enumeration and reliability problems. SIAM Journal on Computing 8, 410–421 (1979)
16. Hintsanen, P., Toivonen, H.: Finding reliable subgraphs from large probabilistic graphs. Data Mining and Knowledge Discovery 17(1), 3–23 (2008)
17. Han, J., Kamber, M.: Data Mining. Concepts and Techniques, 2nd edn. Morgan Kaufmann (2006)
18. Köhler, S., Bauer, S., Horn, D., Robinson, P.: Walking the interactome for prioritization of candidate disease genes. American Journal of Human Genetics 82(4), 949–958 (2008)

(Missing) Concept Discovery in Heterogeneous Information Networks

Tobias Kötter and Michael R. Berthold

Nycomed-Chair for Bioinformatics and Information Mining, University of Konstanz, 78484 Konstanz, Germany
Tobias.Koetter@uni-Konstanz.de

Abstract. This article proposes a new approach to extract existing (or detect missing) concepts from a loosely integrated collection of information units by means of concept graph detection. Thereby a concept graph defines a concept by a quasi bipartite sub-graph of a bigger network with the members of the concept as the first vertex partition and their shared aspects as the second vertex partition. Once the concepts have been extracted they can be used to create higher level representations of the data. Concept graphs further allow the discovery of missing concepts, which could lead to new insights by connecting seemingly unrelated information units.

1 Introduction

The amount of data to which researchers have access is increasing at a breathtaking pace. The available data stems from heterogeneous sources from diverse domains with varying semantics and of various quality. It is a big challenge to integrate and reason from such an amount of data. However by integrating data from diverse domains, relations can be discovered spanning multiple domains, leading to new insights and thus a better understanding of complex systems. In this article we use a network-based approach to integrate data from diverse domains of varying quality. The network consists of vertices that represent information units such as objects, ideas or emotions, whereas edges represent the relations between these information units.

Once the data has been merged into a unifying model it needs to be analyzed. In this article we describe an approach based on concept graphs to extract semantical information from loosely integrated information fragments. This approach was presented at the International Conference on Computational Creativity [6]. Concept graphs allow for the detection of existing concepts, which can be used to create an abstraction of the underlying data. They define a concept by a quasi bipartite sub-graph consisting of two vertex partitions. The first partition contains the members of the concept and the second partition the aspects they have in common. By providing a higher level view on the data the user might obtain a better insight into the integrated data and discover new relations across diverse domains that have been hidden in the noise of the integrated data.

M.R. Berthold (Ed.): Bisociative Knowledge Discovery, LNAI 7250, pp. 230–245, 2012.

Concept graphs also allow for the detection of domain bridging concepts [8] that connect information units from various domains. These domain bridging concepts support creative thinking by connecting seemingly unrelated information units from diverse domains.

Another advantage of concept graphs is that they enable information units to be detected that share common properties but to which no concept has yet been assigned. This might lead to the discovery of concepts that are missing in the data or to the detection of new concepts.

The rest of the chapter is organized as follows: in the next section we will briefly review Bisociative Information Networks [7], which we use for the integration of heterogeneous data sources from diverse domains. We move on to introduce concept graphs and describe their detection, and subsequently discuss the discovery of concept graphs in a real world data set and show some example graphs. Finally we draw conclusions from our discussion and provide an outlook on future work.

2 Bisociative Information Networks

Bisociative Information Networks (BisoNets) [3,7] provide a framework for the integration of semantically meaningful information but also loosely coupled information fragments from heterogeneous data sources. The term *bisociation* [5] was coined by Arthur Koestler in 1964 to indicate the "...joining of unrelated, often conflicting information in a new way...".

BisoNets are based on a k-partite graph structure, whereby the most trivial partitioning consists of two partitions ($k = 2$), with the first vertex set representing units of information and the second set representing the relations among information units. By representing relations as vertices BisoNets support the modeling of relationships among any number of members.

However the role of a vertex is not fixed in the data. Depending on the point of view a vertex can represent an information unit or a relation describing the connection between units of information. Members of a relation are connected by an edge with the vertex describing the relation they share. One example is the representation of documents and authors where documents as well as authors are represented as vertices. Depending on the point of view, a document might play the role of the relation describing authorship or might be a member in the relation of documents written by the same author.

The unified modeling of information units and relations as vertices has many advantages e.g. they both support the assigning of attributes such as different labels. However these attributes do not carry any semantic information. Edges can be further marked as directed to explicit model relationships that are only valid in one direction. Vertices can also be assigned to partitions to distinguish between different domains such as biology, chemistry, etc.

In contrast to ontologies, semantic networks or topic maps, relations are assigned a weight that describes the reliability of the connection. This means that BisoNets support the integration not only of facts but also of pieces of evidence.

Thus units of information and their relations can be extracted from various information sources such as existing databases, ontologies or semantical networks. But also semistructured and noisy data such as literature or biological experiments can be integrated in order to provide a much richer and broader description of the information units. By applying different mining algorithms to the same information source, diverse relations and units of information can be extracted with each mining algorithm representing an alternative view that might highlight a different aspect of the same data.

BisoNets focus solely on the information units and their relations and do not store all the more detailed data underneath the pieces of information. However vertices do reference the detailed data they stem from. This allows BisoNets to integrate huge amounts of data and still be able to show the data from which a vertex originates.

3 Concept Graphs

Once all the data has been integrated, it has to be analyzed in order to find valuable information. We propose a new method to automatically extract semantic information from the loosely integrated collection of information units by means of concept graph detection.

A *concept graph* represents a *concept* that stands for a mental symbol. A concept consists of its *members*, which do not only refer to materialized objects but also to ideas, activities or events, and their common *aspects*, which represent the properties the concept members share. In philosophy and psychology, the concept members are also known as the extension of a concept, which consists of the things to which the concept applies - whereby the aspects are known as the intension of a concept, consisting of the idea or the properties of the concept. An example could be a concept representing birds with specific birds such as eagles or sparrows as members, which in turn are related to their common aspects such as feather, wing, and beak.

Concept graphs base on the assumption that similar information units share more properties than dissimilar information units. Therefore the more similar two information units are, the more properties they share. This assumption bases on the family resemblance proposed by Wittgenstein [12], which states that objects that already share some properties are likely to share further common properties. The theory of basic objects in natural categories from Rosch et al. [9] is also bases on the family resemblance. Rosch et al. define a basic category as the category that carries the most information; the basic categories consist of properties that are mostly connected to the members of the category. Thus family resemblance as well as the basic categories speak in favor of the assumption that a concept does not only possess one property, but many. These properties describe the members of a concept and distinguish the members of a concept from non-members.

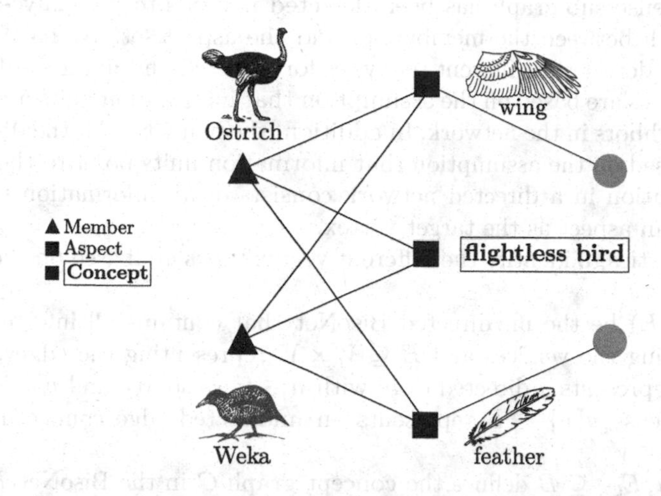

Fig. 1. Example of aconcept graph describing the concept *flightless bird* with its members *Ostrich* and *Weka* and their common aspects *wing* and *feather*

In addition to the concept members and their shared aspects, a concept graph might also contain the symbolic representation of the concept itself. This symbolic representation can be used to generate an abstract view on the data since it represents all members of the corresponding concept graph.

An example of a concept graph that represents the concept of *flightless birds* is depicted in Figure 1. It consists of the two concept members *Ostrich* and *Weka* and their shared aspects *wing* and *feather*. The graph also contains the symbolic representation of the *flightless bird* concept, which can be used as an abstract representation of this particular concept graph.

3.1 Preliminaries

As mentioned above the members of a concept graph are similar in that they share some aspects. In BisoNets the aspects of an information unit are represented by its direct neighbors. The more neighbors two information units share the more similar they are. This leads to the representation of a concept graph as a dense sub-graph in a BisoNet, consisting of two disjoint and fully connected vertex sets. Here the first vertex set represents the concept members and the second vertex set the aspects that are shared by all members of the concept graph. Thus a perfect concept graph would form a complete bipartite graph as depicted in Figure 1 with the concept members as the first partition and the aspects with the concept as the second partition. A concept graph might also contain relations among the vertices within a partition and thus does not necessarily form a perfect bipartite (sub) graph.

Once a dense sub-graph has been detected it needs to be analyzed in order to distinguish between the member set and the aspect set. We have developed heuristics to detect the different set types for directed and undirected networks. Both heuristics are based on the assumption that information units are described by their neighbors in the network. In addition, the heuristics for the directed network are based on the assumption that information units point to their aspects. Hence a relation in a directed network consists of an information unit as the source and an aspect as the target vertex.

The heuristics to identify the different vertex types are based on the following definitions:

Let $B(V, E)$ be the un/directed BisoNet that contains all information with V representing the vertices and $E \subseteq V \times V$ representing the edges. The edge $(u, v) \in E$ represents a directed edge with $u \in V$ as source and $v \in V$ as target vertex whereas $\{u, v\} \in E$ represents an undirected edge connecting the two vertices $u, v \in V$.

$C(V_A, V_M, E_C) \subseteq B$ defines the concept graph C in the BisoNet B. $V_A \subseteq V$ represents the aspect set and $V_M \subseteq V$ the member set of the concept graph C in which $V_A \cap V_M = \emptyset$. $V_C = V_A \cup V_M$ is the set of all vertices within the concept graph. $E_C \subseteq E$ is the subset of all edges that connect vertices within the concept graph $E_C = \{\{u, v\} \in E : u, v \in V_C\}$.

Let

$$N(v) = \{u \in V : \{v, u\} \in E\}$$

be the neighbors of the vertex $v \in V$ in the BisoNet B. Whereby

$$N^+(v) = \{u \in V : (v, u) \in E\}$$

denotes its target neighbors and

$$N^-(v) = \{u \in V : (u, v) \in E\}$$

its source neighbors.

The neighbors of a vertex $v \in V$ in a given vertex set $U \subseteq V$ are denoted by

$$N(v, U) = N(v) \cap U = \{u \in U : \{u, v\} \in E\}.$$

In the directed case

$$N^+(v, U) = N^+(v) \cap U = \{u \in U : (v, u) \in E\}$$

denotes the target neighbors and

$$N^-(v, U) = N^-(v) \cap U = \{u \in U : (u, v) \in E\}$$

the source neighbors of a given vertex $v \in V$ within a given vertex set $U \subseteq V$.

Member Set. The concept members form the first of the two disjoint vertex sets of the concept graph. The heuristic that denotes the probability of a vertex $v \in V_C$ to be part of the member set V_M is denoted by the function $\mathrm{m} : V_C \rightarrow [0, 1]$.

Given the set $V_C \subseteq V$ of all vertices within a concept graph C in an undirected network, $\mathrm{m}(v)$ is defined as the ratio of neighbors inside and outside the concept graph for a given vertex $v \in V_C$

$$\mathrm{m}(v) = \frac{|N(v, V_C)|}{|N(v)|}.$$

In a directed network the heuristic bases on the assumption that concept members point to their aspects. This assumption leads to the computation of the ratio of target neighbors inside and outside the concept graph for a given vertex $v \in V_C$

$$\mathrm{m}(v) = \frac{|N^+(v, V_C)|}{|N^+(v)|}.$$

The set of information units V_M for a given BisoNet B is defined as

$$V_M = \max_{V' \in V_C} \frac{1}{|V'|} \sum_{v \in V'} \mathrm{m}(v).$$

Aspect Set. The aspect set is the second vertex set of the concept graph that describes the members of the concept graph. Each aspect on its own might be related to other vertices as well but the set of aspects is only shared by the members of the concept graph. The vertices of the aspect set might differ considerably in the number of relations to vertices outside of the concept graph depending on their level of detail. More abstract aspects such as animals are likely to share more neighbors outside of the concept graph than more detailed aspects such as bird.

The heuristic that denotes the probability of a vertex $v \in V_C$ to belong to the aspect set V_A is denoted by the function $\mathrm{a} : V_C \rightarrow [0, 1]$.

Given the set $V_C \subseteq V$ of all vertices within a concept graph C in an undirected network, $\mathrm{a}(v)$ is defined as the inverse ratio of neighbors inside and outside the concept graph for a given vertex $v \in V_C$

$$\mathrm{a}(v) = 1 - \frac{|N(v, V_C)|}{|N(v)|} = 1 - \mathrm{m}(v).$$

In a directed network the heuristic is defined as the ratio of the source neighbors inside and outside the concept graph for a given vertex $v \in V_C$

$$\mathrm{a}(v) = \frac{|N^-(v, V_C)|}{|N^-(v)|}.$$

The set of aspects V_A for a given BisoNet B is defined as

$$V_A = \max_{V' \in V_C} \frac{1}{|V'|} \sum_{v \in V'} \mathrm{a}(v).$$

Concepts. The concept $c \in V_A$ is a vertex of the aspect set. A concept differs from the other vertices of the aspect set in that it should only be related to the vertices of the member set within the concept graph. Hence a perfect concept has no relations to vertices outside of the concept graph and can thus be used to represent the concept graph.

The heuristic $\mathrm{c} : V_A \rightarrow [0, 1]$ denotes the probability of a vertex being the concept that can represent concept graph C. The heuristic is based on the cue validity [2] which describes the relevance of an aspect for a given concept. More specific aspects have a higher cue validity than more general ones.

Given the set $V_C \subseteq V$ of all vertices within a concept graph C in an undirected network, the heuristic is defined as the ratio of the neighbors inside and outside the concept graph

$$\mathrm{c}(v) = \frac{|N(v, V_C)|}{|N(v)|} = \mathrm{m}(v), v \in V_A.$$

In a directed network the heuristic considers the ratio of the source neighbors inside and outside the concept graph

$$\mathrm{c}(v) = \frac{|N^-(v, V_C)|}{|N^-(v)|} = \mathrm{a}(v), v \in V_A.$$

The concept c that can represent the concept graph is the vertex $v \in V_A$ with the highest value for $\mathrm{c}(v)$

$$c = \max_{v \in V_A} \mathrm{c}(v).$$

Depending on a user-given threshold we are able to detect a concept graph without a concept. The concept graph lacks a concept if the concept value $\mathrm{c}(v)$ of all vertices of its aspect set is below the given threshold. This might be an indication of an unknown relation among information units that has not been discovered yet and to which no concept has been assigned.

3.2 Detection

In this chapter we use a frequent item set mining algorithm [1] to detect concept graphs in BisoNets. By using frequent item set algorithms we are able to detect concept graphs of different sizes and specificity.

Frequent item set mining has been developed for the analysis of market baskets in order to find sets of products that are frequently bought together. It operates on a transaction database that consists of a transaction identifier and the products that have been bought together in the transaction. Represented as a graph, the overlapping transactions form a complete bipartite graph, which is the basis of our concept graphs.

In order to apply frequent item set mining algorithms to find concept graphs in BisoNets we use the adjacency list of the network as transaction database. Therefore, for each vertex in the BisoNet, we create an entry in the transaction database with the vertex as the identifier and its direct neighbors as the products

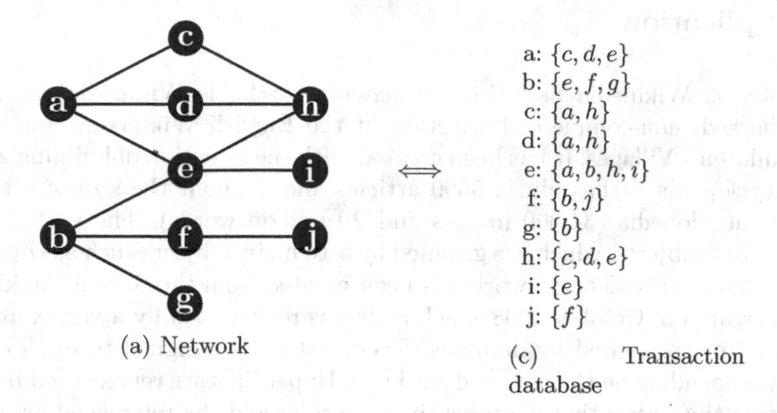

Fig. 2. Network and the corresponding adjacency list which serves as the transaction database for the frequent item set mining algorithm

(Figure 2). Once the database has been created we can apply frequent item set mining algorithms to detect vertices that share some neighbors.

Frequent item set mining algorithms allow the selection of a minimum support that defines the minimum number of transactions containing a given item set in order to make it frequent. They also allow a minimum size to be set for the item set itself in order to discard all item sets that contain fewer items than the given threshold. By setting these two thresholds we are able to define the minimum size of the concept graph.

Since we want to find concept graphs of different specificity we need an additional threshold that takes the general overlap of the transactions into account. To achieve this we used an adaption of the Eclat [13] algorithm called Jaccard Item Set Mining (JIM) [10]. JIM uses the Jaccard index [4] as an additional threshold for pruning the frequent item sets. For two arbitrary sets A and B the Jaccard index is defined as

$$\mathrm{j}(A,B) = \frac{|A \cap B|}{|A \cup B|}.$$

Obviously, $\mathrm{j}(A,B)$ is 1 if the sets coincide (i.e. $A = B$) and 0 if they are disjoint (i.e. $A \cap B = \emptyset$).

By setting the threshold for the JIM algorithm between 0 and 1 we are able to detect concept graphs of different specificity. By setting the threshold to 1 only those vertices that share all of their neighbors are retained by the algorithm. This results in the detection of more specific concept graphs, which contain information units or aspects that exclusively belong to the detected concept graph. Relaxing the threshold by setting a smaller value results in the detection of more general concept graphs where the information units share some but not all of their aspects. Varying thresholds might lead to the detection of overlapping concept graphs. This can be used to create a hierarchy among the concepts.

4 Application

The 2008/09 Wikipedia Selection for schools[1] (Schools Wikipedia) is a free, hand-checked, non-commercial selection of the English Wikipedia[2] funded by SOS Children's Villages. It has been created with the intention of building a child safe encyclopedia. It has about 5500 articles and is about the size of a twenty volume encyclopedia (34,000 images and 20 million words). The encyclopedia contains 154 subjects, which are grouped into 16 main subjects such as countries, religion and science. The network has been created from the Schools Wikipedia version created in October 2008. Each article is represented by a vertex and the subjects are represented by domains. Every article is assigned to one or more domains depending on the assigned subjects. Hyperlinks are represented by links connecting the article that contains the hyperlink and the referenced article.

This example data set and the representation as a hyperlink graph has been chosen since it can be validated manually by reading the Schools Wikipedia articles and inspecting their hyperlinks.

4.1 Results

This section illustrates concept graphs discovered in the Schools Wikipedia data set using the JIM algorithm. The concept graphs consist of the discovered item sets that form the first vertex set and the corresponding root vertices of the transaction that build the second vertex set. Once we have discovered both vertex sets and determined their types we can display them as a graph.

The following graphs display the information units with triangular vertices. Both aspects and the concept are represented by a squared vertex whereas the concept has a box around its label.

Figure 3 depicts such a discovered concept graph that represents the *dinosaur* concept from the biological section of the Schools Wikipedia. The members of the concept graph consist of the orders (e.g. *Ornithischia*) and genera (e.g. *Triceratops*) of the dinosaurs. The members are described by their aspects *Animal*, phylum *Chordate* and the reference to the *Biological classification* as well as the concept *Dinosaur* itself.

Detection and Expansion of Existing Hierarchies. This section demonstrates the ability of concept graphs to detect and expand existing hierarchies in the integrated data. Figure 4 depicts the *Saurischia* order (see Fig. 4a) and one of its suborders *Sauropodomorpha* (see Fig. 4b), which where discovered in the integrated data with the help of the detected concept graphs. These concept graphs benefit from the structure of the Schools Wikipedia pages of the animal section, as they include an information box with the Kingdom, Phylum etc. of the animal.

---
[1] `http://schools-wikipedia.org/`
[2] `http://en.wikipedia.org`

Fig. 3. *Dinosaur* concept graph with the orders and genera of the dinosaurs as members and their biological classification as their common aspects

Figure 5 depicts two different bird categories which were also extracted from the animal section of the Schools Wikipedia data set but which are not part of the standard information box of the corresponding Schools Wikipedia pages.

The concept graph in Figure 5a represents the group of *Waders*. *Waders* are long-legged wading birds such as *Herons*, *Flamingos* and *Plovers*. The concept graph also contains *Terns* even though they are only distantly related to *Waders*. However Schools Wikipedia states that studies in 2004 showed that some of the gene sequences of *Terns* showed a close relationship between *Terns* and the Thinocori, a species of aberrant *Waders*.

The concept graph in Figure 5b represents the *Bird of prey* group. *Birds of prey* or raptors hunt for food on the wing. The graph includes different sub families such as *Hawk*, *Kite* and *Falcon* as well as members of these sub families such as the *Harrier Hawk*. The *Common Cuckoo* is not a *Bird of prey* but is included in the concept graph since it looks like a small *Bird of prey* in flight as stated in its article in Schools Wikipedia. The concept graph contains the *Great Black-backed Gull* which is a gull that behaves more like a *Bird of prey* than a typical gull by frequently hunting any prey that is smaller than itself.

These examples partially benefit from the structure of the Schools Wikipedia pages of the animal section. They all contain an information box specifying the Kingdom, Phylum etc. of the animal. However this demonstrates that our method is able to discover ontologies, such as the biological classification of the dinosaurs (see Fig. 4), if they are available in the integrated data. Furthermore the examples demonstrate the capability of the method to detect further categories such as *Waders* or *Birds of prey* (see Fig. 5) even though they are not part

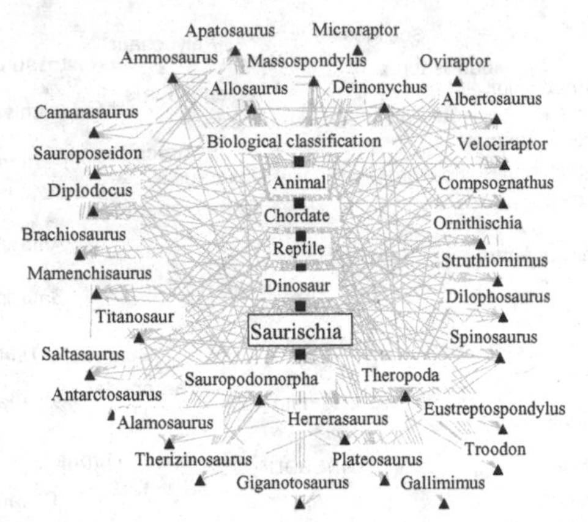

(a) Concept graph of the dinosaur order *Saurischia*.

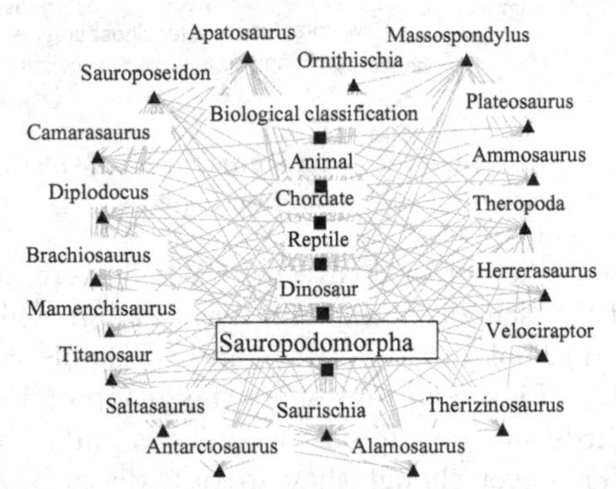

(b) Concept graph of the dinosaur suborder *Sauropodomorpha*.

Fig. 4. Concept graphs of the dinosaur order and one of its suborders

of the ontology structure in Schools Wikipedia. By including the *Great Black-backed Gull* and the *Common Cuckoo* the concept graph for the *Bird of prey* also demonstrates the ability to detect information units that are not typically related to the concept.

Missing Concept Detection. This section demonstrates the ability of concept graphs to detect groups of information units that share common aspects but to which no concept has been assigned. These concept graphs might be the result of incomplete or erroneous data. They might also be a hint of groups of information units that share certain aspects that have not been discovered yet.

The concept graph in Figure 6 is an example of such a concept graph, which lacks an appropriate concept. The graph describes battles between the *United States* and the *Imperial Japanese Navy* during *World War II*. The various battles represent the information units of the concept graph whereas the combatants and

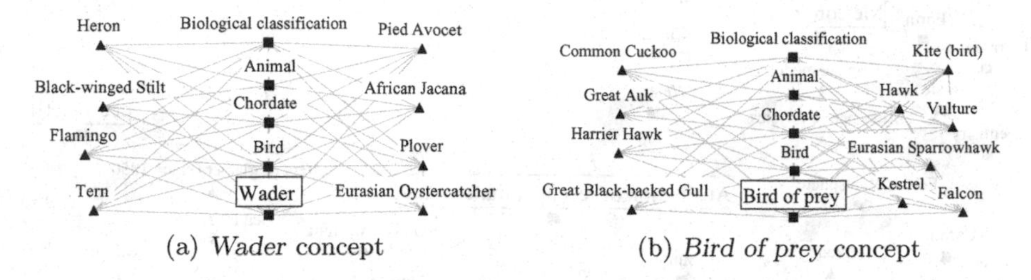

(a) *Wader* concept (b) *Bird of prey* concept

Fig. 5. Concept graphs that expand the existing birds hierarchy

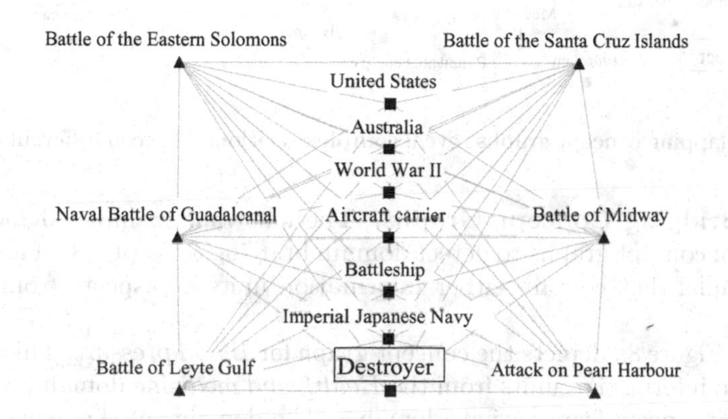

Fig. 6. Example concept graph with a low concept confidence

some of their warships form its aspects. The concept is missing since the Schools Wikipedia data set does not contain an article that groups the described battles.

The following concept graphs contain more aspects than information units. The information units represented by triangular vertices are therefore depicted in the center whereas the aspects and the concepts represented by squared vertices form the outer circle.

Overlapping Concept Graphs. This section describes concept graphs that overlap by sharing common vertices. These shared vertices might have different types since their type depends on the role they possess in a given concept graph. These vertices, which belong to several concept graphs and possess different types, are represented by circular vertices in the following figures.

Figure 7 depicts the connection among the concepts of *Meteorology*, *Greenhouse effect*, *Attribution of recent climate change* and the *Ice sheet* of *Glaciers*. The concept graphs demonstrate the ability of the discussed method to assign different types to the same vertex depending on a given concept graph. The vertex *Earth's atmosphere* as an example is an information unit of the *Greenhouse effect* but an aspect of the concept *Meteorology*. The vertices *Global warming* and *Climate change* are information units of the concept of *Attribution of recent climate change* but aspects of the *Ice sheet* concept.

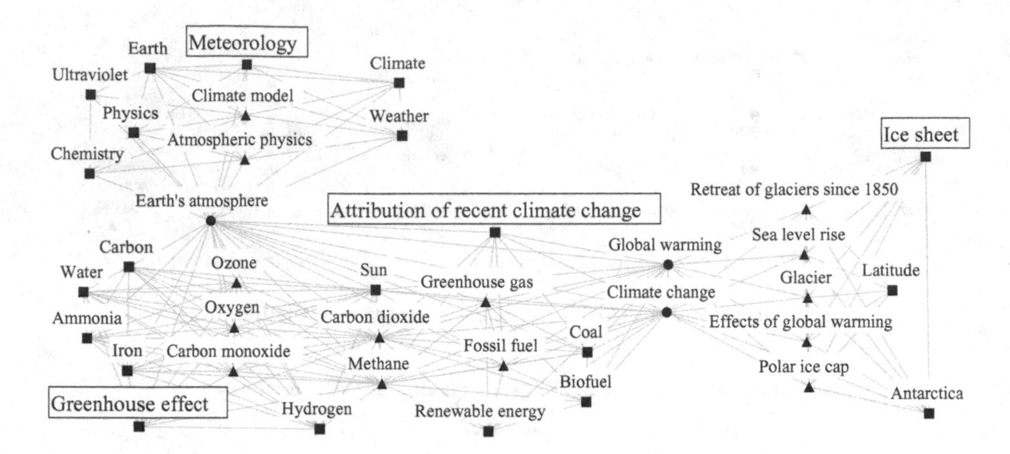

Fig. 7. Overlapping concept graphs reveal complex relations between different concepts

Domain Bridging Concept Graphs. The following examples demonstrate the ability of concept graphs to detect domain bridging concepts [8]. They depict concept graphs that contain either information units or aspects from diverse domains.

The first Figure 8a depicts the concept graph for *Blood pressure*. This concept contains two information units from the *Health and medicine* domain, which are described by aspects from various domains. The domains of the aspects range from *Health and medicine* (e.g. *Blood*), to *Chemical compounds* (e.g. *Glucose*), to *Sports* (e.g. *Sumo*) and *Recreation* (e.g. *Walking*) as well as *Plants* (e.g. *Garlic*).

The second Figure 8b groups information units from diverse domains such as *Cartoons* and *Military History and War*. The vertex *Donald Duck* represents a famous character from the *Cartoons* domain that was used as propaganda

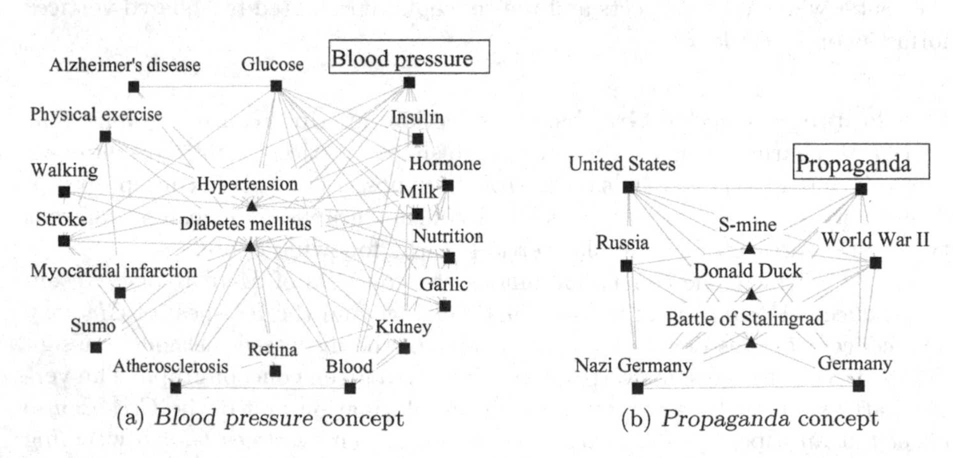

(a) *Blood pressure* concept (b) *Propaganda* concept

Fig. 8. Two domain bridging concept graphs that connect aspects and information units from different domains

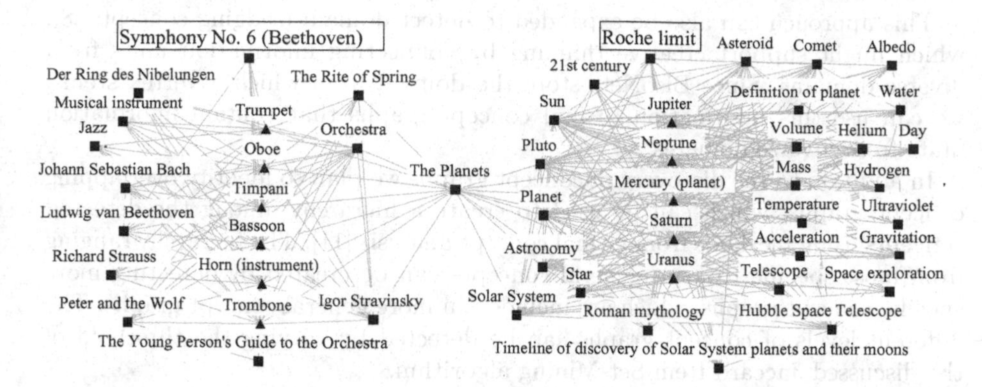

Fig. 9. Example of an overlapping concept graph that connects concepts from heterogeneous domains

whereas *S-mine* describes a weapon from the *Military History and War* domain that was used during *World War II*.

The last Figure 9 depicts not a single concept graph connecting vertices from diverse domains but two overlapping concept graphs that describe concepts from diverse domains. The first concept graph from the *Musical recordings and compositions* domain describes the concept of the *Symphony No. 6*, a famous symphony written by Ludwig van Beethoven. The second concept graph stems from the *Space(Astronomy)* domain describing the concept of the *Roche limit*, which defines the distance within a celestial body such as a planet held together by its own gravity alone. Both concepts are connected by their shared aspect, *The Planets*, which is an orchestral suite that consists of seven movements named after planets (e.g. *Jupiter*, *Neptune*, etc.).

5 Conclusion and Future work

In this chapter we have discussed a new approach to detect existing or missing concepts from a loosely integrated collection of information fragments that can lead to deeper insight into the underlying data. We have discussed concept graphs as a way to discover conceptual information in BisoNets. Concept graphs allow for the abstraction of the data by detecting existing concepts and producing a better overview of the integrated data. They further support the detection of missing concepts by discovering information units that share certain aspects but which have no concept and could be a hint of a previously unknown and potentially novel concept.

By using information networks as input data and retrieving existing as well as unknown concepts from the integrated data, the discussed approach supports creative thinking by improving the understanding of complex systems and the discovery of interesting and unexpected relationships.

This approach can also be expanded to detect domain bridging concepts [8], which might support creative thinking by connecting information units from diverse domains. Since BisoNets store the domain from which a vertex stems, we can use this information to find concept graphs that contain information units from diverse domains.

In addition to the discovery of concept graphs we plan to identify overlapping concept graphs, which can be used to create a hierarchy among the detected concepts using methods from formal concept analysis [11]. The hierarchy ranging from most specific to most general concepts can be created by detecting more specific concept graphs, which are included in more general concept graphs. The different levels of concept graphs can be detected by varying the threshold of the discussed Jaccard Item Set Mining algorithm.

References

1. Agrawal, R., Srikant, R.: Fast algorithms for mining association rules in large databases. In: Proceedings of the 20th International Conference on Very Large Data Bases (1994)
2. Beach, L.R.: Cue probabilism and inference behavior. Psychological Monographs: General and Applied 78, 1–20 (1964)
3. Berthold, M.R., Dill, F., Kötter, T., Thiel, K.: Supporting Creativity: Towards Associative Discovery of New Insights. In: Washio, T., Suzuki, E., Ting, K.M., Inokuchi, A. (eds.) PAKDD 2008. LNCS (LNAI), vol. 5012, pp. 14–25. Springer, Heidelberg (2008)
4. Jaccard, P.: Étude comparative de la distribution florale dans une portion des alpes et des jura. Bulletin de la Société Vaudoise des Sciences Naturells 37, 547–579 (1901)
5. Koestler, A.: The Act of Creation. Macmillan (1964)
6. Kötter, T., Berthold, M.R.: (Missing) concept discovery in heterogeneous information networks. In: Proceedings of the 2nd International Conference on Computational Creativity, pp. 135–140 (2011)
7. Kötter, T., Berthold, M.R.: From Information Networks to Bisociative Information Networks. In: Berthold, M.R. (ed.) Bisociative Knowledge Discovery. LNCS (LNAI), vol. 7250, pp. 33–50. Springer, Heidelberg (2012)
8. Kötter, T., Thiel, K., Berthold, M.R.: Domain bridging associations support creativity. In: Proceedings of the International Conference on Computational Creativity, pp. 200–204 (2010)
9. Rosch, E., Mervis, C.B., Gray, W.D., Johnson, D.M., Boyes-Braem, P.: Basic objects in natural categories. Cognitive Psychology 8, 382–439 (1976)
10. Segond, M., Borgelt, C.: Item Set Mining Based on Cover Similarity. In: Huang, J.Z., Cao, L., Srivastava, J. (eds.) PAKDD 2011, Part II. LNCS, vol. 6635, pp. 493–505. Springer, Heidelberg (2011)

11. Wille, R.: Restructuring lattice theory: An approach based on hierarchies of concepts. In: Ordered Sets, pp. 314–339 (1982)
12. Wittgenstein, L.: Philosophical Investigations. Blackwell, Oxford (1953)
13. Zaki, M.J., Parthasarathy, S., Ogihara, M., Li, W.: New algorithms for fast discovery of association rules. In: Proceedings of the 3rd International Conference on Knowledge Discovery and Data Mining (1997)

Node Similarities from Spreading Activation

Kilian Thiel and Michael R. Berthold

Nycomed Chair for Bioinformatics and Information Mining
Dept. of Computer and Information Science
University of Konstanz, Konstanz, Germany
Kilian.Thiel@Uni-Konstanz.DE

Abstract. In this paper we propose two methods to derive different kinds of node neighborhood based similarities in a network. The first similarity measure focuses on the overlap of direct and indirect neighbors. The second similarity compares nodes based on the structure of their possibly also very distant neighborhoods. Both similarities are derived from spreading activation patterns over time. Whereas in the first method the activation patterns are directly compared, in the second method the relative change of activation over time is compared. We applied both methods to a real world graph dataset and discuss some of the results in more detail.

1 Introduction

It is essential for many experts of various fields to consider all or at least the bigger part of their accessible data before making decisions, to make sure that no important pieces of information are ignored or underestimated. In many areas the amount of available data grows rapidly and manual exploration is therefore not feasible. Many of these datasets consist of units of information, e.g. genes, or proteins in biomedical datasets, or terms and documents in text datasets, as well as relations between these units, and thus can be represented as networks, with units of information represented as nodes or vertices and their relations as edges.

For analysts and experts it can be interesting to find nodes in such networks that are directly or indirectly connected to given query nodes in order to find a community, or a dense subgraph located around a given query node. In terms of biomedical networks, proteins can be found interacting with a query protein, or sharing certain properties. In social networks a circle of friends or acquaintances of a person can be determined, and in textual networks frequently shared terms or documents, according to a query can be discovered. To identify and extract closely connected nodes to certain query nodes, often methods based on spreading activation are used, especially in the field of information retrieval [9,10,4].

Besides finding nodes, which are part of the community of a query node and thereby closely positioned to it, the discovery of structurally similar nodes can be desirable, too. Nodes are structurally similar if the connection structure of their neighborhoods is similar. An overlap of neighborhoods is not required, which means that the nodes can be located far away from each other. For instance structurally similar individuals in social networks may play the same role in their community. In biomedical networks, for

M.R. Berthold (Ed.): Bisociative Knowledge Discovery, LNAI 7250, pp. 246–262, 2012.

example, proteins can be found playing the same role in their metabolic pathways. Additionally the comparison of communities of the query node and its structurally similar result nodes, can lead to new insights as well. For instance information such as the number and size of sub-communities, the number and connectedness of central nodes, and the structural position of the query, and result nodes in their corresponding community, can be interesting.

Experts and analysts do not always know exactly what to look for, or where. Thus the usage of classical information retrieval systems, requiring specific queries, is often not sufficient. Methods that suggest unknown, interesting and potentially relevant pieces of information around a certain topic can help to find a focus, induce new ideas, or support creative thinking. In [11,18] these pieces of information are described as domain bridging associations, or *bisociations*. The underlying data is thereby organized in a *bisociative network* or *BisoNet*, consisting of units of information and their relations [6,17]. A bisociation pattern based on the structural similarity of two subgraphs from different knowledge domains is defined in [17,18].

In contrast to nodes of bisociations based on bridging concepts and bridging graphs, nodes of bisociations based on structural similarity do not necessarily have to be positioned close to each other. These patterns of bisociation link domains, which may not have any direct connections by means of the abstract concepts they have in common, are represented by the structure of their node neighborhood. A prodrug that passes the blood-brain barrier by carrier-mediated transport, and soldiers who pass the gate of Troy hidden in a wooden horse are examples of the kind of bisociation, described in [18]. Both the prodrug as well as the soldiers cannot pass the barrier or gate without the help of a carrier. The abstract concept of using a carrier in order to pass a barrier is represented by the structure of the corresponding subgraph. Finding nodes that are structurally similar to a query node, extracting densely connected, direct and indirect neighbor nodes, and comparing these subgraphs can lead to the discovery of structural bisociations.

Therefore two different kinds of node similarities can be used, structural and spatial similarity. We propose two methods to derive these two kinds of similarities between nodes in a graph from spreading activation processes. The first method is based on the comparison of activation vectors, yielding a spatial similarity. The second method is based on the comparison of change of activation, the *velocity*, yielding a structural similarity. The focus of this article is the definition and explanation of these two similarities and their application to a real world dataset in order to estimate their suitability.

The article is organized as follows. The next section concerns related work about spreading activation and node similarities. Section 3 defines the preliminaries of spreading activation processes on graphs and the underlying framework. The concept of signature vectors, which can be derived from spreading activation processes to represent nodes is introduced in Section 4. In Section 5 we introduce two kinds of node similarities based on the comparison of activation vectors and signature vectors. This is followed by Section 6, which describes the application of these similarities on the Schools-Wikipedia[1] (2008/09) data. Finally Section 7 concludes the article.

[1] `http://schools-wikipedia.org/`

2 Related Work

In the field of graph analysis different kinds of node equivalences, especially in the context of role assignments have been made [8]. Thereby nodes can be considered equivalent based on different properties, such as neighborhood identity, neighborhood equivalence, automorphic mapping, equal role assignments to neighboring nodes, and others. Networks from real world data are usually noisy and irregular, which makes finding equivalent nodes unlikely. Thus a relaxed notion of equivalence, in the sense that nodes are defined similarly to a certain extent, based on certain properties, is useful for a robust comparison of nodes [20].

Approaches which are conceptually similar to the comparison of activation pattern of nodes, from spreading activation processes are given in [22,23,19]. These approaches, like spreading activation, base on an iterative process, consider nodes to be more similar the more their direct and indirect neighborhood overlaps. The aim of these approaches is to detect dense clusters and communities. Since they also take into account an overlap of indirect node neighborhoods as well, they are more robust than measures comparing only the direct neighborhoods, such as e.g. the *Jaccard index* [14].

Each of these approaches suffers from different drawbacks. In [22] the characteristic node vectors of the corresponding normalized adjacency matrix are projected onto a lower dimensional space. Then the values of the projected vectors of each node are replaced iteratively by the mean of their neighbor values. Due to the projection into a lower dimensional space, information can get missing. In [23] node distances are determined, based on random walks, which are iterative processes as well. Nodes are similar if the probability of reaching other nodes in a specified number of iterations is similar. Here only walks of a certain length are considered when computing the distances. However, a more general variant of the algorithm considers walks of different lengths as well. Taking into account all computed iterations, as in [19] may yield to higher accuracy. In [19] all iteration results are accumulated with a decay to decrease the impact of the global node neighborhood. Since the accumulated and normalized activation values are used as similarities the method may yield asymmetric similarities on directed graphs.

In our approach we compare the computed activation pattern by means of a well known similarity measure, the *cosine similarity*, yielding symmetric values. Thus our method can be applied to directed graphs as well. We do not use a lower dimensional node representation by means of a projection into a lower dimensional space and hence may not lose information. We consider all iteration results up to a maximal number of iterations and not only walks of a certain length.

Additionally we propose a second node similarity derived from the comparison of activation changes in each iteration. Based on this method nodes are similar if the structure of their neighborhood is similar, although the neighborhood does not need to overlap at all. This yields a completely different similarity compared to those mentioned above.

Originally spreading activation was proposed by Quillian [24,25] and Collins et al. [9] to query information networks. The method facilitates the extraction of subgraphs, nodes and edges directly and indirectly related to a given query. Initially the nodes representing the query are activated. The activation is than spread iteratively to adjacent nodes, which are activated with a certain level as well until a termination criterion is reached or the process converges. The subset of activated nodes, their level of activation, as well as the induced subgraph compose the final result. The level of activation of nodes is often used as relevancy heuristic.

Spreading activation has been applied in many fields of research from semantic networks [25], associative retrieval [27], to psychology [9,1,2], information retrieval [26,4,10,3] and others [13,28,21,16,12]. Most of these approaches use a set of common heuristic constraints [10] in order to restrict the dynamics of the process, such as distance constraints to terminate the process after a certain number of iterations, or fan out constraints to avoid excessive spreading. In [5] it is shown that pure (constraint free) spreading activation with a linear activation function on a connected and not bipartite graph always converges to the principal eigenvector of the adjacency matrix of the graph.

Usually the level of activation itself, which is sometimes normalized or accumulated over the iterations, represents the relevancy or similarity of nodes to a given query. We propose the comparison of (accumulated) activation patterns, as well as the change of activation patterns to determine similarities between nodes of the underlying network.

In the next section the preliminaries of spreading activation and its framework, that we use in this work are defined.

3 Spreading Activation

Activation is spread on a graph $G = (V, E, w)$, with V as the set of nodes $V = \{1, \dots, n\}$, $E \subseteq V \times V$ as the set of edges and $w(u, v)$ as the weight of the edge connecting u and v, with $u, v \in V$, $w(u, v) = 0$ if $(u, v) \notin E$. For an ease of exposition we assume that the graph G is undirected, however our results easily generalize to directed graphs. The activation state at a certain time k is denoted by $\mathbf{a}^{(k)} \in \mathbb{R}^n$ with $\mathbf{a}_v^{(k)}$ as the activation of node $v \in V$. Each state $\mathbf{a}^{(k)}$ with $k > 0$ is obtained from the previous state $\mathbf{a}^{(k-1)}$ by the three families of functions described below.

- *Input function*: combines the incoming activation from adjacent nodes.
- *Activation function*: determines the state of activation based on the incoming activation.
- *Output function*: determines the outgoing activation based on the current activation.

The initial state $\mathbf{a}^{(0)}$ defines the activation of nodes representing the query. In each iteration activation is spread to adjacent nodes activating them with a certain level as well. The process is usually terminated after a certain number of iterations, activated nodes or convergence.

3.1 Linear Standard Scenario

In our approach we use a linear standard scenario described in [5] for which convergence is shown for non-bipartite connected graphs. The input, activation, and output function can be combined to one function. Given a graph $G = (V, E, w)$ and an activation state $\mathbf{a}^{(k-1)}$ at time $k-1$, the activation of a certain node v at time k is defined by

$$\mathbf{a}_v^{(k)} = \sum_{u \in N(v)} w(u,v) \cdot \mathbf{a}_u^{(k-1)}, \forall v \in V, \tag{1}$$

with $N(v) = \{u \mid \{u,v\} \in E\}$ as the set of neighbors of v. Furthermore the spreading activation process can be described in matrix notation. With $W \in \mathbb{R}^{n \times n}$ as the weight matrix defined by $(W)_{uv} = w(u,v)$ a single iteration can be stated as $\mathbf{a}^{(k)} = W\mathbf{a}^{(k-1)}$ leading to

$$\mathbf{a}^{(k)} = W^k \mathbf{a}^{(0)} . \tag{2}$$

Note that this holds for undirected graphs only. In general an iteration can be stated as $\mathbf{a}^{(k)} = (W^T)^k \mathbf{a}^{(0)}$, holding for directed graphs as well. In order to prevent the activation values from increasing heavily or vanishing, the activation vector is normalized by its Euclidean length after each iteration.

$$\mathbf{a}^{(k)} = \frac{W^k \mathbf{a}^{(0)}}{\left\| W^k \mathbf{a}^{(0)} \right\|} . \tag{3}$$

Rescaling does not change the direction of the activation vector, so convergence to the principal eigenvector $\mathbf{v}_1$ of W is still ensured since $\lim_{k \to \infty} \mathbf{a}^{(k)} = \frac{\mathbf{v}_1}{\|\mathbf{v}_1\|}$.

4 Node Signatures

Convergence of the spreading activation process yields to query independent results. No matter from which node(s) spreading processes have been started initially, the activation state becomes equal after a sufficient number of iterations. From iteration to iteration, activation vectors change their directions towards the direction of the principal eigenvector of the weight matrix W. How quickly a process converges can be described by its velocity and depends on the node(s) from which it was started. For each node the corresponding convergence speed can be determined and represented as a vector, called *signature vector*.

The velocity represents the change of direction of activation patterns between each subsequent iterations. A velocity vector at time k of a spreading process started at $v \in V$ is defined as

$$\delta^{(k)}(v) = \begin{cases} \mathbf{0} & \text{, if } k = 0 \\ \mathbf{a}^{(k)}(v) - \mathbf{a}^{(k-1)}(v) & \text{, else} \end{cases} , \tag{4}$$

with $\mathbf{0}$ as a vector of all 0 and $\mathbf{a}^{(k)}(v)$ as the activation vector at iteration k of a spreading process started at node v, whereas

$$\mathbf{a}_i^{(0)}(v) = \begin{cases} 1 & \text{, if } i = v \\ 0 & \text{, else} \end{cases} ,$$

for all $i \in V$. A norm of a velocity vector represents the amount of change, the step size of the process towards the principal eigenvector of the adjacency matrix. In this work we use the l_2 norm as step size $\|\cdot\|$. Based on the step sizes of each iteration k up to a maximum number of iterations k_{max}, with $0 \leq k \leq k_{max}$, the signature vector of each node is defined. This vector provides information about the convergence speed of a spreading process, starting from a certain node v and is defined as

$$\tau_k(v) = \left\|\delta^{(k)}(v)\right\|, \tag{5}$$

with $\tau(v) \in \mathbb{R}^{k_{max}}$.

5 Node Similarities

Two kinds of node similarities can be derived based on the comparison of activation and convergence behaviors of spreading activation processes starting from each node. On the one hand nodes can be considered similar if their activation vectors are similar (*activation similarity*). On the other hand nodes can be considered similar if the change of activation from one iteration to another is similar (*signature similarity*).

These two kinds of similarities compare nodes based on two different properties, (direct and indirect) neighborhood overlap or neighborhood similarity. A neighborhood overlap between two nodes means that a part of the neighborhood of these two nodes is identical. This consequently means, the larger the overlap the closer the nodes are in the graph. This property yields a spatial similarity measure and is taken into account when activation vectors are compared (activation similarity). A neighborhood similarity of two nodes means that their neighborhood is structurally equivalent to a certain degree but not necessarily identical [20], which can be determined when comparing the change of activation vectors (signature similarity). This property yields a structural similarity measure.

Two node partitionings based on these two different properties are illustrated in Figure 1. The partitioning is indicated by the shading of the nodes. Nodes with the same shade are considered maximally similar (with a similarity value of 1) w.r.t. an equivalent (Figure 1a) or identical (Figure 1b) neighborhood. In Figure 1a the white as well as the black nodes are structurally equivalent since they are automorphic images of each other [8]. In Figure 1b the leaf nodes $\{4,5,6,7\}$, $\{8,9,10,11\}$ and $\{12,13,14,15\}$ are the most similar nodes, due to their identical neighborhood, depicted by the shading gray, black, and white. Even if the leaf nodes are structurally equivalent only those with an identical neighborhood are highly similar. Furthermore the three nodes in the middle $\{1,2,3\}$ are not equal based on the comparison of their neighborhood. Node 3 is more similar to $\{12,13,14,15\}$ than to 1 or 2 when comparing their pattern of activation.

The two different similarity measures derived from spreading activation processes allow on the one hand for the identification of structurally similar nodes to a given query node, even if they are located far apart in the graph via the signature similarity. On the other hand a densely connected subgraph of direct and indirect neighbors can be extracted for each node applying the activation similarity measure. In the following, these two node similarities are formalized and described in detail.

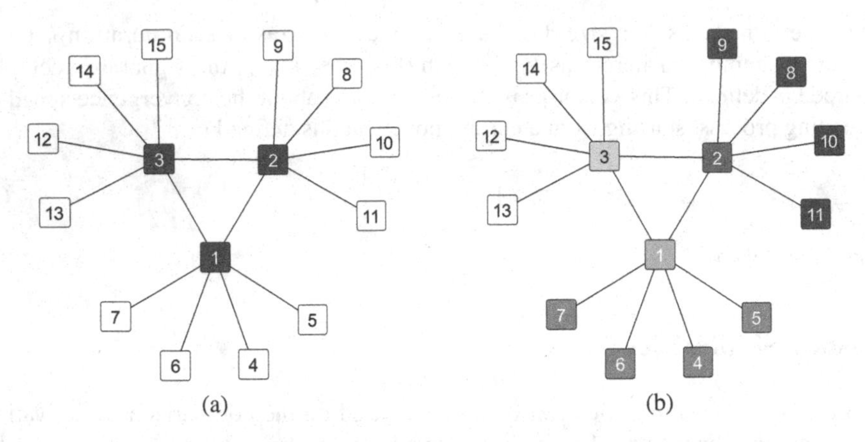

Fig. 1. Two node partitionings, indicated by the shading based on two different node properties, equivalent and identical neighborhood. In 1a the white nodes are structurally equivalent as well as the black nodes, which can be determined by the comparison of the signature vectors (signature similarity). In 1b the leaf nodes are divided into three partitions white, gray, and black, since their neighborhood is only partially identical. In addition node 3 is more similar to the white nodes, node 2 to the black nodes, and node 1 to the gray nodes than to others, which can be determined by the comparison of the accumulated activation vectors (activation similarity).

5.1 Activation Similarity

The first similarity described is based on the comparison of activation vectors and named *activation similarity*. The sequence of activation states of a spreading process started from a certain node describes the node relative to its local and global neighborhood in the graph. Dependent on its neighborhood many or few nodes will be activated and activation will spread fast or slow. Nodes close to the initially activated node will get activated sooner than nodes further apart from this node. Furthermore nodes will get activated to a higher level, at least in the primary iterations, if many walks of different lengths exist, connecting these nodes with the initially activated node. Nodes that are similarly connected to a shared neighborhood will induce similar activation states.

The level of activation $\mathbf{a}_i^{(k)}(v)$ of a node $i \in V$ at a time k, induced by a spreading process started at node v, reflects the reachability of i from node v along (weighted) connecting walks of length k. The more (highly weighted) walks of length k exist connecting i and v, the higher the level of activation. A query node u inducing a similar level of activation $\mathbf{a}_i^{(k)}(u)$ at node i at iteration k is consequently similarly connected to i along (weighted) connecting walks of length k.

Comparing the activation pattern of iterations $k \geq 1$ allows for the determination of the direct and indirect neighborhood overlap of nodes, whereas measures like the *cosine similarity*

$$\sigma_{\cos}(u, v) = \frac{|N(u) \cap N(v)|}{\sqrt{|N(u)|\,|N(v)|}}$$

or *Jaccard index* [14]

$$\sigma_{\text{jaccard}}(u, v) = \frac{|N(u) \cap N(v)|}{|N(u) \cup N(v)|}$$

based on the characteristic vectors of nodes allow for a comparison of the direct node neighborhood only. Figure 2 depicts a graph for which the cosine node similarities have been computed and indicated by the node shading. The nodes 1 and 3 have a cosine similarity of 1, since their direct neighborhood (node 2) is identical. The nodes 1 and 2 have a similarity of 0, as well as nodes 2 and 3. Although they are direct neighbors, the related similarity is 0 since the particular direct neighborhoods do not overlap. Consideration of the indirect neighborhood via connecting walks of lengths greater than 1 ($k > 1$) by applying activation similarity (with 5 iterations) still yields a similarity of 1 for nodes 1 and 3 due to their identical neighborhood, but a similarity greater than 0 for nodes 1 and 2 as well as for nodes 2 and 3. For the detection of dense subgraphs, comparison of the direct node neighborhoods only is too strict. Not all of the nodes in a dense subregion necessarily share a direct neighborhood. Taking into account the indirect k-neighborhoods yields a more robust similarity measure.

Fig. 2. Nodes 1 and 3 (white) have a cosine node similarity of 1, since their direct neighborhood is identical. Although nodes 1 and 2, as well as 2 and 3 are direct neighbors their cosine similarity is 0 since their direct neighborhood is not overlapping.

In [23] it is stated that in terms of random walks of length k starting from a node v the probability is high for other nodes to be reached if they are located in the same dense subgraph or community as node v. For an additional node u, the probability of reaching these nodes is high as well if it is located in the same community. Since random walks are driven by power iterations of the transition matrix of a graph they can be seen as spreading activation processes on a normalized weight matrix.

Considering not only walks of a certain length k as in [23] but all connecting walks of different lengths as in [19] provides a more detailed representation of the local and global neighborhood of a node. Accumulating all activation vectors $\mathbf{a}^{(k)}(v)$ from a spreading process starting from v with a decay α results in a final activation vector $\mathbf{a}^*(v)$ defined by

$$\mathbf{a}^*(v) = \sum_{k=1}^{k_{max}} \alpha^k \mathbf{a}^{(k)}(v) \tag{6}$$

with $0 < \alpha < 1$. The decay α decreases the impact of longer walks and ensures convergence for $k_{max} \to \infty$ for l_2 normalized systems [5]. It is reasonable to decrease the contribution of longer walks in order to keep more information about the local neighborhood of v. The above mentioned form is closely related to the centrality index of Katz [15]. We do not want to let the series fully converge since activation vectors of latter iterations do not contribute much to the final activation based on the decay α, and become more and more similar due to convergence of the spreading processes. We chose k_{max} based on the convergence behavior of the underlying graph as well as the decay factor α.

Before a similarity on the final activation vectors is defined it needs to be considered that nodes with very high degrees will be activated to a higher level. They are more likely to be reached even if they are not located in the same dense region as the node from which activation has spread initially. To take this into account we normalize, related to [23], the final activation by the degree of the corresponding node. The degree normalized final activation vector is thereby denoted as

$$\hat{\mathbf{a}}^*(v) = D^{-\frac{1}{2}}\mathbf{a}^*(v) = D^{-\frac{1}{2}}\left(\sum_{k=1}^{k_{max}} \alpha^k \mathbf{a}^{(k)}(v)\right) \tag{7}$$

with D as the (weighted) degree matrix defined by $(D)_{ii} = d(i), (D)_{ij} = 0$ for $i \neq j, \forall i$ and $d(i) = \sum_{j=1}^{n}(W)_{ij}$. Based on these normalized final activation vectors we define the activation similarity between two nodes u and v

$$\begin{aligned}\sigma_{\text{act}}(u,v) &= \cos(\hat{\mathbf{a}}^*(v), \hat{\mathbf{a}}^*(u)) \\ &= \frac{\langle \hat{\mathbf{a}}^*(u), \hat{\mathbf{a}}^*(v)\rangle}{\|\hat{\mathbf{a}}^*(u)\| \, \|\hat{\mathbf{a}}^*(v)\|} \\ &= \frac{\sum_{i=1}^{n} \mathbf{a}_i^*(u)\mathbf{a}_i^*(v)d(i)^{-1}}{\|\hat{\mathbf{a}}^*(u)\| \, \|\hat{\mathbf{a}}^*(v)\|},\end{aligned} \tag{8}$$

with $\langle \mathbf{x}, \mathbf{y}\rangle$ as the inner product between vectors $\mathbf{x}, \mathbf{y} \in \mathbb{R}^n$. The more nodes are similarly activated in both spreading processes, one starting at node u and one at v, the more similar u and v are. This measure allows for a detection of dense communities and requires a direct and indirect neighborhood overlap, as can be seen in Figure 1b. Node 1 is more similar to $\{4,5,6,7\}$ than to 2 or 3 even if 1 is automorphically equivalent to 2 and 3. In [20] this kind of node similarity is categorized as closeness similarity.

The computation of node similarities proposed in [19] can be seen in terms of spreading activation as well. The accumulated and normalized activation values themselves represent the similarities between the activated nodes and the node at which the spreading process started. As stated, their method is applicable only on undirected graphs. For directed graphs the activation values are not necessarily symmetric, yielding asymmetric similarities.

5.2 Signature Similarity

The second similarity is based on the comparison of the amount of activation changes during spreading activation processes and named *signature similarity*. For each node a signature vector can be determined, consisting of velocity vector norms (see Section 4). The direction of the velocity vectors represent the change of direction of the activation patterns and their norms represent the step size between subsequent iterations towards the principal eigenvector of the weight matrix W. By the comparison of the signature vectors a structural similarity can be derived. In this work we use the cosine measure to compare the signature vectors, thus the signature similarity is denoted as

$$\begin{aligned}\sigma_{\text{sig}}(u,v) &= \cos(\tau(u),\tau(v)) \\ &= \frac{\langle \tau(u),\tau(v)\rangle}{\|\tau(u)\|\,\|\tau(v)\|} \\ &= \frac{\sum_{k=1}^{k_{max}} \|\delta^{(k)}(u)\|\,\|\delta^{(k)}(v)\|}{\|\tau(u)\|\,\|\tau(v)\|}.\end{aligned} \tag{9}$$

Nodes that are similar due to the activation similarity have to be close to each other in the graph, since the same direct and indirect neighbor nodes need to be activated similarly. The signature similarity is not based on the activation pattern itself but on the amount of change of these patterns. If the structure of the neighborhood of two nodes is similar, the change of activation will be similar too, and thus the signature similarity will yield higher values as if the structure is different.

A similar step size between two subsequent iterations yields from a similar structure, i.e. the nodes $\{1,2,3\}$ (black) of Figure 1a are not distinguishable by their signature vectors, since they are automorphic images from each other. Whereas the activation vectors of these nodes are different, as well as the corresponding velocity vectors, the amount of change of direction of the activation vectors in each iteration is equal. Nodes do not necessarily have to be located in the same densely connected region to have a high signature similarity. This makes the signature similarity not a closeness but a structural similarity measure. Nodes with a structurally similar neighborhood have a high signature similarity even if they are located far apart from each other. An overlapping neighborhood is thereby not necessary, which can be seen in Figure 1a, where all the leaf nodes (white) have a signature similarity value of 1, even if their direct neighborhood is not overlapping at all.

6 Experiments

To demonstrate our approach we apply the two kinds of node similarities to the Schools-Wikipedia[2] (2008/09) dataset. The first aim is to find result nodes that are structurally similar to given query nodes by using the signature similarity. Secondly we want to find nodes that are closely connected (directly or indirectly) to the query nodes or interesting result nodes, respectively, using the activation similarity, and extract the corresponding subgraphs. Since the extraction of communities is not the aim of this work we do not focus on this issue. Instead we consider the induced subgraph of the k most similar nodes based on the activation similarity according to a query node, as dense local neighborhood, or community of that query.

Once structurally similar nodes have been detected and the corresponding communities have been extracted and illustrated by means of centrality layouts, we manually compare these subgraphs in order to find structural coherences. We are thereby interested in the status or rank of the result nodes in their community and the most central

[2] http://schools-wikipedia.org/

nodes. Our assumption is that the communities of the result nodes are similar, in these terms, to the community of the query node.

6.1 Schools-Wikipedia

The Schools-Wikipedia (2008/09) dataset consists of a subset of the English Wikipedia[3] dataset, with around 5500 articles. The articles are grouped into 154 different categories, consisting of 16 main or top-level categories, where each article is assigned to at least one category. As in Wikipedia, articles can reference other articles via hyperlinks. In Schools-Wikipedia external links have been filtered.

To create the graph, each article is considered as a unit of information and modeled as a node. Each hyperlink that connects articles is considered as a relation between two units of information and represented as an undirected edge connecting the corresponding nodes. The resulting graph consists of four connected components, whereas three of the components consist only of one node and are also filtered. Convergence of all spreading activation processes on the filtered graph is ensured by connectedness, non-bipartiteness and undirectedness. Table 1 lists some basic properties of the remaining graph.

Table 1. Basic graph properties of the filtered Schools-Wikipedia graph

Schools-Wikipedia graph properties	
Number of nodes	5536
Number of edges	190149
Minimal node degree	1
Maximal node degree	2069
Average node degree	68.7
Diameter	5

We applied spreading activation processes as described in Section 3 to the graph, in order to compute the activation and signature similarities between all nodes, defined in Section 5. Since the spreading activation processes converge quickly due to the underlying graph structure, indicated e.g. by the small diameter, we only computed the first 10 iterations of each spreading process to compute the similarities. Concerning the activation similarity we used a decay value of $\alpha = 0.3$ to compute the accumulated activation vectors in order to focus on the local neighborhood of nodes. The choice of parameters is not discussed in this work. Here it is sufficient to mention that further iterations (> 10) do not contribute significantly to both similarities due to the small decay as well as the small diameter and thus fast convergence.

In our experiment we wanted to find well-known, scholarly persons from different areas of research, which play similar roles in their communities. Our focus is on well-known people, since the results can be reasonably evaluated based on general knowledge. The query consists of the node of the well-known Italian physicist *Galileo Galilei*.

[3] http://en.wikipedia.org/wiki/Main_Page

To find the structurally most similar persons, all nodes, which are assigned to the *People* category are sorted based on their corresponding signature similarity to the query. Since we focused only on people with a similar structural position, we filtered out all nodes not belonging to the People category. Additionally we were interested in the nodes belonging to the community around Galileo Galilei. Therefore we considered all nodes, not only those assigned to the People category and sorted them according to their activation similarity to the query. Table 2 lists the 10 most similar nodes, as well as the 16th and 17th nodes of the People category, based on the signature similarity, and the 10 most similar nodes, as well as the 16th and 17th nodes of all categories based on the activation similarity, compared to Galileo Galilei.

Table 2. 10 most similar nodes to Galileo Galilei and the 16th and 17th nodes; *left* assigned to the *People* category, based on the signature similarity; *right* of all categories, based on the activation similarity.

Galileo Galilei		
Rank	Signature similarity	Activation Similarity
1	Galileo Galilei	Galileo Galilei
2	Isaac Newton	Johannes Kepler
3	Johannes Kepler	Heliocentrism
4	Aristotle	Nicolaus Copernicus
5	Leonhard Euler	Isaac Newton
6	Mary II of England	Phil. Nat. Principa Mathematica
7	James Clerk Maxwell	Kepler's laws of planetary motion
8	Anne of Great Britain	Classical mechanics
9	James I of England	HIstory of physics
10	Henry VII of England	Astronomy
⋮	⋮	⋮
16	Plato	Newton's laws of motion
17	Euclid	General relativity

It can be seen that Galileo himself is the most similar node, which makes sense in terms of the cosine similarity used on activation and signature vectors. Nodes such as *Heliocentrism*, *Astronomy*, *History of physics*, etc. are part of his closer community, reasonably, since he worked primarily in these fields and played a major role in them. Among others Galileo is called "the father of modern physics". Other important scientists who played a major role in these areas as well, such as *Nicolaus Copernicus*, *Johannes Kepler*, and *Isaac Newton* are also part of his community.

On inspecting the structurally similar nodes, the names *Plato* and *Euclid* attract our attention; they are the 16th and 17th structurally most similar nodes of the People category. Both men played a major role in their areas of research too, philosophy and mathematics, respectively, which are different to those of Galileo. Plato contributed significantly to the foundations of Western philosophy and Euclid is said to be the "father of geometry". Newton and Kepler, have a high signature similarity as well and played - like Galilei - a major role in their areas of research too. However, their areas of

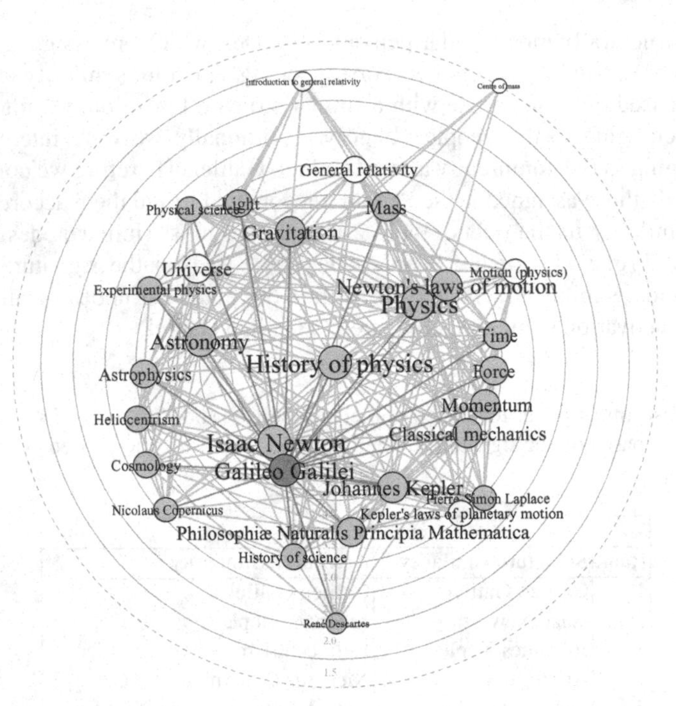

Fig. 3. The subgraph of the 30 most similar nodes to Galileo Galilei based on activation similarity. The used layout is a centrality layout, based on eigenvector centrality.

research do not differ to Galilei's as much as those of Plato and Euclid do. In terms of structural bisociations, structurally similar nodes, that represent units of information in unrelated fields of knowledge are potentially more interesting, than those in the same or similar fields. As a result of this fact and due to their high degree of popularity, Plato and Euclid were chosen in order to compare their communities.

We extracted the induced subgraphs of their communities consisting of the 30 most similar nodes based on the corresponding activation similarities. We used 30 nodes, since the subgraphs of this size can be visualized in a reasonable manner and both structural similarities as well as differences can be shown. Figure 3 shows the community around Galileo, Figure 4a that around Plato and Figure 4b that around Euclid.

Nodes are represented as circles, whereas the corresponding size and the size of the label is proportional to their degree. The nodes of the corresponding persons are emphasized by dark gray; their direct neighbors are light gray and all other nodes white. The layout of all graphs is a centrality layout based on eigenvector centrality [7]. The eigenvector centrality is, like other centrality indices, a measure to quantify the status of nodes in a graph. The higher the value compared to others, the more central or important the node, and the more central its position in the visualization. In contrast, the lower the value, the lower its status or importance and the more peripherical the position.

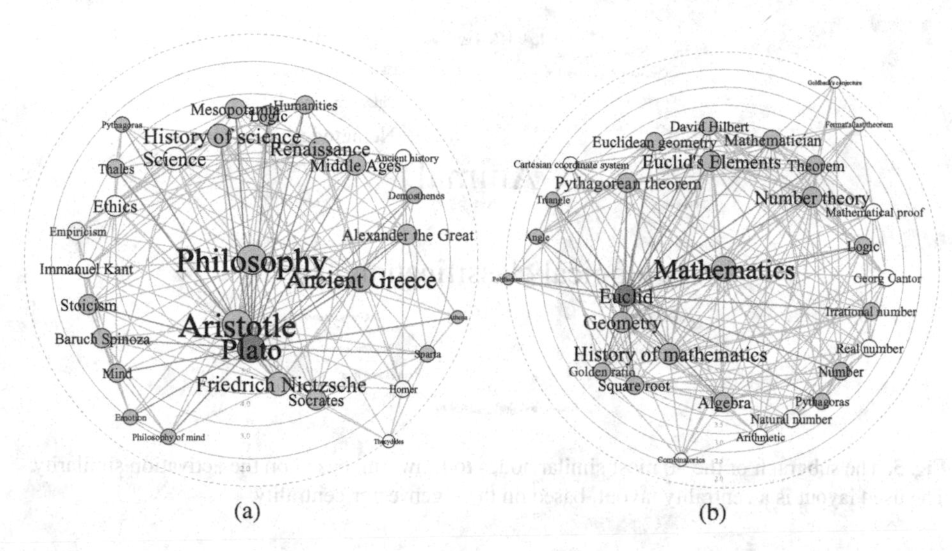

Fig. 4. Two subgraphs of the 30 most similar nodes to Plato (4a) and Euclid (4b), based on the activation similarity. The used layout is a centrality layout, based on the eigenvector centrality.

It can be seen that Galileo, Plato, and Euclid are connected to most of the nodes of their communities. Galileo has 23 direct neighbors, Plato 22 and Euclid 21. Their neighborhoods can roughly be partitioned into three semantic groups: the fields they worked in, topics and issues important in these particular fields and other important persons who contributed significantly to these fields as well. In the case of Galileo, the fields of research are *Physics*, *Astronomy*, *Classical mechanics* etc. Important topics and issues in these fields are e.g. *Gravitation*, *Mass*, and *Force* and other important persons who worked in these fields are e.g. *Nicolaus Copernicus*, *Isaac Newton* or *Johannes Kepler*. In the case of Plato, the fields of research are *Philosophy* and *Philosophy of mind*. Important topics are e.g. *Emotion* and *Logic* and other important persons are *Aristotle* and *Socrates*. In the case of Euclid, the fields of research are *Mathematics* and *Euclidean geometry*. Important issues are e.g. *Angle* and *Triangle* and other important persons are *Pythagoras* and *David Hilbert*. Even if Galileo, Plato, and Euclid are directly connected to most of the nodes in their community, the most central nodes are, however the fields for which (among others) they are famous for: *History of physics*, *Philosophy*, and *Mathematics*. Nevertheless their status is very central compared to all other nodes in the corresponding communities.

In all three subgraphs there exist other nodes with a similar centrality. In Galiliei's community these nodes are *Isaac Newton*, *Johannes Kepler*, *Physics*, *Astronomy*, and *Gravitation*, in Plato's *Aristotle*, and *Anchient Greece* and in Euclid's *Geometry*, *Euclids Elements*, and *History of mathematics*. All of these nodes, except *Euclids Elements* and *Gravitation* have a high signature similarity according to Galilei, even though some nodes, such as *Aristotle* are not part of his community and thus do not have a high degree of activation similarity. However, the signature similarity of *Euclids Elements* and *Gravitation* is also not very low. The nodes are part of the 270 most similar nodes of all categories. Additionally it can be seen that in the case of Galileo and Plato, there

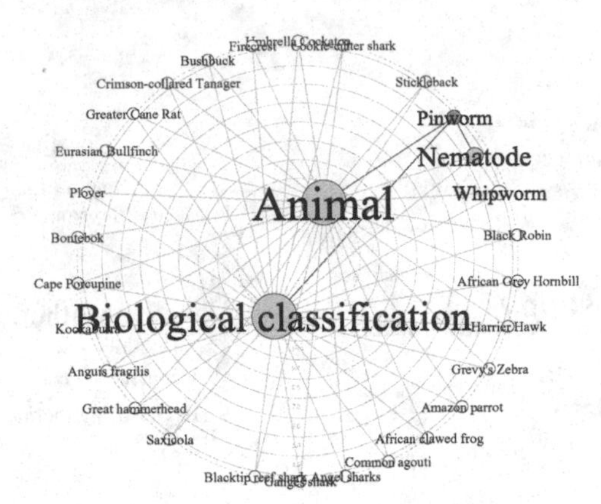

Fig. 5. The subgraph of the 30 most similar nodes to Pinworm, based on the activation similarity. The used layout is a centrality layout, based on the eigenvector centrality.

is one slightly more central node in the People category: *Isaac Newton* and *Aristotle*, respectively. In all three communities the most peripheric nodes are not connected to Galileo, Plato and Euclid.

In a nutshell, structural coherences of nodes with high signature similarity and their corresponding communities can be seen based on various aspects, such as their own status or centrality and those of other nodes, their connectedness, their degree as well as the density of their community, etc.

On the one hand very similar nodes to a certain query, based on signature similarity, are interesting to show structural coherences. On the other, very dissimilar nodes are also interesting to show structural differences. The most dissimilar node to Galileo is *Pinworm*. Again we extracted the 30 nodes most similar to Pinworm, based on activation similarity, and illustrated the induced subgraph in Figure 5 using the centrality layout.

Nodes are represented as circles, whereas size and labelsize are proportional to their degree, up to a certain maximum size. The Pinworm node is emphasized by dark gray, its direct neighbors are light gray and all other nodes white. The structural differences of the Pinworm node as well as its community can be seen clearly. Pinworm has, as well as almost all other nodes a very peripheric position. Additionally the most central position is shared by the two nodes *Animal* and *Biological classification*, which are connected to all nodes of the community. In addition the density of the community is much lower then those of the communities of Galilei, Plato, or Euclid.

7 Conclusion

In this work we have shown how two kinds of similarities to compare nodes in a graph can be derived from spreading activation processes. The activation similarity is based on the comparison of accumulated activation vectors and yields a spatial or closeness similarity. The signature similarity is based on the comparison of norms of velocity vectors and yields a structural similarity. By applying both kinds of similarities we can find structurally similar nodes on the one hand, which are not necessarily located close together, and dense subgraphs or communities around nodes on the other. We applied this procedure to the Schools-Wikipedia (2008/09) dataset and preliminary results are very encouraging: the nodes of Euclid, and Plato for example, are structurally similar to that of Galileo Galilei. By comparing their communities structural similarity was able to be confirmed manually. The experiments suggested that the combination of these two kinds of similarities is a promising tool in terms of identification and extraction of structural bisociations.

References

1. Anderson, J.R.: A spreading activation theory of memory. Journal of Verbal Learning and Verbal Behavior 22(3), 261–295 (1983)
2. Anderson, J.R., Pirolli, P.L.: Spread of activation. Journal of Experimental Psychology: Learning, Memory, and Cognition 10(4), 791–798 (1984)
3. Aswath, D., Ahmed, S.T., D'cunha, J., Davulcu, H.: Boosting item keyword search with spreading activation. In: Web Intelligence, pp. 704–707 (2005)
4. Belew, R.K.: Adaptive information retrieval: using a connectionist representation to retrieve and learn about documents. In: SIGIR 1989: Proceedings of the 12th Annual International ACM SIGIR Conference on Research and Development in Information Retrieval, pp. 11–20. ACM Press, New York (1989)
5. Berthold, M.R., Brandes, U., Kötter, T., Mader, M., Nagel, U., Thiel, K.: Pure spreading activation is pointless. In: Proceedings of the CIKM the 18th Conference on Information and Knowledge Management, pp. 1915–1919 (2009)
6. Berthold, M.R., Dill, F., Kötter, T., Thiel, K.: Supporting Creativity: Towards Associative Discovery of New Insights. In: Washio, T., Suzuki, E., Ting, K.M., Inokuchi, A. (eds.) PAKDD 2008. LNCS (LNAI), vol. 5012, pp. 14–25. Springer, Heidelberg (2008)
7. Bonacich, P.: Factoring and weighting approaches to status scores and clique identification. Journal of Mathematical Sociology 2, 113–120 (1972)
8. Brandes, U., Erlebach, T.: Network Analysis: Methodological Foundations. Springer (2005)
9. Collins, A.M., Loftus, E.F.: A spreading-activation theory of semantic processing. Psychological Review 82(6), 407–428 (1975)
10. Crestani, F.: Application of spreading activation techniques in information retrieval. Artif. Intell. Rev. 11(6), 453–482 (1997)
11. Dubitzky, W., Kötter, T., Schmidt, O., Berthold, M.R.: Towards Creative Information Exploration Based on Koestler's Concept of Bisociation. In: Berthold, M.R. (ed.) Bisociative Knowledge Discovery. LNCS (LNAI), vol. 7250, pp. 11–32. Springer, Heidelberg (2012)
12. Duch, W., Matykiewicz, P., Pestian, J.: Neurolinguistic approach to natural language processing with applications to medical text analysis. Neural Networks 21, 1500–1510 (2008)
13. Hopfield, J.J.: Neurons with graded response have collective computational properties like those of two-state neurons. Proceedings of the National Academy of Sciences 81(10), 3088–3092 (1984)
14. Jaccard, P.: Étude comparative de la distribution florale dans une portion des alpes et des jura. Bulletin del la Société Vaudoise des Sciences Naturelles 37, 547–579 (1901)

15. Katz, L.: A new status index derived from sociometric analysis. Psychometrika 18(1), 39–43 (1953)
16. Kosinov, S., Marchand-Maillet, S., Kozintsev, I.: Dual diffusion model of spreading activation for content-based image retrieval. In: MIR 2006: Proceedings of the 8th ACM International Workshop on Multimedia Information Retrieval, pp. 43–50. ACM, New York (2006)
17. Kötter, T., Berthold, M.R.: From Information Networks to Bisociative Information Networks. In: Berthold, M.R. (ed.) Bisociative Knowledge Discovery. LNCS (LNAI), vol. 7250, pp. 33–50. Springer, Heidelberg (2012)
18. Kötter, T., Thiel, K., Berthold, M.R.: Domain bridging associations support creativity. In: Proceedings of the International Conference on Computational Creativity, pp. 200–204 (2010)
19. Leicht, E.A., Holme, P., Newman, M.E.J.: Vertex similarity in networks. Physical Review E 73(2), 026120 (2006)
20. Lerner, J.: Structural Similarity of Vertices in Networks. PhD thesis, Universität Konstanz, Universitätsstr. 10, 78457 Konstanz (2007)
21. Liu, W., Weichselbraun, A., Scharl, A., Chang, E.: Semi-automatic ontology extension using spreading activation. Journal of Universal Knowledge Management, 50–58 (2005)
22. Moody, J.: Peer influence groups: identifying dense clusters in large networks. Social Networks 23(4), 261–283 (2001)
23. Pons, P., Latapy, M.: Computing communities in large networks using random walks. Journal of Graph Algorithms and Applications 10(2), 191–218 (2006)
24. Quillian, M.R.: A revised design for an understanding machine. Mechanical Translation 7, 17–29 (1962)
25. Quillian, M.R.: Semantic memory. In: Minsky, M. (ed.) Semantic Information Processing, pp. 227–270. The MIT Press, Cambridge (1968)
26. Salton, G., Buckley, C.: On the use of spreading activation methods in automatic information retrieval. Technical report, Dept. Computer Science, Cornell Univ., Ithaca, NY (1988)
27. Salton, G.: Automatic Information Organization and Retrieval. McGraw Hill (1968)
28. Ziegler, C.N., Lausen, G.: Spreading activation models for trust propagation. In: 2004 IEEE International Conference on e-Technology, e-Commerce and e-Service, EEE 2004, pp. 83–97. IEEE Computer Society, Washington, DC (2004)

Towards Discovery of Subgraph Bisociations

Uwe Nagel*, Kilian Thiel, Tobias Kötter,
Dawid Piątek, and Michael R. Berthold

Nycomed-Chair for Bioinformatics and Information Mining
Dept. of Computer and Information Science
University of Konstanz
firstname.lastname@uni-konstanz.de

Abstract. The discovery of surprising relations in large, heterogeneous information repositories is gaining increasing importance in real world data analysis. If these repositories come from diverse origins, forming different domains, domain bridging associations between otherwise weakly connected domains can provide insights into the data that are not accomplished by aggregative approaches. In this paper, we propose a first formalization for the detection of such potentially interesting, domain-crossing relations based purely on structural properties of a relational knowledge description.

1 Motivation

Classical data mining approaches propose (among others) two major alternatives to exploit data collections. One scenario tries to fit a model to the given data and thereby to predict the behavior of some underlying system. Another approach describes all or part of the given data by patterns such as clusters or frequent itemsets to provide an insight or highlight mechanisms that led to such patterns. Both variants have in common that some hypothesis about the considered data is involved and that the processing is motivated by a concrete question. A necessity for such a motivated processing is some a priori knowledge or decision about either the involved data (i.e. what type of model could be fitted) or the form in which findings are described (e.g. clusters or frequent item sets). While in the first case the possible findings are narrowed by the aspect of the systems behavior that is modelled, in the latter case the choice of patterns limits the possible findings. In short, one could say that in those approaches the problem to be solved is simplified by narrowing it down through the investment of a priori knowledge or by specifying the form of outcome.

Alternatively, Explorative (or Visual) Data Mining attempts to overcome this problem by creating a more abstract overview of the entire data together with subsequent drill-down operations. Thereby it additionally enables the search for arbitrary interesting patterns on a structural level, detached from the semantics of the represented information. However, such overviews still leave the entire

* Corresponding author.

M.R. Berthold (Ed.): Bisociative Knowledge Discovery, LNAI 7250, pp. 263–284, 2012.

search for interesting patterns to the user and therefore often fail to actually point to interesting and truly novel details.

We propose a different approach: instead of considering all available data or large parts of it at once, we concentrate on the identification of interesting, seldom appearing details. In that, integrated data is explored by finding unexpected and potentially interesting connections that hopefully trigger the user's interest, ultimately supporting creativity and outside-the-box thinking. Our assumption is that such connections qualify as unexpected by connecting seemingly unrelated domains. As already pointed out by Henri Poincaré [14]: "Among chosen combinations the most fertile will often be those formed of elements drawn from domains which are far apart... Most combinations so formed would be entirely sterile; but certain among them, very rare, are the most fruitful of all." A historical example for such a combination is provided by the theory of electromagnetism by Maxwell [11], connecting electricity and magnetism. Consequently, we embrace the diversity of different data sources and domains of knowledge. We thus do not fuse those into one large homogeneous knowledge base by e.g. mapping them into a common feature space. Instead, we model the given data sparsely, try to identify (possibly hidden) domains and search for rare instead of frequent and weak instead of strong patterns, i.e. exclusive, domain crossing connections. Though for technical integration we still need a certain homogeneity in the data representation.

With respect to this demand, we assume a knowledge representation fulfilling only very few conditions: the representation of *information units* and links between them without any further attributes. Based on that, we address two sub-problems: the identification of domains and the assessment of the potential interestingness of connections between these domains.

2 Networks, Domains and Bisociations

In this section, we transfer the theoretical concept of domain crossing associations which are called *bisociations* [10] (to emphasize the difference to associations within a single domain) to a setting where a relational description of knowledge is given. We will explain the model that incorporates our knowledge base, narrow down the concepts underlying *domains* and *bisociations* and identify properties that allow to assess the interestingness of a bisociation.

2.1 Knowledge Modeling

As a preliminary, we assume that the available knowledge is integrated into a unifying data model. We model this as an undirected, unweighted graph structure with nodes representing units of information and edges representing their relations. Examples for information units are terms, documents, genes or experiments. Relations could arise from references, co-occurrences or explicitly encoded expert knowledge. The only semantic assumption we do make is, that the relation expressed by the links is of positive nature, i.e. is to be interpreted as similarity not dissimilarity.

A graph is described as $G = (V, E)$ with node set V, edge set $E \subseteq \binom{V}{2}$ and $n = |V|$ the number of nodes. The degree of a node, i.e. the number of incident edges, is denoted as $d(v)$ and its neighboring nodes in G as $N(v)$. We further access the structure of G via its adjacency matrix A, with $(A)_{uv} = 1$ if $\{u, v\} \in E$ and 0 otherwise. Finally, for a set of nodes $U \subseteq V$, $G[U]$ denotes the subgraph of G induced by the nodes of U, i.e. the nodes of U and all edges of E that connect two nodes of U.

The presented model does not contain any hints on the semantics behind the incorporated units of information except for their relations. In practice, such semantics would be provided by additional attributes attached to nodes and links. Our approach will, however, not employ them in any automatic processing, thereby ensuring maximal flexibility. Yet supporting attributes are helpful and necessary in the process of manual result interpretation and should consequently not be removed. In contrast they should be preserved completely: in the phase of manual interpretation they provide the necessary link to the semantic layer of the considered data. The fact that they are not employed in any automatic processing rules out the demand for homogeneity and thereby the necessity to convert them into a common format, leaving freedom to attach arbitrary information that might be useful.

While we ignore additional information attached to nodes and details about link interpretations, the general interpretation of links is an important aspect since the structural information provided by them is the sole basis of reasoning about the connected objects. In general, we consider different types of links expressing different relations and we allow some inhomogeneity within the set of links, such as different sources that led to the formation of links. However, we assume that within an individual data set all links obey roughly the same interpretation. Consider as an example a knowledge collection consisting of scientific articles. We would interpret the articles as information units and therefore model them as nodes. For the derivation of links we can choose between alternative semantic relations. As an example we could derive similarities by text analysis and derive links from these. Alternatively, we could exploit the fact that scientific articles reference each other and introduce a link for each reference. Both approaches have their assets and drawbacks and surely do not represent all possibilities of link derivation. In the identification of domains as described in the following, these two interpretations have to be handled differently which makes their distinction very important. We do not fix any decision about the type of links, but consider it as an important design decision in the process of data modeling and stress that it has to be considered carefully in the whole process. In the remainder we restrict our considerations to the two interpretations described above and point out where the method has to be adapted to the type of link interpretation.

2.2 Domains

As indicated before, in this context a *domain* is a set of information units from the same field or area of knowledge. Domains exist with different granularity and thus can be partially ordered in a hierarchical way from specific to general. As an example consider the domains chemistry, biology, biochemistry and the science domain in general. While the first three are clearly subsets of the science domain, biochemistry is neither a proper subset of biology nor chemistry but overlaps with both of them. Furthermore, this distinction of domains may be sufficient in the context of common knowledge while scientists working in these fields would surely subdivide them. Consequently, the granularity of a domain depends on a specific point of view, which can be a very local one. In addition, information units may belong to several domains which are not necessarily related. The eagle for example belongs to the domain of animals and in addition to the coat of arms domain.

Relation to Graph Structure. Intuitively, a set of highly interconnected nodes indicates an intense interrelation that should be interpreted as a common domain. While this is a sound assumption when connections express similarities between the involved concepts, it is not necessarily true when links express other semantic relations. In the example of scientific articles a collection of papers approaching a common problem would signify an example domain. Yet the similarity of these articles is not necessarily reflected by mutual references, especially if they were written at the same time. However, they will very likely share a number of references. Consequently, we derive domains from common neighborhoods instead of relying on direct connections between information units. This allows domains to be identified when the connections express either references or similarities since nodes in a densely connected region also have similar neighborhoods. Two information units that share all (or - more realistically - almost all) their connections to other information units should therefore belong to a common domain. Since they are in this respect indistinguishable and their relations form the sole basis for our reasoning about them, all possibly identifiable domains have to contain either both or none of them. We will show a concrete node similarity that expresses this property and relaxes the conditions in Section 3. This similarity will then be used as a guidance in the identification of domains.

As mentioned before, the discussed guidelines are necessarily tailored to the considered link semantics and have to be adapted if links in the graph representation are derived differently. The interface between link interpretation and the process of domain identification is here established by a node similarity hinting at common domain affiliation. For the adaption to different link interpretations it is thus only necessary to adapt the node similarity correspondingly, ensuring that highly similar nodes tend to belong to identical domains on all levels while decreasing similarity indicates that nodes share fewer and thus only upper level domains.

Domain Recovery. Assuming a node similarity with the described properties, recursive merging of the most similar nodes leads to a merge tree as produced by hierarchical clustering. In the following, we consider the inner nodes of such a merge tree as candidates for domains.

The resulting domains form a hierarchy on the information units which is similar to an ontology but is not able to render all possible domain assignments. That is, any two domains resulting from this process overlap completely (one is contained in the other) or are completely disjoint. A number of domains could remain unidentified since cases of partially overlapping domains are excluded by the procedure, as the domain of biochemistry from the example above, given that biology and chemistry are identified as domains. We consider this as an unavoidable approximation for now, posing the extraction of domains as a separate problem.

2.3 Bisociations

A connection - usually indirect - between information units from multiple, otherwise unrelated domains is called *bisociation* in contrast to associations that connect information units within the same domain. The term was introduced by Koestler [9] in a theory to describe the creative act in humor, science and art.

Up to now, three different patterns of bisociation have been described in this context [10]: bridging concepts, bridging graphs and structural similarity. Here we focus on the discovery of bridging graphs, i.e. a collection of information units and connections providing a "bisociative" relation between diverse domains.

Among the arbitrary bisociations one might find, not all are going to be interesting. To assess their interestingness, we follow Boden [2] defining a creative idea in general as *new*, *surprising*, and *valuable*. All three criteria depend on a specific reference point: A connection between two domains might be long known to some specialists but new, surprising, and hopefully valuable to a specific observer, who is not as familiar with the topic. To account for this, Boden [2] defines two types of creativity namely H-creativity and P-creativity. While H-creativity describes globally (historical) new ideas, P-creativity (psychological) limits the demand of novelty to a specific observer. Our findings are most likely to be P-creative since the found connections have to be indicated by the analyzed data in advance. However, a novel combination of information sources could even lead to H-creative bisociations. Analogous to novelty, the value of identified bisociations is a semantically determined property and strongly depends on the viewers' perspective. Since both novelty and value cannot be judged automatically, we leave their evaluation to the observer. In contrast, the potential surprise of a bisociation can be interpreted as the unlikeliness of a connection between the corresponding domains. We will express this intuition in more formal terms and use it as a guideline for an initial evaluation of possible bisociations.

Identifying Bisociations. Based on these considerations, we now narrow down properties that are in our view essential for two connected domains to form a bisociation. Despite the discussion above, we have not yet given a technical definition of what a domain is. We will return to this problem in Section 3.1 and assume for now that a domain is simply a set of information units. In the graph representation, two domains are connected either directly by edges between their nodes or more generally by nodes that are connected to both domains - the *bridging nodes*. Analogous to these more or less direct connections, of course connections spanning larger distances in the graph are possibly of interest. However, for the simplicity of description and due to the complexity of their determination, we will reduce the following considerations to the described simple cases. Such connecting nodes or edges bridge the two domains and together with the connected domains they form a *bisociation candidate*:

Definition 1 (Bisociation Candidate). *A* bisociation candidate *is a set of two domains and their connection within the network. That is, the subgraph induced by the nodes of the two domains* δ_1, δ_2 *and any further nodes that are connected to both domains:*

$$G\left[\delta_1 \cup \delta_2 \cup \{v \in V : N(v) \cap \delta_1 \neq \emptyset \wedge N(v) \cap \delta_2 \neq \emptyset\}\right]$$

Since it is impossible to precisely define what a surprising bisociation is, we develop three properties that distinguish promising bisociation candidates: *exclusiveness*, *size*, and *balance*. These technical demands are derived from an information-scientific view as e.g. expressed in [6]. In Ford's view, the creativity of a connection between two domains is related to (i) the dissimilarity of the connected domains and (ii) the level of abstraction on which the connection is established. In the following, we transfer these notions into graph theoretic terms by capturing them in properties that relate to structural features which can be identified in our data model.

We begin with the dissimilarity of two domains which we interpret as their connectedness within the graph. Maximal dissimilarity is rendered by two completely unconnected domains, closely followed by "minimally connected" domains. While the former case does not yield a bridging graph based bisociation (i.e. the connection itself is missing) the latter is captured by the exclusiveness property. Exclusiveness of a bisociation candidate states that the two involved domains are only sparsely connected, thereby expressing the demand of dissimilarity. On a more technical level it also excludes merely local exclusivity caused by nodes of high degree which connect almost everything, even unrelated domains, without providing meaningful connections.

Property (Exclusiveness)
A bisociation candidate is exclusive iff the connection between the two domains is

1. *small: the number of nodes connected to both domains (bridging nodes) is small in relation to the number of nodes in the adjacent domains;*
2. *sparse: the number of links between either the two domains directly or the domains and the nodes connecting them is small compared to the theoretical maximum;*
3. *concentrated: neighbors of the bridging nodes are concentrated in the adjacent domains and not scattered throughout the rest of the graph.*

Alternatively, this could be described in terms of probabilities: In a bisociation candidate two nodes from different domains are *linked* when they share an edge or have a common neighbor. Then exclusiveness describes the fact that two such nodes, randomly chosen from the two domains, are linked only with a low probability.

Directly entangled with this argument is the demand for size: a connection consisting of only a few nodes and links becomes less probable with growing domain sizes. In addition, a relation between two very small domains is hard to judge without knowledge about the represented semantic. It could be an expression of their close relation being exclusive only due to the small size of the connected domains. In that case the larger domains containing these two would show even more relations. It could also be an exclusive link due to domain dissimilarity. However, this situation would in turn be revealed when considering the larger domains, since these would also be exclusively connected. In essence, the exclusiveness of such a connection is pointless if the connected domains are very small, while it is amplified by domains of larger size. We capture this as follows:

Property (Size)
The size of a bisociation candidate is the number of nodes in the connected domains.

In terms of [6], the demand for size relates to the level of abstraction. A domain is more abstract than its subdomains since it includes more information units and thus an exclusive link between larger (i.e. more abstract) domains is a more promising bisociation candidate than a link between smaller domains.

Finally, the balance property assures that we avoid the situation of a very small domain attached to a large one:

Property (Balance)
A bisociation candidate is balanced iff the connected domains are of similar size.

In addition, we assume that domains of similar size tend to be of similar granularity and are thus likely to be on comparable levels of abstraction. Again, this is an approximation based on the assumption that domains are covered in comparable density. Thereby the demand for balance avoids exclusive links to small subdomains that are actually part of a broader connection between larger ones.

Following a discussion of the domain extraction process in Section 3.1, we will turn these three properties into a concrete measure for the quality of a bisociation candidate in Section 3.2.

3 Finding and Assessing Bisociations

In this section, we translate the demands described in the last section into an algorithm for the extraction and rating of bisociations. Therein we follow the previously indicated division of tasks: (i) domain extraction and (ii) scoring of bisociation candidates.

3.1 Domain Extraction

As described in Section 2, domain affiliation of nodes is reflected by similar direct and indirect neighborhoods in the graph. Thus comparing and grouping nodes based on their neighborhoods yields domains. In the following, we establish the close relation of a node similarity measure called *activation similarity* [16] to the above described demands. Based on this similarity, we show in a second part how domains can be found using hierarchical clustering.

Activation Similarity. The employed node similarity is based on *spreading activation* processes in which initially one node is activated. The activation spreads iteratively from the activated node, along incident edges, to adjacent nodes and activates them to a certain degree as well. Given that the graph is connected, not bipartite, and activation values are normalized after each step, the process converges after sufficient iterations. The final activation states are determined by the principal eigenvector of the adjacency matrix of the underlying graph as shown in [1]. They differ, however, by their initial state and those following it. Adopting the notation of [1], activation states of all nodes at a certain time k can be represented by the activation vector $\mathbf{a}^{(k)} \in \mathbb{R}^n$ given by

$$\mathbf{a}^{(k)} = A^k \mathbf{a}^{(0)} / \left\| A^k \mathbf{a}^{(0)} \right\|_2 .$$

The value $\mathbf{a}^{(k)}$ at index v, i.e. $\mathbf{a}_v^{(k)}$, is the activation level of node v and the initial activation levels of all nodes are determined by $\mathbf{a}^{(0)}$. The denominator in the equation further ensures that the overall activation levels do not grow unrestricted with the dominating eigenvalue. We add a parameter u to denote that the spreading activation was started by activating node u with unit value. The level of activation $\mathbf{a}_v^{(k)}(u)$ of a certain node $v \in V$ at a time k, induced by a spreading activation process started at node u, reflects the reachability of node v from node u via walks of length k. More precisely, it represents the fraction of weighted walks of length k from u to v among all walks of length k started at u. The more walks end at v the better is v reachable from u and the higher its activation level $\mathbf{a}_v^{(k)}(u)$ will be. To consider more than just walks of a certain length, the activation vectors are normalized and accumulated. In this accumulation an additional decay $\alpha \in [0, 1)$ serves to decrease the impact of longer walks. The *accumulated activation vector* of node u is then defined by

$$\hat{\mathbf{a}}^*(u) = D^{-\frac{1}{2}} \left(\sum_{k=1}^{k_{\max}} \alpha^k \mathbf{a}^{(k)}(u) \right) ,$$

with $D = \operatorname{diag}(d(v_1), \ldots, d(v_n))$ being the degree matrix and $k_{\max}$ the number of spreading iterations. Using $D^{-\frac{1}{2}}$ for degree normalization accounts for nodes of a very high degree: these are more likely to be reached and would thus distort similarities by attracting the activation if not taken care of. The value $\hat{\mathbf{a}}^*_v(u)$ represents the (normalized) sum of weighted walks of different lengths $1 \leqslant k \leqslant k_{\max}$ from u to v proportional to all weighted walks of different length starting at u and thus the relative reachability from u to v. $\hat{\mathbf{a}}^*(u)$ consequently serves as a description of the relations of node u to all other nodes in the graph.

Our basic assumption was, that nodes of similar domain are strongly connected and have a strong overlap of direct and indirect neighborhood. Hence, their reachability among each other is higher than that to other nodes. A comparison of the accumulated activation vectors of nodes compares the reachability of all other nodes from the specific nodes. On this basis we define the activation similarity $\sigma_{\text{act}} : V \times V \to \mathbb{R}$ between nodes $u, v \in V$ as

$$\sigma_{\text{act}}(u, v) = \cos(\hat{\mathbf{a}}^*(u), \hat{\mathbf{a}}^*(v))$$

and use it as node similarity for domain identification. For usual reasons we use the corresponding distance $1 - \sigma_{\text{act}}(u, v)$ for hierarchical clustering.

Domain Identification. Based on the distance described above, we apply hierarchical clustering for domain identification. To decide which subsets are to be merged we use Ward's linkage method [17], which minimizes the sum of squared distances within a cluster. This corresponds well with the notion of a domain since it tends to produce compact clusters and to merge clusters of similar size. First of all, we would expect a certain amount of similarity for arbitrary information units within a domain and thus a compact shape. Further, clusters of similar size are likely to represent domains on the same level of granularity and thus merging those corresponds to building upper-level domains. The resulting *merge tree* is formalized as follows:

Definition 2 (Merge tree). *A merge tree $T = (V_T, E_T)$ for a graph $G = (V, E)$ is a tree produced by a hierarchical clustering with node set $V_T = V \cup \Lambda$ where Λ is the set of clusters obtained by merging two nodes, a node and a cluster or two clusters. E_T describes the merging structure: $\{u\lambda, v\lambda\} \subseteq E_T$ iff the nodes or clusters u and v are merged into cluster $\lambda \in \Lambda$.*

However, not all clusters in the hierarchy are good domain candidates. If a cluster is merged with a single node, the result is unlikely to be an upper-level domain. Most likely, it is just an expansion of an already identified domain resulting from agglomerative clustering. These considerations lead to the domain definition:

Definition 3 (Domain). *A cluster δ_1 is a domain iff it is merged with another cluster in the corresponding merge tree:*

$$\delta_1 \in \Lambda \textit{ is a domain} \Leftrightarrow \exists \delta_2, \kappa \in \Lambda \textit{ such that } \{\{\delta_1, \kappa\}, \{\delta_2, \kappa\}\} \subseteq E_T \, .$$

Note that in this definition δ_2 is also a domain.

3.2 Scoring Bisociation Candidates

In the next step, we iterate over all pairs of disjoint domains and construct a bisociation candidate for each pair. We then try to assess the potential of each candidate using the properties shown in Section 2. A first step in this assessment is the identification of the bridging nodes:

Definition 4 (Bridging nodes). *Let δ_1 and δ_2 be two domains derived from the merge tree of the graph $G = (V, E)$. The set of bridging nodes* $\text{bn}(\delta_1, \delta_2)$ *containes all nodes that are connected to both domains:*

$$\text{bn}(\delta_1, \delta_2) = \{v \in V : \exists \{v, u_1\}, \{v, u_2\} \in E \text{ with } u_1 \in \delta_1, u_2 \in \delta_2\} .$$

Note that this definition includes nodes belonging to one of the two domains, thereby allowing direct connections between nodes of these domains.

Using this concept we can define the *b-score*, which combines the properties described in Section 2 into a single index that can be used to compare bisociation candidates directly. We therefore consider each property separately and combine them into an index at the end.

Exclusiveness could be directly expressed by the number of nodes in $\text{bn}(\delta_1, \delta_2)$. However, this is not a sufficient condition. Nodes of high degree are likely to connect different domains, maybe even some of them exclusively. Nevertheless, such nodes are unlikely to form good bisociations since they are not very specific. On the other hand, bridging nodes providing only a few connections at all (and thus a large fraction of them within δ_1 and δ_2) tend to express a very specific connection. This interpretation of node degrees is of course an unproved assumption, yet we consider it as necessary and reasonable. Since we are only interested in the case of specific connection, we assess exclusiveness by using the inverse of the sum of the bridging nodes' degrees: $2/\sum_{v \in \text{bn}(\delta_1, \delta_2)} d(v)$. The 2 in the numerator ensures that this quantity is bound to the interval $[0, 1]$, with 1 being the best possible value. The balance property is accounted for by relating the domain sizes in a fraction: $\min\{|\delta_1|, |\delta_2|\}/\max\{|\delta_1|, |\delta_2|\}$, again bound to $[0, 1]$ with one expressing perfect balance. Finally, the size property is integrated as the sum of the domain sizes.

As described above, a combination of all three properties is a necessary prerequisite for an interesting bisociation. Therefore, our bisociation score is a product of the individual quantities. Only in the case of $\text{bn}(\delta_1, \delta_2) = \emptyset$ is our measure undefined. However, this situation is only possible if the domains are unconnected, so we define the score to be 0 in this case. For all non-trivial cases the score has strictly positive values and is defined as follows:

Definition 5 (b-score). *Let δ_1 and δ_2 be two domains, then the b-score of the corresponding bisociation candidate is*

$$\textit{b-score}(\delta_1, \delta_2) = \frac{2}{\sum\limits_{v \in bn(\delta_1, \delta_2)} d(v)} \cdot \frac{\min\{|\delta_1|, |\delta_2|\}}{\max\{|\delta_1|, |\delta_2|\}} \cdot (|\delta_1| + |\delta_2|).$$

This combination acts comparably to a conjunction of the involved properties. In our opinion, an ideal bisociation is represented by two equally sized domains connected directly by a single edge or indirectly by a node connected to both domains. This optimizes the b-score, leaving the sum of the domain sizes as the only criterion for the assessment of this candidate. Further, every deviation from this ideal situation results in a deterioration of the score. In addition, the calculation of the b-score only involves information about the two domains and their neighborhoods and not the whole graph, which is important when the underlying graph is very large.

3.3 Complexity and Scalability

To determine the complexity and scalability of the complete algorithm, the process can be split into three parts: similarity computation, clustering, and scoring of domain pairs. In the following, we examine the complexity of each of these three parts.

To compute the pairwise activation similarities, the accumulated activation vectors for all nodes need to be determined. For each node several matrix vector multiplications, normalizations, scalings and additions are necessary. The computational complexity is dominated by that of the matrix vector multiplication with a complexity of $\mathcal{O}(n^2)$. Repeating this process for all nodes leads to an overall complexity of $\mathcal{O}(n^3)$.

Note, that this is a worst case result which can be improved substantially by exploiting the graph structure and the characteristics of the convergence of the spreading activation process: First of all, in a large, sparsely connected network the activation is only spread over existing edges. This alone speeds up the matrix multiplication depending on the network density which should usually be low, since otherwise the considered information units are connected to most other information units which is not very informative. Further, a large network diameter yields strongly localized activation vectors (i.e. most nodes have zero activation) in the first few iterations, since activation can only spread to nodes that are adjacent to already activated nodes in each step. This can be exploited, when only the first few activation vectors are used to approximate the activation similarity. In addition, the convergence rate of the power iteration itself is exponentially related to the ratio of the first two eigenvalues and the additional decay factor (c.f. [1]). The latter guarantees that only a few iterations of activation spreading are necessary and together with the sparsity and large diameter of the underlying network, only a small part has to be considered in the computation for an individual node. Unfortunately, the assumption of a large diameter is contradicted by many observed real-world networks and our own application example. A possible counter measure could be the removal of high-degree nodes, which should result in a larger diameter and only minimal information loss since the information provided by these nodes is most likely of highly general nature.

The crucial part of the domain identification process is the clustering of the nodes based on the computed similarity. Here, the complexity is dominated by the ward clustering which involves $\mathcal{O}(n^2 \log^2 n)$ steps (c.f. [5]). This could be further reduced by the employment of other clustering algorithms (i.e. density based approaches) or a completely different domain identification process.

The final step is the determination of the b-scores for all domain pairs. Since in the hierarchical clustering $n-1$ merge steps are executed, $|V_T|$ and thus the number of domains is bound by $n-1$. Consequently, the number of domain pairs to be analyzed is less than $\binom{n}{2}$, i.e. in $\mathcal{O}(n^2)$. For the determination of the b-score of an individual domain pair, the domain sizes can already be prepared in the domain identification process, avoiding additional time consumption. The complexity of the b-score computation for two domains δ_1, δ_2 is therefore determined completely by the calculation of $\sum_{v \in bn(\delta_1,\delta_2)} d(v)$ - the sum of the bridging nodes' degrees. This calculation is again dominated by the determination of the elements of $bn(\delta_1, \delta_2)$. Considering a domain as a set of its contained nodes these common neighbors can be determined in $\mathcal{O}\left(\max_{\delta \in \{\delta_1,\delta_2\}} \sum_{v \in \delta} d(v)\right)$ as shown by Algorithm 1.

Algorithm 1. b-score computation

```
Input: domains δ1, δ2, graph G = (V, E)
Result: b-score(δ1, δ2)
for v ∈ δ1 do
    for u ∈ N(v) do
        m_u := true;
s := 0;
for v ∈ δ2 do
    for u ∈ N(v) do
        if m_u then
            s := s + d(u);
            m_u := false;
return 2/s · min(|δ1|,|δ2|)/max(|δ1|,|δ2|) · (|δ1| + |δ2|);
```

With $m_v = \text{false}\ \forall v \in V$ being initialized only once for the whole computation and cleaned up after each candidate evaluation, the complexity of the procedure is directly related to the loops over the neighbors of each node in either domain.

Besides the clustering process, the determination of b-scores is an important aspect in the total time spent in the analysis of a dataset. To speed up this process, we propose pruning of the set of domains and candidate domain pairs. Recall our definition of a domain based on the merge tree: it is sufficient that a cluster from Λ is merged with another cluster from Λ in contrast to merging with a single element from V. Firstly, this produces a large number of small domains: e.g. two node domains which are in turn merged with other elements from Λ. Secondly, this procedure yields a number of largely overlapping clusters that differ only in a small numer of nodes, e.g. when a large cluster is merged with a very small one. This is illustrated by the distribution of domain sizes in

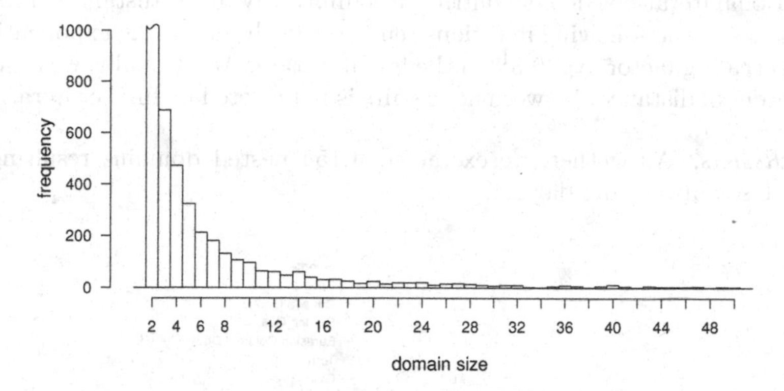

Fig. 1. Distribution of domain sizes for domains of size 0 to 50 (the number of 2-node domains is not depicted completely)

our evaluation example of Section 4 shown in Figure 1. It can be observed that a large number of domains consist of only two or three nodes. Considering the demand for size, balance and the exclusiveness of a connection between such small domains it can be seen that a large gain in efficiency could be obtained by pruning small domains or bisociation candidates involving very small domains.

In addition, a further reduction of the number of domain pairs to be considered may be achieved by filtering highly unbalanced candidates, though in that case a threshold needs to be chosen cautiously.

4 Preliminary Evaluation

To demonstrate our approach, we applied our method to the Schools-Wikipedia[1] (2008/09) dataset, which is described in more detail in [16]. Due to the lack of a benchmark mechanism we manually explored the top rated bisociation candidates and describe some of them to demonstrate the reasonability of the results.

The dataset consists of a subset of the English Wikipedia with about 5500 articles. For our experiment, we consider each article as a separate unit of information and model it as a node. We interpret cross-references as relations and introduce an undirected edge whenever one article references another. The resulting graph is connected except for two isolated nodes which we removed beforehand. For the remaining nodes we extracted the domains using the procedure described above.

Parameter Choices. To focus on the local neighborhood of nodes we used the decay value $\alpha = 0.3$ for the activation similarity. Due to this decay and the graph structure the activation processes converged quickly allowing a restriction to $k_{\max} = 10$ iterations for each process. This choice seems arbitrary, but we ensured

[1] `http://schools-wikipedia.org/`.

that additional iterations do not contribute significantly to the distances. First of all, the values of the following iterations tend to vanish due to the exponentially decreasing scaling factor, e.g. 0.3^{10} in the last iteration. Additionally, we ensured that the order of distances between node pairs is not altered by further iterations.

Domain Results. Altogether, we extracted 4,154 nested domains resulting in 8,578,977 bisociation candidates.

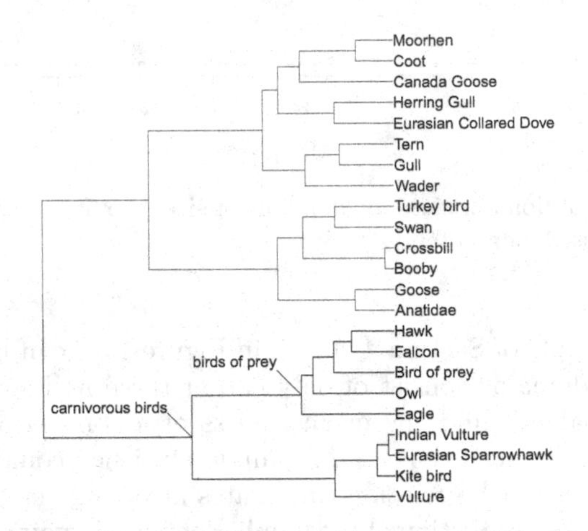

Fig. 2. Part of merge tree with articles about birds

A part of the merge tree involving birds is shown in Figure 2. This small excerpt illustrates that the hierarchical clustering can yield conceptually well defined domains, though we could not verify the complete result manually. In the example, birds of prey such as hawk, falcon, eagle etc. end up in the same cluster with carnivorous birds such as e.g. vulture and are finally combined with non-carnivorous birds to a larger cluster. This example further illustrates that the nodes of a good domain are not necessarily connected, as there are few connections within the sets of birds, and yet they share a number of external references.

Bisociation Results. The b-scores of the best 200 bisociation candidates are shown in Figure 3. It can be observed that the scores quickly decrease from some exceptionally high rated examples (b-score 1.5 and more) to the vast majority of candidates rated lower than 1. This indicates that - using the b-score as basis of judgement - the dataset contains some outstanding bisociations while most candidates are uninteresting. Since the individual candidates have to be assessed manually, this encourages the decision to concentrate on the first few pairs. Note, that due to the design of the b-score the best rated candidates often exhibit only a single bridging node. In addition, these bridging nodes appear repeatedly in bisociation candidates that differ only slightly in the composition of the connected domains which usually shrink along with decreasing b-scores.

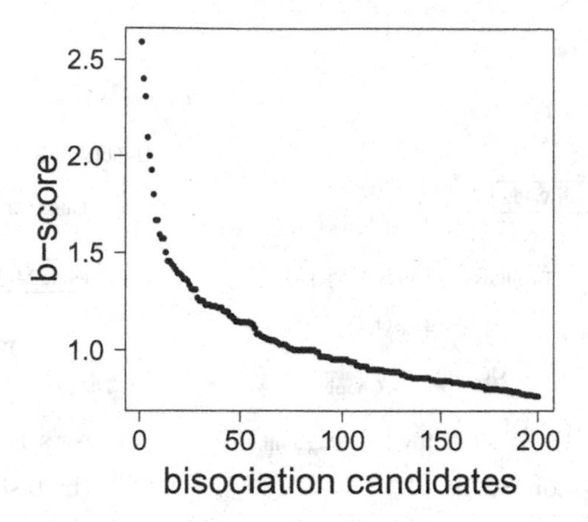

Fig. 3. Distribution of the b-score for the 200 top rated bisociation candidates

In such cases we focused on the first appearance and ignored the lower rated ones. Finally, due to the employed distance in the domain extraction process the resulting domains are not necessarily connected.

Result Evaluation. Since a comprehensive description and illustration of results would quickly exceed the scope of this article, we only show the three top-rated candidates and emulate a realistic result evaluation by additionally presenting interesting candidates found under the top-rated ones. In our visualizations of the individual candidates, we show the nodes of the individual candidate together with the link structure of the extracted network. Domain affiliation of the nodes is indicated by their background color (white or gray) while bridging nodes are highlighted by a black node border.

The overall best rated bisociation candidate, shown in Figure 4a, is composed of a domain of classical music composers such as *Schumann*, *Schubert*, *Beethoven*, *Mozart*, etc. connected to a domain incorporating operating systems and software such as *Microsoft Windows*, *Linux*, *Unix* and the *X window system*. Intuitively these domains - composers and operating systems - are highly separated and a direct connection seems to be unlikely. However, a connection between both is provided by *Jet Set Willy* which is a computer game for the *Commodore 64*.

The unusual connection to the domain of composers is explained by its title music, which was adapted from the first movement of *Beethoven*'s Moonlight Sonata. On the other side, a level editor developed for *Microsoft Windows* connects to the domain of operating systems. To us, this connection was new and surprising, though one might argue its value. Besides that, the formalized demands are met well. The connection itself is very specific, since *Jet Set Willy* provides only a few links and the two domains are far apart, i.e. not connected otherwise. The sizes of the two domains are not exactly equal but with 5 nodes

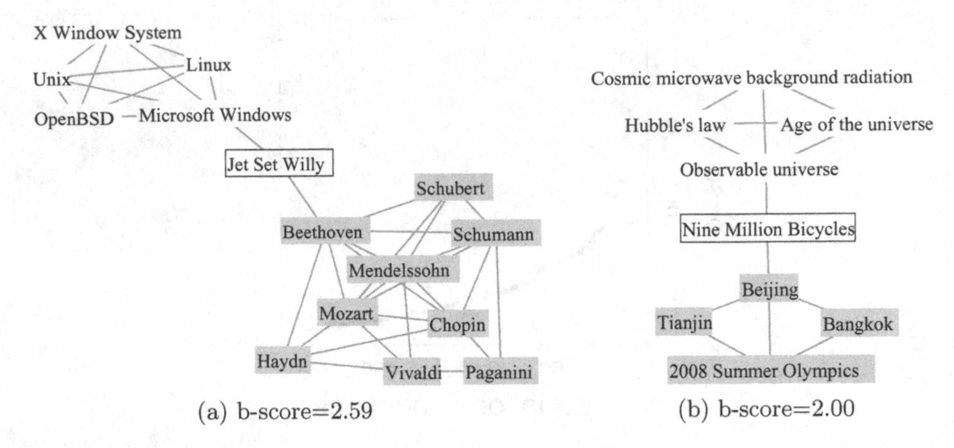

Fig. 4. The two top rated bisociations and their b-score (see text for details)

in one domain and 9 nodes in the other they are comparable. Finally, in view of the small size of the dataset and its wide distribution of topics, the absolute domain sizes are reasonable.

Figure 4b depicts the next best candidate where the node *Nine Million Bicycles* connects a geography with an astronomy domain. Excluding repetitions of *Jet Set Willy*, this is the second best rated bisociation candidate and it also appears in several variants within the best rated candidates. The bridging node *Nine Million Bicycles* refers to a song which was inspired by a visit in *Beijing*, China. The connection to the astronomy section is established by the second verse which states that the *Observable universe* is twelve billion years old. The actual link in the data set is established by the report of a discussion about the correctness of this statement in the article. As in the first example, the value of this connection is at least arguable, while the formal criteria are met well.

Figure 5a shows the third best rated candidate. Here, the node *Tesseract* connects a geometry domain with the domain of a famous BBC series called *Doctor Who*. A *tesseract* is a special geometrical shape also known as hypercube, which has a natural connection to the domain of geometry. In the context of the TV-series it is used to describe the form of *Doctor Who*'s spaceship called *TARDIS*.

In the following, we present some hand-picked samples that appeared as interesting to us. These should illustrate, that despite the limits imposed by the analyzed data some interesting, though not always valuable, proposals where made by the presented method. Figure 5b shows a bisociation candidate, where the node *Sequence alignment* connects domains from computer science and chemistry. The connection to the computer science domain results from reports about open source software programs implementing some of the involved algorithms. *NMR spectroscopy*, providing the connection to a domain about chemistry, is an analysis technique with applications in organic chemistry.

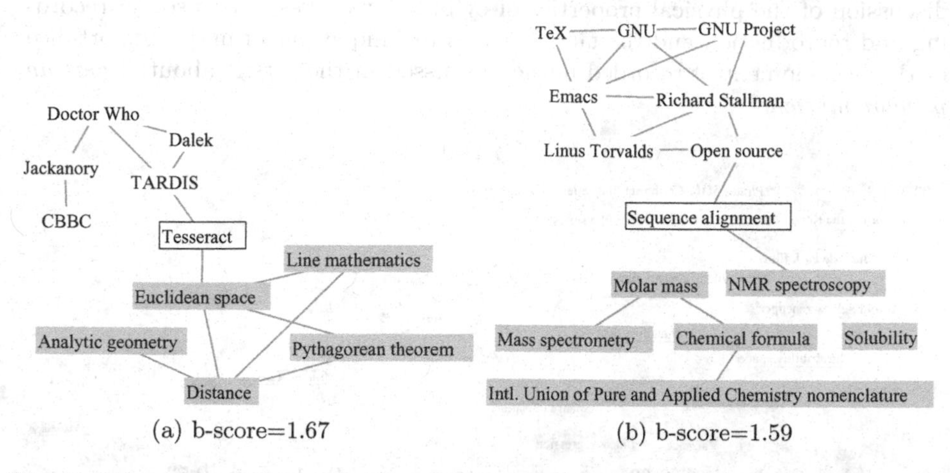

(a) b-score=1.67 (b) b-score=1.59

Fig. 5. Third best rated bisociation candidate and a bisociation candidate connecting open source related articles with articles about chemistry

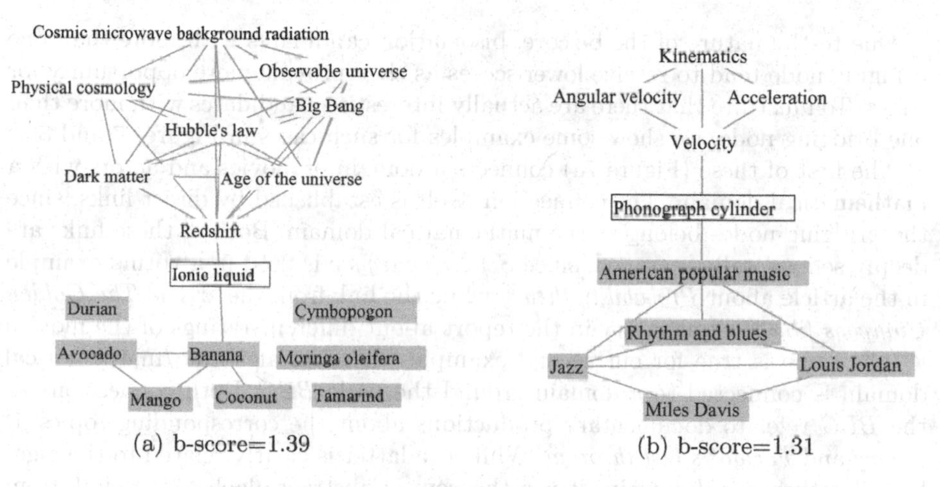

(a) b-score=1.39 (b) b-score=1.31

Fig. 6. Two interesting bisociation candidates with good b-scores

A quite surprising example - at least for us - is given in Figure 6a, where *ionic liquid* provides a nearly direct connection between the *Redshift* effect and *Banana.* While the relation to the *Redshift* effect is due to applications of ionic liquids in infra-red imaging in telescopes, the link to *Banana* is produced by an example for an application in food science

An example for a historical bisociation is shown in Figure 6b, where a music domain is connected to some physical notions in an article about the *phonograph-cylinder*. The *phonograph-cylinder* was - quoting the article - the "earliest method of recording and reproducing sound" and thus a historically new connection between physics and music. The concrete connections are established by a discussion of the physical properties of cylinders with respect to sound recording and reproduction and the fact that this technique was of major importance in the development of recorded music, discussed in the article about *American popular music.*

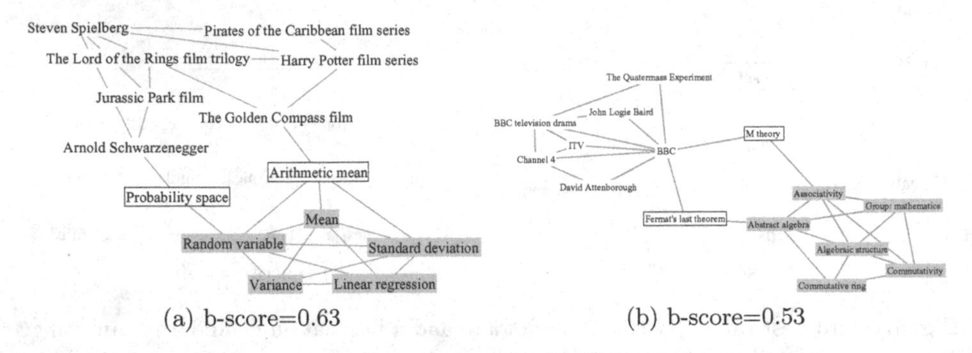

(a) b-score=0.63 (b) b-score=0.53

Fig. 7. Bisociation candidates with two bridging nodes

Due to the nature of the b-score, bisociation candidates with more than one bridging node tend to receive lower scores as they provide more opportunity for edges. To illustrate that there are actually interesting candidates with more than one bridging node, we show some examples for such cases in Figures 7 and 8a.

The first of these (Figure 7a) connects a domain of movies and actors with a mathematical domain. The connection itself is established by direct links, since the bridging nodes belong to the mathematical domain. Both of these links are deeply semantically reasoned, since *Schwarzenegger* is used in a voting example in the article about *Probability theory* while the link from the article *The Golden Compass film* is explanatory in the report about different ratings of the movie.

The same is true for our second example where an algebraic/mathematical domain is connected to a domain around the node *BBC*. Both connections to the *BBC* refer to documentary productions about the corresponding topics *M theory* and *Fermat's last theorem.* While the latter is clearly related to the algebraic/mathematical domain (it is a theorem of abstract algebra), the link from *M theory* to *Associativity* appears in the explanation of some details about *M theory.*

An example for a bisociation candidate with three bridging nodes and a b-score of 0.35 is shown in Figure 8a. It is basically an extension of the example shown in Figure 7a. The bridging nodes *Chaos theory, Probability space* and *Arithmetic mean* connect a statistics domain with nodes like *Variance* or *Mean* and a movie domain with nodes like *Arnold Schwarzenegger* or *The Lord of the Rings.*

For completeness, we additionally evaluated very low rated candidates. A negative example of a bisociation can be seen in Figure 8b. A football domain consisting of football clubs *Celtic F.C., Rangers F.C.*, etc. is connected to an arctic domain containing *Arctic, Arctic Ocean, Polar bear,* etc. The bridging nodes

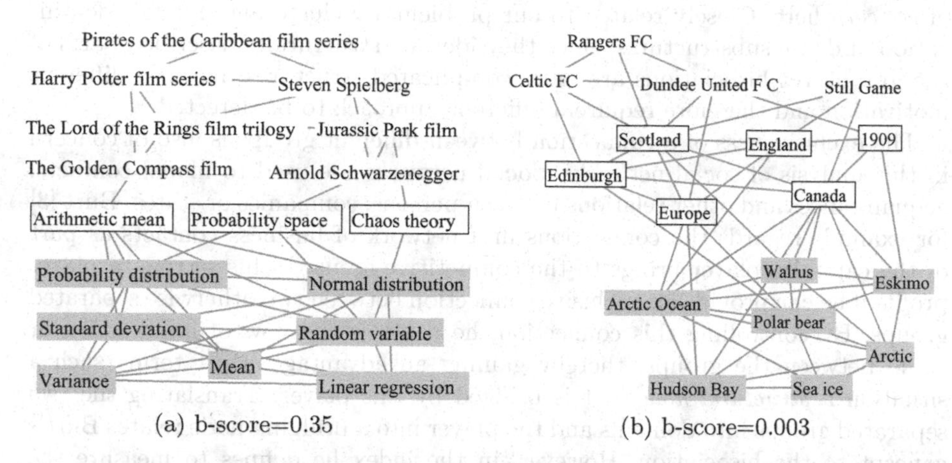

Fig. 8. A bisociation candidate (b-score=0.35) involving three bridging nodes (8a) and a bad bisociation (b-score=0.003) due to non exclusiveness and missing balance (8b)

are countries such as *Canada*, *Europe*, *England*, etc. and *Edinburgh*. Clearly, this bisociation candidate is not exclusive since the number of connecting nodes is high (proportional to the domain sizes) and the degree of these nodes is high as well (countries have a very high degree in Schools-Wikipedia).

The above examples illustrate that our index discriminates well with respect to exclusiveness and balance. A detailed examination showed in addition that size is negatively correlated with both other index components. This and the limited size of the dataset could explain the small sizes of the best rated candidates.

Our preliminary evaluation indicates the potential of the presented method to detect bisociations based on the analysis of the graph structure. Even though Schools-Wikipedia is a reasonable dataset for evaluation purposes, one cannot expect to find valuable or even truly surprising bisociations therein since it is limited to handpicked, carefully administrated common knowledge, suitable for children. We opted to manually evaluate the results since the value of a bisociation is a highly subjective semantic property, which inhibits automatic evaluation. An evaluation using synthetic data is complicated by the difficulty of realistic simulation and could introduce an unwanted bias on certain types of networks, distorting the results. Finally, manually tagged datasets for this purpose are not available.

5 Related Work

Although a wealth of techniques solving different graph mining problems already exist (see e.g. [4] for an overview), we found none to be suitable for the problem addressed here. Most of them focus on finding frequent subgraphs, which is not of concern here. Closely related to our problem are clustering and the identification of dense substructures, since they identify structurally described parts of the graph. Yet bisociations are more complicated structures due to a different motivation and therefore require a different approach to be detected.

The exclusiveness of a connection between different groups is also of concern in the analysis of social networks. Social networks are used to model contacts, acquaintances and other relations between persons, companies or states. Burt [3] for example regards the connections in a network of business contacts as part of the capital a player brings to the competitive arena. In his setting, a player profits if he can provide an exclusive connection between two otherwise separated groups. By controlling this connection, he controls the flow of information or value between the groups, thereby gaining an advantage. Burt terms such a situation a *structural hole* that is bridged by this player. Translating the two separated groups into domains and the player into a bridging node relates Burt's concept to the bisociation. However, in the index he defines to measure the presence of a structural hole only the very local view of the player himself is integrated. Further, his index would implicitly render domains a product of only direct connections between concepts, whereas we showed earlier that a more general concept of similarity is advisable.

A global measure for the amount of control over connections between other players is provided by *betweenness* [7]. Betweenness measures the fraction of shortest paths between all other nodes that employ an individual node. Intuitively, the shortest paths in a network are the preferred channel of transportation. Consequently, if a node appears in a large fraction of these shortest connections, it can exert a certain amount of control on the flow of goods or information. The translation to our setting is again straightforward, but it provides no explanation of what a domain is. However, this approach leads to the variant of hierarchical graph clustering proposed in [12]. Girvan and Newman develop a top-down approach in which the edges of highest betweenness are removed recursively until the graph splits up into several components. Still, only a subset of the properties we demand from a bisociation is considered.

Strongly related to bisociations is the notion of serendipity [15] which describes accidental discoveries. Serendipitous discoveries strongly overlap with bisociations since the involved fortuitousness is often caused by the connection of dissimilar domains of knowledge. A number of approaches (e.g. [13,8]) were developed to implement this concept in recommender systems balancing between the suggestion of strongly related versus loosely related surprising suggestions of content which lead the user into new directions not too far from his original interests. In a sense this work is parallel to ours, but targets a different setting - users and their preferences - and thus follows different criteria of optimality.

None of these approaches provide a coherent, formal setting for the description of domains and potentially interesting links between these. Note further, that our approach is additionally distinguished from all of the mentioned variants in that the process of community or domain detection is guided by a node similarity tailored to the identification of domains of knowledge.

6 Conclusion

We presented an approach for the discovery of potentially interesting, domain crossing associations, so-called bisociations. For this purpose we developed a formal framework to describe potentially interesting bisociations and corresponding methods to identify domains and rank bisociations according to interestingness. Our evaluation on a well-understood benchmark data set has shown promising first results. We expect that the ability to point the user to potentially interesting, truly novel insights in data collections will play an increasingly important role in modern data analysis.

The presented method is, however, not intended to be directly applicable to real world problems. Instead, we presented a framework that can be used as a guideline and benchmark for further developments in this direction. Conceptually, we divide the presented approach in two parts: (i) basic considerations about the expression of domains and bisociations in a structural knowledge description and (ii) a framework for the identification of the described concepts. To demonstrate the soundness of our considerations and their general applicability, we then filled this framework with a number of heuristics to solve the resulting subproblems. Clearly, the choice of these heuristics is to some extend arbitrary and can be improved, especially in light of additional experience with more realistic data. However, by using them in a first instantiation of the described framework, we demonstrated that the underlying assumptions lead to promising results. Finally, we hope that further improvements of this framework will ultimately lead to systems that are applicable in practical settings.

Acknowledgements. This research was supported by the DFG under grant GRK 1042 (Research Training Group „Explorative Analysis and Visualization of Large Information Spaces“) and the European Commission in the 7th Framework Programme (FP7-ICT-2007-C FET-Open, contract no. BISON-211898).

References

1. Berthold, M.R., Brandes, U., Kötter, T., Mader, M., Nagel, U., Thiel, K.: Pure spreading activation is pointless. In: Proceedings of the CIKM the 18th Conference on Information and Knowledge Management, pp. 1915–1919 (2009)
2. Boden, M.A.: Précis of the creative mind: Myths and mechanisms. Behavioral and Brain Sciences 17(03), 519–531 (1994)
3. Burt, R.S.: Structural holes: the social structure of competition. Harvard University Press (1992)
4. Cook, D.J., Holder, L.B.: Mining graph data. Wiley-Interscience (2007)

5. Eppstein, D.: Fast hierarchical clustering and other applications of dynamic closest pairs. In: Proceedings of the Ninth Annual ACM-SIAM Symposium on Discrete Algorithms, SODA 1998, pp. 619–628. Society for Industrial and Applied Mathematics, Philadelphia (1998)
6. Ford, N.: Information retrieval and creativity: Towards support for the original thinker. Journal of Documentation 55(5), 528–542 (1999)
7. Freeman, L.C.: A set of measures of centrality based upon betweenness. Sociometry 40, 35–41 (1977)
8. Kamahara, J., Asakawa, T., Shimojo, S., Miyahara, H.: A community-based recommendation system to reveal unexpected interests. In: Proceedings of the 11th International Multimedia Modelling Conference (MMM 2005), pp. 433–438. IEEE (2005)
9. Koestler, A.: The Act of Creation. Macmillan (1964)
10. Kötter, T., Thiel, K., Berthold, M.R.: Domain bridging associations support creativity. In: Proceedings of the International Conference on Computational Creativity, Lisbon, pp. 200–204 (2010)
11. Maxwell, J.C.: A treatise on electricity and magnetism. Nature 7, 478–480 (1873)
12. Newman, M.E.J., Girvan, M.: Finding and evaluating community structure in networks. Physical Review E 69(2), 026113 (2004)
13. Onuma, K., Tong, H., Faloutsos, C.: Tangent: a novel, 'surprise me', recommendation algorithm. In: Proceedings of the 15th ACM SIGKDD International Conference on Knowledge Discovery and Data Mining, KDD 2009, pp. 657–666. ACM, New York (2009)
14. Poincaré, H.: Mathematical creation. Resonance 5(2), 85–94 (2000); reprinted from Science et méthode (1908)
15. Roberts, R.M.: Serendipity: Accidental Discoveries in Science. Wiley-VCH (1989)
16. Thiel, K., Berthold, M.R.: Node similarities from spreading activation. In: Proceedings of the IEEE International Conference on Data Mining, pp. 1085–1090 (2010)
17. Ward Jr., J.H.: Hierarchical grouping to optimize an objective function. Journal of the American Statistical Association 58(301), 236–244 (1963)

Exploration: Overview

Andreas Nürnberger

Data and Knowledge Engineering Group,
Faculty of Computer Science, Otto-von-Guericke-University, Germany
andreas.nuernberger@ovgu.de

1 Introduction

In the previous chapters of this book quite different approaches to create networks based on existing data collections (Part II) have been discussed and diverse methods for network analysis have been proposed (Part III). All these methods provide powerful means in order to obtain different insights into the properties of huge information networks or graphs. However, one disadavantage of these individual approaches is that each approach provides only specific facets of information to the end user. Integrated exploration tools that enable the interactive exploration of huge graphs and data collections using data viusalization, aggregation and mining methods could be much more beneficial for the (interactive) task of finding bisociations. Therefore, we present and discuss in this part of the book methods for interactive exploration that bring these fields together.

2 Contributions

This part starts with a more general discussion of methods for interactive data exploration. Therefore, the chapter of Gossen et al. [1] provides first a brief review of methods and tools for the interactive exploration of graphs with a focus on approaches to support bisociative discoveries. Furthermore, a critical discussion of the challenges of evaluating the performance and quality of exploration tools is given.

In the second chapter of Haun et al. [2] the Creative Exploration Toolkit (CET) is presented, which was developped as part of the BISON project. CET is a user interface for graph visualization designed towards explorative tasks. It supports the integration and communication with external data sources and mining tools, especially the data-mining platform KNIME. Besides the tool itself, the chapter also presents the results of a small case study, in which the applicability of the tool for bisociative discoveries is analyzed.

In the chapter "Bisociative Knowledge Discovery by Literature Outlier Detection", Petrič et al. [4] discuss the role of outliers in literature-based knowledge discovery. They show that outlier documents can be successfully used as means of detecting bridging terms that connect documents of two different literature sources.

The chapter "Exploring the Power of Outliers for Cross-Domain Literature Mining" of Sluban et al. [5] proposes an approach to find outliers that are potential candidates of documents that discuss bisociative discoveries bridging different scientific domains. The discussed approach aims at finding cross-domain

M.R. Berthold (Ed.): Bisociative Knowledge Discovery, LNAI 7250, pp. 285–286, 2012.

links by mining for bridging concepts or terms (b-terms) and can be used for an exploration task using iterative search and filter methods.

An example of an exploration tool that is based on the idea of Sluban et al. [5] is presented in the final chapter by Juršič et al. [3]. Juršič et al. introduce a system for bisociative literature mining called *CrossBee*. This system is focussed on b-term identification and ranking and supports the search for hidden links connecting two different domains. The chapter contains a detailed description of the proposed methodology that is implemented in *CrossBee*. Furthermore, an experimental evaluation on two datasets, Migraine-Magnesium and Autism-Calcineurin, is reported.

3 Conclusions

Interactive exploration methods provide powerful means to support bisociative discoveries in huge data collections and networks. The contributions presented in this part of the book, provide some insights into their capabilities and the challenges these approaches have to face. The tools presented can – together with the critical discussions – serve as a basis for the development of advanced exploration methods that can enable interactive bisociative discoveries.

References

1. Gossen, T., Nitsche, M., Haun, S., Nürnberger, A.: Data Exploration for Knowledge Discovery: A brief Overview of Tools and Evaluation Methods. In: Berthold, M.R. (ed.) Bisociative Knowledge Discovery. LNCS (LNAI), vol. 7250, pp. 287–300. Springer, Heidelberg (2012)
2. Haun, S., Gossen, T., Nürnberger, A., Kötter, T., Thiel, K., Berthold, M.R.: On the Integration of Graph Exploration and Data Analysis: The Creative Exploration Toolkit. In: Berthold, M.R. (ed.) Bisociative Knowledge Discovery. LNCS (LNAI), vol. 7250, pp. 301–312. Springer, Heidelberg (2012)
3. Juršič, M., Cestnik, B., Urbančič, T., Lavrač, N.: Bisociative Literature Mining by Ensemble Heuristics. In: Berthold, M.R. (ed.) Bisociative Knowledge Discovery. LNCS (LNAI), vol. 7250, pp. 338–358. Springer, Heidelberg (2012)
4. Petrič, I., Cestnik, B., Lavrač, N., Urbančič, T.: Bisociative Knowledge Discovery by Literature Outlier Detection. In: Berthold, M.R. (ed.) Bisociative Knowledge Discovery. LNCS (LNAI), vol. 7250, pp. 313–324. Springer, Heidelberg (2012)
5. Sluban, B., Juršič, M., Cestnik, B., Lavrač, N.: Exploring the Power of Outliers for Cross-Domain Literature Mining. In: Berthold, M.R. (ed.) Bisociative Knowledge Discovery. LNCS (LNAI), vol. 7250, pp. 325–337. Springer, Heidelberg (2012)

Data Exploration for Bisociative Knowledge Discovery: A Brief Overview of Tools and Evaluation Methods

Tatiana Gossen, Marcus Nitsche, Stefan Haun, and Andreas Nürnberger

Data and Knowledge Engineering Group,
Faculty of Computer Science, Otto-von-Guericke-University, Germany
`http://www.findke.ovgu.de`

Abstract. In this chapter we explain the definition of the term *(data) exploration*. We refine this definition in the context of browsing, navigating and searching. We provide a definition of *bisociative exploration* and derive requirements on user interfaces, which are designed to support bisociative knowledge discovery. We discuss how to support subtasks of bisociative data exploration with appropriate user interface elements. We also present a set of exploratory tools, which are currently available or in development. Finally, we discuss the problem of usability evaluation in the context of exploratory search. Two main issues - complexity and comparability - are explained and possible solutions proposed.

Keywords: exploration, exploratory search, tools, usability evaluation.

1 Introduction

A lot of data in different domains, e.g. biology, astronomy, geography, and other sciences were gathered and became available during the last decades. Much useful and interesting knowledge is hidden in these data sets. Therefore, experts in different knowledge domains explore the data in order to make new discoveries and, thus, data exploration becomes one of the standard user tasks.

Unfortunately, the well-known phrase of the futurist John Naisbitt "We are drowning in information but starved for knowledge" [26] is still relevant. One man alone is not able to examine even small parts of the available data. Knowledge discovery tools are a way out and show a promising direction to support users by data exploration. Finding implicit links between given data (sets) from different domains is an even more challenging task. This is what bisociative knowledge discovery tools are supposed to support. This chapter addresses the issue of data exploration for bisociative knowledge discovery. Further details on bisociative knowledge discovery in general can be found in [6].

The structure of this work is as follows. In Sect. 2 we explain the meaning of (data) exploration, which, although often used, has no formal definition. Then, we discuss how to support bisociative data exploration trough user interface

M.R. Berthold (Ed.): Bisociative Knowledge Discovery, LNAI 7250, pp. 287–300, 2012.

elements and give a short overview of tools for data discovery and exploration in Sect. 3. In Sect. 4 we discuss the problem of evaluating the performance and usability of knowledge discovery tools and propose a possible solution. The chapter ends with a conclusion and a discussion of future work in Sect. 5.

2 Bisociative Data Exploration

Many authors use the term *exploration* without giving a well-formed explanation. Since this term has many meanings and just some of them are related to computer science, it is important to clarify what exploration in the context of this chapter refers to. Therefore, we explain the meaning of the term *exploration* in the context of information retrieval, human computer interaction, visual data analysis and bisociative knowledge discovery. We also elaborate on the implications for user interface design.

2.1 Different Meanings of Exploration

A general definition of *exploring* is found in The American Heritage Dictionary of the English Language. Here, exploring is defined as "to search into or travel in for the purpose of discovery" [3].

In his book "Interaction Design" Michael Herczeg provides a more specific explanation of the term *exploration* in the context of human-computer interaction (translated from German into English from [19], p. 81):

> "Exploration is a method for searching in new, unknown information spaces. It is similar to browsing with the important difference that exploration is defined by sounding the user's interest in the information space. Further exploration is connected to the user's wish to get to know almost the entire structure and the content of the information space. After a successful exploration of the information space users mainly navigate in it. The user builds up an overview map while exploring. The entry point for exploratory search is [...] in most cases given by accident."

Contrary to Herczeg's definition we see exploration not as a specific method of information access. More likely we see it as an enhancement of other methods: The basic information access methods *browsing*, *navigating* and *searching* can be enhanced by aspects of exploration. Hence, there exist *exploratory browsing*, *exploratory navigation* and *exploratory search*. Furthermore, the information space does not necessarily need to be *unknown*. For example, in exploratory navigation users are already familiar with certain navigation points and are able to make more sophisticated decisions where to go next.

Searching is often an integral part of exploration. Searching means "to look through (a place, records, etc.) thoroughly in order to find someone or something" [2]. Many users on computer systems are engaged in searching. Search systems usually provide an input box for entering search keywords to describe the user's information need. Gary Marchionini [24] calls this *lookup* which is a

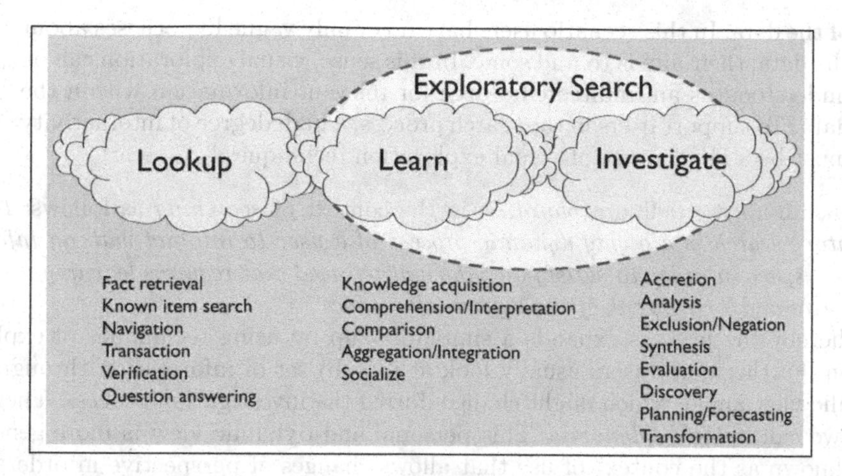

Fig. 1. Categorization of information access methods according to Marchionini [24]

summarization for procedures like fact retrieval, known item search, navigation, transaction, verification and question answering (see Figure 1).

Furthermore, Marchionini distinguishes between *lookup activity* and *exploratory search*. Exploratory search is, according to Marchionini, divided into *learning* and *investigating*. The first describes methods like knowledge acquisition, comprehension and interpretation, comparison, aggregation, integration and socializing. Investigative tasks contain accretion, analysis, exclusion and negation, synthesis, evaluation, discovery, planning, forecasting and transformation. Therefore the general task of exploration can be defined as exploring data sets in order to filter or extract relevant information from it, to (re)define the user's information need and to find associations between displayed information items.

Exploratory data analysis (EDA) is an approach to analyze data in order to formulate hypotheses worth testing and to complement programs of conventional testing for testing hypotheses [22]. We distinguish between this definition based on statistics and definitions found in human-computer interaction. EDA is a term named by John Tukey [38], which is strongly related to suggesting hypotheses about causes of observed phenomena in data and therefore has a specific use case in data analysis, while exploratory search is more about supporting the task in general.

Furthermore, there is a concept of *visual exploration*. Kreuseler [21] describes exploration as an undirected search for interesting features in a data set. According to John Tuckey [37], users engaged in exploratory data analysis are searching interactively and usually undirected for structures and trends in the data. Their goal is to find implicit but potentially useful information and they do not have an initial hypothesis about the data.

A definition for *visual exploration* is given by Tominski [35]:

> "The aim pursued with visual exploration is to give an overview of the data and to allow users to interactively browse through different portions

of the data. In this scenario users have no or only vague hypotheses about the data; their aim is to find some. In this sense, visual exploration can be understood as an undirected search for relevant information within the data. To support users in the search process, a high degree of interactivity must be a key feature of visual exploration techniques."

To sum up, we define *exploration* in the context of *searching* as follows: *Exploratory search is a highly dynamic process of a user to interact with an information space in order to satisfy an information need that requires learning about structure and/or content of the information space.*

Thereby the process expands a simple lookup by using techniques of exploration. Furthermore, users usually look at a (sub) set of information through a specific view angle, which might change during the investigation process. Therefore we call it *highly dynamic.* This personal and dynamic view is more generally known as the context of use that allows changes of perspective in order to (re)formulate or to refine an initial query. User's overall goals are to learn, to investigate, to understand, or to conceptualize (about) their initial information need by building up a personal mental map or model. Thereby the acts of explorative searching, browsing and navigation are often more important than the actual find, i.e. success in this context does not necessarily mean to find a certain piece of information. This makes evaluation of exploration tasks rather difficult, as we discuss later.

2.2 Definition of Bisociative Exploration

In the context of this book we are talking about *bisociative knowledge discovery.* That means that people are engaged in the *creative* discovery of previously unknown information, in particular relationships that were before overlooked in-between different data sets. To find those *bisociations* users explore the data sets in a creative way: "Creative information exploration refers to a new information exploration paradigm that aims to facilitate the generation of creative insight or solutions." *Bisociative information exploration* is an approach where elements from two or more "incompatible concepts" or information spaces are combined to generate creative solutions and insight [10]. Thus, we can define bisociative data exploration as follows:

> Exploration is *bisociative*, if the data set consists of two or more habitually incompatible domains and the user is presented unusual, but interesting domain-crossing connections with the aim of finding relevant and new relationships between those domains.

2.3 Implications for User Interface Design

During exploration, the user is interactively exploring a given data set. His task is to build up a mental model - an overview map - in order to get a structured view on unknown data. Furthermore, the user needs to be able to follow connected items in order to understand the context of the explored items. Another task, a user might try to fulfil, is to follow the boundaries of information he or she is already aware of to gain new insights and to establish further relations between the information items that are already known.

Since association is an important capacity in human communication, visual information addresses patterns of understanding [9]. Furthermore, exploration can be seen as a creative act that requires the user to recognize important parts of the presented information and build connections to the facts he or she already knows while discarding irrelevant knowledge or postpone it for later reference in a different context. This process can be easily disturbed by external influences such as distractions from the environments or inconvenient means of navigation through the information space.

As the user can only keep the most important seven chunks of information in his working memory [25] anything requiring one's immediate attention leads to a loss of just gained insight, which most probably has not been retained yet. Constantly losing ideas due to interruptions will make the user anxious and less productive. Not only one will lose information, but the higher probability of user's work being in vain will lower the motivation. Users may settle with worse results than the one initially intended to find. Supporting exploration in an ergonomic way means taking into account a lot of psychological and physical properties of human beings. Overall the requirements for bisociative knowledge discovery tools are:

1. Supporting *dynamic* exploration within information spaces.
2. Supporting users in deriving connections between different *domains*.
3. Supporting the human *creativity* processes.
4. It should incorporate *online mining tools* that are capable to return the results to the user interface in real-time, see e.g. [15].

Here we use the domain and creativity definitions from [10]. We believe that a tool that only visualize the whole data set, e.g. gives a large graph view, would not appropriatly support bisociative knowledge discovery. The mentioned implications should be reflected in appropriate user interfaces. In the next section we give a brief overview of possible widgets for data exploration and discuss their applicability for bisociative knowledge discovery. We also present a short state of the art on existing exploration tools.

3 Supporting Bisociative Data Exploration

In this section, we first elaborate on the interface elements (widgets) which could be used to support data exploration. We analyse their applicability for bisociative data exploration. The second part of the section contains the state of the art on tools for data exploration.

There are several ways of providing users with appropriate user interface elements in order to support subtasks of exploration like *getting an overview*. Most common solutions are *lists*, *item sets* and *graphs*.

1. Lists are good to present accurate rankings, but due to their structure they only support sequential item observation. It is not possible to discover items in parallel or even generate an overview. Furthermore, relations between listed items are hard to identify without prior knowledge. Therefore lists do not provide good means for exploration tasks.
2. Item sets, often represented as Venn diagrams, provide a good overview about a data set. By using different colors and sizes, relations and groups can be easily recognized by users. Unfortunately, item sets do not provide users with an understanding where certain connections between single item sets exist.
3. Graphs consist of nodes and edges. While nodes often symbolize data entries or information items, edges provide users with understanding which connections / relations do exist between single items. Also groups of items can be identified by using different colors or sizes of nodes or by identifying separated sub graphs. One big disadvantage in this visualization method might be that users find it difficult to find an entry point to start from to explore the data set. Navigation might be also harder than in lists because there are multiple directions to go to. Especially for unknown data sets users might have difficulties to decide which path they like to follow. On the other hand exactly these characteristics support data sets analysis in an unconventional way and offer the possibility to find something new and unexpected, to find new insights in a certain topic.

Lists, item sets and graphs can be utilized to visualize the whole data set. They may also be incorporated within a tool for dynamic exploration of huge data, e.g. users can explore different levels of a collection which is visualized as a hierarchical list. But only graphs support users in deriving connections between different domains. These arguments and the requirements for bisociative knowledge discovery tools (from the previous section) lead us to the conclusion that a graph structure is the most promising approach for bisociative data exploration.

3.1 Tools for Data Exploration

The *Jigsaw* [12] system for investigative analysis across collections of text documents, the *Enronic* [18] tool for a graph based information exploration in emails (see Figures 2,3) and the *CET* [15,16] for efficient exploration and analysis of complex graph structures are some examples of exploration tools.

Since data exploration is an interdisciplinary topic at the intersection of knowledge discovery, visualization techniques and human-computer-interaction, we would like to structure the following overview into these subsections:

Data Set Analysis. By exploring data sets users want to locate anomalies, manage information and understand trends. It is very hard for users to deal with large, high-dimensional and heterogeneous data sets. Therefore, users need to be supported in their data analysis tasks by powerful data analysis methods [34]. Meanwhile, there exist a large amount of powerful algorithms for data analysis which deal with [14].

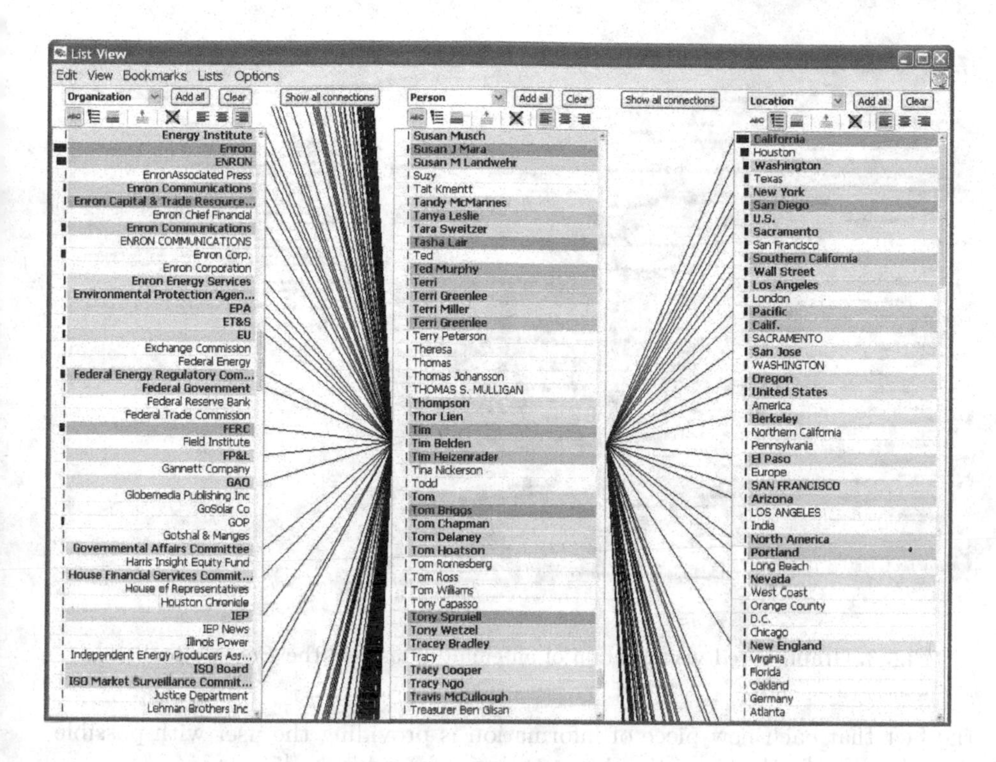

Fig. 2. List based visualization of an email dataset in the *Jigsaw* tool [12] showing connections of "Tim Belden"

Visualization Techniques Supporting Knowledge Discovery. Visualization, as a kind of external memory, supports users cognitive abilities [8] and enhances her or his memory while performing complex cognitive operations [28]. Tominski et al. [36] proposed an interactive graph visualization system *CGV*. This system is supposed to support users in visual exploration of linked information items. Ten years before Eick and Williams [11] already proposed a similar tool called *HierNet* for network-based visualization, which allows grouping, expanding and collapsing of information items. In Jigsaw [12] the authors present a system for investigative analysis across collections of text documents, exemplarily demonstrated on the ENRON data set. The same data set has been used in [18] to build up a graph based information exploration tool *Enronic*. While *ChainGraph* [23] is a further example for a graph-based exploration tool, designers of the *Tiara* tool [39] show that there are also other opportunities for supporting exploratory search than using graphs. Here, a visual summary based on different layers, which are organized in a coordinate system, is used to support exploratory search.

Human-Computer Interaction (HCI) for Explorative Tasks. Users interact with exploration tools in order to formulate problems and to solve them [32]. When considering HCI aspects in exploratory search, we need to take into account

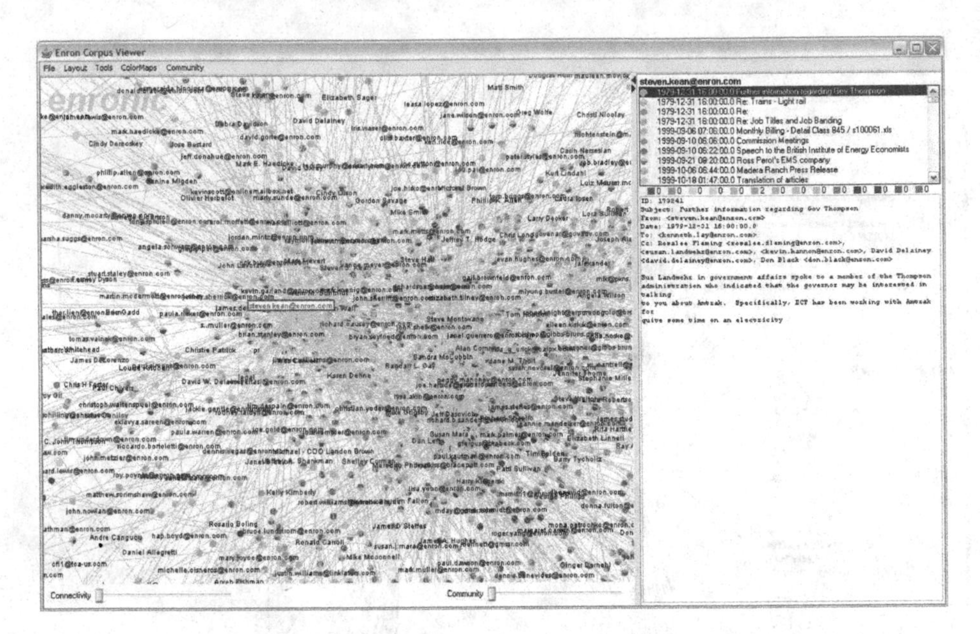

Fig. 3. Graph based visualization of an email dataset in the *Enronic* tool [18]

the fact that each new piece of information is providing the user with possible directions for further investigation, new insights and ideas [5].

4 Evaluation of Knowledge Discovery Tools

Usability evaluation is an integral part of user interface software development. With tool evaluation designers prove that their software fulfils its purpose and satisfies the needs of the target user, i.e. the software is "usable" [27]. With an evaluation it should be proven that, using exploration tools, users are able to make discoveries: effectively, efficiently and with positive attitude.

Our starting point are systems designed for exploration of large, heterogeneous and high-dimensional data sets. The research question that we target is how to evaluate such systems. The most important functionality of knowledge discovery tools is to support users in the creative discovery of new information and relations that were overlooked before in data sets. Thus, users of such tools usually have complex information needs.

Evaluation methods which can be used vary and consist of formal usability studies in the form of controlled experiments and longitudinal studies, benchmark evaluation of the underlying algorithms, informal usability testing and large-scale log-based usability testing [17]. There is also some research in the area of automatic evaluation of user interfaces [33]. Here the idea is using simulation to reflect the way a user is exploring a document collection. We consider an automatic approach, but it is not clear if this would work for biociative exploration which requires creativity.

In the following, we discuss how to apply existing evaluation methods to exploration tools for biociative knowledge discovery. Evaluation of such complex systems is very challenging and requires collaboration with domain experts for creating scenarios and participation. Furthermore, complex information needs are usually vaguely defined and require much user time to be solved. In order to evaluate these tools more efficiently four components are essential: a standardized evaluation methodology, benchmark data sets, benchmark tasks and clearly defined evaluation measures [13].

4.1 Evaluation Challenges

Since knowledge discovery tools are complex systems [30], evaluation of them is very challenging. The first challenge is to create an appropriate scenario for evaluation. The tasks must be complex enough to represent a realistic situation. Such realistic exploratory tasks might require much time (sometimes weeks or even months) to be solved. Lab experiments are limited in time, therefore a "good balance" between time and the right level of complexity is crucial for lab user studies. Longitudinal studies, i.e. research studies that observe users' usage of a system over long periods of time (e.g. months), overcome lab experiments drawbacks like strong time limitation and artificial environment. Researchers motivate the community to conduct long-term user studies because they can be well applied for studying the creative activities that users of information visualization systems engage in. [31]

Controlled lab studies and longitudinal studies require an involvement of target users. Unfortunately, knowledge discovery tools are often designed to be used by experts with domain-specific knowledge, e.g. molecular biologists, who are more difficult to find than participants without special skills or knowledge. Thus, the second challenge is recruiting the participants. This should be a group of people which represents the end users. It requires either collaboration with scientific institutions or some incentive (like money) to engage their participation [30]. In the study preparation step collaboration with domain experts is also needed to help the researchers in creation of appropriate scenarios.

4.2 Open Issues

By evaluating knowledge discovery tools we can either focus on the tool examination or carry out a comparative evaluation. Most researchers concentrate on evaluating their own tool to gain a deeper understanding of user interactions with it. However, the results do not provide such important information if or under what conditions their tool outperforms alternative tools for the same purpose. We found only one publication [20] that proposed an experimental design and a methodology for a comparative user study of complex systems.

To be able to compare and rank a tool among similar ones, benchmark data sets and tasks for user studies are essential [29]. Suppose we wanted to repeat the study conducted in [20] to compare our tool to theirs, we would need the document collection and the task solution used by the authors. However, this data is not available to the public, so we cannot compare the results. A promising direction here is the *Visual Analytics Science and Technology* (VAST) contest[1] which offers data sets of different application domains with description and open-ended domain specific tasks. These tasks should be solved with the help of specific software within the contest. After the contest the solutions are made public, making the data available to evaluations. Thus, the data can be used for evaluations.

Additionally, clearly defined evaluation measures are also important in order to evaluate exploration tools more efficiently. These could be measures from different domains, e.g. information retrieval and human computer interaction, but new measures are still necessary in order to capture the amount of discoveries in document collections or how creative a solution is. The solution of a task itself can be very complex, so we need a way to account for answers which are only partially correct or complete.

The well established three usability aspects from HCI which are usually evaluated in user studies, are *effectiveness*, *efficiency* and *satisfaction* [1,17]. Each of these aspects can be expressed in various measures. In the context of discovery tools evaluation, one can express effectiveness in the amount of discovered information, efficiency in time to find new facts or in importance of the made discovery and satisfaction in the user's rating of the tool's comfort and acceptability [7]. All the three aspects should be ideally measured when evaluating a discovery user interface. Depending on the use case scenario, some of the criteria can be more important than others. If the exploration tool, for example, is primarily designed to support creative discovery of earlier unseen relations among data, the focus may lay more on user satisfaction and less on efficiency. It is not crucial to find the relations fast, but user's satisfaction by using the tool may directly influence his creativity, which is very important by bisociation discovery. A positive attitude helps the user to keep an open mind or play around with the information. User satisfaction, in general, is important, because if the user does not like the interface, one is not likely to use the tool any more again.

One can draw an analogy between user evaluation of exploration tools and automated benchmark evaluation of ranking algorithms in information retrieval. The latter requires a set of test queries, a document collection with labels according to relevancies (e.g. TREC) and a measure (e.g. Average Precision) [17], while discovery tools user evaluation requires a benchmark data set, a benchmark task with a standard solution and an evaluation measure.

4.3 Benchmark Evaluation for Discovery Tools

In the following we propose an evaluation method for discovery tools, consisting of two parts: The first part is a "small" controlled experiment with about 5–10 participants. The purpose of this is to collect qualitative data using user observations like audio/video recording and interviewing the participants afterwards. We actually do not need a special task to be solved by the participants. The assignment can be to discover new information using the software. From this

[1] `http://hcil.cs.umd.edu/localphp/hcil/vast11/`

study we collect data about learnability improvements and user satisfaction. We also get feedback about the users' favourite features and software drawbacks.

The second part is an online study, in which the software is provided to the participants as an online application. This makes it possible to overcome the time limitation found in lab experiments. The participants can access the tool from their own working environment and spend as much time as they like with the tool, even working discontinuously. After that they can use an online questionnaire to provide the task solution and usability feedback. Participants are motivated to solve an interesting task using the tool. We assume that the VAST benchmark data with an investigative task (from IEEE VAST 2006 Contest) can be used as a benchmark data set and a benchmark task. The tool interactions of each participant are logged on the server side. Each participant can spend arbitrarily much time to solve the quest.

We can analyze the log files to get the time spent by participants, to get the solution and interaction patterns. The outcome of the study also contains the number of participants who succeeded in solving the task in comparison to all participants who tried. Each participant is motivated to answer an online questionnaire to provide the task solution and usability feedback. It is beneficial to get the user feedback during the study as it may forget some important issues due to the extended duration. This can be done in the form of a diary. The purpose of the second part is to collect quantitative data.

The described method is only the first step in the creation of a good methodology. It still has several limitations. The first problem is to get an appropriate number of participants. It is not easy to stimulate the participation even with money and if it would work the study becomes cost consuming. One possible solution lies in automatic evaluation (see, e.g., [4]). We could simulate exploration process on different levels and for diverse tasks. However it is not clear how to model a *creative* exploration process, which is important in the case of tasks like creative information discovery, e.g. of relations that were overlooked before in data sets. We also do not have a clear understanding how to judge the success of the search given a complex information need. Thus, the question about evaluation measures remains.

5 Conclusion and Future Work

In this chapter we tried to elaborate on a definition for the term *data exploration*, which—although often used—has no formal description. Our findings, however, can only serve as a starting point for more thorough research on contexts and tools which are recognized as *exploratory* and therefore should be covered by a formal definiton of the term. A selection of tools for graph and data exploration has been provided in Section 3. We propose a simple classification, however creating a taxonomy on these tools would on the one hand require a more specific context, on the other hand we would need a definitive formal specification of what exactly an exploratory tool would be. Finally, we have discussed the problem of evaluating the performance and usability of exploratory tools and identified two main issues: First, evaluation scenarios for those tools are much more complex, resulting in longer sessions and more effort. Second, there is no benchmark against which a tool can be tested, i.e. each evaluation needs to come up with a reference scenario of itself.

With the VAST data set we proposed a starting point for finding such benchmark, which still needs to be specified and tested. However, several aspects are yet unclear. This applies to evaluation methodology, in particular the possibility to evaluate the discovery tools automatically, and evaluation measures. We would like to motivate the community and make the researchers pay attention to the fact that evaluation of knowledge discovery tools should be carried out using a standardized evaluation methodology in combination with benchmark data sets, tasks and measures. Only then discovery tools designers can evaluate their tools more efficiently.

Acknowledgement. The work presented here was supported by the European Commission under the 7th Framework Programme FP7-ICT-2007-C FET-Open, contract no. BISON-211898 and the German Ministry of Education and Science (BMBF) within the ViERforES II project, contract no. 01IM10002B.

References

1. ISO 9241-11: Ergonomic requirements for office work with visual display terminals (VDTs). Part 11 - guidelines for specifying and measuring usability. Geneva: International Standards Organisation. Also available from the British Standards Institute, London (1998)
2. Collins English Dictionary - Complete and Unabridged. HarperCollins Publishers (2003)
3. The American Heritage Dictionary of the English Language. Houghton Mifflin (2009)
4. Azzopardi, L., Järvelin, K., Kamps, J., Smucker, M.: Proc. of SIGIR 2010 Workshop on the Simulation of Interaction: Automated Evaluation of Interactive IR (SimInt 2010). ACM Press (2010)
5. Bates, M.J.: The design of browsing and berrypicking techniques for the online search interface. Online Review 13(5), 407–424 (1989)
6. Berthold, M.R. (ed.): Bisociative Knowledge Discovery. LNCS (LNAI), vol. 7250. Springer, Heidelberg (2012)
7. Bevan, N.: Measuring usability as quality of use. Software Quality Journal 4(2), 115–130 (1995)
8. Card, S., Mackinlay, J., Shneiderman, B.: Using visualization to think. Readings in Information Visualization (1999)
9. Dengel, A., Agne, S., Klein, B., Ebert, A., Deller, M.: Human-centered interaction with documents. In: Proceedings of the 1st ACM International Workshop on Human-centered Multimedia, HCM 2006, pp. 35–44. ACM, New York (2006)

10. Dubitzky, W., Kötter, T., Berthold, M.R.: Towards Creative Information Exploration Based on Koestler's Concept of Bisociation. In: Bisociation, Part I. Springer (2011)
11. Eick, S.G., Wills, G.J.: Navigating large networks with hierarchies. Readings in Information Visualization Using Vision to Think, 207–214 (1999)
12. Görg, C., Stasko, J.: Jigsaw: investigative analysis on text document collections through visualization. In: DESI II: Second International Workshop on Supporting Search and Sensemaking for Electronically Stored Information in Discovery Proceedings. University College London, UK (2008)
13. Gossen, T., Haun, S., Nuernberger, A.: How to Evaluate Exploratory User Interfaces? In: Proceedings of the SIGIR 2011 Workshop on "Entertain Me": Supporting Complex Search Tasks, pp. 23–24. ACM Press (2011)
14. Han, J., Kamber, M.: Data mining: concepts and techniques. Morgan Kaufmann (2006)
15. Haun, S., Gossen, T., Nürnberger, A., Kötter, T., Thiel, K., Berthold, M.R.: On the Integration of Graph Exploration and Data Analysis: The Creative Exploration Toolkit. In: Berthold, M.R. (ed.) Bisociative Knowledge Discovery. LNCS (LNAI), vol. 7250, pp. 301–312. Springer, Heidelberg (2012)
16. Haun, S., Nürnberger, A., Kötter, T., Thiel, K., Berthold, M.R.: CET: A Tool for Creative Exploration of Graphs. In: Balcázar, J.L., Bonchi, F., Gionis, A., Sebag, M. (eds.) ECML PKDD 2010. LNCS, vol. 6323, pp. 587–590. Springer, Heidelberg (2010)
17. Hearst, M.: Search user interfaces. Cambridge University Press (2009)
18. Heer, J.: Exploring Enron: Visualizing ANLP results (2004), `http://hci.stanford.edu/jheer/projects/enron/v1/`
19. Herczeg, M.: Interaktionsdesign. Oldenbourg Wissenschaftsverlag (2006)
20. Kang, Y., Goerg, C., Stasko, J.: How can visual analytics assist investigative analysis? design implications from an evaluation. IEEE Transactions on Visualization and Computer Graphics (2010)
21. Kreuseler, M., Nocke, T., Schumann, H.: A history mechanism for visual data mining. In: Proceedings of the IEEE Symposium on Information Visualization (Infovis 2004). IEEE Computer Society (2004)
22. Leinhardt, G., Leinhardt, S.: Exploratory data analysis: New tools for the analysis of empirical data. Review of Research in Education 8, 85–157 (1980)
23. Lohmann, S., Heim, P., Tetzlaff, L., Ertl, T., Ziegler, J.: Exploring Relationships between Annotated Images with the ChainGraph Visualization. In: Chua, T.-S., Kompatsiaris, Y., Mérialdo, B., Haas, W., Thallinger, G., Bailer, W. (eds.) SAMT 2009. LNCS, vol. 5887, pp. 16–27. Springer, Heidelberg (2009)
24. Marchionini, G.: Exploratory search: From finding to understanding. Communications of the ACM - Supporting Exploratory Search 49(4) (April 2006)
25. Miller, G.A.: The magical number seven, plus or minus two: Some limits on our capacity for processing information. Psychological Science (63), 81–97 (1956)
26. Naisbitt, J.: Megatrends. Ten New Directions Transforming Our Lives. Warner Books (1982)
27. Nielsen, J.: Usability engineering. Morgan Kaufmann (1993)
28. Norman, D.: Visual representations. Things that Make us Smart: Defending Human Attributes in the Age of the Machine (1994)
29. Plaisant, C.: The challenge of information visualization evaluation. In: Proceedings of the Working Conference on Advanced Visual Interfaces, pp. 109–116. ACM (2004)

30. Redish, J.: Expanding usability testing to evaluate complex systems. Journal of Usability Studies 2(3), 102–111 (2007)
31. Shneiderman, B., Plaisant, C.: Strategies for evaluating information visualization tools: multi-dimensional in-depth long-term case studies. In: Proceedings of the 2006 AVI Workshop on BEyond Time and Errors: Novel Evaluation Methods for Information Visualization, pp. 1–7. ACM (2006)
32. Spence, R.: Information Visualization. Addison Wesley/ACM Press, Harlow (2001)
33. Stober, S., Nürnberger, A.: Automatic evaluation of user adaptive interfaces for information organization and exploration. In: SIGIR Works. on SimInt 2010, pp. 33–34 (July 2010)
34. Thomas, J.J., Cook, K.A.: Illuminating the path –The research and development agenda for Visual Analytics. IEEE Computer Society Press (2005)
35. Tominski, C.: Event-Based Visualization for User-Centered Visual Analysis. PhD thesis, Institute for Computer Science, Department of Computer Science and Electrical Engineering, University of Rostock (2006)
36. Tominski, C., Abello, J., Schumann, H.: Cgv—an interactive graph visualization system. Computers & Graphics 33(6), 660–678 (2009)
37. Tukey, J.: Exploratory data analysis. Addison-Wesley, Massachusetts (1977)
38. Tukey, J.W.: Exploratory Data Analysis. Addison-Wesley (1977)
39. Wei, F., Liu, S., Song, Y., Pan, S., Zhou, M.X., Qian, W., Shi, L., Tan, L., Zhang, Q.: Tiara: A visual exploratory text analytic system. In: Proceedings of the 16th ACM SIGKDD International Conference on Knowledge Discovery and Data Mining, KDD 2010, pp. 153–162. ACM, New York (2010)

On the Integration of Graph Exploration and Data Analysis: The Creative Exploration Toolkit

Stefan Haun[1], Tatiana Gossen[1], Andreas Nürnberger[1],
Tobias Kötter[2], Kilian Thiel[2], and Michael R. Berthold[2]

[1] Data and Knowledge Engineering Group,
Faculty of Computer Science, Otto-von-Guericke-University, Germany
`http://www.findke.ovgu.de`
[2] Nycomed-Chair for Bioinformatics and Information Mining
University of Konstanz, Germany
`http://www.inf.uni-konstanz.de/bioml`

Abstract. To enable discovery in large, heterogenious information networks a tool is needed that allows exploration in changing graph structures and integrates advanced graph mining methods in an interactive visualization framework. We present the Creative Exploration Toolkit (CET), which consists of a state-of-the-art user interface for graph visualization designed towards explorative tasks and support tools for integration and communication with external data sources and mining tools, especially the data-mining platform KNIME. All parts of the interface can be customized to fit the requirements of special tasks, including the use of node type dependent icons, highlighting of nodes and clusters. Through an evaluation we have shown the applicability of CET for structure-based analysis tasks.

1 Introduction

Today's search is still concerned mostly with keyword-based searches and the closed discovery of facts. Many tasks, however, can be solved by mapping the underlying data to a graph structure and searching for structural features in a network, e.g. the connection between certain pages in Wikipedia[1] or documents closely related to a specific document, which may be defined by the exploration task itself, i.e. documents mentioning each other, documents which are term-related, etc. Exploring a hyperlink structure in a graph representation enables these tasks to be fulfilled much more efficiently. On the other hand, graph visualization can handle quite large graphs, but is rather static, i.e. the layout and presentation methods calculate the graph visualization once and are not well suited for interactions, such as adding or removing nodes. For example, one of the well known graph layout methods, the *Spring Force Layout*, can yield very

[1] `http://www.wikipedia.org`

M.R. Berthold (Ed.): Bisociative Knowledge Discovery, LNAI 7250, pp. 301–312, 2012.

chaotic results when it comes to small changes in the graph, leading to a completely different layout if just one node is removed [13]. Since a user's memory is strongly location-based [16] and relies on the node positions during interaction with the graph, such behavior is not desirable.

With the *Creative Exploration Toolkit (CET)*, we present a user interface with several distinct features: Support of interactive graph visualization and exploration, integration of a modular open source data analytics system, and easy configuration to serve specific user requirements.

In the following sections, we describe these features in more detail. We start with a short overview on the state of the art in graph interaction and visualization (Sect. 2), describe the explorative user interface (Sect. 3) and the XMPP communication (Sect. 4.1), discuss the integrated (graph)mining methods (Sect. 4), present a first evaluation of the tool by a user study (Sect. 5) and finally discuss some future work.

2 State of the Art in Graph Interaction and Visualization

Related work can be found in the field of graph visualization and graph layouting. Cook and Holder, although mostly concerned with graph mining, provide a good overview on the state of the art and current systems [3]. For a general overview there are several surveys on graph visualization available (c.f. [14], [4], [18]). According to [5], there are three major methods for graph layouting: *force-directed*, *hierarchical* and *topology-shape-metrics*, where the force directed method introduced by [7] is most used today, despite its disadvantageous behavior in interactive systems [14]. Special visualizations can be used to accommodate data specific features such as time lines: [6] introduces a 2.5D time-series data visualization, which uses stacks to represent time-dependent advances in data series. A large number of visualization systems is available. Approaches tailored to web searching and the visualization of hypermedia structures can be found among the web meta-search clustering engines (Vivsmo[2], iBoogie[3], SnakeT[4], WhatsOnWeb[5]) and in the field of semantic wikis (iMapping Wiki [10]).

However, existing layout and visualization methods do not take continuous graph development in the exploration scenario or the heterogeneousity of visualized information networks and their data sources into account. Besides the grave differences between data mining tools and human-computer interaction (see [11]) and the aforementioned shortcomings in continuous visualization of changing graph structures, a loosly-coupled, but efficient integration between network-providing services and visualization tools is often not available.

[2] `http://vivismo.com/`
[3] `http://www.iboogie.com/`
[4] `http://snaket.di.unipi.it`
[5] `http://whatsonweb.diei.unipg.it`

3 The Creative Exploration Toolkit

The Creative Exploration Toolkit (CET) is a user interface that visualizes the graph—derrived from an information network—and allows interaction with it. The global design, shown in Figure 1, consists of

- a *dashboard* at the top, where the controls are located,
- a *logging area*, below, to show information on running processes and the tool status,
- a *sidebar* on the right which displays detailed information about a node,
- and the *workspace* in the center, which is used for visualization.

We currently use the *Stress Minimization Layout* [15] to determine the initial graph layout, followed by an overlap removal [8]. Nodes can be moved to create certain arrangements, selected for further action, and expanded by double-clicking them. On expansion the surrounding graph structure—obtained from the data provider—is added to the visualization. Additionally, the user may issue keyword-based queries. The corresponding results consists of graphs and can be visualized as well. Subsequent query results are added to the graph, enabling the user to explore the graph itself and the structures between the query results. Additionally, there is support for node handling such as a list of all available or all marked nodes, a keyword search for specific nodes and an attribute editor for the selected node, allowing to manually add, change and delete properites.

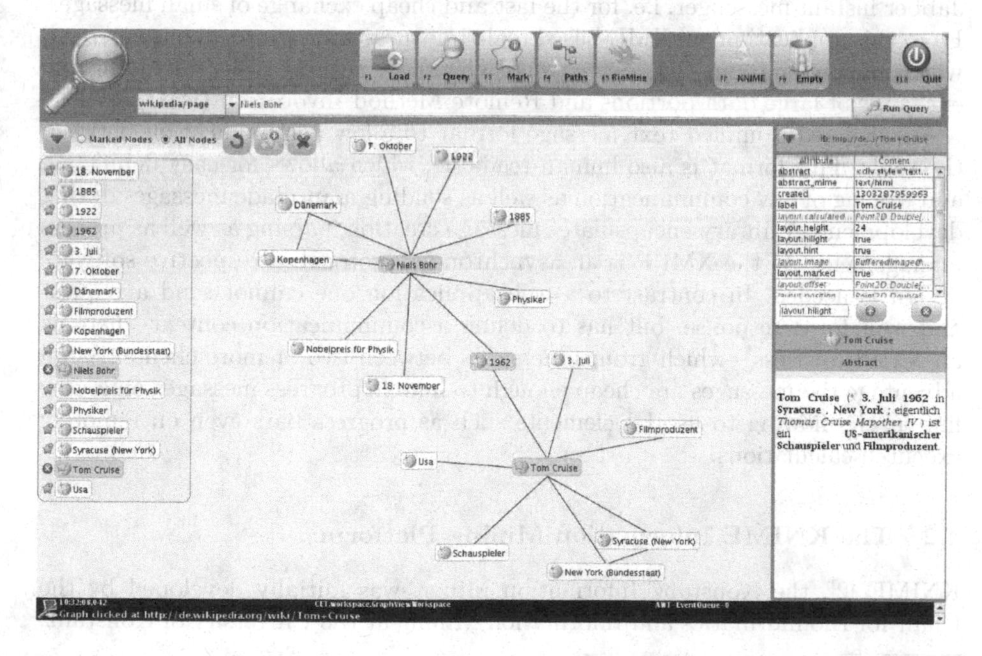

Fig. 1. Screenshot of the Creative Exploration Toolkit (CET), showing an exploration subgraph from the Wikipedia provider

While CET takes care of graph visualization and presentation, domain-specific semantics are not supported. To give an example: The shortest path between two nodes is displayed by highlighting all nodes on the path. However, the user interface is not aware of the path-property, but only displays the highlight attribute of the nodes, while the actual calculation takes place in the underlying data analysis platform described in the next section. The user interface is therefore very flexible when it comes to tasks from different domains.

4 Network and Algorithm Providers

While the CET provides interaction and visualization, it relies on external tools to provide graphs and algorithms. We are working on a selection of providers, including the KNIME platform and Wikipedia—both presented here—as well as a provider for the MusicBrainz[6] network and Personal Information Management (PIM) data.

4.1 Communication between CET and Other Tools

When an interactive process is spread over several nodes, i.e. databases and computation services, it is necessary to keep track of the current status and to be able to propagate changes in the request or outcome very quickly. The Extensible Messaging and Presence Protocol (XMPP)[7] has originally been developed for the Jabber instant messenger, i.e. for the fast and cheap exchange of small messages. From this application, an XML-based real-time messaging protocol has emerged, which now offers numerous extensions (XEPs) for several tasks, including the exchange of large data portions and Remote Method Invocation (RMI) [17].

We definied a unified text message format to allow communication between the tools. This format is also human-readable, which allows for easy debugging and tracing of any communication as well as sending man-made messages during development. A library encapsulates message creation/parsing as well as process management. As the XMPP is an asynchronous protocol, a respective software design is needed. In contrast to a web application one cannot send a request and wait for a response, but has to define a communication context—here as an XMPP process—which groups messages between two or more clients. As an advantage the messages are cheap enough to handle progress messages on a very fine level, allowing to use UI elements such as progress bars even on remotely executed calculations.

4.2 The KNIME Information Mining Platform

KNIME [2], the Konstanz Information Miner, was initially developed by the Chair for Bioinformatics and Information Mining at the University of Konstanz,

[6] `http://www.musicbrainz.org`
[7] `http://www.xmpp.org`

Germany. KNIME is released under an open source license (GPL v3[8]) and can be downloaded free of charge[9]. KNIME is a modular data exploration platform that enables the user to visually create data flows (often referred to as pipelines), selectively execute some or all analysis steps, and later investigate the results through interactive views on data and models. The KNIME base version already incorporates hundreds of processing nodes for data I/O, preprocessing and cleansing, modeling, analysis and data mining as well as various interactive views, such as scatter plots, parallel coordinates and others. It integrates all analysis modules of the well known Weka data mining environment and additional plugins allow, among others, R-scripts[10] to be run, offering access to a vast library of statistical routines.

KNIME has been extended to allow for the flexible processing of large networks via a network plugin, which is available on the KNIME Labs homepage[11]. The network plugin provides new data types and nodes to process (un)weighted and (un)directed multigraphs as well as hypergraphs within KNIME. It further supports the handling of vertex, edge and graph features. Networks can either be processed in memory or in a relational database which enables large networks to be handled within KNIME.

The plugin provides nodes to create, read and write networks. It also provides nodes for filtering and analyzing networks. The created networks can be visualized directly in KNIME or in other existing network visualization tools e.g. visone[12]. Nodes to convert a network into other data types such as matrices and various data tables allow already existing nodes to be used within KNIME. Due to this seamless integration, KNIME can be applied to model complex network processing and analysis tasks.

CET offers a very generic access to KNIME, enabling the user to make arbitrary calls without adapting the user interface. CET can be configured to directly call a KNIME workflow via a configuration and execution dialog. which provides a list of all available workflows and parameters for a selected workflow, which can be edited by the user. Essentially, all information that would be sent by the user interface can be provided to start a KNIME workflow. The result is then visualized in the graph. New analysis methods can therefore be integrated easily into CET by simply adding a new workflow providing the corresponding functionality.

Figure 2 shows an example of a workflow computing the network diameter. In this workflow, first all nodes with a certain feature value are filtered to take only those into account that have been selected and marked by the user. Second, degree filters are applied on nodes and edges to filter unconnected nodes. The shortest paths of all node pairs are subsequently computed and a feature is

---

[8] http://www.gnu.org/licenses/gpl.html
[9] http://www.knime.org
[10] http://www.r-project.org
[11] http://tech.knime.org/knime-labs
[12] http://visone.info

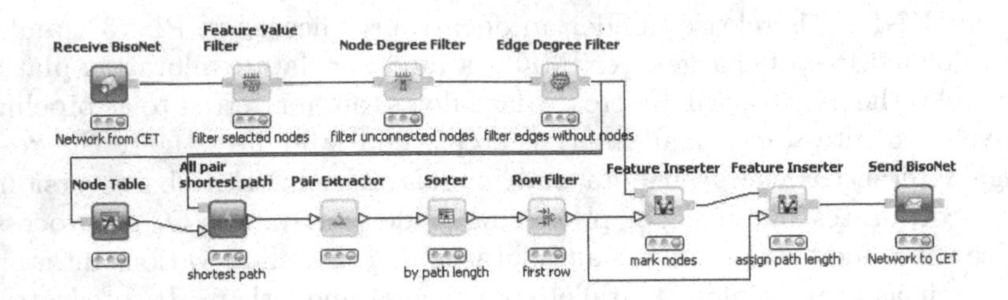

Fig. 2. An example KNIME workflow for calculating the network diameter which is called from CET

assigned consisting of the path length to those nodes of the longest of shortest paths. Finally the graph is sent back to the CET.

4.3 Wikipedia

The Wikipedia client provides query methods needed to explore the Wikipedia structure graph. To obtain the structure, a Wikipedia database dump[13] has been crawled and its structure stored into a MySQL database containing

- the pages—which constitute the nodes in the graph structure—with the page URL, their title and a flag stating whether a page is a redirect to another page. For pages with content, the text part before the first heading is stored as an abstract. However, this only holds the source code from the Wiki page, not any rendered contents, i.e. to display the abstract as it can be seen from Wikipedia, a MediaWiki conversion must be applied.
- links between the pages, including the source page, the destination and an alternative text given for a link, as the MediaWiki markup allows to state any text to be displayed for the link content.

An XMPP service has been set up to provide access to the Wikipedia structure. There are commands to query pages by their name, get the list of all pages connected to a specific page (expansion) and find all links connecting a list of pages (completion). The completion step is needed to fill a graph after nodes have been added, as a simple expansion only views outgoing links with respect to a node. To find incoming links, i.e. links which are outgoing from another node, all pages in the displayed subgraph have to be revisited. Additionally, the abstract of a page can be acquired to be displayed in the side-bar.

5 Evaluation

In this section we describe the evaluation of the CET in the form of a small user study. With this study we prove the applicability of the CET for graph exploration tasks and show how basic functionality of KNIME can support the user by these activities. In this case study we concentrate on one possible scenario

[13] `http://en.wikipedia.org/wiki/Wikipedia:Database_download`

for graph exploration which is knowledge discovery, ex. *bisociations* discovery or discovery of new unexpected relations in the graph data (see [1]).

The case study is carried out in the form of a controlled comparative lab experiment. Our research question is, whether graph based navigation outperforms hypertext navigation in terms of effectiveness, efficiency and user satisfaction. Thus, the target of this study is to compare knowledge discovery, using the CET, which is based on graph navigation with hypertext navigation when users are searching online using web resources e.g. Wikipedia. For a general discussion about the evaluation of exploratory user interfaces see [9].

Hypothesis: Users can make more novel discoveries or make them faster when using our graph-based interface in comparison to exploration based on hypertext navigation.

5.1 Study Design

The study consisted of a lab experiment combined with a questionnaire. By the questionnaire we collected the participants' demographic data, their computer skills and search experience, results of the search experiment using the user interface and Wikipedia, and usability assessment. For the experiment we used the German version of Wikipedia.

We designed several search tasks to reflect knowledge discovery. Especially we concentrated on bisociations discovery. There are three types of bisociations: bridging concepts, bridging graphs and structural similarity (see [1]). We concentrated on bisociations of type "bridging graphs", which can be found in Wikipedia. These bisociations contain named entities that have many "connecting domains" in common. As example, Indira Gandhi and Margaret Thatcher are concepts from the same domain (person) connected through such domains as: university (both attended Oxford), career (Female heads of government), political position (Cold War leaders), sex and century (Women in 20th century warfare). Here, the "connecting domains" are Wikipedia categories. However, the direct link between Indira Gandhi and Margaret Thatcher was missing in Wikipedia.

To be able to find such bisociations the user interface should support searching for similarities between concepts. Therefore we considered the following independent searching tasks for our study:

- Participants should find what the two concepts have in common.
- Participants should build the association chain between two concepts[14].

In the lab experiment the participants used the CET to solve two tasks of the types described above. They also did two similar tasks using online web search on Wikipedia. We employed a Latin Square blocking design [12] in our lab study to get rid of the effects like the order the users use the discovery tools which can bias the results. That means one group of our participants started with the first task set and used the CET and after that used online Wikipedia to solve

[14] To avoid confusion we omitted the term "bisociation" in the study and used "association" instead.

the second task set. Second half of the participants started with the first task set and used online Wikipedia interface and after that used the CET to solve the second task set. Our toolkit supports the user discovery process with the following features:

- *Querying the graph:* exact match and its neighbour nodes can be found.
- *Graph visualisation:* with Wikipedia articles as nodes and links between them if there is a hypertext link in the abstract of one of the articles.
- *Explanation of relations between graph nodes:* for each node a corresponding article abstract can be seen.
- *Graph navigation:* each node can be expanded by its neighbours.
- *Shortest path:* indication of the minimum intermediate concepts/nodes between two or more concepts/nodes in the graph.

Before conducting the lab experiment, the participants were given instructions on how to operate the CET. They were also given some time to try out the toolkit. When using online Wikipedia the participants used the *Firefox*[15] browser and were allowed to perform Wikipedia search as they usually do, e.g. open several tabs and use the internal search feature. The participants were allowed to use only the text in the abstract of an article. They could also follow the hypertext links found only within an abstract. This limitation arose from the limited experiment time (otherwise the users could spend much time reading and understanding information) and from the construction of our graph-based interface. Each participant was told that the target answer, he or she was supposed to find in the lab experiment, should be derived from article abstracts.

5.2 Results of the Study

Twelve users participated in our study: 66.7% (8) men and 33.3% (4) women. Their average age was about 26. The majority of participants had informatics-related profession like engineering, IT-research assistant or student. As we employed a Latin Square blocking design there were 50% of the users (6) in the first group and 50 % (6) in the second with equal percentage of women in each group. All participants categorized themselves as professional users of computer programs. Almost all participants (91.7%) used search engines (e.g. Google) every day to make their investigations. One participant uses search engine several times a week. The majority used Wikipedia to make their investigations several times a week.

The experiment consisted of two similar sets, each with two unrelated tasks (see Table 1). The participants were equally successful solving the first task set independent of the tool they used: based on hypertext navigation or graph based navigation. The second task set was more complicated than the first one. Especially the task about the association chain between amino acids and Gerardus Johannes Mulder showed that graph-based tool better supports users by knowledge discovery. All the participants managed the task using the CET while only

[15] `http://www.mozilla.com/firefox`

one third did it based on Wikipedia. One third of the participants who were supposed to solve the task based on Wikipedia even gave up and presented no solution. The participants also spent less time on task solution if using the CET in comparison to hypertext navigation with exception on one task (Table 2). One important note is that the participants mainly did not know the answers in advance[16]. The proportion of people who knew the answer before the search experiment was equal comparing two configuration groups. This information is important because it would not make sense to compare the programs if the users already knew the answers as then they could find the right answer not because they used one of the tools.

Table 1. Task solving statistic. Success rate in %.

Task set	Task description	Wikipedia			CET		
		Right answer	Wrong answer	Not solved	Right answer	Wrong answer	Not solved
1	What do Tom Cruise and Niels Bohr have in common?	83.3	16.7	0	83.3	16.7	0
	Build an association chain between computer science and Netherlands	100	0	0	100	0	0
2	What do Jean-Marie Lehn and Thomas Mann have in common?	83.3	16.7	0	100	0	0
	Build an association chain between amino acids and Gerardus Johannes Mulder	33.3	33.3	33.3	100	0	0

Table 2. Task solving statistic. Mean time spent on solving (in minutes).

Task set	Task description	Wikipedia	CET
1	What do Tom Cruise and Niels Bohr have in common?	4.00	1.33
	Build an association chain between computer science and Netherlands	1.50	1.67
2	What do Jean-Marie Lehn and Thomas Mann have in common?	2.33	1.50
	Build an association chain between amino acids and Gerardus Johannes Mulder	3.83	2.67

[16] For the task about association chain between computer science and Netherlands, two participants, who used Wikipedia, and two participants, who used CET, knew the answer before the search. This information was learned from the questionnaire.

To summarize, our hypothesis that users can make more new discoveries or achieve them faster using our graph-based interface in comparison to online web search based on hypertext navigation was supported by the study[17]. We also observed on user actions during the experiment. Participants experienced difficulties analyzing even small portions of text without the support of CET (see the example of *Tom Cruise* and *Niels Bohr* in Figure 3).

CET, which has a graph-based interface, helps users to see the connections between concepts at once (see the example of *Tom Cruise* and *Niels Bohr* in Figure 1). That is why our tool is better for knowledge discovery.

We analyzed the participants' opinion on our tool to improve it. The overall rate of the program support of the functions for information discovery was good (see Table 3). The best mean assessment (nearly very good) was for finding relations between topics. The study results show that the program does not sufficiently support the search for topics and we should work in the direction to better support this functionality.

Furthermore we evaluated the usability of the user interface. This statistic is summarized in Table 4. The overall usability assessment was good. The best mean assessment was for user support by solving the searching tasks which was nearly very good. The participants again confirmed our hypothesis that

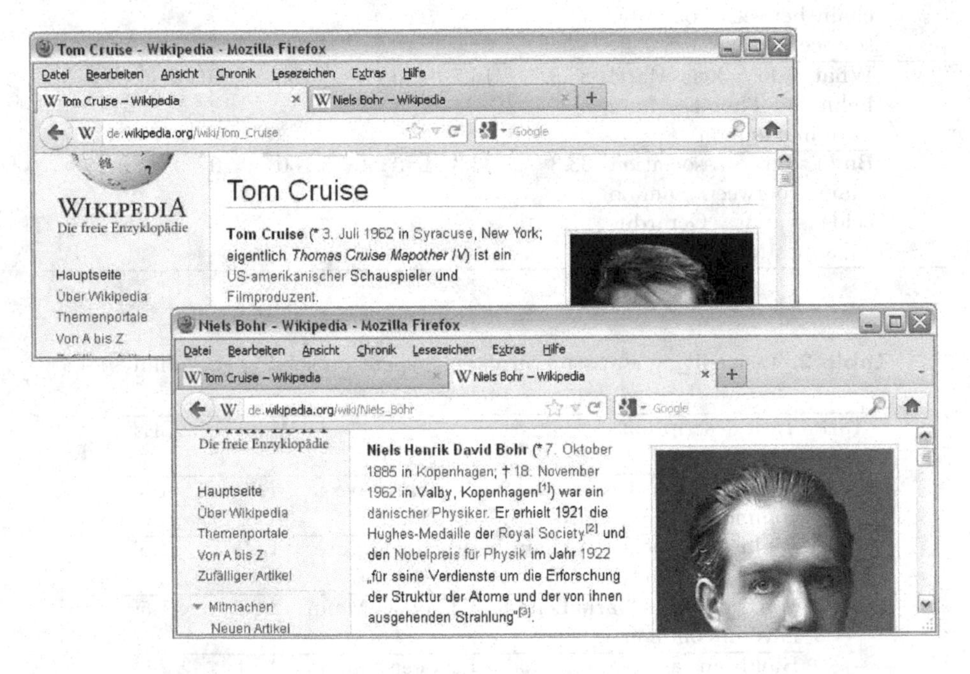

Fig. 3. Screenshots of abstracts of the Wikipedia articles on Tom Cruise and Niels Bohr. Tom Cruise was born in the year Niels Bohr died.

[17] As the participants group was relatively small we do not have a statistical proof. Further studies would be beneficial.

Table 3. User assessment how well the program supports them by knowledge discovery with a scale from 1 (very bad) to 5 (very good)

Functionality	Mean	Min	Max	St. Dev.
Topic search	3.67	2	5	0.78
Navigation between topics	4.25	3	5	0.75
Finding relations between topics	4.67	4	5	0.49
Understanding the relations between topics	4.08	2	5	1.08
Knowledge discovery	4.08	3	5	0.67

Table 4. Usability assessment with a scale from 1 (very bad) to 5 (very good)

Usability criteria	Mean	Min	Max	St. Dev.
Intuitive operation	4.00	3	5	0.74
User support by task solving	4.75	3	5	0.62
User support of knowledge discovery vs. Wikipedia	4.17	3	5	0.84

graph-based interface in comparison to online web search based on hypertext navigation better supports knowledge discovery[18].

6 Conclusion and Future Work

We presented a user interface for generic, exploratory graph visualization with special emphasis on extensibility by integration with data and graph analysis methods provided by KNIME. The presented interface allows for easy interaction with the visualized graphs. This setup is particularly interesting for researchers in the area of Data Mining and Network Analysis, as it is very simple to plug in new approaches and visualize the results, even if there is interaction involved. With a case study we proved the CET applicability for knowledge discovery on tasks requiring structural analysis of data sets.

Future work includes the enhancement of available interaction elements, eventually being able to plug in arbitrary control widgets, improvements on the communication facilities—with extensions of our XMPP library—and the integration of more data sources.

Acknowledgement. The work presented here was supported by the European Commission under the 7th Framework Programme FP7-ICT-2007-C FET-Open, contract no. BISON-211898.

[18] One participant wrote a note, that he did not use Wikipedia for discovery. That is why he could not compare these two tools. But he admitted he knew no alternative to our graph-based tool.

References

1. Berthold, M.R. (ed.): Bisociative Knowledge Discovery. LNCS (LNAI), vol. 7250. Springer, Heidelberg (2012)
2. Berthold, M.R., Cebron, N., Dill, F., Gabriel, T.R., Kötter, T., Meinl, T., Ohl, P., Sieb, C., Thiel, K., Wiswedel, B.: KNIME: The Konstanz Information Miner. In: Data Analysis, Machine Learning and Applications - Proceedings of the 31st Annual Conference of the Gesellschaft für Klassifikation e.V. Studies in Classification, Data Analysis, and Knowledge Organization, pp. 319–326. Springer, Heidelberg (2007)
3. Cook, D.J., Holder, J.B. (eds.): Mining Graph Data. John Wiley & Sons, Inc., Hoboken (2007)
4. Battista, G.D., Eades, P., Tamassia, R., Tollis, I.G.: Algorithms for drawing graphs: An annotated bibliography. Computational Geometry: Theory and Applications 4(5), 235 (1994)
5. Didimo, W., Liotta, G.: Mining Graph Data (Graph Visualization and Data Mining), ch.3, pp. 35–63. John Wiley & Sons, Inc., Hoboken (2007)
6. Dweyer, T., Rolletschek, H., Schreiber, F.: Representing experimental biological data in metabolic networks. In: 2nd Asia-Pacific Bioinformatics Conference (APBC 2004). CRPIT, vol. 29, pp. 13–20. ACS, Sydney (2004)
7. Eades, P.: A heuristic for graph drawing. Congr. Numer. 42, 149–160 (1984)
8. Gansner, E.R., Hu, Y.: Efficient, proximity-preserving node overlap removal. J. Graph Algorithms Appl. 14(1), 53–74 (2010)
9. Gossen, T., Nitsche, M., Haun, S., Nürnberger, A.: Data Exploration for Knowledge Discovery: A brief Overview of Tools and Evaluation Methods. In: Berthold, M.R. (ed.) Bisociative Knowledge Discovery. LNCS (LNAI), vol. 7250, pp. 287–300. Springer, Heidelberg (2012)
10. Haller, H., Kugel, F., Völkel, M.: iMapping Wikis - Towards a Graphical Environment for Semantic Knowledge Management. In: SemWiki (2006)
11. Haun, S., Nürnberger, A.: Supporting exploratory search by user-centered interactive data mining. In: SIGIR Workshop Information Retrieval for E-Discovery (SIRE) (2011)
12. Hearst, M.: Search user interfaces. Cambridge University Press (2009)
13. Herman, I., Melançon, G., de Ruiter, M.M., Delest, M.: Latour – A Tree Visualisation System. In: Kratochvíl, J. (ed.) GD 1999. LNCS, vol. 1731, pp. 392–399. Springer, Heidelberg (1999)
14. Herman, I., Melancon, G., Marshall, M.S.: Graph visualization and navigation in information visualization: A survey. IEEE Transactions on Visualization and Computer Graphics 6(1), 24–43 (2000)
15. Koren, Y., Çivril, A.: The Binary Stress Model for Graph Drawing. In: Tollis, I.G., Patrignani, M. (eds.) GD 2008. LNCS, vol. 5417, pp. 193–205. Springer, Heidelberg (2009)
16. Payne, S.J.: Mental models in human-computer interaction. In: Sears, A., Jacko, J.A. (eds.) The Human-Computer Interaction Handbook, pp. 63–75. Lawrence Erlbaum Associates (2008)
17. Saint-Andre, P.: Streaming XML with Jabber/XMPP. IEEE Internet Computing 9(5), 82–89 (2005)
18. Tamassia, R.: Advances in the theory and practice of graph drawing. Theoretical Computer Science 17, 235–254 (1999)

Bisociative Knowledge Discovery by Literature Outlier Detection

Ingrid Petrič[1], Bojan Cestnik[2,3], Nada Lavrač[3,1], and Tanja Urbančič[1,3]

[1] University of Nova Gorica, Nova Gorica, Slovenia
{ingrid.petric,tanja.urbancic}@ung.si
[2] Temida, d.o.o., Ljubljana, Slovenia
bojan.cestnik@temida.si
[3] Jožef Stefan Institute, Ljubljana, Slovenia
nada.lavrac@ijs.si

Abstract. The aim of this chapter is to present the role of outliers in literature-based knowledge discovery that can be used to explore potential bisociative links between different domains of expertise. The proposed approach upgrades the RaJoLink method which provides a novel framework for effectively guiding the knowledge discovery from literature, based on the principle of rare terms from scientific articles. This chapter shows that outlier documents can be successfully used as means of detecting bridging terms that connect documents of two different literature sources. This linking process, known also as closed discovery, is incorporated as one of the steps of the RaJoLink methodology, and is performed by using publicly available topic ontology construction tool OntoGen. We chose scientific articles about autism as the application example with which we demonstrated the proposed approach.

Keywords: outliers, bisociations, literature mining, knowledge discovery.

1 Introduction

In statistics, an outlier is described as an observation that is numerically distant from the rest of the data, or more formally, it is an observation that falls outside the overall pattern of a distribution [1]. While in many data sets outliers may be due to data measurement errors (therefore it would be best to discard them from the data), there are also several examples where outliers actually led to important discoveries of intriguing information.

In this chapter we explore the potential of outliers for guiding bisociative knowledge discovery from literature. We present an approach to outliers-based knowledge discovery from text documents that can be used to explore implicit relationships across different domains of expertise, indicating interesting cross-domain connections, called *bisociations* [2], [3]. The development of this approach was conducted in three phases that were described in a comprehensive report [4]. The approach upgrades the RaJoLink method [5] for knowledge discovery from literature,

M.R. Berthold (Ed.): Bisociative Knowledge Discovery, LNAI 7250, pp. 313–324, 2012.

where the hypotheses generation phase is based on the principle of rare terms from scientific articles, with the notion of bisociation.

The motivation for work has grounds in the associationist creativity theory [8]. Mednick [8] defines creative thinking as the faculty of generating new combinations of distant associative elements (e.g. words). He explicates how thinking of concepts that are not strictly related to the elements under research inspires unforeseen useful connections between elements. In this manner, bisociations considerably improve the knowledge discovery process. This chapter pays special attention to the category of context-crossing associations, called bisociations [3].

RaJoLink is intended to support experts in their overall process of open knowledge discovery, where hypotheses have to be generated, followed by the closed knowledge discovery process, where hypotheses are tested. It was demonstrated in [5], [6], and [7] that this method can successfully support the user-guided knowledge discovery process.

The RaJoLink methodology has been applied to a challenging medical domain: the set of records for our study was selected from the domain of autism. Autism belongs to a group of pervasive developmental disorders that are portrayed by an early delay and abnormal development of cognitive, communication and social interaction skills of a person [9]. It is a very complex and not yet sufficiently understood domain, where precise causes are still unknown, hence we have chosen it as our experimental testing domain.

This chapter is organized as follows. Section 2 presents the related work in the area of literature mining. Section 3 introduces the literature-based knowledge discovery process and further explores rarity as a principle for guiding the knowledge discovery in the upgraded RaJoLink method. Section 4 presents the RaJoLink approach by focusing on outliers in the closed discovery process. Section 5 illustrates the application of outlier detection to the autism literature. Section 6 provides discussion and conclusions.

2 Related Work in Literature Mining

Novel interesting connections between disparate research findings can be extracted from the published literature. Analysis of implicit associations hidden in scientific literature can guide the hypotheses formulation and lead to the discovery of new knowledge. To support such literature-based discoveries in medical domains, Swanson has designed the *ABC model* approach [10] that investigates whether an agent *A* influences a phenomenon *C* by discovering complementary structures via interconnecting phenomena *B*. Two literatures are complementary if one discusses the relations between *A* and *B*, while a disparate literature investigates the relations between *B* and *C*. If combining these relations suggests a previously unknown meaningful relation between *A* and *C*, this can be viewed as a new piece of knowledge that might contribute to a better understanding of phenomenon *C*.

Weeber and colleagues [11] defined the hypothesis generation approach as an open discovery process and the hypothesis testing as a closed discovery process. In the open discovery process only the phenomenon under investigation (*C*) is given in advance, while the target agent *A* is still to be discovered. In the closed discovery process, both *C* and *A* are known and the goal is to search for bridging phenomena *B* in order to support the validation of the hypothesis about the connection between *A* and *C*. Smalheiser and Swanson [12] developed an online system named ARROWSMITH, which takes as input two sets of titles from disjoint domains *A* and *C* and lists terms *b* that are common to literature *A* and *C*; the resulting terms *b* are used to generate novel scientific hypotheses.[1] As stated by Swanson [13], his major focus in literature-based discovery has been on the closed discovery process, where both *A* and *C* have to be specified in advance.

Several researchers have continued Swanson's line of research. Most of them have made literature-based discoveries in the field of biomedicine. In biomedicine, huge literature databases and well structured knowledge based-systems provide effective supports for literature mining tasks. An on-line literature-based discovery tool called BITOLA has been designed by Hristovski [14]. It uses association rule mining techniques to find implicit relations between biomedical terms. Weeber and colleagues [15] developed Literaby, the concept-based Natural Language Processing tool. The units of analysis that are essential for their approach are UMLS Metathesaurus concepts. The open discovery approach developed by Srinivasan and colleagues [16], on the other hand, relies almost completely on Medical Subject Headings (MeSH). Yetisgen-Yildiz and Pratt [17] proposed a literature-based discovery system called LitLinker. It mines biomedical literature by employing knowledge-based and statistical methods. All the pointed systems use MeSH descriptors [18] as a representation of scientific medical documents, instead of using title, abstract or full-text words. Thus, problems arise since MeSH indexers normally use only the most specific vocabulary to describe the topic discussed in a document [19] and therefore some significant terminology from the documents' content may not be covered. The Swanson's literature-based discovery approach has been extended also by Lindsay and Gordon [20], who used lexical statistics to determine relative frequencies of words and phrases. In their open discovery approach they search for words on the top of the list ranked by these statistics. However, their approach fails when applied to Swanson's first discoveries and extensive analysis has to be based on human knowledge and judgment.

Unlike related work, we put an emphasis on rare terms. Since rare terms are considered to be special terms, not characteristic for a particular domain context, they are more informative than frequent terms. For this reason, rare terms are very likely to be relevant for crossing the boundaries of domains and leading to some interesting observations.

3 The Upgraded RaJoLink Knowledge Discovery Process

The aim of knowledge discovery presented in this chapter is to detect the previously unnoticed concepts (chances) at the intersections of multiple meaningful scenarios. As a consequence, tools for indicating rare events or situations prove to play a significant

[1] Here we use the notations *A*, *B*, and *C* that are written in uppercase letter symbols to represent a set of terms (e.g., literature or collection of titles, abstracts or full texts of documents), while with *a*, *b*, and *c* (lowercase symbols) we represent a single term.

role in the process of research and discovery [21]. From this perspective curious or rare observations of phenomena can provide novel possible opportunities for reasoning [22]. Regarding this, the use of data mining tools is essential to support experts, in choosing meaningful scenarios.

Outliers actually attract a lot of attention in the research world and are becoming increasingly popular in text mining applications as well. Detecting interesting outliers that rarely appear in a text collection can be viewed as searching for the needles in the haystack. This popular phrase illustrates the problem with rarity since identifying useful rare objects is by itself a difficult task [22].

The rarity principle that we apply in the first (open discovery) step of the RaJoLink literature-based discovery is a fundamental difference from the previously proposed methods and represents a unique contribution of the RaJoLink method. In our earlier work [5], [6], and [7] we presented the idea of extending the Swanson's ABC model to handle the open discovery process with rare terms from the domain literature. For that purpose we employed the Txt2Bow utility from the TextGarden library [23] in order to compute total frequencies of terms in the entire text corpus/corpora.

The entire RaJoLink method involves three principal steps, *Ra*, *Jo* and *Link*, which have been named after the key elements of each step: Rare terms, Joint terms and Linking terms. Note that the steps *Ra* and *Jo* implement the open discovery, while the step *Link* corresponds to the closed discovery. The methodological description of the three steps has been provided in our previous publications [5], [6], and [7].

We developed a software tool that implements the RaJoLink method and provides decision support to experts. It can be used to find scientific articles in MEDLINE database [24], to compute statistics about the data, and to analyze them to discover eventually new knowledge. By such exploration, massive amounts of textual data are automatically collected from databases, and text mining methods are employed to generate and test hypotheses. In the step *Ra*, a specified number (set by user as a parameter value) of interesting rare terms in literature about the phenomenon C under investigation are identified. In the step *Jo*, all available articles about the selected rare terms are inspected and interesting joint terms that appear in the intersection of the literatures about rare terms are identified and selected as the candidates for A. In order to provide explanation for hypotheses generated in the step *Jo*, our method searches for links between the literature on joint term a and the literature on term c.

The upgraded RaJoLink methodology for bisociative knowledge discovery consists of the following steps.

- The crucial step in the RaJoLink method is to identify rare elements within scientific literature, i.e., terms that rarely appear in articles about a certain phenomenon.
- Sets of literature about rare terms are then identified and considered together to formulate one or more initial hypotheses in the open discovery process.
- Next, in the closed discovery process, RaJoLink focuses on outlying and their neighbouring documents in the documents' similarity graphs. We construct such graphs with the computational support of a semi-automatic tool for topic ontology construction, called OntoGen [25].

- Outlier documents are then used as a heuristic guidance to speed-up the search for the linking terms (bridging terms, also called b-terms) between different domains of expertise and to alleviate the burden from the expert in the process of hypothesis testing. In this way, the detection of outlier documents represents an upgrade to our previous method that results in significant improvements of the closed discovery process. This step of the upgraded RaJoLink methodology is the focus of research presented in this chapter.

4 Outlier Detection in the RaJoLink Knowledge Discovery Process

This chapter focuses on the steps of the closed discovery process, where two domains of interest *A* and *C* have already been identified prior to starting the knowledge discovery process. The closed discovery process is supported by using the OntoGen tool [25]. One of its features is its capacity of visualizing the similarity between the selected documents of interest. The main novelty of the upgraded RaJoLink methodology is the visualization of outlier documents in the documents' similarity graph (Figure 1) which enables us to find bisociations in the combined set of literatures *A* and *C*. Our argumentation is that outlier documents of two implicitly linked domains can be used to search for relevant linking terms (bridging terms or b-terms) between the two domains. The idea of representing instances of literature *A* together with instances of literature *C* in the same similarity graph with the purpose of searching for their bisociative links is a unique aspect of our method in comparison to the literature-based discovery investigated by other researchers.

When investigating whether disjoint domains *A* and *C* can be connected by domain bridging concepts *B*, we take as input two sets of documents from disjoint domains *A* and *C* and visualize them in the documents' similarity graph. The goal of constructing such graphs is to discover complementary structures that are common to both literatures, *A* and *C* via domain bridging concepts *B*. These domain bridging terms can be found in similarity graphs in those outlying documents of literature *A* and/or literature *C* that are not positioned in the mainstream domain literatures but are relatively distinct from a prototypical document of each domain literature, where a prototypical/average document is, technically speaking, computed as the centroid of the selected domain. Such outlying documents are most frequently observed at the intersection between literatures *A* and *C* as shown in Figure 1.

In the closed discovery process of the RaJoLink method, text documents containing terms *b* that bridge the literature *A* and the literature *C* can be expected to be present in outlier documents. Therefore, in our approach to closed knowledge discovery, outliers are used as heuristic guidance to speed up the search for bridging concepts between different domains of expertise. Having disparate literatures *A* and *C*, both domains are examined by the combined dataset of literatures *A* and *C* in order to assess whether they can be connected by implicit relations. Within the whole corpus of texts consisting of literatures *A* and *C*, which acts as input for step Link (i.e. the closed discovery) of RaJoLink, each text document represents a separate instance/record.

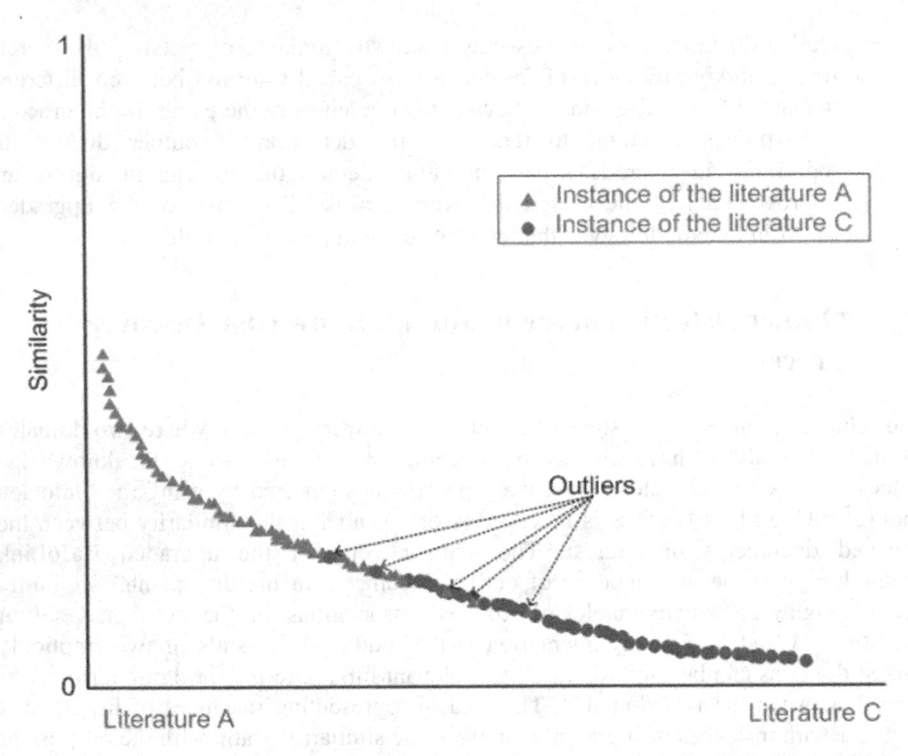

Fig. 1. A graph representing instances (documents) of literature *A* and instances (documents) of literature *C* according to their content similarity to a prototypical document of literature *A*. In this similarity graph, outliers of literature *C* are positioned closer to the typical representatives of the literatures *A* than to the central documents of literature *C*.

Each document from the two literatures is an instance, represented by a set of words using frequency statistics based on the Bag of Words (BoW) text representation [26]. The BoW vector enables to measure content similarity of documents. Content similarity computation is performed with OntoGen, which was designed for interactive data-driven construction of topic ontologies [25]. Content similarity is measured using the standard TF*IDF (term frequency inverse document frequency) weighting method [27], where high frequency of co-occuring words in documents indicates high document similarity. The similarity between documents is visualized with OntoGen in the document's similarity graph, as illustrated in Figure 1.

The cosine similarity measure, commonly used in information retrieval and text mining to determine the semantic closeness of two documents where document features are represented using the BoW vector space model, is used to position the documents according to their similarity to the representative document (centroid) of a selected domain. Documents positioned based on the cosine similarity measure can be visualized in OntoGen by a similarity graph with cosine similarity values that fall within the [0, 1] interval. Value 0 means extreme dissimilarity, where two documents (a given document and the centroid vector of its cluster) share no common words, while value 1 represents the similarity between two semantically identical documents in the BoW representation.

The method uses domains *A* and *C*, and builds a joint document set *AC* (i.e. A∪C). For this intention, two individual sets of documents (e.g. titles, abstracts or full texts of scientific articles), one for each term under research (namely, literature *A* and literature *C*), are automatically retrieved from bibliographic databases or extracted from other document sources. The documents from the two individual sets are loaded as a single text file (i.e. a joint document set *AC*) where each line represents a document with the first word in the line being its name. We consider all the terms and not just the medical ones. A list of 523 English stop words is then used to filter out meaningless words, and English Porter stemming is applied.

From a joint document set A∪C, a similarity graph (Figure 1) between two document sets *A* and *C* is constructed with OntoGen by ranking and visualizing all the documents from *AC* in terms of their similarity to centroid *a* of document set *A*. The OntoGen tool can then be used to build two document clusters, *A'* and *C'* (where A'∪C'=*AC*) in an unsupervise manner, using OntoGen's 2-means clustering algorithm. Cluster *A'* consists mainly of documents from *A*, but may contain also some documents from *C*. Similarly, cluster *C'* consists mainly of documents from *C*, but may contain also some documents from *A*.

Each cluster is further divided into two document subclusters based on domains *A* and *C* with the aim to identify outlying documents. For each individual document cluster we proceed as follows: cluster *A'* is divided into subclusters A'∩A and A'∩C, while cluster *C'* is divided into C'∩A and C'∩C. In this manner, subclusters A'∩C (outliers of *C*, consisting of documents of domain *C* only) and C'∩A (outliers of *A*, consisting of documents of domain *A* only) are the two document sets that consist of outlying documents.

5 Application of Outlier Detection in the Autism Literature

This section is dedicated to a practical application of the upgraded RaJoLink methodology to text analysis of biomedical scientific documents. We present how text mining and link analysis techniques, which are implemented in our approach, can be performed and show how they can be applied to a biomedical domain. For the experimental field we chose autism, for which causes and risk factors are still poorly recognized although it is known, that both genetic and environmental factors influence this disorder. When exploring the literature on autism, the collaborating medical expert has proposed to take the NF-kappaB literature as one of the most promising potential target domains for further focused studies [7]. For a given hypothesis of NF-kappaB and autism relationship we automatically extracted abstracts of MEDLINE articles that could connect the domain of autism with the knowledge gained through the studies of the transcription factor NF-kappaB. In fact, according to the semantic similarity measure we identified some articles on NF-kappaB in the group of articles on autism. Technically speaking, when autism literature was selected as domain *A* and when from the joint domain *AC* (joining autism and NF-kappaB literatures) OntoGen's 2-means clustering method was applied to obtain document groups (clusters) *A'* and *C'*, some documents from domain C (NF-kappaB literature) appeared as members of document cluster *A'* containing mostly articled from domain *A* (autism). Similarly, there were also some articles on autism in the group of articles on NF-kappaB (article group *C'*). It turned out that indeed these exceptional documents contain uncommon and therefore potentially bridging terms. In particular, terms Bcl-2, cytokines, MCP-1, oxidative stress and other meaningful linking terms between the literature on autism and the literature on NF-kappaB were detected in these outlier documents.

Here we present finding of an abstract of MEDLINE articles that makes logical connection between the specific autism observations and the NF-kappaB findings across the bridging term Bcl-2, a regulatory protein for control of programmed brain cell death. Figure 2 shows the similarity graph representing instances of literature *A* (autism context) among instances of literature *C* (nf-kappab+ context) according to their content similarity, where *A* denotes a set of documents containing term autism, *A'* denotes the group of documents constructed from the *AC* document set where most documents are documents on autism (i.e., the so-called autism+ context, where + autism being the majority document class in this document group), and *C* denotes a set of documents containing term NF-kappaB (i.e., the so-called nf-kappab+ context).

The presented bisociative linking approach suggests a novel way to improve the evidence gathering phase when analyzing individual terms appearing in literature *A* in terms of their potential for connecting with terms from literature *C*. In fact, even

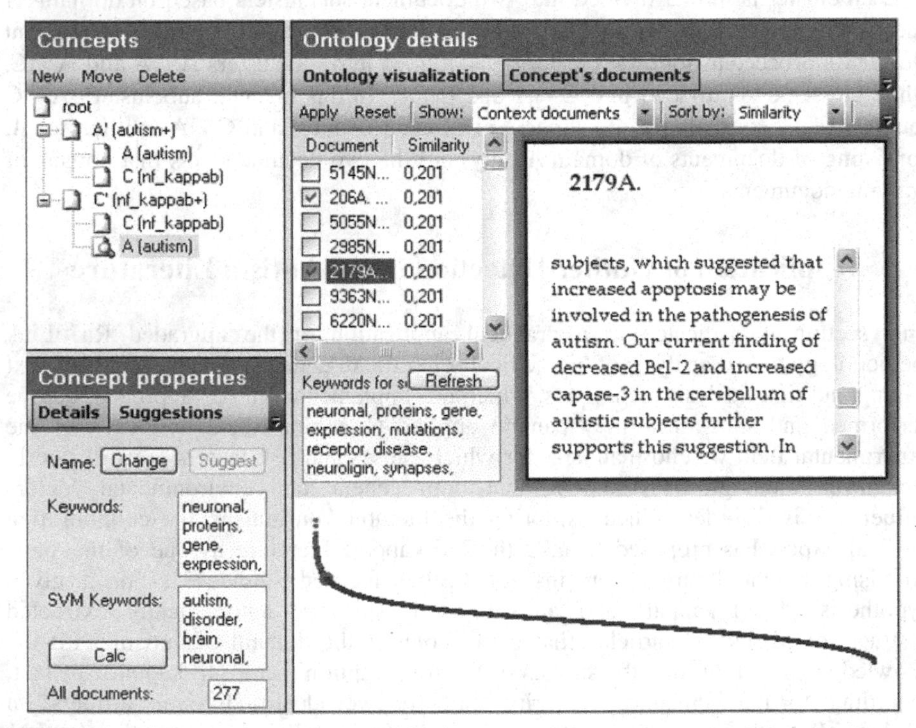

Fig. 2. OntoGen's similarity graph representing instances of literature *A* (*autism+ context*) among instances of the literature *C* (*nf-kappab+ context*) according to their content similarity. The distinctive article about the substance Bcl-2 in relation to autism (*2179A*) is visualized among the nf-kappab+ context documents.

Srinivasan and colleagues, who declared to have developed the algorithms that require the least amount of manual work in comparison with other studies [16], still need significant time and human effort for collecting evidence relevant to the hypothesized connections. In the comparable upgraded RaJoLink approach, the domain expert should be involved only in the conclusive actions of the *Link* step to accelerate the choice of significant linking terms. In this step, similarity graph visualization proves to be extremely beneficial for speeding the process of discovering the bridging concepts. Not only that the documents detected as outliers are visualized and their contents presented on the screen by simply clicking on the pixel representing the document (see Figure 2), but also the keywords are listed, explicitly indicating a set of potential bridging concepts (terms) to be explored by the domain experts.

6 Conclusions

Current literature-based approaches depend strictly on simple, associative information search. Commonly, literature-based association is computed using measures of similarity or co-occurrence. Because of their 'hard-wired' underlying criteria of co-occurrence or similarity, these methods often fail to discover relevant information, which is not related in obvious associative ways. Especially information related across separate contexts is hard to identify with the conventional associative approach. In such cases the context-crossing connections, called bisociations, can help generate creative and innovative discoveries. The RaJoLink method has the potential for bisociative relation discovery as it allows switching between contexts and for discovering interesting terms in the intersections between contexts.

Similar to Swanson's closed discovery approach [10], the search for bridging terms consists of looking for terms *b* that can be found in the intersection of two separate sets of records, namely in the literature *A* as well as in the literature *C*. However, our focusing is on outliers from the two sets of records and their neighbouring documents. Thus we show how outlying documents in the similarity graphs yield useful information in the closed discovery, where bridging concepts have to be found between the literatures *A* and *C*. In fact, such visual analysis can show direction to the previously unseen relations like bisociations, which provide new knowledge. This is an important aspect and significant contribution of our method to literature-based discovery research.

Most of the data analysis research is focused on discovering mainstream relations. These relations are well statistically supported; findings usually confirm the conjectured hypothesis. However, this research provides insight into the relationship between outliers and the literature-based knowledge discovery. An important feature of our approach is the way of detecting the bridging concepts connecting unrelated literatures, which we have performed by the OntoGen's similarity graphs. We used them for representing instances of the literature *A* together with instances of the literature *C* according to their content similarity with the goal to identify outliers from the two sets of literatures and their neighbouring documents. We showed that with the similarity graphs that enable the visual analysis of the literature it is easier to detect the documents, which are very interesting for a particular link analysis investigation, for the reason that such outlying documents often represent particularities in domain literature. Therefore, to test whether the hypothetical observation could be related to the phenomenon under investigation or not, we compare the sets of literature about the initial phenomenon with the literature about the hypothetically related one in the documents' similarity graphs. By our original discovery of linking terms between the literature on autism and the literature on calcineurin we proved that such combination of two previously unconnected sets of literatures in a single content similarity graph can be very effective and useful [5] and [6]. In the autism domain we also discovered a relation between autism and transcription factor NF-kappaB, which has been evaluated by a medical expert as relevant for better understanding of autism [7]. From the similarity graphs that we drew with OntoGen we could quickly notice, which documents from the observed domain are semantically more related to another context. They were positioned in the middle portions of the similarity curves. In the present autism experiment we found a document about the anti-apoptotic protein Bcl-2 [28] that presents a bridging concept among disjoint sets of scientific articles about autism on one hand, and NF-kappaB on the other hand. In fact, Sheikh and colleagues [28] found reduction of Bcl-2, the important marker of apoptosis, in the cerebellum of autistic subjects. Some years before them also Araghi-Niknam and Fatemi showed the reduction of Bcl-2 in superior frontal and cerebellar cortices of autistic individuals [29]. On the other hand, Mattson [30] reported in his review that activation of NF-kappaB in neurons can promote their survival by inducing the expression of genes encoding antiapoptotic proteins such as Bcl-2. However, further research about timing, maturational differences in brain development, and other determinants of NF-kappaB involvement in autism would be needed to substantiate the hypotheses generated by our literature-based experiments.

Acknowledgments. This work was partially supported by the project Knowledge Technologies (grant P2-0103) funded by the Slovene National Research Agency, and the EU project FP7-211898 BISON (Bisociation Networks for Creative Information Discovery).

References

1. Moore, D.S., McCabe, G.P.: Introduction to the Practice of Statistics, 3rd edn. W.H. Freeman, New York (1999)
2. Berthold, M.R. (ed.): Bisociative Knowledge Discovery, 1st edn. LNCS(LNAI), vol. 7250. Springer, Heidelberg (2012)
3. Koestler, A.: The act of creation. MacMillan Company, New York (1964)
4. Petrič, I., Cestnik, B., Lavrač, N., Urbančič, T.: Outlier detection in cross–context link discovery for creative literature mining. Comput. J., 15 (2010)

5. Petrič, I., Urbančič, T., Cestnik, B., Macedoni–Lukšič, M.: Literature mining method RaJoLink for uncovering relations between biomedical concepts. J. Biomed. Inform. 42(2), 219–227 (2009)
6. Petrič, I., Urbančič, T., Cestnik, B.: Discovering hidden knowledge from biomedical literature. Informatica 31(1), 15–20 (2007)
7. Urbančič, T., Petrič, I., Cestnik, B., Macedoni-Lukšič, M.: Literature Mining: Towards Better Understanding of Autism. In: Bellazzi, R., Abu-Hanna, A., Hunter, J. (eds.) AIME 2007. LNCS (LNAI), vol. 4594, pp. 217–226. Springer, Heidelberg (2007)
8. Mednick, S.A.: The associative basis of the creative process. Psychol. Rev. 69(3), 220–232 (1962)
9. American Psychiatric Association: Diagnostic and Statistical Manual of Mental Disorders, 4th edn. Text Revision, Washington, DC (2000)
10. Swanson, D.R.: Undiscovered public knowledge. Library Quarterly 56(2), 103–118 (1986)
11. Weeber, M., Vos, R., Klein, H., de Jong–van den Berg, L.T.W.: Using concepts in literature–based discovery: Simulating Swanson's Raynaud–fish oil and migraine–magnesium discoveries. J. Am. Soc. Inf. Sci. Tech. 52(7), 548–557 (2001)
12. Smalheiser, N.R., Swanson, D.R.: Using ARROWSMITH: a computer–assisted approach to formulating and assessing scientific hypotheses. Comput. Methods Programs Biomed. 57(3), 149–153 (1998)
13. Swanson, D.R., Smalheiser, N.R., Torvik, V.I.: Ranking indirect connections in literature–based discovery: The role of Medical Subject Headings (MeSH). J. Am. Soc. Inf. Sci. Tech. 57(11), 1427–1439 (2006)
14. Hristovski, D., Peterlin, B., Mitchell, J.A., Humphrey, S.M.: Using literature–based discovery to identify disease candidate genes. Int. J. Med. Inform. 74(2-4), 289–298 (2005)
15. Weeber, M.: Drug Discovery as an Example of Literature-Based Discovery. In: Džeroski, S., Todorovski, L. (eds.) Computational Discovery 2007. LNCS (LNAI), vol. 4660, pp. 290–306. Springer, Heidelberg (2007)
16. Srinivasan, P., Libbus, B.: Mining MEDLINE for implicit links between dietary substances and diseases. Bioinformatics 20(suppl. 1), I290–I296 (2004)
17. Yetisgen–Yildiz, M., Pratt, W.: Using statistical and knowledge–based approaches for literature–based discovery. J. Biomed. Inform. 39(6), 600–611 (2006)
18. Nelson, S.J., Johnston, D., Humphreys, B.L.: Relationships in Medical Subject Headings. In: Bean, C.A., Green, R. (eds.) Relationships in the Organization of Knowledge, pp. 171–184. Kluwer Academic Publishers, New York (2001)
19. Principles of MEDLINE Subject Indexing, `http://www.nlm.nih.gov/bsd/disted/mesh/indexprinc.html`
20. Lindsay, R.K., Gordon, M.D.: Literature–based discovery by lexical statistics. J. Am. Soc. Inf. Sci. 50(7), 574–587 (1999)
21. Ohsawa, Y.: Chance discovery: the current states of art. Chance Discoveries in Real World Decision Making 30, 3–20 (2006)
22. Magnani, L.: Chance Discovery and the Disembodiment of Mind. In: Khosla, R., Howlett, R.J., Jain, L.C. (eds.) KES 2005. LNCS (LNAI), vol. 3681, pp. 547–553. Springer, Heidelberg (2005)
23. Grobelnik, M., Mladenić, D.: Extracting human expertise from existing ontologies. EU–IST Project IST–2003–506826 SEKT (2004)
24. MEDLINE Fact Sheet, `http://www.nlm.nih.gov/pubs/factsheets/medline.html`

25. Fortuna, B., Grobelnik, M., Mladenić, D.: Semi–automatic data–driven ontology construction system. In: Bohanec, M., Gams, M., Rajkovič, V., Urbančič, T., Bernik, M., Mladenić, D., Grobelnik, M., Heričko, M., Kordeš, U., Markič, O., Musek, J., Osredkar, M.J., Kononenko, I., Novak Škarja, B. (eds.) Proceedings of the 9th International Multi-Conference Information Society, Ljubljana, Slovenia, pp. 223–226 (2006)
26. Sebastiani, F.: Machine learning in automated text categorization. ACM Comput. Surv. 34(1), 1–47 (2002)
27. Salton, G., Buckley, C.: Term Weighting Approaches in Automatic Text Retrieval. Information Processing and Management 24(5), 513–523 (1988)
28. Sheikh, A.M., Li, X., Wen, G., Tauqeer, Z., Brown, W.T., Malik, M.: Cathepsin D and apoptosis related proteins are elevated in the brain of autistic subjects. Neuroscience 165(2), 363–370 (2010)
29. Araghi–Niknam, M., Fatemi, S.H.: Levels of Bcl–2 and P53 are altered in superior frontal and cerebellar cortices of autistic subjects. Cellular and Molecular Neurobiology 23(6), 945–952 (2003)
30. Mattson, M.P.: NF–kappaB in the survival and plasticity of neurons. Neurochemical Research 30(6-7), 883–893 (2005)

Exploring the Power of Outliers for Cross-Domain Literature Mining

Borut Sluban[1], Matjaž Juršič[1], Bojan Cestnik[1,2], and Nada Lavrač[1,3]

[1] Jožef Stefan Institute, Ljubljana, Slovenia
[2] Temida d.o.o., Ljubljana, Slovenia
[3] University of Nova Gorica, Nova Gorica, Slovenia
{borut.sluban,matjaz.jursic,bojan.cestnik,nada.lavrac}@ijs.si

Abstract. In bisociative cross-domain literature mining the goal is to identify interesting terms or concepts which relate different domains. This chapter reveals that a majority of these domain bridging concepts can be found in outlier documents which are not in the mainstream domain literature. We have detected outlier documents by combining three classification-based outlier detection methods and explored the power of these outlier documents in terms of their potential for supporting the bridging concept discovery process. The experimental evaluation was performed on the classical migraine-magnesium and the recently explored autism-calcineurin domain pairs.

1 Introduction

Scientific literature serves as the basis of research and discoveries in all scientific domains. In literature-based creative knowledge discovery one of the interesting goals is to identify terms or concepts which relate different domains, as these terms may represent germs of new scientific discoveries.

The aim of this chapter[1] is to present an approach which supports scientists in their creative knowledge discovery process when analyzing scientific papers of their interest. The presented research follows Mednick's *associative creativity theory* [9] defining creative thinking as the capacity of generating new combinations of distinct associative elements (e.g. words), and Koestler's book *The act of creation* [7] stating that scientific discovery requires creative thinking to connect seemingly unrelated information. Along these lines, Koestler explores domain-crossing associations, called *bisociations*, as a crucial mechanism for progressive insights and paradigm shifts in the history of science.

Based on the definition of bisociations—defined by Koestler [7] and further refined by Dubitzky et al. [3]—our work addresses the task of supporting the search for bisociative links that cross different domains. We consider a simplified setting, where a scientist has identified two domains of interest (two different scientific areas or two different contexts) and tries to find concepts that represent potential links between the two different contexts. This simplified cross-context

[1] This chapter is an extension of our short paper [16].

M.R. Berthold (Ed.): Bisociative Knowledge Discovery, LNAI 7250, pp. 325–337, 2012.

link discovery setting is usually referred to as the *closed discovery* setting [23]. Like Swanson [19] and Weeber et al. [23], we address the problem of literature mining, where papers from two different scientific areas are available, and the task is to support the scientist in cross-context literature mining. By addressing this task, our aim is to contribute to a methodology for semi-automated cross-context literature mining, which will advance both the area of computational creativity as well as the area of text mining.

We investigate the role of *outliers* in literature mining, and explore the utility of outliers in this non-standard text mining task of *cross-context link discovery*. We provide evidence that outlier detection methods can contribute to literature-based cross-domain scientific discovery based on the notion of bisociation.

This chapter is organized as follows. Section 2 presents the related work in literature mining and outlier detection. In Section 3 we present the experimental datasets, and the method for transforming a set of documents into a format required for text processing and outlier detection. Section 4 presents the methodology for outlier document detection in cross-domain knowledge discovery, together with its evaluation in two medical problem settings: in the classical migraine-magnesium cross-domain discovery problem and in the autism-calcineurin domain pair. Section 5 concludes by a discussion and directions for further work.

2 Related Work

The motivation for new scientific discoveries from disparate literature sources grounds in Mednick's *associative creativity theory* [9] and in the literature on domain-crossing associations, called *bisociations*, introduced by Koestler [7]. Furthermore, we are inspired by the work of Weeber et al. [23] who followed the work of creative literature-based discovery in medical domains introduced by Swanson [19]. Swanson designed the *ABC model* approach that investigates whether an agent A is connected with a phenomenon C by discovering complementary structures via interconnecting phenomena B (see Figure 1)[2]. Two literatures are complementary if one discusses the relations between A and B, while a disparate literature investigates the relations between B and C. If combining these relations suggests a previously unknown meaningful relation between A and C, this can be viewed as a new piece of knowledge that may contribute to a better understanding of phenomenon C.

In a *closed discovery process*, where domains A and C are specified by the expert at the beginning of the discovery process, the goal is to search for bridging concepts (terms) b in B in order to support the validation of the hypothesized connection between A and C (see Figures 1 and 2). Smalheiser and Swanson [17] developed an online system ARROWSMITH, which takes as input two sets of titles from disjoint domains A and C and lists b-terms that are common to literature A and C; the resulting bridging terms (b-terms) are used to generate

[2] Uppercase letter symbols A, B and C are used to represent sets of terms, and lowercase symbols a, b and c to represent single terms.

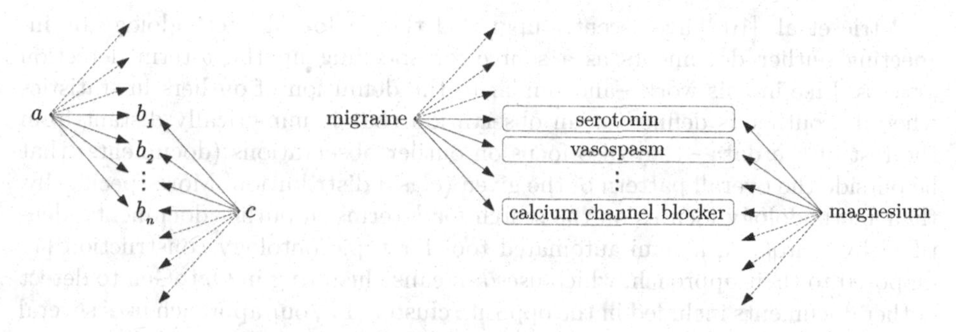

Fig. 1. Closed discovery process as defined by Weber et al. [23]

Fig. 2. Closed discovery when exploring migraine and magnesium documents, with *b*-terms as identified by Swanson et al. [21]

novel scientific hypotheses. As stated by Swanson et al. [21], the major focus in literature-based discovery has been on the closed discovery process, where both A and C are specified in advance.

Srinivasan [18] developed an algorithm for bridging concept identification that is claimed to require the least amount of manual work in comparison with other literature-based discovery studies. However, it still needs substantial time and human effort for collecting evidence relevant to the hypothesized connections. In comparison, one of the advantages of the approach presented in this chapter is that the domain expert needs to be involved only in exploring the potential *b*-terms in outlier documents, instead of exploring all the most frequent potential *b*-terms in all the documents.

In a closely related approach, rarity of terms as means for knowledge discovery has been explored in the RaJoLink system [13,22], which can be used to find interesting scientific articles in the PubMed database with the aim to discover new knowledge. The RaJoLink method involves three principal steps, Ra, Jo and Link, which have been named after the key elements of each step: Rare terms, Joint terms and Linking terms, respectively. In the Ra step, interesting rare terms in literature about the phenomenon *A* under investigation are identified. In the Jo step, all available articles about the selected rare terms are inspected and interesting joint terms that appear in the intersection of the literatures about rare terms are identified as the candidates for *C*. This results in a candidate hypothesis that *C* is connected with *A*. In order to provide explanation for hypotheses generated in the Jo step, in the Link step the method searches for *b*-terms, linking the literature on joint term *c* from *C* and the literature on term *a* from *A*. Note that steps Ra and Jo implement the open discovery, while step Link corresponds to the closed discovery process, searching for *b*-terms when *A* and *C* are already known, as illustrated in Figure 1. Figure 2 illustrates the closed discovery process for a real-life case of exploring the migraine-magnesium domain pair.

Petrič et al. [10] have recently upgraded the RaJoLink methodology by inspecting outlier documents as a source for speeding up the b-term detection process. Like in this work—and similar to the definition of outliers in statistics where an outlier is defined as an observation that is numerically distant from the rest of the data—they also focus on outlier observations (documents) that lie outside the overall pattern of the given (class) distribution. More specifically, their methodology focuses on the search for b-terms in outlier documents identified by OntoGen, a semi-automated tool for topic ontology construction [4]. Opposed to their approach, which uses k-means clustering in OntoGen to detect outlier documents included in the opposite cluster [14], our approach uses several classification algorithms to identify misclassified documents as domain outliers, which are inspected for containing domain bridging terms.

Since outlier mining has already proved to have important applications in fraud detection and network intrusion detection [1], we focused on outliers as they may actually have the potential to lead to the discovery of intriguing new information. Classification noise filters and their ensembles, recently investigated by the authors [15], are used for outlier document detection in this chapter. Documents of a *domain pair* dataset (i.e., the union of two different domain literatures) that are misclassified by a classifier can be considered as domain outliers, since these instances tend to be more similar to regular instances of the opposite domain than to instances of their own domain. The utility of domain outliers as relevant sources of domain bridging terms is the topic of study of this chapter.

3 Experimental Datasets

This section shortly describes two datasets which were used to evaluate the proposed outlier detection approach for cross-domain literature mining. Along with the descriptions of datasets we also provide the description of our preprocessing techniques and some basic statistics for the reader to get a better idea of the data.

The first dataset - the *migraine-magnesium* domain pair - was previously well researched by different authors [13,19,20,21,23]. In the literature-based discovery process Swanson managed to find more than 60 pairs of articles connecting the migraine domain with the magnesium deficiency via several bridging concepts. In this process Swanson identified 43 b-terms connecting the two domains of the *migraine-magnesium* domain pair [21].

The second dataset - the *autism-calcineurin* domain pair - was introduced and initially researched by Petrič et al. in [11,12,22] and later also in [8,10,13]. Autism belongs to a group of pervasive developmental disorders that are portrayed by an early delay and abnormal development of cognitive, communication and social interaction skills of a person. It is a very complex and not yet sufficiently understood domain, where precise causes are still unknown. Alike Swanson, Petrič et al. [13] also provide b-terms, 13 in total, whose importance in connecting autism to calcineurin (a protein phosphatase) is discussed and confirmed by the domain expert.

Table 1. Bridging terms – b-terms identified by Swanson et al. [21] and Petrič et al. [13] for the *migraine-magnesium* and *autism-calcineurin* domain pair, respectively

migraine-magnesium	autism-calcineurin
serotonin, spread, spread depression, seizure, calcium antagonist, vasospasm, paroxysmal, stress, prostaglandin, reactivity, spasm, inflammatory, anti inflammatory, 5 hydroxytryptamine, calcium channel, epileptic, platelet aggregation, verapamil, calcium channel blocker, nifedipine, indomethacin, prostaglandin e1, anticonvulsant, arterial spasm, coronary spasm, cerebral vasospasm, convulsion, cortical spread depression, brain serotonin, 5 hydroxytryptamine receptor, epilepsy, antimigraine, 5 ht, epileptiform, platelet function, prostacyclin, hypoxia, diltiazem, convulsive, substance p, calcium blocker, prostaglandin synthesis, anti aggregation	synaptic, synaptic plasticity, calmodulin, radiation, working memory, bcl 2, type 1 diabetes, ulcerative colitis, asbestos, deletion syndrome, 22q11 2, maternal hypothyroxinemia, bombesin

We use the b-terms, which were identified in each of the two domain pair datasets, as the gold standard to evaluate the utility of domain outlier documents in the cross-context link discovery process. Table 1 presents the b-terms for the *migraine-magnesium* and the *autism-calcineurin* domain pair datasets used in our experiments.

Both datasets were retrieved from the PubMed database[3] using the keyword query; however, we used additional filtering condition for selection of migraine-magnesium dataset. It was necessary to select only the articles published before the year 1988 as this was the year when Swanson published his research about this dataset and thus making an explicit connection between migraine and magnesium domain. Preprocessing was done in a standard text mining way, using the preprocessing steps described in [6]: (a) text tokenization, (b) stopword removal, (c) word stemming/lemmatization using LemmaGen lemmatizer for English [5], (d) construction of N-grams which are terms defined as a concatenation of 1 to N words than appear consecutively in text with minimum supporting frequency, (e) creation of standard bag-of-words (BoW) representation of text using term-frequency-inverse-document-frequency (tf-idf) or binary (depends on classification algorithm) term weights. Besides this standard workflow we additionally removed from the dataset all terms (N-grams) containing words which were used as query terms during document selection. Experiments showed that the correlation between the domain class and the query terms is too high for an outlier detection algorithm to find a reasonable number of high quality outliers. A summary of statistics on the datasets used in our experiments is presented in Table 2.

The 43 b-terms identified by Swanson in the standard *migraine-magnesium* dataset were retrieved from article titles only [21]. Therefore, we also used only article titles in our experiments. In the preprocessing of this dataset we constructed 3-grams to obtain more features for each document despite a relatively

[3] PubMed: `http://www.ncbi.nlm.nih.gov/pubmed`

Table 2. Some basic properties and statistics of the domain pair datasets used in the experimental evaluation

Dataset name	migraine-magnesium	autism-calcineurin
Document source	PubMed	PubMed
Query terms	"migraine" |"magnesium" (condition: year<1988)	"autism" |"calcineurin"
Number of retrieved doc.	8,058 (2,425 |5,633)	15,243 (9,365 |5,878)
Part of document used (text)	title	abstract
Average text length	11 words, 12 terms	180 words, 105 terms
Term definition	3-grams, min. freq. 2	1-grams, min. freq. 15
Number of distinct terms	13,524	5,255
Number of b-terms	43	13
Num. of doc. with b-terms	394 = 4.89%	1672 = 10.97%

low average word count. On the other hand, for the *autism-calcineurin* dataset, which contains titles and the abstracts, we had to limit ourselves to 1-grams and had to set the minimum supporting frequency of terms higher to reduce the number of features due to computational limitations.

4 Detecting Outlier Documents

This research aims at supporting the search for cross-domain links between concepts from two disparate literatures A and C, based on exploring outlier articles of the two domains. Our method assumes that by exploring outlier documents it will be easier to discover linking b-terms (bridging concepts) that establish previously unknown links between literature A and literature C, as the hypothesis of this work is that most bridging concepts occur in outlier documents. This section first presents the algorithms used for outlier detection, followed by the experimental validation of our hypothesis that outlier documents contain a relatively higher number of bridging terms than other documents.

4.1 Classification Noise Filters for Outlier Detection

The novelty of our work is to use noise detection approaches for findinging outlier documents containing cross-domain links (bridging terms – b-terms) between different domains. When exploring a domain pair dataset we searched for a set of outlier documents with different classification noise filtering approaches [2], implemented and adapted for this purpose.

Classification noise filtering is based on the idea of using a classifier as a tool for detecting noisy and outlier instances in data. In this work the simple classifiers used in [2] were replaced by new, better performing classifiers, as the noise filter should, as much as possible, trust the classifiers that they will be able to correctly predict the class of a data instance. In this way the incorrectly classified instances are considered to be noise/outliers. In other words, if an instance of class A is classified in the opposite class C, we consider it to be an

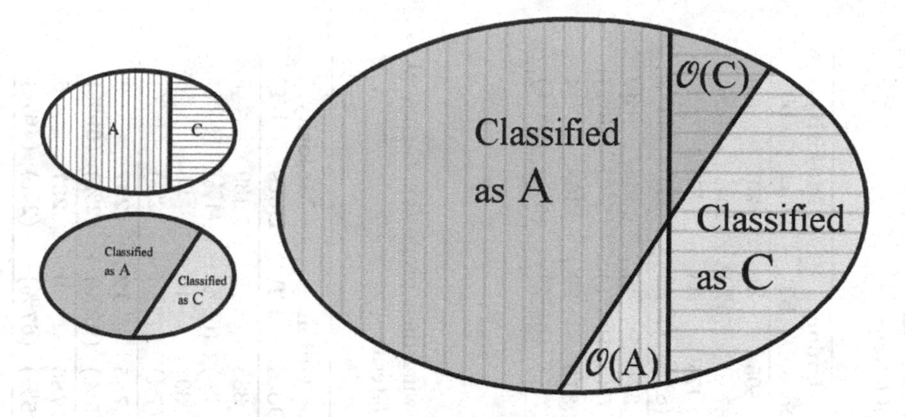

Fig. 3. Detecting outliers of a domain pair dataset with classification filtering

outlier of domain A, and vice versa. We denote the two sets of domain outlier documents with $O(A)$ and $O(C)$, respectively. Figure 3 depicts this principle.

The proposed outlier detection method works in a 10-fold cross-validation manner, where repeatedly nine folds are used for training the classifier and on the complementary fold the misclassified instances are denoted as noise/outliers. Instances of a domain pair dataset that are misclassified by a classifier can be considered as domain outliers, since these instances tend to be more similar to regular instances of the opposite domain than to instances of their own domain.

4.2 Experimental Evaluation

The goal of this section is to provide experimental evidence for the hypothesis that outliers can be used as the focus of exploration to speed-up the search for bridging concepts between different domains of expertise [10,14]. Therefore, our experiments are designed to validate that sets of outlier documents are rich on b-terms and contain significantly more b-terms than sets of arbitrary documents.

We implemented three classification noise detection algorithms, using three different classifiers: Naïve Bayes (abbreviated: Bayes), Random Forest (RF) and Support Vector Machine (SVM). In addition to the outlier sets obtained by these three classification filters, we examined also the union of these outlier sets and the so called "Majority" outlier set containing outlier documents that were detected by at least two out of three classification filters.

Our experiments were performed on the migraine-magnesium and the autism-calcineurin domain pair datasets, described in Section 3. To measure the relevance of the detected outlier documents in terms of their potential for containing domain bridging terms, we inspected 43 terms known as bridging terms appearing in the migraine-magnesium domain pair and 13 known b-terms in the autism-calcineurin domain pair. Tables 3 and 4 present the size of all examined sets of outlier documents and the amount of b-terms they contain, for the migraine-magnesium and autism-calcineurin dataset, respectively.

Table 3. Numbers of documents and *b*-terms (bT) in different outlier sets, with percentages showing their proportion compared to all documents in the migraine-magnesium domain. The bT percentages can be interpreted as recall of *b*-terms.

Class	All docs		Bayes		RF		SVM		Union		Majority	
	Docs	bT	Docs	bT	Docs	bT	Docs	bT	Docs	bT	Docs	bT
MIG	2,425 (100%)	43 (100%)	248 (10%)	23 (53%)	772 (32%)	26 (60%)	192 (8%)	12 (28%)	895 (37%)	34 (79%)	237 (10%)	20 (47%)
MAG	5,633 (100%)	43 (100%)	335 (6%)	27 (63%)	124 (2%)	19 (44%)	170 (3%)	24 (56%)	475 (8%)	33 (77%)	131 (2%)	21 (49%)
Total	8,058 (100%)	43 (100%)	**583 (7%)**	**32 (74%)**	**896 (11%)**	**36 (84%)**	**362 (4%)**	**29 (67%)**	**1,370 (17%)**	**40 (93%)**	**368 (5%)**	**30 (70%)**
Randomly sampled	8,058 (100%)	43 (100%)	**583 (7%)**	**19 (44%)**	**896 (11%)**	**24 (56%)**	**362 (4%)**	**14 (33%)**	**1,370 (17%)**	**29 (68%)**	**368 (5%)**	**14 (33%)**

Table 4. Numbers of documents and *b*-terms (bT) in different outlier sets, with percentages showing their proportion compared to all documents in the autism-calcineurin domain. The bT percentages can be interpreted as recall of *b*-terms.

Class	All docs		Bayes		RF		SVM		Union		Majority	
	Docs	bT	Docs	bT	Docs	bT	Docs	bT	Docs	bT	Docs	bT
AUT	9,365 (100%)	13 (100%)	349 (4%)	8 (62%)	77 (1%)	7 (54%)	147 (2%)	6 (46%)	388 4%)	9 (69%)	139 (1%)	7 (54%)
CAL	5,878 (100%)	13 (100%)	316 (5%)	7 (54%)	65 (1%)	5 (38%)	145 (2%)	7 (54%)	397 (7%)	10 (77%)	98 (2%)	6 (46%)
Total	15,243 (100%)	13 (100%)	**665 (4%)**	**9 (69%)**	**142 (1%)**	**9 (69%)**	**292 (2%)**	**9 (69%)**	**785 (5%)**	**12 (92%)**	**237 (2%)**	**9 (69%)**
Randomly sampled	15,243 (100%)	13 (100%)	**665 (4%)**	**8 (63%)**	**142 (1%)**	**5 (35%)**	**292 (2%)**	**6 (46%)**	**785 (5%)**	**9 (67%)**	**237 (2%)**	**6 (46%)**

Columns of Tables 3 and 4 present the numbers of outlier documents (and contained b-terms) identified by different outlier detection approaches, together with percentages showing their proportion compared to the given dataset. The rows present these numbers separately for each class, for both classes together, and—for the needs of results validation explained below—for a random sample of documents in the size of the detected outlier set.

These results show that all five outlier subsets[4] of each of the two domain pairs contain from 70% to over 90% (for the "Union" subset) of b-terms, on average in less than 10% of all documents from the migraine-magnesium dataset and in less than 5% of all documents of the autism-calcineurin dataset. This means that by inspecting outlier documents, which represent only a small fraction of the datasets, a great majority of b-terms can be found, which substantially reduces the time and effort needed by the domain expert to discover cross-domain links.

To confirm that these results are not due to chance (do not hold for just any arbitrary subset that has the same size as an outlier set), we have randomly sampled 1,000 subsets for each of the five outlier sets (all of them having the same size as their corresponding outlier set) in order to present the average b-term occurrences in randomly sampled subsets. The last row of Tables 3 and 4 shows that the sets of outlier documents contain on average more than 30% more of all b-terms in the migraine-magnesium dataset and more than 20% more of all b-terms in the autism-calcineurin dataset than randomly sampled sets of the same size.

A comparison of the above discussed results relative to the whole migraine-magnesium and autism-calcineurin datasets is summarized in Figures 4 and 5, respectively.

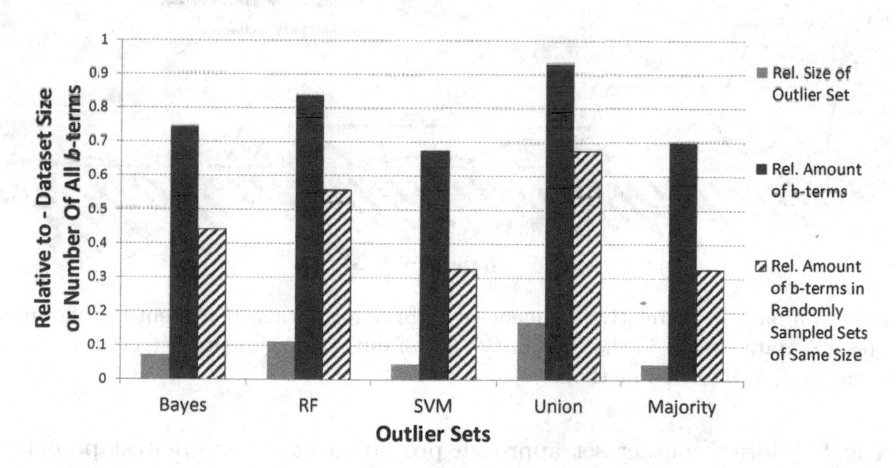

Fig. 4. Relative size of outlier sets and the amount of b-terms for the migraine-magnesium dataset

[4] *Outlier subset* is used instead of *outlier set* to emphasize its relation to the entire dataset of documents. The terms are used interchangeably, however they always refer to a set of detected outlier documents that belong to a certain domain pair.

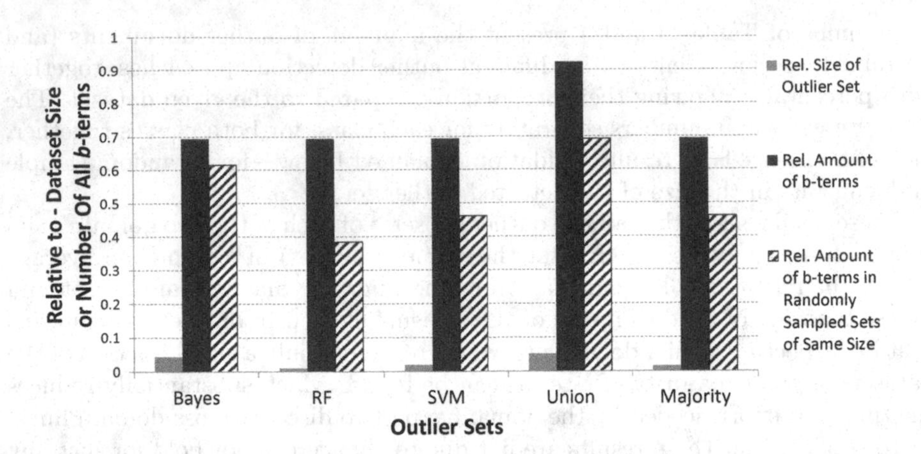

Fig. 5. Relative size of outlier sets and the amount of *b*-terms for the autism-calcineurin dataset

Additionally, we compared relative frequencies of *b*-terms in the detected outlier sets to their relative frequencies in the whole dataset, i.e. the fraction of documents containing a certain *b*-term among the documents of a chosen set. In Figure 6 we present the increase of relative frequencies of *b*-terms in the "Majority" outlier set detected on the migraine-magnesium dataset.

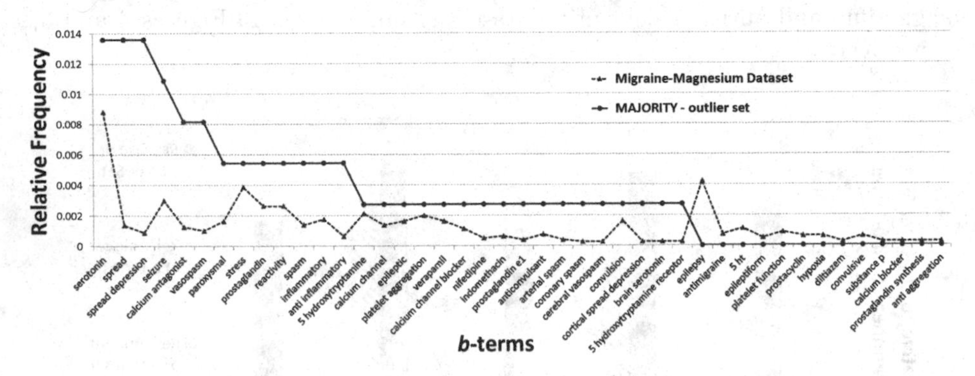

Fig. 6. Comparison of relative frequencies of bridging terms in the entire migraine-magnesium dataset and in the "Majority" set of outlier documents detected by three different outlier detection methods

The "Majority" outlier set approach proved to have the greatest potential for bridging concept detection. Firstly, because of the best ratio among the proportion of the size of the outlier subset and the proportion of *b*-terms which are present in that outlier subset (see Table 4 and Figure 4), and secondly, because the relative frequency of all the *b*-terms present in the "Majority" outlier set is higher compared to the entire migraine-magnesium dataset, as can be clearly seen from Figure 6.

Similarly, encouraging results for the "Majority" outlier set detected on the autism-calcineurin dataset can be observed in Figure 7.

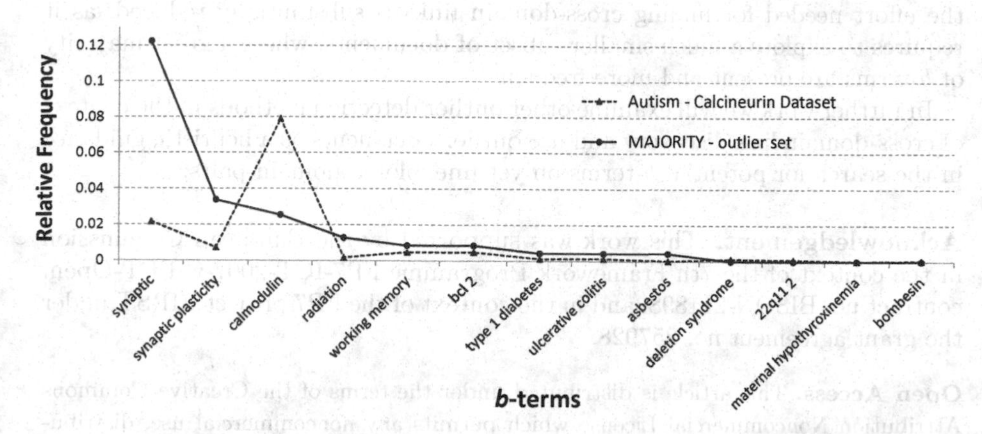

Fig. 7. Comparison of relative frequencies of bridging terms in the entire autism-calcineurin dataset and in the "Majority" set of outlier documents detected by three different outlier detection methods[5]

All *b*-terms that are present in the "Majority" outlier set, except for one ("*calmodulin*"), have a higher relative frequency in the outlier set compared to the relative frequency in the entire dataset. Although (1) the RF outlier set is best in terms of the ratio among the proportion of the size of the outlier subset and the proportion of *b*-terms which are present in that outlier subset and (2) the "Majority" outlier set is second best (for the autism-calcineurin dataset), in general we prefer the "Majority" outlier set for bridging concept detection. The majority approach is more likely to give quality outliers on various datasets, in contrast to a single outlier detection approach, since it reduces the danger of overfitting or bias to a certain domain by requiring the agreement of at least two outlier detection approaches for a document to declare it as an domain outlier.

5 Conclusions

In our research we investigated the potential of outlier detection methods in literature mining for supporting the discovery of bridging concepts between disparate domains.

We retrieved articles for the migraine-magnesium and the autism-calcineurin domain pairs from the PubMed database. In our experiments we obtained five sets of outlier documents for each domain pair by three different outlier detection methods, their union and a majority voting approach. Experimental results

[5] Note that the scale of the chart in Figure 7 is different from the scale of the chart in Figure 6.

show that inspecting outlier documents considerably contributes to the bridging concept discovery process, since it enables the expert to focus only on a small fraction of documents which is rich on concept bridging terms (b-terms). Thus, the effort needed for finding cross-domain links is substantially reduced, as it requires to explore a much smaller subset of documents, where a great majority of b-terms are present and more frequent.

In further work we will examine other outlier detection methods in the context of cross-domain link discovery and use outlier documents as a heuristic guidance in the search for potential b-terms on yet unexplored domain-pairs.

Acknowledgement. This work was supported by the European Commission in the context of the 7th Framework Programme FP7-ICT-2007-C FET-Open, contract no. BISON-211898, and in the context of the FP7 project FIRST under the grant agreement no. 257928.

References

1. Aggarwal, C.C., Yu, P.S.: Outlier detection for high dimensional data. In: Sellis, T. (ed.) Proceedings of the 2001 ACM SIGMOD International Conference on Management of Data, pp. 37–46 (2001)
2. Brodley, C.E., Friedl, M.A.: Identifying mislabeled training data. Journal of Artificial Intelligence Research 11, 131–167 (1999)
3. Dubitzky, W., Kötter, T., Schmidt, O., Berthold, M.R.: Towards Creative Information Exploration Based on Koestler's Concept of Bisociation. In: Berthold, M.R. (ed.) Bisociative Knowledge Discovery. LNCS (LNAI), vol. 7250, pp. 11–32. Springer, Heidelberg (2012)
4. Fortuna, B., Grobelnik, M., Mladenic, D.: OntoGen: Semi-automatic Ontology Editor. In: Smith, M.J., Salvendy, G. (eds.) HCII 2007. LNCS, vol. 4558, pp. 309–318. Springer, Heidelberg (2007)
5. Juršič, M., Mozetič, I., Erjavec, T., Lavrač, N.: Lemmagen: Multilingual lemmatisation with induced ripple-down rules. Journal of Universal Computer Science 16(9), 1190–1214 (2010)
6. Juršič, M., Sluban, B., Cestnik, B., Grčar, M., Lavrač, N.: Bridging Concept Identification for Constructing Information Networks from Text Documents. In: Berthold, M.R. (ed.) Bisociative Knowledge Discovery. LNCS (LNAI), vol. 7250, pp. 66–90. Springer, Heidelberg (2012)
7. Koestler, A.: The act of creation. MacMillan Company, New York (1964)
8. Macedoni-Lukšič, M., Petrič, I., Cestnik, B., Urbančič, T.: Developing a deeper understanding of autism: Connecting knowledge through literature mining. Autism Research and Treatment (2011)
9. Mednick, S.A.: The associative basis of the creative process. Psychological Review 69, 219–227 (1962)

10. Petrič, I., Cestnik, B., Lavrač, N., Urbančič, T.: Outlier detection in cross-context link discovery for creative literature mining. The Computer Journal (2010)
11. Petrič, I., Urbančič, T., Cestnik, B.: Literature mining: Potential for gaining hidden knowledge from biomedical articles. In: Bohanec, M., et al. (eds.) Proceedings of the 9th International Multiconference Information Society, pp. 52–55 (2006)
12. Petrič, I., Urbančič, T., Cestnik, B.: Discovering hidden knowledge from biomedical literature. Informatica 31, 15–20 (2007)
13. Petrič, I., Urbančič, T., Cestnik, B., Macedoni-Lukšič, M.: Literature mining method RaJoLink for uncovering relations between biomedical concepts. Journal of Biomedical Informatics 42(2), 220–232 (2009)
14. Petrič, I., Cestnik, B., Lavrač, N., Urbančič, T.: Bisociative Knowledge Discovery by Literature Outlier Detection. In: Berthold, M.R. (ed.) Bisociative Knowledge Discovery. LNCS (LNAI), vol. 7250, pp. 313–324. Springer, Heidelberg (2012)
15. Sluban, B., Gamberger, D., Lavrač, N.: Performance analysis of class noise detection algorithms. In: Ågotnes, T. (ed.) Proceedings of the 5th Starting AI Researchers Symposium - STAIRS at ECAI 2010, pp. 303–314 (2011)
16. Sluban, B., Juršič, M., Cestnik, B., Lavrač, N.: Evaluating Outliers for Cross-Context Link Discovery. In: Peleg, M., Lavrač, N., Combi, C. (eds.) AIME 2011. LNCS, vol. 6747, pp. 343–347. Springer, Heidelberg (2011)
17. Smalheiser, N.R., Swanson, D.R.: Using ARROWSMITH: a computer-assisted approach to formulating and assessing scientific hypotheses. Comput. Methods Programs Biomed. 57(3), 149–153 (1998)
18. Srinivasan, P.: Text mining: Generating hypotheses from MEDLINE. Journal of the American Society for Information Science and Technology 55, 396–413 (2004)
19. Swanson, D.R.: Undiscovered public knowledge. Library Quarterly 56(2), 103–118 (1986)
20. Swanson, D.R.: Medical literature as a potential source of new knowledge. Bulletin of the Medical Library Association 78(1), 29–37 (1990)
21. Swanson, D.R., Smalheiser, N.R., Torvik, V.I.: Ranking indirect connections in literature-based discovery: The role of medical subject headings (mesh). Journal of the American Society for Information Science and Technology 57(11), 1427–1439 (2006)
22. Urbančič, T., Petrič, I., Cestnik, B., Macedoni-Lukšič, M.: Literature Mining: Towards Better Understanding of Autism. In: Bellazzi, R., Abu-Hanna, A., Hunter, J. (eds.) AIME 2007. LNCS (LNAI), vol. 4594, pp. 217–226. Springer, Heidelberg (2007)
23. Weeber, M., Vos, R., Klein, H., de Jong-van den Berg, L.T.W.: Using concepts in literature-based discovery: Simulating Swanson's Raynaud–fish oil and migraine–magnesium discoveries. Journal of the American Society for Information Science and Technology 52, 548–557 (2001)

Bisociative Literature Mining by Ensemble Heuristics

Matjaž Juršič[1], Bojan Cestnik[1,2], Tanja Urbančič[1,3], and Nada Lavrač[1,3]

[1] Jožef Stefan Institute, Ljubljana, Slovenia
[2] Temida d.o.o., Ljubljana, Slovenia
[3] University of Nova Gorica, Nova Gorica, Slovenia
{matjaz.jursic,bojan.cestnik,tanja.urbancic,nada.lavrac}@ijs.si

Abstract. In literature mining, the identification of bridging concepts that link two diverse domains has been shown to be a promising approach for finding bisociations as distinct, yet unexplored cross-domain connections which could lead to new scientific discoveries. This chapter introduces the system CrossBee (on-line Cross-Context Bisociation Explorer) which implements a methodology that supports the search for hidden links connecting two different domains. The methodology is based on an ensemble of specially tailored text mining heuristics which assign the candidate bridging concepts a bisociation score. Using this score, the user of the system can primarily explore only the most promising concepts with high bisociation scores. Besides improved bridging concept identification and ranking, CrossBee also provides various content presentations which further speed up the process of bisociation hypotheses examination. These presentations include side-by-side document inspection, emphasizing of interesting text fragments, and uncovering similar documents. The methodology is evaluated on two problems: the standard migraine-magnesium problem well-known in literature mining, and a more recent autism-calcineurin literature mining problem.

Keywords: Bisociative Literature Mining, Term Ranking, Ensemble Heuristics, Bisociation Score.

1 Introduction

One of the prevailing trends in research and development is professional over-specialization, resulting in islands of deep, but relatively isolated knowledge. On the other hand, many complex problems require knowledge from different domains to be combined. Due to huge amounts of information available on-line it has become difficult to follow even specific literature limited to a single specialization. Searching for cross-domain scientific connections is even harder, as also scientific literature all too often remains closed and cited only in professional sub-communities. As a promising solution to this problem, literature mining offers methods and software tools which support the experts in their knowledge discovery process, especially in searching for yet unexplored connections between different domains. The notion of such connections is closely related to bisociations as defined by Koestler [8] and further refined by Dubitzky et al. [3].

M.R. Berthold (Ed.): Bisociative Knowledge Discovery, LNAI 7250, pp. 338–358, 2012.

A specific type of knowledge discovery problems, addressed in this chapter, is *closed discovery* introduced by Weeber et al., [21] which has been explored previously in literature mining. In closed discovery we start with a hypothesis that two particular concepts usually investigated in separate literatures are connected. We search for supportive evidence for this by investigating available literatures about these two concepts. As suggested already by Swanson [16], this can be done by identifying interesting bridging terms (b-terms) appearing in both literatures and bearing a potential of indirectly connecting the two concepts under investigation. Although being time-consuming, searching for terms appearing in both literatures is not the main problem. The main issue which also motivated the research presented in this chapter is the fact that a list of terms shared by the two literatures can be very long. Estimating which of the terms have higher potential for interesting discoveries is an interesting research question, important for practical applications.

Narrowing the list of candidate bridging terms can be done in different ways. For example, in the RaJoLink methodology presented by Petrič et al. [11] the list of interesting terms is effectively filtered according to MeSH (Medical Subject Headings) categories; in the next step the expert checks which of the remaining terms seem to be promising. In spite of MeSH filtering, the list of interesting terms can still be long and estimating the potential of a particular bridging term candidate to lead to useful bisociations is based on the expert's knowledge and intuition. The expert's involvement assures that the search is guided towards promising bridging concepts which are meaningful and interesting for the expert [11]. Therefore, we believe that experts' involvement should remain an important part of the process. However, in order to ensure that the expert's inspection of the list of candidate bridging terms is made easier, our main motivation was to automatically estimate the bisociation potential of term candidates and rank the terms.

In the methodology proposed in this work, we estimate the bisociation potential of a term by calculating its *bisociation score*. To this end, different heuristics were developed (see [7]), which are summarized in this chapter. As the experiments described in this chapter show the choice of the right heuristic for a particular domain is far from being trivial. A solution, proposed in this work, is to combine multiple heuristics into an ensemble heuristic which is less sensitive to the variability of domain characteristics.

Ensemble learning is a known approach used in machine learning for combining predictions of multiple models into one final prediction. It is well known [2] that the resulting ensemble model is more accurate than any of the individual models used to build it as long as the models are similarly accurate, are better than random, and their errors are uncorrelated. There is a wide variety of known and well tested ensemble techniques, e.g., Bagging, Boosting, Majority voting, Random forest, Naïve bayes, etc. (see [14]). However, these approaches are usually used for the problem of classification while the core problem presented in this work is ranking. Nevertheless, as information retrieval and especially ranking of web pages by search engines are becoming more and more popular also the ensemble ranking is gaining research attention, e.g., [4, 6].

To evaluate the proposed methodology, implemented in the on-line CrossBee (Cross-Context Bisociation Explorer) system, we applied it to two problems. The first one is the well-known migraine-magnesium example [16, 17] which represents a gold standard in literature mining and served as a testing dataset also in more recent studies [20]. To prevent overfitting the given literature pair and to show the performance in a more complex case, we performed the evaluation of the proposed methodology in the autism-calcineurin problem introduced in [19, 12, 13].

This chapter is structured as follows. In Section 2 the problem of ranking potential b-terms according to their bisociation potential is defined in detail. Section 3 describes the newly introduced ranking methodology and deals with heuristics, the main emphasis being on the proposed ensemble heuristic. In Section 4, the proposed methodology is evaluated through the migraine-magnesium and autism-calcineurin experiments. Section 5 presents a new on-line software tool CrossBee which implements the methodology and provides additional functionalities, making expert's knowledge discovery process easier and more efficient. The chapter concludes with a discussion and plans for further work.

2 Problem Description

The problem addressed in this work is to help the domain expert to effectively find bisociations between two domains presented by two sets of text documents. The main inputs to this task are two sets of documents – one for each of the examined domains. The top-level problem is split, mainly for the reason of evaluation, in two subproblems as follows:

- Develop a *methodology for identifying bisociations* which (among various patterns of bisociation, identified by Dubitzky et al. in [3]) identifies and ranks the key bridging concepts (also named bridging terms or b-terms) that provide the expert with clues about the potential bisociations. The evaluation of this subproblem is based on defining quality values of different solutions which can then be compared to each other. In this way one is able to evaluate the improvements made over the previously existing solutions.
- Create a *system which can support the expert* not only by providing results of the b-term identification methodology but also by adding multiple layers of information to plain data (documents). The added information can be used for human exploration and judgment whether the connections suggested by b-terms are indeed bisociations. The evaluation of this subproblem is slightly less clear, however, by setting an experiment and observing the effectiveness of the expert using the system, one can approximately estimate the quality of different solutions.

3 Methodology for Bridging Concept Identification and Ranking

This section describes the methodology for identifying and ranking of terms according to their potential for being b-terms. The basics of our methodology was developed with the purpose of using potential bridging concepts in the construction of information networks from text documents (see [7] for details), as well as for b-term identification and ranking in our new CrossBee system described in more detail in Section 5 below.

The input to the procedure for b-term identification and ranking consists of two sets of documents – one for each domain. Input documents can be either in the standard form of running text, e.g., titles and abstracts of scientific documents or in the form of partly preprocessed text, e.g., text with already recognized named entities. The output of the procedure is a ranked list of all identified interesting terms. The output list of terms is ordered according to terms' *bisociation score* which is the estimate of a potential that the evaluated term is indeed a b-term which can trigger a bisociation. Our solution to the presented problem of b-term identification and ranking is based on the following three procedural steps:

1. *Preprocess input documents:* Employ state of the art approaches for text preprocessing to extract the most of useful information present in raw texts. Documents are transformed into the bag-of-words [5] feature vector representation, where features represent the terms or concepts. The extracted concepts are *identified as candidate b-terms* and ranked in the next step. More details on text preprocessing are presented in [7].
2. Score *candidate b-terms:* Take the list of candidate b-terms generated in the document preprocessing step and evaluate their b-term potential by calculating the *bisociation score* for each term from the list. This is performed in two steps:
   a. *Employ the base heuristics*: Based on the feature vector representation and some other properties of documents and terms, use specially designed base heuristic functions to score the terms. The output of a base heuristic (the term's score) evaluates the term's potential of being a b-term (see [7] for details).
   b. *Employ the ensemble heuristic*: Scores of base heuristics are integrated into one ensemble heuristic score which represents the final output of the scoring candidate b-terms step and is used as the estimate of the term's bisociation potential. The exact procedure for calculating the ensemble bisociation score is explained in more detail below in this section.
3. *Output the ranked list of b-terms:* Order the list of terms according to the descending order of the calculated bisociation score and return the ranked list of terms with their bisociation scores. This step is elementary and does not need to be presented in detail.

The rest of this section deals with the second step sketched above.

3.1 Base Heuristics

We use the term "heuristic" or "heuristic function" to name a function that numerically evaluates term's quality in the view of its bisociation potential. Ranking all the terms using the scores calculated by an ideal heuristic should result in finding all the b-terms together at the top of such a sorted list. This ideal scenario is generally not realistic; however, ranking by heuristic scores (either ascending or descending) should still increase the proportion of b-terms at the top of the term list.[1]

In [7] we defined a heuristic as a function with two inputs: (a) a set of documents labeled with two domain labels and (b) a term t appearing in these documents; and one output, i.e., a score that estimates the term's bisociation potential. We here list the heuristics with short descriptions only, while the detailed heuristics definition along with their equations are provided in [7].

BoW Heuristics

The heuristics in the group of BoW (bag-of-words) work in a similar way – they manipulate the data present in document vectors to derive the terms' bisociation score. They can be divided into three subgroups:

Term frequency based

(1) $\boldsymbol{freqTerm(t)}$: dataset term frequency,
(2) $\boldsymbol{freqDoc(t)}$: dataset document frequency,
(3) $\boldsymbol{freqRatio(t)}$: dataset term to document frequency ratio,
(4) $\boldsymbol{freqDomnRatioMin(t)}$: minimum of domain term frequencies ratio,
(5) $\boldsymbol{freqDomnProd(t)}$: product of domain term frequencies,
(6) $\boldsymbol{freqDomnProdRel(t)}$: product of domain term frequencies relative to a dataset term frequency.

Tf-idf based

(7) $\boldsymbol{tfidfSum(t)}$: sum of document tf-idf weights of a term in a dataset,
(8) $\boldsymbol{tfidfAvg(t)}$: average of document tf-idf weight of a term in a dataset,
(9) $\boldsymbol{tfidfDomnProd(t)}$: product of domain centroid tf-idf weights of a term,
(10) $\boldsymbol{tfidfDomnSum(t)}$: sum of domain centroid tf-idf weights of a term.

Similarity based

(11) $\boldsymbol{simAvgTerm(t)}$: similarity of a term to an average dataset document – the distance of a term to the dataset centroid,
(12) $\boldsymbol{simDomnProd(t)}$: product of similarities of a term to domain centroids,
(13) $\boldsymbol{simDomnRatioMin(t)}$: min of similarities of a term to domain centroids.

Outlier Heuristics

The outlier heuristics focus on outlier documents since they frequently embody new information that is often hard to explain in the context of existing knowledge. We concentrate on a specific type of outliers, i.e., domain outliers, which are the documents that tend to be more similar to the documents of the opposite domain than to those of their own domain. In the definition of outlier heuristics we used three outlier sets of documents corresponding to the three different underlying document

[1] Note that regardless of the choice, all the heuristics give score 0 to all the terms which appear only in one of the two domains, as these terms have zero potential for bisociation between the two domains.

classification algorithms used for outlier detection: Centroid Similarity classifier (CS), Random Forest classifier (RF), and Support Vector Machine classifier (SVM). Research focused in detecting the outlier documents was performed in [15] and two of the sets, namely RF and SVM were provided by that research. The detection of CS outlier documents was implemented directly in CrossBee using the principles described in [15] but using the Centroid Similarity classifier. The resulting heuristics are:

Based on absolute term frequency in outlier sets

(14)***outFreqCS***(t): term frequency in CS outlier set,
(15)***outFreqRF***(t): term frequency in RF outlier set,
(16)***outFreqSVM***(t): term frequency in SVM outlier set,
(17)***outFreqSum***(t): sum of term frequencies in all three outlier sets.

Based on relative term frequency in outlier sets

(18)***outFreqRelCS***(t): relative frequency in CS outlier set,
(19)***outFreqRelRF***(t): relative frequency in RF outlier set,
(20)***outFreqRelSVM***(t): relative frequency in SVM outlier set,
(21)***outFreqRelSum***(t): sum of relative term frequencies in all three outlier sets.

Baseline Heuristics

We defined two heuristics which are supplementary and serve as baselines:

(22)***random***(t): random number in the interval [0,1),
(23)***appearInAllDomn***(t): a better baseline heuristic which separates two classes of terms, the ones that appear in both domains and the ones that appear in one domain only. The terms that appear in one domain only have a strictly lower heuristic score that those appearing in both. The inner scores of terms inside these two classes are still random numbers.

3.2 Ensemble Heuristic

An ensemble heuristic is a heuristic which combines results of multiple base heuristics into one aggregated result. This work extends the methodology presented in our previous work [7] with an ensemble heuristic due to identified problematic aspect of using a single heuristic for final ranking. The problem arises from the fact that the process of selection of a single heuristic is prone to overfitting the training dataset which results in heuristics' performance instability across other datasets. As long as our experiments were performed only on a single dataset, i.e., the migraine-magnesium dataset, the results of the selected single heuristic, i.e., the (21)outFreqRelSum which proved to be the best heuristic on that dataset were stable, even if we used various modifications of data preprocessing, removed random documents from the set, randomly deleted words from documents or did some other data perturbations.

One possible approach to designing an ensemble heuristic from a set of base heuristics consists of two steps. In the first step the task is to select member heuristics for the ensemble heuristic using standard data mining approaches like feature selection. In the second step equation discovery is used to obtain an optimal combination of member heuristics. The advantage of such approach is that the ensemble creation does not require manual intervention. Therefore, we performed several experiments with such approach; however, the results were even more overfitted to the training domain used in our study. Consequently, we decided to manually – based on experiences and experimentation – select appropriate base heuristics and construct an ensemble heuristic. As the presentation of numerous experiments which support our design decisions is beyond the scope of this chapter, we only describe the final solution, presented in the following subsections.

Ensemble Construction

The ensemble heuristic results in the *ensemble score*, constructed from two parts: the *ensemble voting score* and the *ensemble position score* which are summed together to give the final ensemble score.

- The *ensemble voting score* (s_t^{vote}) of a given term t is an integer which denotes how many base heuristics voted for the term. Each selected base heuristic h_i gives one vote ($s_{t_j,h_i}^{vote} = 1$) to each term which is in the first third[2] in its ranked list of terms and zero votes to all the other terms ($s_{t_j,h_i}^{vote} = 0$). Formally, the ensemble voting score of a term t_j that is at position p_j in the ranked list of n terms is computed as a sum of individual heuristics' voting scores:
$$s_{t_j}^{vote} = \sum_{i=1}^{k} s_{t_j,h_i}^{vote} = \sum_{i=1}^{k} \begin{cases} 1: p_j < n/3, \\ 0: otherwise \end{cases}.$$
Therefore, each term can get a score $s_{t_j}^{vote} \in \{0, 1, 2, \ldots, k\}$, where k is the number of base heuristics used in the ensemble.
- The *ensemble position score* (s_t^{pos}) is calculated as an average of position scores of individual base heuristics. For each heuristic h_i, the term's *position score* s_{t_j,h_i}^{pos} is calculated as $(n - p_j)/n$, which results in position scores being in the interval $[0,1)$. For an ensemble of k heuristics, the ensemble position score is computed as an average of individual heuristics' position scores:
$$s_{t_j}^{pos} = \frac{1}{k}\sum_{i=1}^{k} s_{t_j,h_i}^{pos} = \frac{1}{k}\sum_{i=1}^{k} \frac{(n - p_j)}{n}.$$
- The final *ensemble score* is computed as:
$$s_t = s_t^{vote} + s_t^{pos}.$$

Using the proposed construction we make sure that the integer part of the ensemble score always presents the ensemble vote score, while the ensemble score's fractional part always presents the ensemble position score. An ensemble position score is

[2] The voting threshold is one third (1/3) of the terms which appear in both domains (not one third of all the terms). It was set empirically based on the evaluation of the ensemble heuristic on the migraine-magnesium domain.

strictly lower than 1, therefore, a term with a lower ensemble voting score can never have a higher final ensemble score than a term with a higher ensemble voting score.

Note that at the first sight our method of constructing the ensemble score looks rather intricate. An obvious way to construct an ensemble score of a term could be simply to sum together individual base heuristics scores; however, the calculation of the ensemble score by our method is well justified by extensive experimental results on the migraine-magnesium dataset.

The described method for ensemble score calculation is illustrated in Example 1. In the upper left table the base heuristics scores are shown for each term. The next table presents terms ranked according to the base heuristics scores. From this table, the voting and position scores are calculated for every term based on its position, as shown in the upper right table. For example, all terms at position 2, i.e., t_1, t_6, and t_6, get voting score 1 and position score 4/6. The central table below shows the exact equation how these individual base heuristics' voting and position scores are combined for each term. The table at the bottom displays the list of terms ranked by the calculated ensemble scores.

	Base scores		
Term	h_1	h_2	h_3
t_1	0.93	0.46	0.33
t_2	0.26	0.15	0.10
t_3	0.51	0.22	0.79
t_4	0.45	0.84	0.73
t_5	0.41	0.15	0.11
t_6	0.99	0.64	0.74

Base heuristic scores

	Base ranking		
Pos.	h_1	h_2	h_3
1	t_6	t_4	t_3
2	t_1	t_6	t_6
3	t_3	t_1	t_4
4	t_4	t_3	t_1
5	t_5	t_2	t_5
6	t_2	t_5	t_2

Terms ranked by base heuristics

Pos.	Voting score $s^{vote}_{t_j,h_i}$	Position score $s^{pos}_{t_j,h_i}$
1	1	(6-1)/6=5/6
2	1	(6-2)/6=4/6
3	0	(6-3)/6=3/6
4	0	(6-4)/6=2/6
5	0	(6-5)/6=1/6
6	0	(6-6)/6=0/6

Voting and position scores based on positions in the ranked lists

Voting score sum	+ Pos. score average	= Ensemble score
$(s^{vote}_{t_j,h_1} + s^{vote}_{t_j,h_2} + s^{vote}_{t_j,h_3})$	$+ (s^{pos}_{t_j,h_1} + s^{pos}_{t_j,h_2} + s^{pos}_{t_j,h_3})/k =$	$s^{vote}_{t_j} + s^{pos}_{t_j} = s_{t_j}$
$s_{t_1} = ($ 1 + 0 + 0 $)$	$+ ($ 4/6 + 3/6 + 2/6 $)/3 =$	1 + 9/18 = 1.50
$s_{t_2} = ($ 0 + 0 + 0 $)$	$+ ($ 0/6 + 1/6 + 0/6 $)/3 =$	0 + 1/18 = 0.06
$s_{t_3} = ($ 0 + 0 + 1 $)$	$+ ($ 3/6 + 2/6 + 5/6 $)/3 =$	1 + 10/18 = 1.56
$s_{t_4} = ($ 0 + 1 + 0 $)$	$+ ($ 2/6 + 5/6 + 3/6 $)/3 =$	1 + 10/18 = 1.56
$s_{t_5} = ($ 0 + 0 + 0 $)$	$+ ($ 1/6 + 0/6 + 1/6 $)/3 =$	0 + 2/18 = 0.11
$s_{t_6} = ($ 1 + 1 + 1 $)$	$+ ($ 5/6 + 4/6 + 4/6 $)/3 =$	3 + 13/18 = 3.72

Calculation of ensemble heuristic score

t_6 (3.72), [t_2, t_3] (1.56), t_1 (1.50), t_5 (0.11), t_2 (0.06)

Ranked list of terms produced by the ensemble

Example 1. Ensemble construction illustrated on a simple example with six terms and three heuristics. The last table states the result – the ranked list of terms.

Selecting Base Heuristics for the Ensemble

Another important decision when constructing the ensemble is the selection of base heuristics. Table 1 shows the results that influenced our decision which base heuristics to select. The measure used for heuristic performance comparison is the AUC (area under ROC) presented and discussed already in [7]. Our final set of heuristics included in the ensemble is the following:

- (19)outFreqRelRF
- (20)outFreqRelSVM
- (18)outFreqRelCS
- (17)outFreqSum
- (10)tfidfDomnSum
- (3)freqRatio

Table 1. Comparison of the results (presented and discussed already in [7]) for the base heuristics ordered by the quality – AUC. The first column states the name of the heuristic; the second displays the AUC. The heuristics chosen for the ensemble are shown in italics.

Heuristic	AUC	(16)outFreqSVM	94,70%	(5)freqDomnProd	93,42%
(21)outFreqRelSum	95,33%	(14)outFreqCS	94,67%	(3)*freqRatio*	93,35%
(19)*outFreqRelRF*	95,24%	(4)freqDomnRatioMin	94,36%	(23)appearInAllDomn	93,31%
(20)*outFreqRelSVM*	95,06%	(10)*tfidfDomnSum*	93,85%	(12)simDomnProd	93,27%
(18)*outFreqRelCS*	94,96%	(6)freqDomnProdRel	93,71%	(1)freqTerm	93,20%
(17)*outFreqSum*	94,96%	(13)simDomnRatioMin	93,58%	(2)freqDoc	93,19%
(8)tfidfAvg	94,87%	(7)tfidfSum	93,58%	(11)simAvgTerm	92,71%
(15)outFreqRF	94,73%	(9)tfidfDomnProd	93,47%	(22)random	50,00%

Our initial idea was to choose one (possibly the best performing) heuristic form each set. The rationale behind this idea was to include the top performing heuristics that are as independent as possible. In such a way, the combined information provided by the constructed ensemble was expected to be higher than the information contributed by the individual heuristics. However, certain additional decisions were made to maximize ensemble performance on the migraine-magnesium dataset as well as due to trying not to overfit this dataset:

- The first observation (see Table 1) is that all outlier heuristics based on relative term frequency, i.e., (19)outFreqRelRF, (20)outFreqRelSVM, and, (18)outFreqRelCS perform very well. Actually the only heuristic that is better is the (21)outFreqRelSum which is the combination of all these three. As we want to emphasize the power of this best performing set, we include all three heuristics into the ensemble instead of only (21)outFreqRelSum. So they get more votes and a chance to over-vote some other – not so well performing – heuristics.
- A representative heuristic of the second outlier heuristic set, based on absolute term frequency, is (17)outFreqSum which is not only the best performing of this set, but also integrates the votes of other three heuristics from this set and is therefore the best candidate.
- Representatives of BoW heuristics based on frequency and tf-idf were chosen in a way which tries to avoid overfitting the migraine-magnesium dataset. We chose (3)freqRatio and (10)tfidfDomnSum with the reasoning that they are not among the best performing on the training dataset (but we expect them to

perform better on other datasets) and will therefore act as a counterweight to prevent overfitting.
– We completely discarded all the heuristics of the type similarity, as their performance is in the range of the baseline heuristic $^{(23)}$appearInAllDomn.

Table 2. B-terms for the autism-calcineurin dataset identified by Petrič et al. [11]

1	synaptic	6	bcl 2	11	22q11 2
2	synaptic plasticity	7	type 1 diabetes	12	maternal hypothyroxinemia
3	calmodulin	8	ulcerative colitis	13	bombesin
4	radiation	9	asbestos		
5	working memory	10	deletion syndrome		

4 Evaluation of the Methodology

This section presents the evaluation of the presented base and ensemble heuristics. The key result of this evaluation is the assessment how well the proposed ensemble heuristic performs when ranking the terms from the perspective of the domain expert who acts as the end-user of the CrossBee system. From the expert's point of view, the ROC curves and AUC statistics (as used and described in [7]) are not the most crucial information about the quality of a single heuristic – even though, in general, a better ROC curve reflects a better heuristic. Usually the user is interested in questions like: (a) how many b-terms are likely to be found among the first *n* terms in a ranked list (where *n* is a selected number of terms the expert is willing to inspect, e.g., 5, 20 or 100), or (b) how much one can trust a heuristic if a new dataset is explored. This section provides the evaluation of the heuristics in terms of their performance on a training dataset as well as on a new experimental dataset.

4.1 Experimental Setting

The experimental setting is related to the one in [7] and [15]. The evaluation was performed based on two datasets (or two domain pairs, since each dataset consists of two domains), which can be viewed as a training and test dataset. The training dataset is the dataset we employed when developing the methodology, i.e., for creating a set of base heuristics in [7], as well as for creating the ensemble heuristic presented in this work. The results of the evaluation on the training dataset are important, but needs to be interpreted carefully due to a danger of overfitting the dataset. The test dataset is used for the evaluation of the methodology in a broader (non-dataset biased) scenario.

As the training data we used the well-researched *migraine-magnesium* domain pair which was introduced by Swanson in [16] and was later explored in [17, 18, 20, 11] and others. In the literature-based discovery process Swanson managed to find more than 60 pairs of articles connecting the migraine domain with the magnesium deficiency via 43 bridging concepts (b-terms). Using the developed methodology we tried to rank these 43 b-terms (listed in Table 1 in [7]) as high as possible among other terms which are not marked as b-terms. Since Swanson does not state that this is an exclusive list, there may be also other important bridging terms which he did not list. Consequently, there are two obvious reasons for our results not showing 43 b-terms as the first 43 terms on the ensemble's ranked list. The first reason is a non-optimal ensemble performance and the second reason is that some other terms – not listed by Swanson – may be equally important for bridging the two domains.

For the training dataset we used the *autism-calcineurin* domain pair which was introduced and initially researched by Urbančič et al. [19] and later also in [11, 12]. Like Swanson, Petrič et al. [11] also provide b-terms, 13 in total (listed in Table 2), whose importance in connecting autism to calcineurin (a protein phosphatase) is discussed and confirmed by the domain expert. In the view of searching for b-terms, this dataset has a relatively different dimensionality compared to the migraine-magnesium dataset. On the one hand it has only approximately one fourth of the b-terms defined, while on the other hand, it contains more than 40 times as many potential b-term candidates. Therefore, the ratio between b-terms and candidate terms is substantially lower – approximately by factor 160, i.e., the chance to find a b-term among the candidate terms if picking it at random is 160 times lower in the autism-calcineurin dataset then in the magnesium-migraine dataset. Consequently, finding the actual b-terms in the autism-calcineurin dataset is much more difficult compared to the migraine-magnesium dataset.

Both datasets, retrieved from the PubMed database using the keyword query, are formed of titles or abstracts of scientific papers returned by the query; however, we used an additional filtering condition for selecting the migraine-magnesium dataset. We needed to select only the articles published before the year 1988 as this was the year when Swanson published his research about this dataset and consequently making an explicit connection between the migraine and magnesium domains.

Table 3 states some properties for comparing the two datasets used in the evaluation. One of the major differences between the datasets is the length of an average document since only the titles were used in the migraine-magnesium dataset, while the full abstracts were used in the autism-calcineurin case – due to matching the properties of experiments of original research [16, 19] on these two datasets. Consequently, also the number of distinct terms and b-term candidates is much larger in

Table 3. Comparison of statistical properties of the two datasets used in the experiments

		Migraine-magnesium	Autism-calcineurin
Retrieval	Source	PubMed	PubMed
	Query terms	"migraine"-"magnesium"	"autism"-"calcineurin"
	Additional conditions	Year < 1988	/
	Part of paper used	Title	Abstract
Docum. Statistics	Number	8,058 (2,415-5,633)	15,243 (9,365-5,878)
	Doc. with b-term	394 (4.89%)	1672 (10.97%)
	Avg. words per doc.	11	180
	Outliers (CS-SVM-RF)	(505 - 362 - 896)	(377 - 292 - 142)
Term statistic	Avg. term per doc.	7	173
	Distinct terms	13,525	322,252
	b-term candidates	1,847	78,805
	Defined b-terms	43	13

the case of the autism-calcineurin dataset. Nevertheless, the preprocessing of both datasets was the same with the exception of outlier document identification. For the needs of RF and SVM outlier based heuristics we used the outlier documents identified by Sluban et al. [15] since we did not implement RF and SVM classifiers ourselves. Thus, our outlier heuristics results are completely aligned with the results provided in [15] for both datasets; however, Sluban et al. used slightly different document preprocessing for each of the two datasets. Table 3 also shows the exact number of outliers identified in each dataset. We can inspect higher numbers in the migraine-magnesium dataset which points to the problem of harder classification of documents in this dataset – this is also partly due to shorter texts.

4.2 Results in the Migraine-Magnesium Dataset

Fig. 1 shows the comparison of ranking performance for the ensemble and all the base heuristics on the migraine-magnesium dataset. The heuristics are ordered by their AUC. Black dots along with percentages show the heuristic's AUC performance. Gray bars around AUC central point shows the interval of a heuristics' AUC result, explained below.

The property of heuristics having AUC on the interval and not as a fixed value is due to the fact that some heuristics do not produce unambiguous ranking of all the terms. Several heuristics assign the same score to a set of terms – including both the actual

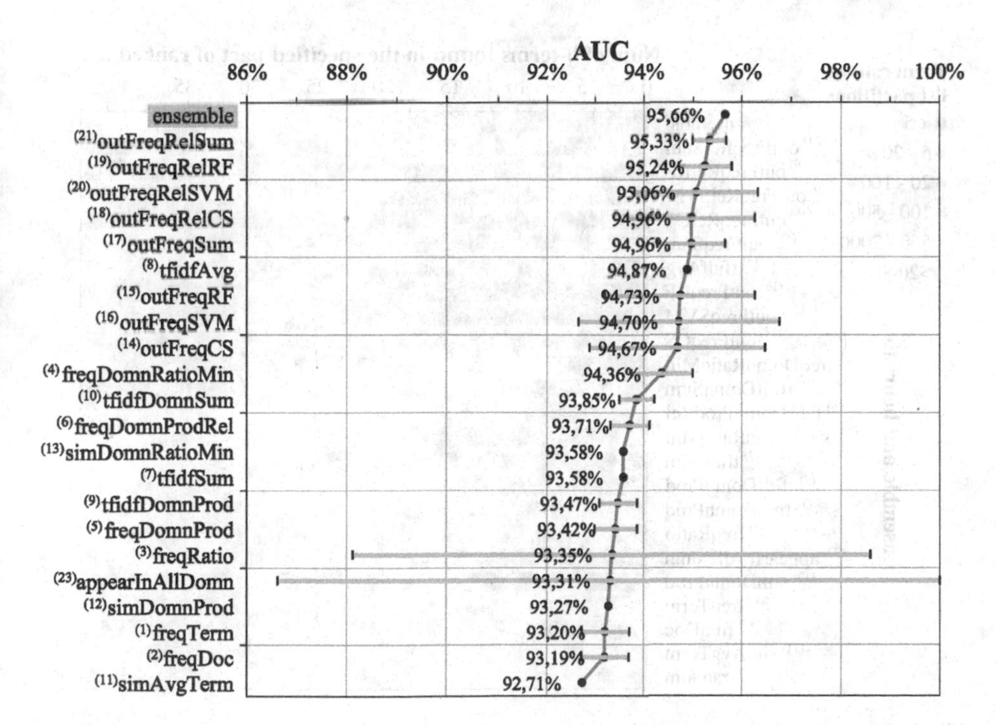

Fig. 1. Graphical representation of the AUC measure for all the individual heuristics and the ensemble heuristic on the migraine-magnesium dataset

b-terms as well as non b-terms – which results in a fact that unique sorting is not possible (i.e., see equal ensemble scores for terms t_2 and t_3 in Example 1). In such cases, the AUC calculation can either maximize the AUC by sorting all the b-terms in front of all the other terms inside equal scoring sets or minimize it by putting the b-terms at the back. The AUC calculation can also achieve many AUC values in between these two extremes by using different (e.g., random) sorts of equal scoring sets. Therefore, an interval bar of AUC shows the interval which contains all the possible AUC values and a black dot shows the interval's middle point which represents the average AUC over a large number of random sorts of equal scoring sets.

Fig. 1 shows no surprises among the base heuristics, since the results are equal to those presented in our previous work (see [7]), however, when focusing on the ensemble heuristic, we notice that it is better in both, higher AUC value and lower AUC interval compared to all the other heuristics. We constructed the ensemble using also two not so well performing heuristics ((10)tfidfDomnSum and (3)freqRatio) in order to avoid overfitting on the training domain. This could have a negative effect to the ensemble performance, however, the ensemble performance was not seriously affected which signals an evidence on the right decisions when designing the ensemble.

As stated in the introduction of this section, we are mostly interested in the heuristics quality from the end user's perspective. Such evaluation of heuristics quality is shown in Fig. 2, where the length of colored bars tells how many b-terms were found among the first 5, 20, 100, 500 and 2000 terms on the ranked list of terms

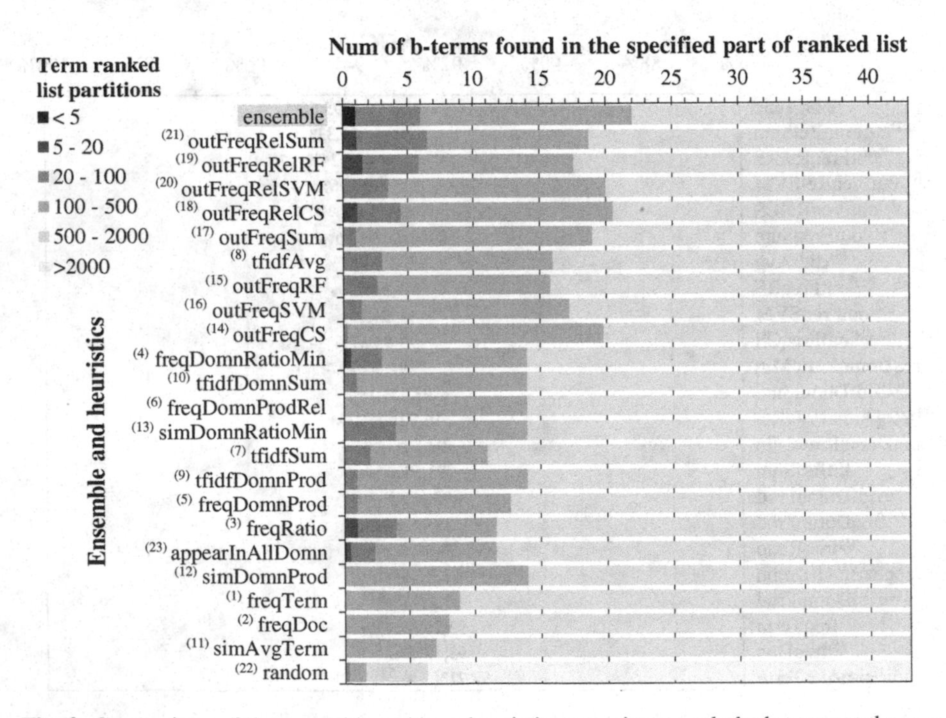

Fig. 2. Comparison of the ensemble and base heuristics capacity to rank the b-terms at the very beginning of the terms list for the migraine-magnesium dataset

produced by a heuristic. We can see that the ensemble finds one b-term among the first 5 terms (the darkest gray bar), one b-term – no additional b-terms – among the first 20 terms (no bar), 6 b-terms – 5 additional – among the first 100 terms (lighter gray bar), 22 b-terms – 16 additional – among first 500 terms (even lighter gray bar) and all the 43 b-terms – 21 additional – among the first 2000 terms (the lightest gray bar). Thus, if the expert limits himself to inspect only the first 100 terms, he will find 6 b-terms in the ensemble list, slightly more than 6 in the (21)outFreqRelSum list, 6 in the (19)outFreqRelRF, and so on. Results in Fig. 2 also give us the confirmation that the ensemble is among the best performing heuristics also from the user's perspective. Even though a strict comparison depends also on the threshold of how many terms an expert is willing to inspect, the ensemble is always among the best.

4.3 Results in Autism-Calcineurin Dataset

Fig. 3 shows how our methodology works on a new independent test dataset which was not used in the development of our methodology. As discussed, the dimensionality of the autism-calcineurin dataset is considerably different and less favorable compared to the migraine-magnesium dataset. This is evident also when observing Fig. 3, since the performance of individual base heuristics significantly changes. Some of the originally best performing heuristics, e.g., based on relative frequency in outlier sets are now among the worst and the other types, e.g., tf-idf based that were not performing well before, are now among the best. The most important observation is

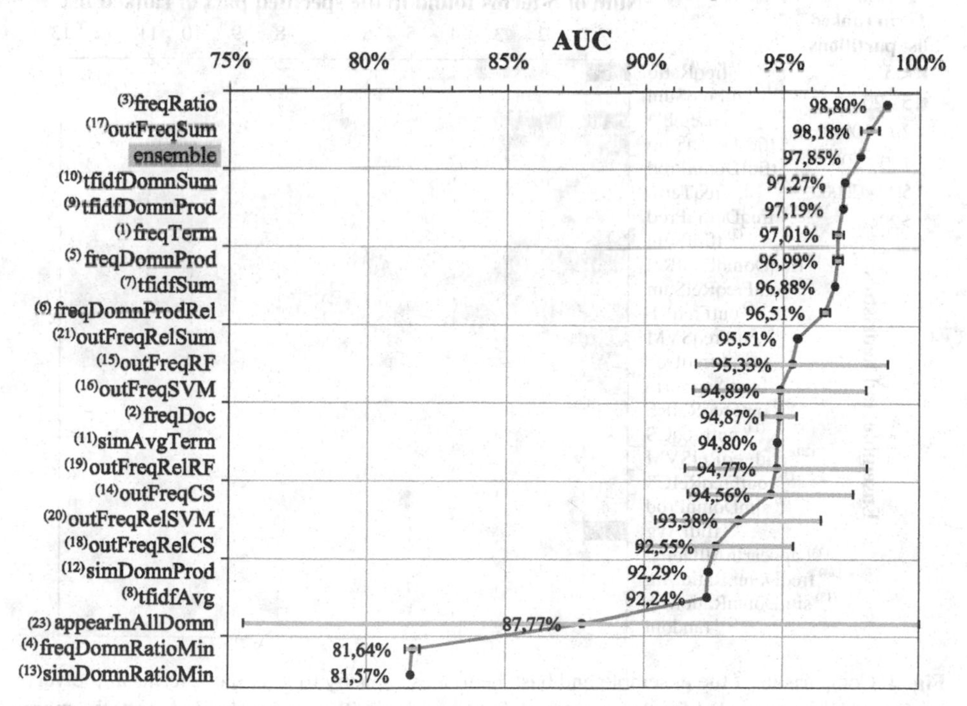

Fig. 3. Graphical representation of the AUC measure for all the individual heuristics and the ensemble heuristic on the autism-calcineurin dataset

that the ensemble heuristic is still among the best (placed after (3)freqRatio and (17)outFreqSum) and preserves a zero AUC interval. Otherwise, we can notice a slight AUC increase of the best performing heuristics which is very positive since the candidate term list is much longer now and we expect we will find the same number of b-terms much later in the candidate term list compared to the migraine-magnesium dataset.

The last result in this section is the user oriented visualization of heuristics performance shown in Fig. 4. This gives us the final argument for the quality of the ensemble heuristic since it outperforms or at least equals to all the other heuristics on the most interesting ranked list lengths (up to 20, 100, 500 terms). The ensemble finds one b-term among 20 ranked terms, 2 among 100 and 3 among 500 ranked terms. At a first sight, this may seem a bad performance, but, note that there are 78,805 candidate terms which the heuristics have to rank. The evidence of the quality of the ensemble can be understood if we compare it to the (23)appearInAllDomn heuristic which is the baseline heuristic and represents the performance which is achievable without developing the methodology presented in this work. The (23)appearInAllDomn heuristic discovers in average only approximately 0.33 b-terms before position 2000 in the ranked list while the ensemble discovers 5 – not to mention the shorter term lists where the ensemble is relatively even better compared to the (23)appearInAllDomn heurisitc.

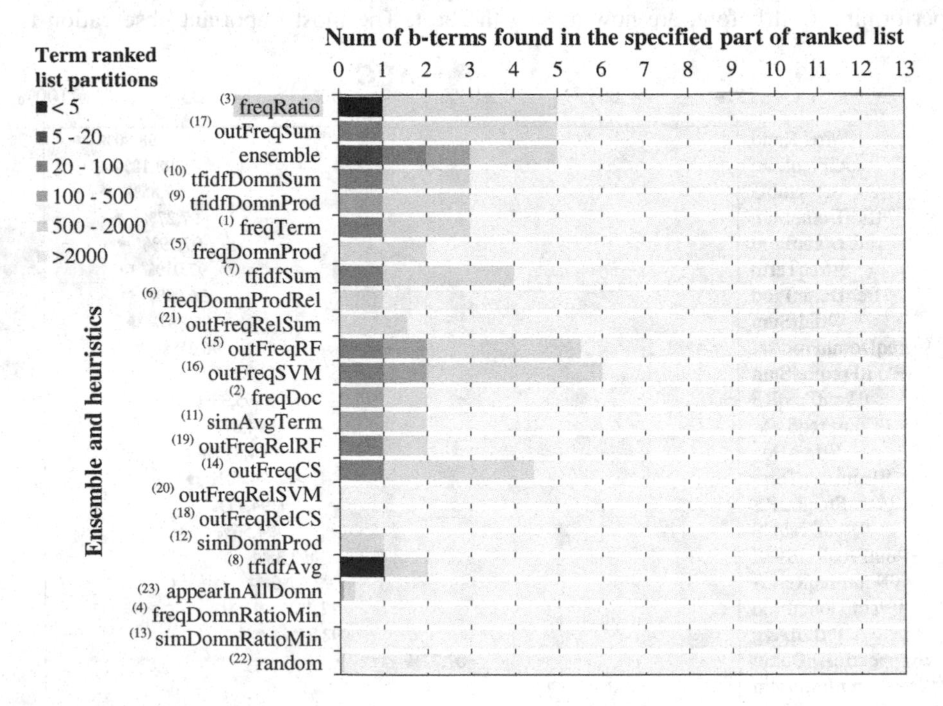

Fig. 4. Comparison of the ensemble and base heuristics capacity to rank the b-terms at the very beginning of the terms list for the autism-calcineurin dataset. The longer the dark part, the more b-terms a heuristic ranks at the specified partition of the ranked list.

5 The CrossBee System

Besides introducing the methodology for identifying bisociations, our aim was to create a system which helps the experts when searching for hidden links that connect two seemingly unrelated domains. We designed and implemented an online system CrossBee – a *Cross-Context Bisociation Explorer*. Initially, the system was designed as an online implementation of the ensemble ranking methodology presented in this chapter. To this core functionality, we have added various supplementary functionalities and content presentations which effectively turn CrossBee into a user-friendly tool for ranking and exploration of perspective cross-context links. This enables the user not only to spot but also to efficiently investigate cross-domain links pointed out by our ensemble ranking methodology.

This section presents CrossBee by describing its most important functionality and a typical use case example. The CrossBee system is built on top of the TexAs (Text Assistant) library created for the needs of our previous work [7]. From this perspective CrossBee is firstly, a functional enhancement (ensemble functionality) of TexAs and secondly, a wrapping of this functionality into a practical web user interface especially designed for the requirements of bisociation discovery (available online at the web address http://crossbee.ijs.si). Note that future versions of CrossBee might not be visually identical as the current version presented here. Nevertheless, the core ensemble ranking algorithm, the two datasets and the results presented in this work will remain available in the future by providing a link to the application, data and settings compatible with this chapter in order to ensure the repeatability of the experiments.

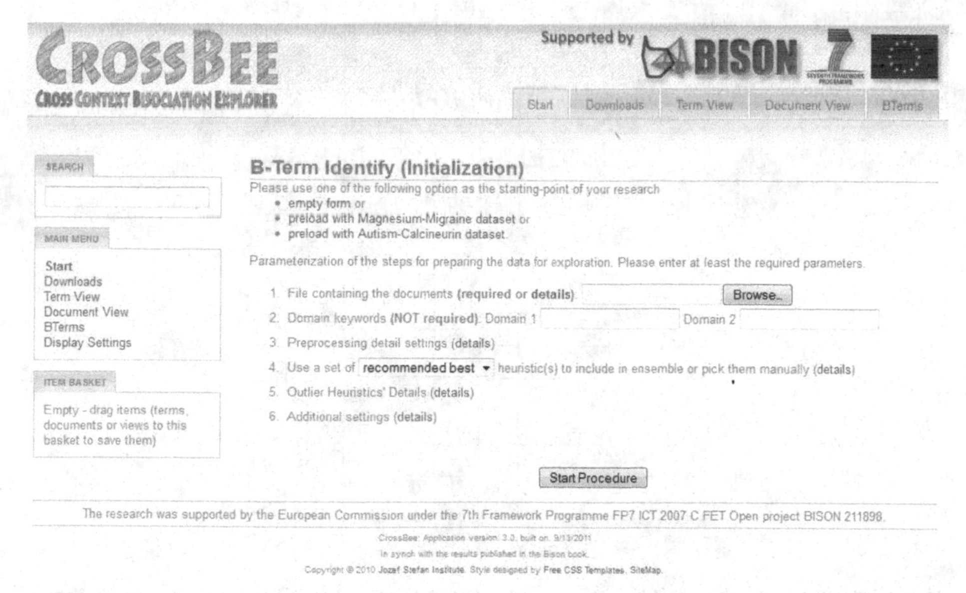

Fig. 5. Home page of the CrossBee system. The user starts an exploration at this point by inputting documents of interest and by tuning the parameters of the system.

5.1 A Typical Use Case

The most standard use case – as envisioned by CrossBee authors – is the following:

1. Prior to starting the process of bisociation exploration, the user needs to prepare the input file of documents. The prescribed format of the input file is kept simple to enable all users, regardless of their computing skills, to prepare the file of documents of their interest. Each line of this file contains exactly three tab-separated entries: (a) the document identification number, (b) the domain acronym, and (c) the document text.
2. The user starts at the initialization page (see Fig. 5) which is the main entry point to the system. The minimal required user's input at this point is a file of documents from two domains. The other options available to the user at this point include specifying the exact preprocessing options, specifying the base heuristics to be used in the ensemble, specifying outlier documents identified by an external outlier detection software (e.g., [15]), defining the already known

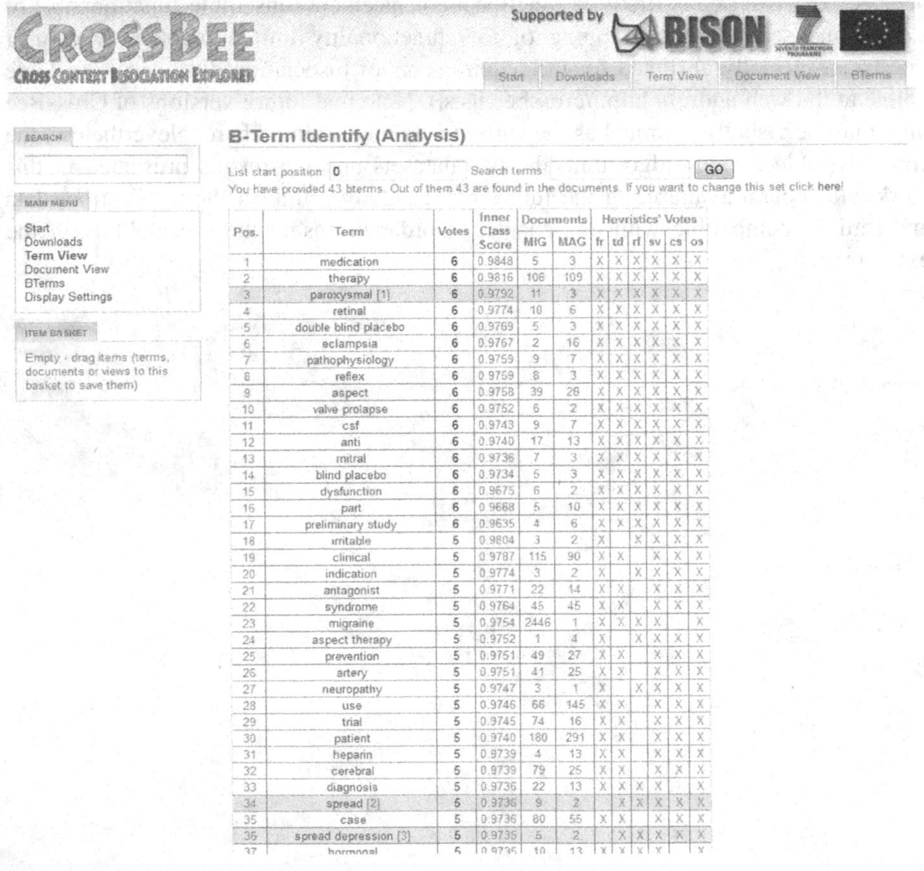

Pos.	Term	Votes	Inner Class Score	Documents		Hevristics' Votes					
				MIG	MAG	fr	td	rf	sv	cs	os
1	medication	6	0.9848	5	3	X	X	X	X	X	X
2	therapy	6	0.9816	106	109	X	X	X	X	X	X
3	paroxysmal [1]	6	0.9792	11	3	X	X	X	X	X	X
4	retinal	6	0.9774	10	6	X	X	X	X	X	X
5	double blind placebo	6	0.9769	5	3	X	X	X	X	X	X
6	eclampsia	6	0.9767	2	16	X	X	X	X	X	X
7	pathophysiology	6	0.9759	9	7	X	X	X	X	X	X
8	reflex	6	0.9759	8	3	X	X	X	X	X	X
9	aspect	6	0.9758	39	28	X	X	X	X	X	X
10	valve prolapse	6	0.9752	6	2	X	X	X	X	X	X
11	csf	6	0.9743	9	7	X	X	X	X	X	X
12	anti	6	0.9740	17	13	X	X	X	X	X	X
13	mitral	6	0.9736	7	3	X	X	X	X	X	X
14	blind placebo	6	0.9734	5	3	X	X	X	X	X	X
15	dysfunction	6	0.9675	6	2	X	X	X	X	X	X
16	part	6	0.9668	5	10	X	X	X	X	X	X
17	preliminary study	6	0.9635	4	6	X	X	X	X	X	X
18	irritable	5	0.9804	3	2	X		X	X	X	X
19	clinical	5	0.9787	115	90	X	X		X	X	X
20	indication	5	0.9774	3	2	X		X	X	X	X
21	antagonist	5	0.9771	22	14	X	X		X	X	X
22	syndrome	5	0.9764	45	45	X	X		X	X	X
23	migraine	5	0.9754	2446	1	X	X	X	X		X
24	aspect therapy	5	0.9752	1	4	X		X	X	X	X
25	prevention	5	0.9751	49	27	X	X		X	X	X
26	artery	5	0.9751	41	25	X	X		X	X	X
27	neuropathy	5	0.9747	3	1	X		X	X	X	X
28	use	5	0.9746	66	145	X	X		X	X	X
29	trial	5	0.9745	74	16	X	X		X	X	X
30	patient	5	0.9740	180	291	X	X		X	X	X
31	heparin	5	0.9739	4	13	X	X		X	X	X
32	cerebral	5	0.9739	79	25	X	X		X	X	X
33	diagnosis	5	0.9736	22	13	X	X	X	X		X
34	spread [2]	5	0.9736	9	2		X	X	X	X	X
35	case	5	0.9736	80	55	X	X		X	X	X
36	spread depression [3]	5	0.9735	5	2		X	X	X	X	X
37	hormonal	5	0.9735	10	13	X	X	X	X		X

Fig. 6. Candidate b-term ranking page as displayed by CrossBee after the preprocessing is done. This example shows CrossBee's term list output as ranked by the ensemble heuristic in the migraine-magnesium dataset. The terms marked yellow are the Swanson's b-terms.

b-terms, and others. When the user selects all the desired options he proceeds to the next step.

3. CrossBee starts a computationally very intensive step in which it prepares all the data needed for the fast subsequent exploration phase. During this step the actual text preprocessing, base heuristics, ensemble, bisociation scores and rankings are computed in the way presented in Section 3. This step does not require any user's intervention.
4. After computation, the user is presented with a ranked list of b-term candidates as seen in Fig. 6. The list provides the user some additional information including the ensemble's individual base heuristics votes (columns 7-12) and term's domain occurrence statistics in both domains (columns 5 and 6). If the user defines the actual b-terms during the initialization (which is not a realistic scenario when exploring a new domain for the first time) then these b-terms are marked throughout the whole CrossBee session, as seen in Fig. 6 and Fig. 7. The user then browses through the list and chooses the term he believes to be promising for finding meaningful connections between domains.

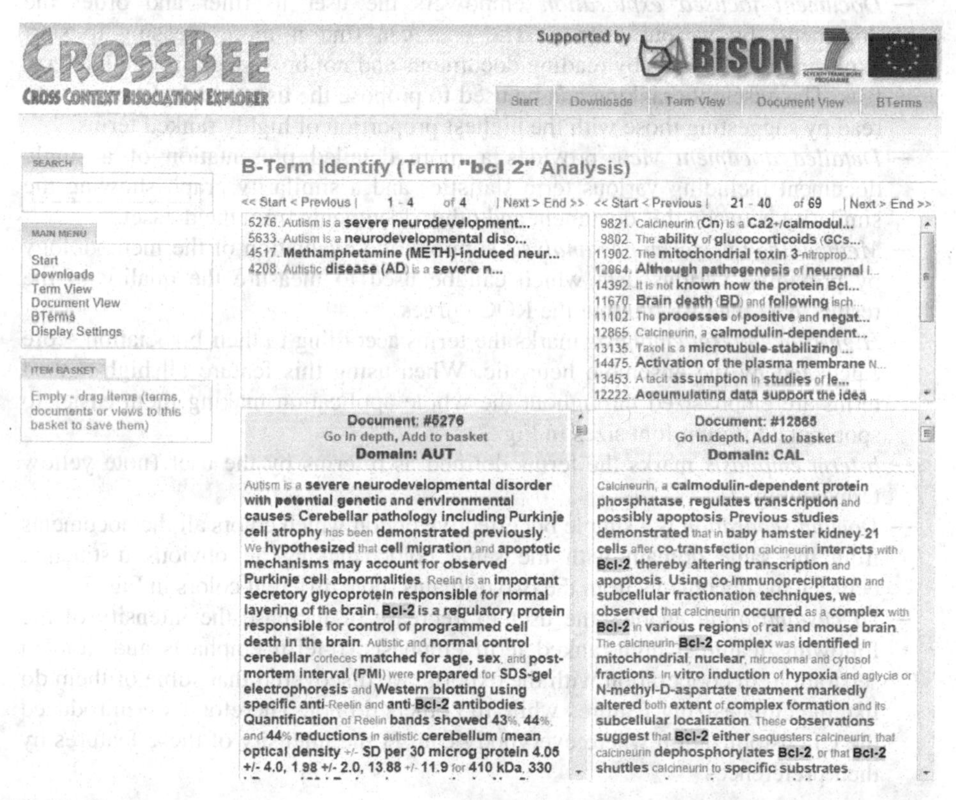

Fig. 7. CrossBee supports the inspection of potential cross-domain links by a side-by-side view of documents from the two domains under investigation. The figure presents an example from the autism-calcineurin dataset, showing the analysis of the *bcl 2* term. The presented view enables efficient comparison of documents, the left one from the autism and the right one from the calcineurin domain. The actual displayed documents were reported by Macedoni-Lukšič et al. [10] as relevant for exploring the relationship between autism and calcineurin.

5. At this point, the user inspects the actual appearances of the selected term in both domains, using the side-by-side document inspection as shown in Fig. 7. In this way, he can verify whether his rationale behind selecting this term as a bridging term can be justified based on the contents of the inspected documents.
6. Afterwards, the user continues with the exploration by returning to step 3 or by choosing another term in step 4, or concludes the session.

The most important result of the exploration procedure is a proof for a chosen term to be an actual bridge between the two domains, based on supporting facts from the documents. As experienced in sessions with the experts, the identified documents are an important result as well, as they usually turn out to be a valuable source of information providing a deeper insight into the discovered cross-domain relations.

5.2 Other CrossBee Functionalities

Below we list the most important additional functionalities of the CrossBee system:

- *Document focused exploration* empowers the user to filter and order the documents by various criteria. The user can find it more pleasing to start exploring the domains by reading documents and not browsing through the term lists. The ensemble ranking can be used to propose the user which documents to read by suggesting those with the highest proportion of highly ranked terms.
- *Detailed document view* provides a more detailed presentation of a single document including various term statistics and a similarity graph showing the similarity between this document and other documents from the dataset.
- *Methodology performance analysis* supports the evaluation of the methodology by providing various data which can be used to measure the quality of the results, e.g., data for plotting the ROC curves.
- *High-ranked term emphasis* marks the terms according to their bisociation score calculated by the ensemble heuristic. When using this feature all high-ranked terms are emphasized throughout the whole application making them easier to spot (note different font sizes in Fig. 7).
- *b-term emphasis* marks the terms defined as b-terms by the user (note yellow terms in Fig. 7).
- *Domain separation* is a simple but effective option which colors all the documents from the same domain with the same color, making an obvious distinction between the documents from the two domains (note different colors in Fig. 7).
- *UI customization* enables the user to decrease or increase the intensity of the following features: high-ranked term emphasis, b-term emphasis and domain separation. In cooperation with the experts, we discovered that some of them do like the emphasizing features while the others do not. Therefore, we introduced the UI customization where everybody can set the intensity of these features by their preferences.

6 Discussion and Further Work

This work presents a methodology and a system for bisociative literature mining focusing on b-term identification and ranking by using an ensemble heuristic. First, a detailed description of the proposed methodology and its experimental evaluation are provided, followed by the overview of the implemented system CrossBee.

In the experimental evaluation we tested a set of base heuristics and the proposed ensemble heuristic on two datasets: migraine-magnesium and autism-calcineurin. While the first dataset was used to develop the b-term ranking methodology, the second dataset was used as an independent test to validate the findings.

The comparison of the results on both datasets has shown that the performances of individual heuristics vary substantially. This indicates that there are differences between datasets which influence the performance of individual heuristics; while some base heuristic can be more adapted to one dataset, the others might be better suited to another. The proposed ensemble heuristic, which is among the best performing heuristics in both datasets, is therefore suggested as a dataset independent methodology.

The results of the heuristics were evaluated from two perspectives: (a) using the AUC measure, and (b) by counting the number of b-terms found in the first *n* term candidates. While the first measure (a) is used to estimate the quality of heuristics as a single number, which is good for ranking the heuristics, the second measure (b) is used to illustrate the heuristics quality from the end-user's perspective. In a typical scenario, the end-user appreciates reducing the burden of exploration by browsing through as few b-term candidates as possible to find the b-terms bridging the two domains. The comparison of baseline heuristics results with the constructed ensemble heuristic results confirms that the proposed methodology substantially reduces the end-user burden in this respect.

The CrossBee System has proved to be a user-friendly implementation of the presented methodology. Its visualization functionalities, in particular its presentation of pairs of documents which can be inspected in more detail for meaningful relations, is very helpful. An obvious extension planned for the near future is automatic download of documents from a selected bibliographic database, such as MEDLINE.

Investigation of more general connections between properties of domains and the best choice of selected heuristics combined into ensemble heuristic remains an important issue for further work, together with a more systematic study and comparison with other ensemble approaches known from the literature.

Acknowledgements. The work presented in this chapter was supported by the European Commission under the 7th Framework Programme FP7-ICT-2007-C FET-Open project BISON-211898, and the Slovenian Research Agency grant Knowledge Technologies (P2-0103).

References

1. Albert, R., Barabasi, A.L.: Statistical mechanics of complex networks. Rev. Mod. Phys. 74(1), 47–97 (2002)
2. Dietterich, T.G.: Ensemble Methods in Machine Learning. In: Kittler, J., Roli, F. (eds.) MCS 2000. LNCS, vol. 1857, pp. 1–15. Springer, Heidelberg (2000)

3. Dubitzky, W., Kötter, T., Schmidt, O., Berthold, M.R.: Towards Creative Information Exploration Based on Koestler's Concept of Bisociation. In: Berthold, M.R. (ed.) Bisociative Knowledge Discovery. LNCS (LNAI), vol. 7250, pp. 11–32. Springer, Heidelberg (2012)
4. Dwork, C., Kumar, R., Naor, M., Sivakumar, D.: Rank Aggregation Methods for the Web. In: Proc. of the 10th int. Conference on World Wide Web, pp. 613–622 (2001)
5. Feldman, R., Sanger, J.: The Text Mining Handbook: Advanced Approaches in Analyzing Unstructured Data. Cambridge University Press (2007)
6. Hoi, S.C.H., Jin, R.: Semi-Supervised Ensemble Ranking. In: Proc. of the 23rd National Conference on Artificial Intelligence, vol. 2. AAAI Press (2008)
7. Juršič, M., Sluban, B., Cestnik, B., Grčar, M., Lavrač, N.: Bridging Concept Identification for Constructing Information Networks from Text Documents. In: Berthold, M.R. (ed.) Bisociative Knowledge Discovery. LNCS (LNAI), vol. 7250, pp. 66–90. Springer, Heidelberg (2012)
8. Koestler, A.: The Act of Creation. The Macmillan Co. (1964)
9. Li, D., Wang, Y., Ni, W., Huang, Y., Xie, M.: An Ensemble Approach to Learning to Rank. In: 5th Int. Conf. on Fuzzy Systems and Knowledge Discovery, pp. 101–105 (2008)
10. Macedoni Lukšič, M., Petrič, I., Cestnik, B., Urbančič, T.: Developing a Deeper Understanding of Autism: Connecting Knowledge through Literature Mining. In: Autism Research and Treatment (2011)
11. Petric, I., Urbancic, T., Cestnik, B., Macedoni-Luksic, M.: Literature mining method RaJoLink for uncovering relations between biomedical concepts. Journal of Biomedical Informatics 42(2), 219–227 (2009)
12. Petrič, I., Cestnik, B., Lavrač, N., Urbančič, T.: Outlier Detection in Cross-Context Link Discovery for Creative Literature Mining. Comput. J., November 2 (2010)
13. Petrič, I., Cestnik, B., Lavrač, N., Urbančič, T.: Bisociative Knowledge Discovery by Literature Outlier Detection. In: Berthold, M.R. (ed.) Bisociative Knowledge Discovery. LNCS (LNAI), vol. 7250, pp. 313–324. Springer, Heidelberg (2012)
14. Rokach, L.: Ensemble-based classifiers. Art. Int. Review 33(1-2), 1–39 (2010)
15. Sluban, B., Juršič, M., Cestnik, B., Lavrač, N.: Exploring the Power of Outliers for Cross-domain Literature Mining. In: Berthold, M.R. (ed.) Bisociative Knowledge Discovery. LNCS (LNAI), vol. 7250, pp. 325–337. Springer, Heidelberg (2012)
16. Swanson, D.R.: Migraine and magnesium: Eleven neglected connections. Perspectives in Biology and Medicine 31(4), 526–557 (1988)
17. Swanson, D.R.: Medical literature as a potential source of new knowledge. Bull. Med. Libr. Assoc. 78(1), 29–37 (1990)
18. Swanson, D.R., Smalheiser, N.R., Torvik, V.I.: Ranking Indirect Connections in Literature-Based Discovery: The Role of Medical Subject Headings (MeSH). Journal of the American Society for Inf. Science and Technology 57, 1427–1439 (2006)
19. Urbančič, T., Petrič, I., Cestnik, B., Macedoni-Lukšič, M.: Literature Mining: Towards Better Understanding of Autism. In: Bellazzi, R., Abu-Hanna, A., Hunter, J. (eds.) AIME 2007. LNCS (LNAI), vol. 4594, pp. 217–226. Springer, Heidelberg (2007)
20. Urbančič, T., Petrič, I., Cestnik, B., Macedoni Lukšič, M.: RaJoLink: A Method for Finding Seeds of Future Discoveries in Nowadays. In: Proceedings of the 18th Symposium on Methodologies for Intelligent Systems, Prague, pp. 129–138 (2009)
21. Weeber, M., Vos, R., Klein, H., de Jong-van den Berg, L.T.W.: Using concepts in literature-based discovery: Simulating Swanson's Raynaud–fish oil and migraine-magnesium discoveries. J. Am. Soc. Inf. Sci. Tech. 52(7), 548–557 (2001)

Applications and Evaluation: Overview

Igor Mozetič and Nada Lavrač

Jožef Stefan Institute, Ljubljana, Slovenia

1 Introduction

This part of the book presents several applications which were motivated by the concept of bisociation, and to some extent exploited the notions of heterogeneous information networks, explicit contextualization and/or context crossing.

The main goals of these applications are:

- to verify if the principles of heterogeneous information networks and bisociation, and their computational realization, can lead to new discoveries,
- to test the software platforms developed for bisociative knowledge discovery, and
- to find actual new discoveries in at least some application domains.

Most of the applications are in the area of biology, but in addition there are interesting digressions to finance, improvements of business processes, and music recommendations.

2 Contributions

Eronen at al. [1] discusses Biomine as a BisoNet which integrates heterogeneous biological databases. It consists of over 1 million nodes, representing biological entities (genes, proteins, ontology terms, ...), and over 8 million edges, representing weighted relations of different types. Biomine search algorithms implement link discovery between distant nodes in the graph, and can be exploited for context crossing in bisociative reasoning.

Biomine is an essential component of SegMine, described by Mozetič et al. [8], which implements a form of bisociative reasoning for the analysis of microarray data. SegMine first performs explicit contextualization by subgroup discovery (implemented by the SEGS algorithm), where sets of enriched genes are found. Context crossing is then triggered by queries to Biomine which discovers long range links between distant sets of genes. In the analysis of senescence in human stem cells [10] (not described in this book), a biology expert used SegMine to formulate three new hypotheses which can improve the understanding of the underlying mechanisms in senescence.

A novel application of SegMine, extended to plant biology, is described by Langohr et al. [5]. The problem addressed is the analysis of plant response to a virus attack, from a series of microarray datasets. All human related databases and ontologies in SEGS and Biomine were replaced by plant related data, and

M.R. Berthold (Ed.): Bisociative Knowledge Discovery, LNAI 7250, pp. 359–363, 2012.

subgroup discovery is used again in the later stage of analysis to characterize contrasts between different microarray datasets. The bisociative component in this study consists of the transfer of knowledge about a well-understood plant, namely A. thaliana, to investigate a less well-understood plant, in this case the potato.

Bisociations between related organisms are also exploited in the approach by Kimming and Costa [4]. They investigate metabolic pathways with the goal of automatic pathway curration by link and node prediction, similar to Biomine. However, they make use of information on related organisms to suggest filling of incomplete pathways.

An exploration of textual resources to get an insight into a biological domain is described by Miljković et al. [7]. The ultimate goal is to develop a dynamic model of plant defense to a virus attack. Scientific literature read by domain experts is automatically analyzed to extract triplets of the form <node1, edge, node2> which then form a heterogeneous information network. The network can be explored to find cross-context links between different bodies of human expertise, or eventually (not described in this book) to find novel cross-talk links between different submodels of the plant defense response.

Another analysis of textual resources is described by Schmidt et al. [11]. The goal is to better understand a biological bile acid and xenobiotic system (BAXS) by bisociative hints from a drastically different domain of finance. The idea was to retreive several thousands of scientific papers from both domains, cluster them, and then identify outlier documents. Outliers are biological papers more similar to financial papers than to the rest of biological papers, or vice versa. From the outliers, some interesting bridging terms can be identified which connect the two disparate domains.

An application to business process modelling is presented by Martin and He [6]. The goal is to improve business processes by discovery of process models, their analysis, extension and mining. Process instances concerning repair and call-center data were used to define different contexts, and bisociative reasoning suggested three possible routes which could lead to process improvement.

Finally, Stober et al. [12] present an advanced user interface for music recommendation. Here, the concept of bisociation provides motivation for unexpected and fortunate (serendipitous) recommendations. The article demonstrates how the separation of the similarity measures for projection and distortion makes it possible to link two distinct views on a music collection. As a consequence, it creates a setting where serendipitous recommendations become more likely.

The main conclusions of the applications presented in this volume are the following:

- The concepts of bisociation, explicit contextualization, and context crossing have the potential to help formulate research hypotheses which lead to new discoveries.
- Several software platforms for bisociative reasoning were developed; at least two of them are used regularly by domain experts in human and plant biology (SegMine and Biomine).

- At least in two domains (microarray analysis of human stem cells, autism) biology and medical experts formulated significant new research hypotheses which facilitate novel insights into the domains.

3 Lessons Learned

3.1 The BISON Software for Applications Development

There are numerous lessons learned from the applications described in this volume. In addition to the BISON platform and the software developed within the BISON project, a lesson learned is that in bisociative discovery tasks we were often able to use also other existing software tools, which were used beyond their original scope and purpose. For example, Ontogen [2] is an interactive tool for the construction of topic ontologies, but was used in several applications described in this book for outlier detection and b-term identification. A lesson learned from using Ontogen's similarity graph is, however, that it needs to be used with care. Although two documents appear to be close to each other in the similarity graph, actually they can be distant but at a similar distance from the centroid of the given document cluster. Another tool successfully used in the BISON project was Biomine which was designed for link discovery in biological domains, but in the stem cells microarray analysis its visualization facility enabled the biology expert to identify "gene hubs" (nodes with a large number of edges) and "outlier genes" (nodes with a few edges and of low strength). These concepts are known in social network analysis, but were not exploited in the Biomine context before. Another success, for which Biomine and SEGS were designed but not actually used before, was the relative ease with which we replaced human related databases with plants related databases.

3.2 Application Potential of the BISON Methodology

A major lesson learned from the microarray analysis applications described in this part of the book is that a huge amount of effort is needed to develop a software platform to be used by biology experts. On one hand, the software must match state-of-the-art biological software tools to be competitive. On the other hand, it must address a large number of data management requirements (e.g., different and sometimes inconsistent formats) which are important for routine biological research, but are largely uninteresting and irrelevant from a computer science and knowledge discovery perspective.

The role of bisociations was mainly conceptual but played a crucial role in the formulation of new hypotheses. It made us aware of the need for explicit definition of distinct contexts, for the search of links between them, and for intentional jumps "out-of-the-context". These were initially accomplished by ingenious connection of seemingly unrelated tools (SEGS and Biomine), but later evolved into a novel, interactive, service-oriented platform with a set of SegMine workflows, implemented in a principled way. This led to a natural extension to contrasting coSegMine [5] which opens exciting opportunities for future research.

3.3 Evaluation of the BISON Methodology and the Potential for Triggering Creativity

The problem of cross-context link discovery from scientific papers (presented in Part IV: Exploration) is that in new domains the success criteria are unclear and that only the expert's evaluation is possible. However, in the document analysis case studies on migraine-magnesium and autism-calcineurin [3] this problem did not occur since the task was to evaluate the method by rediscovering known b-terms.

In these cross-context link discovery applications we reused Ontogen in a novel, unforeseen way. The Ontogen approach may not be seen as an approach that triggers creativity, but still it is a useful tool for cross-context discovery. The strongest novelty and lesson learned is that indeed outliers are very useful means of speeding up link discovery in cross-context domains, which was confirmed experimentally in the migraine-magnesium and autism-calcineurin domains [9]. The utility of Ontogen was further proven in the completely new BAXS-finance domain pair as well.

4 The Future of Bisociative Reasoning and Cross-Context Data Mining

Computational creativity community is aware of Koestler's work, but this community can now establish a clear link with the data mining community through the results presented in this book. The BISON project has identified a novel cross-context data mining task which could be of large interest to the data mining community. The investigated research topic, cross-context data mining and knowledge discovery, is not yet part of mainstream data mining research. By further raising awareness of this cross-context/cross-domain knowledge discovery paradigm, the work presented in this book has the potential to ensure that cross-context knowledge discovery will become a recognized topic and thus a first class citizen of major machine learning, data mining and knowledge discovery conferences.

References

1. Eronen, L., Hintsanen, P., Toivonen, H.: Biomine: A Network-Structured Resource of Biological Entities for Link Prediction. In: Berthold, M.R. (ed.) Bisociative Knowledge Discovery. LNCS (LNAI), vol. 7250, pp. 364–378. Springer, Heidelberg (2012)
2. Fortuna, B., Grobelnik, M., Mladenic, D.: OntoGen: Semi-automatic Ontology Editor. In: Smith, M.J., Salvendy, G. (eds.) HCII 2007. LNCS, vol. 4558, pp. 309–318. Springer, Heidelberg (2007)

3. Juršič, M., Sluban, B., Cestnik, B., Grčar, M., Lavrač, N.: Bridging Concept Identification for Constructing Information Networks from Text Documents. In: Berthold, M.R. (ed.) Bisociative Knowledge Discovery. LNCS (LNAI), vol. 7250, pp. 66–90. Springer, Heidelberg (2012)
4. Kimmig, A., Costa, F.: Link and Node Prediction in Metabolic Networks with Probabilistic Logic. In: Berthold, M.R. (ed.) Bisociative Knowledge Discovery. LNCS (LNAI), vol. 7250, pp. 407–426. Springer, Heidelberg (2012)
5. Langohr, L., Podpečan, V., Petek, M., Mozetič, I., Gruden, K.: Subgroup Discovery from Interesting Subgroups. In: Bisociative Knowledge Discovery. LNCS (LNAI), vol. 7250, pp. 390–406. Springer, Heidelberg (2012)
6. Martin, T., He, H.: Bisociation Discovery in Business Process Models. In: Berthold, M.R. (ed.) Bisociative Knowledge Discovery. LNCS (LNAI), vol. 7250, pp. 452–471. Springer, Heidelberg (2012)
7. Miljković, D., Podpečan, V., Grčar, M., Gruden, K., Stare, T., Petek, M., Mozetič, I., Lavrač, N.: Modelling a Biological System: Network Creation by Triplet Extraction from Biological Literature. In: Berthold, M.R. (ed.) Bisociative Knowledge Discovery. LNCS (LNAI), vol. 7250, pp. 427–437. Springer, Heidelberg (2012)
8. Mozetič, I., Lavrač, N., Podpečan, V., Novak, P.K., Motaln, H., Petek, M., Gruden, K., Toivonen, H., Kulovesi, K.: Semantic Subgroup Discovery and Cross-context Linking for Microarray Data Analysis. In: Berthold, M.R. (ed.) Bisociative Knowledge Discovery. LNCS (LNAI), vol. 7250, pp. 379–389. Springer, Heidelberg (2012)
9. Petrič, I., Cestnik, B., Lavrač, N., Urbančič, T.: Bisociative Knowledge Discovery by Literature Outlier Detection. In: Berthold, M.R. (ed.) Bisociative Knowledge Discovery. LNCS (LNAI), vol. 7250, pp. 313–324. Springer, Heidelberg (2012)
10. Podpečan, V., Lavrač, N., Mozetič, I., Kralj Novak, P., Trajkovski, I., Langohr, L., Kulovesi, K., Toivonen, H., Petek, M., Motaln, H., Gruden, K.: SegMine workflows for semantic microarray data analysis in Orange4WS. BMC Bioinformatics 12, 416 (2011)
11. Schmidt, O., Kranjc, J., Mozetič, I., Thompson, P., Dubitzky, W.: Bisociative Exploration of Biological and Financial Literature Using Clustering. In: Berthold, M.R. (ed.) Bisociative Knowledge Discovery. LNCS (LNAI), vol. 7250, pp. 438–451. Springer, Heidelberg (2012)
12. Stober, S., Haun, S., Nürnberger, A.: Bisociative Music Discovery & Recommendation. In: Berthold, M.R. (ed.) Bisociative Knowledge Discovery. LNCS (LNAI), vol. 7250, pp. 472–483. Springer, Heidelberg (2012)

Biomine: A Network-Structured Resource of Biological Entities for Link Prediction

Lauri Eronen, Petteri Hintsanen, and Hannu Toivonen

Department of Computer Science and HIIT, University of Helsinki, Finland
firstname.lastname@cs.helsinki.fi

Abstract. Biomine is a biological graph database constructed from public databases. Its entities (vertices) include biological concepts (such as genes, proteins, tissues, processes and phenotypes, as well as scientific articles) and relations (edges) between these entities correspond to real-world phenomena such as "a gene codes for a protein" or "an article refers to a phenotype". Biomine also provides tools for querying the graph for connections and visualizing them interactively.

We describe the Biomine graph database. We also discuss link discovery in such biological graphs and review possible link prediction measures. Biomine currently contains over 1 million entities and over 8 million relations between them, with focus on human genetics. It is available on-line[1] and can be queried for connecting subgraphs between biological entities.

1 Introduction

Biomine is a large biological graph (or BisoNet [1]) whose entities (vertices) include concrete biological concepts such as genes, proteins and tissues, but also abstract concepts such as biological processes, phenotypes and scientific articles. Relations (edges) between these entities correspond to real-world phenomena such as "a gene codes for a protein" or "an article refers to a phenotype". We are motivated by link discovery in such biological graphs with the primary aim of prioritising putative disease-susceptibility genes.

A generic goal of Biomine is to help users discover and understand relations between biological entities, such as indirect connections between a gene and a disease. In the context of bisociative or creative information exploration [2], our aim is to facilitate discovery of bisociations between biological entities that are not connected within a single existing database. As a concrete and motivating example, consider a gene mapping process for a disease (or other phenotype). Current genome-wide analysis methods produce a large number of candidate genes, i.e., putative disease-susceptibility genes for the disease. A question then is how to prioritize these genes so that further efforts can be focused on the most promising candidates. One approach is to look at what is already known about the putative disease genes and see how they relate to each other and

[1] http://biomine.cs.helsinki.fi

M.R. Berthold (Ed.): Bisociative Knowledge Discovery, LNAI 7250, pp. 364–378, 2012.

to the phenotype under study. This might reveal evidence for the hypothesised association or facilitate a more detailed hypothesis about the mechanisms of the relationship. Due to the lack of automated methods the work is mostly done by manually browsing the databases. This is a slow and laborious process which necessarily limits the extent and coverage of the search. In this chapter we describe a database and methods for (partial) automation of the prioritising task.

Biological graphs can be built from publicly available biological databases. Converting (relational) biological knowledge to a graph form is conceptually simple though not straightforward. For instance, how to map different biological concepts and their attributes into the graph and how to weight edges is non-trivial. In Sections 2 and 3 we consider these issues in the context of Biomine, a relatively large biological graph. In Section 4 we then review some proposed link goodness measures and consider the evaluation of link significance. We briefly review related work in Section 5 and conclude in Section 6.

2 Biomine Database

We now describe in more detail *Biomine*, a large index of various interlinked public biological databases. Biomine offers a uniform view to these databases by representing their contents as a large, heterogeneous graph, with probabilistic edges. Vertices in this graph represent entities (records) in the original databases and edges represent their annotated relationships (cross-references between records). Edges have weights that are interpreted as probabilities. A preliminary version of Biomine has been described by Sevon et al [3]. In this section we take a brief look at the core components of Biomine: its data model and source databases. Edge weighting is considered separately in Section 3.

2.1 Data Model

The choice of *data representation*, or *data model*, is important in link mining [4]. To facilitate wide applicability, the core Biomine data model is deliberately simple: all source database records are represented as vertices in an undirected, labelled and weighted multigraph $G = (V, E)$. The elements of the vertex set V are biological entities such as genes, proteins and biological processes as well as more general objects like article abstracts. They are labelled by a type, such as *gene* or *protein*, from set T_v. We denote the vertex type mapping by $t_v : V \mapsto T_v$.

Edge multiset $E \subset [V]^2$ consists of unordered vertex pairs $\{u, v\}$. As with vertices, edges have labels from edge type set T_e and we denote this mapping by $t_e : E \mapsto T_e$. Edge types depict annotated relations between vertices, such as *codes for* (e.g., gene *codes for* protein) or *refers to* (e.g., article *refers to* gene).

Each edge has a source database where the corresponding relation resides. We denote this source database mapping by $s : E \mapsto \mathcal{D}$ where $\mathcal{D}$ is the set of source databases. For a given graph $G = (V, E)$, we refer to its vertex set V by $V(G)$

and its edge set E by $E(G)$. Finally, we denote the set of neighbouring vertices of v by $N(v) = \{u \in V : \{v,u\} \in E\}$.

Table 1 lists the vertex types used in Biomine; similarly, Table 2 lists the edge types. Some representative examples of typed edges are given in Table 3. All tables refer to the Biomine database built at 4th June 2010.

Table 1. Biomine vertex types T_v, primary source database for each type, and the total amount and mean degrees of corresponding vertices

Type	Primary source database	Amount	Mean degree
Active site	InterPro	89	95.82
Allelic variant	OMIM	19,455	1.44
Article	PubMed	532,675	3.98
Binding site	InterPro	62	111.18
Biological process	GO	19,539	32.47
Cellular component	GO	2,856	122.18
Compound	KEGG	15,879	0.55
Conserved site	InterPro	575	58.29
Domain	InterPro	5,515	69.55
Drug	KEGG	8,846	0.69
Enzyme	KEGG	5,095	10.15
Family	InterPro	12,718	10.61
Gene	Entrez Gene	192,893	18.60
Gene/Phenotype	OMIM	343	82.35
Genomic context	Entrez Gene	11,825	18.68
Glycan	KEGG	2,519	0.92
Homolog group	HomoloGene	25,780	3.18
Molecular function	GO	9,529	49.07
Ortholog group	KEGG	13,067	3.81
Pathway	UniProt	1,875	37.10
Phenotype	OMIM	6,559	16.95
PTM	InterPro	16	82.88
Protein	UniProt	275,292	29.58
Region	InterPro	1,441	20.14
Repeat	InterPro	255	94.84
Tissue	UniProt	1,317	189.10
		total 1,166,020	14.84

2.2 Source Databases

Biomine essentially is an index to several interlinked, publicly available *source databases*. Each database provides different kinds of entities and relations to Biomine, some overlapping. We briefly review the main features of the source databases below.

NCBI's *Entrez Gene* [5,6] provides gene entries for different organisms. Currently, Biomine contains five model organisms: human, mouse, rat, fruit fly and

Table 2. Biomine edge types T_e and amount of edges of each type

Type	Source databases	Amount
affects	Entrez Gene	5,077
belongs to	Entrez Gene, HomoloGene, KEGG, STRING, SwissProt, TrEMBL	689,026
codes for	Entrez Gene, KEGG, STRING	174,480
contains	SwissProt, TrEMBL	454,553
functionally associated to	STRING	2,916,286
has	Entrez Gene, InterPro, KEGG, OMIM, SwissProt, TrEMBL	464,369
has synonym	Entrez Gene	1,666
interacts with	Entrez Gene, SwissProt, TrEMBL	97,361
is a	GO, InterPro, KEGG	51,483
is expressed in	SwissProt, TrEMBL	234,153
is found in	Entrez Gene, InterPro, KEGG, SwissProt, TrEMBL	337,542
is homologous to	HomoloGene	259,390
is located in	Entrez Gene, OMIM	144,495
is part of	GO, InterPro, OMIM	54,196
is related to	GO, HomoloGene, KEGG, OMIM, SwissProt, TrEMBL	25,414
overlaps	OMIM	8,199
participates in	Entrez Gene, InterPro, KEGG, SwissProt, TrEMBL, UniProt	605,237
refers to	Entrez Gene, KEGG, OMIM, SwissProt, TrEMBL	2,216,614
subsumes	Entrez Gene, KEGG, STRING, SwissProt, TrEMBL	140,555
targets	KEGG	4,885
		total 8,884,981

Table 3. Some examples of Biomine edge types, their source databases and the amount of corresponding edges. Observe that a sequence of such edges would constitute a gene–gene path in the graph.

Edge	Source database	Amount
Gene *codes for* Protein	STRING	5,948
Protein *belongs to* Family	SwissProt	30,651
Family *participates in* Biological process	InterPro	5,274
Biological process *is related to* Tissue	GO	13,103
Protein *is expressed in* Tissue	SwissProt	176,034
Enzyme *subsumes* Protein	TrEMBL	7,907
Gene *codes for* Enzyme	KEGG	14,195

nematode (*Caenorhabditis elegans*). Genes are connected to their protein products and other homologous genes (similar genes in different organisms). Homology relations come from an another Entrez database *HomoloGene* [6]

UniProt [7] is the main source of protein-related information. Its core elements are proteins, pathways and tissues. These elements form vertices in the graph. Manually annotated and reviewed proteins are in *Swiss-Prot* subdatabase, while *TrEBML* subdatabase contains automatically annotated and nonreviewed proteins. UniProt contains many relations, such as protein interactions and expressions, and classifications into protein families and pathways.

InterPro [8] is another protein-related database. It indexes protein families and structural elements (domains, regions, sites, etc.), and it has hierarchies for these elements. The third protein database, STRING [9], contains known and predicted protein–protein interactions. The interactions include direct (physical) and indirect (functional) associations. STRING also contains clusters of orthologous groups (COGs) and their interactions, with mappings between proteins and COGs.

Gene Ontology (GO) aims to provide a controlled vocabulary for genes and gene products [10]. Its core domains are cellular components, biological processes and molecular functions. The ontology is structured as a directed acyclic graph and each term has defined relationships to one or more other terms in the same domain and sometimes to other domains. This graph is a subgraph of Biomine, and the term vertices are referred to by other databases such as Entrez Gene and UniProt.

Online Mendelian Inheritance in Man (OMIM) is a catalogue of human genes and genetic disorders. It is the main source of phenotype information in Biomine: most of the OMIM entries are *Phenotype* vertices. The database also contains descriptions of allelic variants, gene locations and a large number of references to biomedical literature.

PubMed [6] is a freely accessible online database of biomedical journal citations and abstracts with approximately 20 million entries at the time of writing. Many biological databases (such as UniProt and OMIM) contain references to PubMed entries, for example to index articles where a particular gene or phenotype is mentioned. In Biomine these cross-referenced PubMed entries are *Article* vertices.

Kyoto Encyclopedia of Genes and Genomes (KEGG) is a large, integrated database resource consisting of 16 main databases broadly categorised into systems information, genomic information and chemical information [11]. Biomine uses a subset of KEGG: its pathway, gene, drug, orthology, compound and glycan databases.

Each of the source databases has its own schema for arranging and formatting data. Raw data files are preprocessed into a uniform intermediate format before integration. Intermediate format files are essentially lists of typed edges, vertex attributes and synonym mappings. These files are then imported into a single database to form a large graph. During the importing process synonyms, invalid references and other anomalies are resolved. The complete conversion and importing process is complicated and out of the scope of this chapter.

3 Edge Goodness in Biomine

One of the goals of Biomine is to allow discovery and evaluation of links between vertices specified by the user. To rank paths or assess the significance of a connection between two vertices we need a measure for edge goodness. Edges sometimes have natural weights in the source databases. For example, a homology between two proteins could have a value denoting the degree of sequence similarity. Biomine extends such domain-specific static weighting by considering edge weight, or *goodness*, as a function of three factors:

1. *Reliability.* How confident are we that the relation (and consequently the edge) really exists? How reliable is the data source, how reliable is the method used to produce or predict the edge and how strong or probable is the connection estimated to be in the data source?
2. *Relevance.* How relevant is the edge with respect to the query? We assume that the investigator can give query-specific weights for vertex and/or edge types according to his or her subjective opinions of the importance of each type for the query at hand.
3. *Rarity* of informativeness. How rare and informative is the edge? As an extreme example, an article [12] that refers to over 18,000 human and mouse genes is not likely to be relevant for a specific gene whereas an article that only refers to few genes is much more likely to be informative. In Biomine edge rarity is directly related to the degrees of its incident vertices.

A distinguishing feature of Biomine is the probabilistic interpretation of the above factors: an edge $e \in E$ is considered to be reliable with probability $r(e)$, relevant with probability $q(e)$ and rare (or *informative*) with probability $d(e)$. These factors are combined to a single probability $g(e)$ so that e is an existing and potentially useful relation if e is at the same time reliable, relevant and informative. In other words, edges are random: e "exists" or "is true" with probability $g(e)$, or "does not exist" or "is not true" with probability $1 - g(e)$. With the probabilistic interpretation G is a random graph that naturally models the uncertainty in the source data and the query-specific relevance. We next give definitions for r, q and d, and we combine them into one goodness g.

Reliability $r(e)$ of an edge $e \in E$ is defined as a product of two (independent) reliabilities: a database reliability $r_d : \mathcal{D} \mapsto [0, 1]$ and a relation (edge) reliability $r_r : E \mapsto [0, 1]$. The database reliability r_d is given by the user, and the interpretation of r_d is the degree of belief the user has for a relation being correctly annotated in the corresponding database. For example, the manually curated Swiss-Prot database could be given a perfect reliability by letting $r_d(\text{Swiss-Prot}) = 1.0$, while the computer-annotated TrEMBL database could be assumed to be less precise by letting $r_d(\text{TrEMBL}) = 0.75$. Relation reliability r_r comes from the source database instead: if there is a separate confidence value c associated to e (that reflects similarity or homology score, for example), we let $r_r(e) = c$, where c is scaled between 0 and 1 if needed. Otherwise we let $r_r(e) = 1$. The interpretation of $r_r(e)$ is the confidence of the data source itself on the relation represented by e.

We define the *edge reliability* $r : E \mapsto [0,1]$ by treating the reliabilities r_d and r_r as probabilities of independent events:

$$r(e) = r_d(s(e)) \cdot r_r(e) \tag{1}$$

where $s(e)$ is the source database of e. The interpretation of $r(e)$ is that e is reliable if both the database (as a whole) and the annotation are considered reliable.

Relevance $q(e)$ of an edge $e \in E$ is the degree of belief that e represents a relevant connection between vertices u and v with respect to the current query. Edge relevance is analogous to edge reliability r but, in contrary to the static database-related reliability, relevance is query-specific.

Relevance values may be sometimes easier to give in terms of vertex types instead of edge types. Hence Biomine uses two relevance functions: $q_v : T_v \mapsto [0,1]$ for vertex types and $q_e : T_e \mapsto [0,1]$ for edge types. Both q_v and q_e are given by the user. A practical implementation could have a default configuration for both q_v and q_e, so only few adjustments would be needed for a typical query.

As in (1), relevance values q_v and q_e are treated as probabilities of independent events. The *edge relevance* $q : E \mapsto [0,1]$ is

$$q(e) = q_e(t_e(e)) \cdot \sqrt{q_v(t_v(u))} \cdot \sqrt{q_v(t_v(v))} \tag{2}$$

where $e = \{u, v\} \in E$. Vertex relevance coefficient $\sqrt{q_v(t_v(x))}$ in (2) decomposes the vertex type specific relevance $q_v(t_v(x))$ of vertex x for each of its adjacent edges. As path relevance will be later defined as a product of edge relevance values this gives the desired outcome: the relevance of any path visiting a vertex of type τ is multiplied by $q(\tau)$.

We want to give lower scores for paths that visit vertices with high degrees: the higher the degree of vertex $v \in V$ the less likely it is that any two neighbours of v actually have an interesting connection through v. Hence we define *rarity* $d_v : V \mapsto [0,1]$ first for vertices. Rarity $d_v(v)$ represents the probability that any two edges incident on v are related to each other and represent a meaningful path; the higher the rarity, the more informative v is. The following ad hoc formula is used as a basis for rarity:

$$d_v(v) = \frac{1}{(\deg(v) + 1)^\alpha} \tag{3}$$

where $0 \leq \alpha \leq 1$ is a *penalising parameter*. It determines how steeply d_v decreases as a function of vertex degree. With $\alpha = 0$ we have $d_v(v) \equiv 1$ so that all vertices are considered equally informative. With $\alpha = 1$ we have $d_v(v) = (deg(v,+)1)^{-1}$ and $d_v(v)$ has the following probabilistic interpretation. Consider a random walker who, at any vertex, is equally likely to follow any edge or stop at the vertex. Given a path $P = (v_1, v_2, \ldots, v_k)$, $v_i \in V$, rarity $d_v(v_i)$ is the probability that the walker who has so far traversed vertices $v_1, \ldots, v_i$ will next stay on the path and visit vertex v_{i+1}.

The simple formula (3) can be too inflexible in practise. Take for example PLA2G7: a widely studied asthma gene that has been referred in 97 articles. Because of these article links (3) would penalise PLA2G7 vertex severely. However, it has only one interaction link and it participates in three biological processes, so PLA2G7 could be informative when the investigator is mostly interested in gene–gene interactions or biological processes. Another issue is that vertex degrees vary wildly between different vertex types (see Table 1) but α is independent of vertex types. This causes unreasonable penalisation for some large-degree vertex types such as GO terms.

To allow more flexibility in degree penalising we replace the single constant α and vertex degree function deg with vertex-type and edge-type specific functions $\alpha : T_v \mapsto [0,1]$ and $\deg : V \times T_e \mapsto \mathcal{N}$ (that is, *deg*$(v, \tau,)$ denotes the number of edges of type τ adjacent to v). Now the vertex *rarity* $d_v : V \times T_e \mapsto [0,1]$ for vertex $v \in V$ is

$$d_v(v, \tau) = \frac{1}{(\deg(v, \tau) + 1)^{\alpha(t_v(v))}}. \tag{4}$$

As with relevance (2), the rarity values are decomposed into edge-specific coefficients. The edge *rarity* $d : E \mapsto [0,1]$ becomes

$$d(e) = \sqrt{d_v(u, t_e(e))} \cdot \sqrt{d_v(v, t_e(e))} = \left[d_v(u, t(e)) \cdot d_v(v, t(e))\right]^{-1/2} \tag{5}$$

where $e = \{u, v\} \in E$.

Now that we have defined all the components of edge goodness, the goodness $g : E \mapsto [0,1]$ itself is simply a product of those factors:

$$g(e) = r(e) \cdot q(e) \cdot d(e) \tag{6}$$

where $r(e)$, $q(e)$ and $d(e)$ are the reliability (1), relevance (2) and rarity (5) of an edge $e \in E$. Under the assumptions that $r(e)$, $q(e)$ and $d(e)$ are probabilities for mutually independent necessary conditions for the edge and that edges are independent of each other, the goodness $g(e)$ is the probability that e exists. We remark that these assumptions of independence are strong and in some cases they are arguably unrealistic. However, independence allows us to calculate path and subgraph probabilities easily; we return to these in Section 4.

4 Link Goodness Measures

A link is a more general concept of connection than a simple relation (edge) between two vertices s and t. Links are useful since they can be used to model indirect, weak or otherwise non-trivial connections. A *path* (a sequence of consecutive edges) is probably the simplest link type, but shared neighbourhoods, connected subgraphs and random walks can also be used to represent links. To discover or predict links, assess their strengths or analyse statistical significances of links we need a measure for link goodness in addition to edge goodness.

We next give a short review of some link goodness measures proposed in the literature. They are presented in the order of increasing generality; more general measures utilise more information to determine the strength of a link. The discussion is not restricted to Biomine graphs, so $G = (V, E)$ refers to an arbitrary directed or undirected graph below. See Liben-Nowell and Kleinberg [13] for an experimental evaluation of many of these measures for link prediction.

4.1 Path and Neighbourhood Level

The shortest s–t-path P is a simple but efficient link type. Its length $w(P)$ is a natural measure for link strength:

$$g_s(s,t) = \min_{P \in \mathcal{P}} w(P) = \min_{P \in \mathcal{P}} \sum_{e \in P} w(e) \tag{7}$$

where $w(e)$ is the length (weight) of an edge $e \in E$ and $\mathcal{P}$ is the set of all s–t-paths in G. This measure is easy and efficient to calculate by any shortest path algorithm.

For random graphs where edge "lengths" are probabilities, (7) does not make much sense. However, if edges are independent of each other, like in Biomine graphs, path "length" or goodness follows in a natural way. Let $P = (e_1, \ldots, e_k)$, $e_i \in E$, be a path in G. The *path goodness* $g_p : \mathcal{P} \mapsto [0, 1]$ is

$$g_p(P) = \prod_{e \in P} g(e). \tag{8}$$

With the interpretation that $g(e)$ is the probability that edge e exists (Section 3) the path goodness $g_p(P)$ is the probability that the whole path P exists in a realisation H of G. A *realisation* of G is a non-random subgraph $H \subset G$ where each edge of G has been randomly and independently decided according to the corresponding edge probability.

With path goodness g_p the shortest path corresponds to the most probable, or *best* path. By combining (7) and (8) we get

$$g_b(s,t) = \max_{P \in \mathcal{P}} g_p(P) = \max_{P \in \mathcal{P}} \prod_{e \in P} g(e). \tag{9}$$

Again, any shortest path algorithm can be applied to find most probable paths by using edge weights $w(e) = -\log\big(g(e)\big)$. Let P be the shortest path found with weight $w(P)$. Then

$$w(P) = \sum_{e \in P} -\log\big(g(e)\big) = -\log\Big(\prod_{e \in P} g(e)\Big) = -\log\big(g_p(P)\big) \tag{10}$$

and since the logarithm function is strictly increasing and $w(P)$ is minimised, $g_p(P)$ is maximised.

Overlapping vertex neighbourhoods may indicate indirect similarity or proximity. The number of overlapping neighbours is the simplest measure in this context:

$$g_n(s,t) = |N(s) \cap N(t)|. \tag{11}$$

This measure has been observed to positively correlate with future collaboration probability in coauthor networks [14]. The normalised form of (11)

$$g_J(s,t) = \frac{|N(s) \cap N(t)|}{|N(s) \cup N(t)|} \tag{12}$$

is the well known *Jaccard index.* Adamic and Adar have proposed [15] a modification of (12) that rewards vertex pairs that share neighbours with low degrees:

$$g_A(s,t) = \sum_{u \in N(s) \cap N(t)} \frac{1}{\log |N(u)|}. \tag{13}$$

4.2 Subgraph Level

The goodness of a single s–t-path, as in (7) and (9), is not necessarily a good measure of the strength of the link between vertices s and t. For example, a link consisting of several parallel paths could be considered to be stronger than a single path even if all of the parallel paths are weak. *Connection subgraphs* take this into account by evaluating connected subgraphs, which can be thought to be a set of paths, containing s and t. Specifically, a connection subgraph between s and t is a connected subgraph $H \subset G$, of a given size, such that $\{s,t\} \subset V(H)$. Subgraph H can be, for example, a set of k shortest paths for some fixed k or it can be chosen to maximise a given connection subgraph goodness function [16].

Faloutsos et al. view G as an electrical network of resistors [16]. They propose an algorithm that extracts a fixed size subgraph H which maximises total delivered current over the subnetwork from s to t when s is assigned a potential of +1 volt and t is grounded (0 volts). Total delivered current has a random walk interpretation [17]. At first, let us define transition probabilities

$$p(u,v) = \frac{g(u,v)}{\sum_{w \in N(u)} g(u,w)} \tag{14}$$

for each $(u,v) \in E$. Next, let p_{esc} denote the escape probability according to (14) from s to t; i.e. the probability that a random walker starting from s will reach t before returning to s. The *effective conductance* between s and t is now

$$g_{\mathrm{EC}}(s,t) = \sum_{u \in N(s)} g(s,u) \cdot p_{\mathrm{esc}} \tag{15}$$

which is the expected number of "successful escapes" when the number of escape attempts is $\sum_{u \in N(s)} g(s,u)$ [18].

Effective conductance is an appealing link goodness measure and it has been used to measure centrality in networks [19]. However, it does not penalise uninformative vertices that have large degrees (cf. (4)). Faloutsos et al. dodge this by introducing a global grounded "sink" vertex that is connected to all vertices $v \in V$ with conductance proportional to $\sum_{u \in N(v)} g(v,u)$. As pointed out by Koren et al. [17], this introduces a counterintuitive size bias where the link goodness can decrease if the connection subgraph is enlarged. They propose a modified version of (15) titled cycle-free effective conductance (CFEC):

$$g_{\text{CFEC}}(s,t) = \sum_{u \in N(s)} g(s,u) \cdot p_{\text{cf-esc}}(s,t) = \sum_{u \in N(s)} g(s,u) \cdot \sum_{P \in \mathcal{P}} \Pr(P), \qquad (16)$$

where $p_{\text{cf-esc}}$ is the escape probability restricted to cycle-free random walks (walks that are simple s–t-paths) and $\mathcal{P}$ is the set of all simple s–t-paths in G. CFEC has two desirable properties: it is monotonically increasing as a function of graph size, and a relatively small connection subgraph consisting of the most probable simple s–t-paths is usually enough to approximate $g_{\text{CFEC}}(s,t)$ [17].

4.3 Graph Level

A link goodness measure can utilise the topology of the whole graph G. Most measures on this scale are based on random walks like (15) and (16), although a measure proposed by Katz [20] considers sets of s–t-paths such that

$$g_K(s,t) = \sum_{l=1}^{\infty} \beta^l |\mathcal{P}_l| \qquad (17)$$

where $\mathcal{P}_l$ is the set of all s–t-paths of length l. Parameter $\beta > 0$ controls the effect of longer paths to the goodness.

Random walk models typically consider a single walker w starting from s or two walkers w_1 and w_2 with one starting from s and the other from t. Walkers traverse G randomly with transition probabilities (14). *Hitting time* $H(s,t)$ considers the expected number of steps w has to take to reach t [21]. Its symmetric variant is *commute time* $C(s,t) = H(s,t) + H(t,s)$. Both can be readily used as distance measures, and they have been used as link goodness (proximity) measures as well [13].

SimRank by Jeh and Widow [22] is based on a recursive definition

$$g_{\text{SR}}(s,t) = \begin{cases} 0 & \text{if } N(s) = \emptyset \text{ or } N(t) = \emptyset, \\ 1 & \text{if } s = t, \\ C/(|N(s)||N(t)|) \cdot \sum_{\substack{u \in N(s) \\ v \in N(t)}} s(u,v) & \text{otherwise} \end{cases} \qquad (18)$$

where $C \in [0,1]$ is a constant. SimRank also has a random walk interpretation: value $g_{\text{SR}}(s,t)$ corresponds to the expected value of $C^{\mathbf{t}}$ where $\mathbf{t}$ is the time (number of steps) when walkers w_1 and w_2 first meet [22].

Liben-Nowell and Kleinberg [13] proposed a *rooted PageRank* measure for link goodness based on the well known *PageRank* measure [23]. In rooted PageRank the random walker w returns to s with probability α in every step, or it continues the walk with probability $1 - \alpha$. The measure is the steady state (stationary) probability of t.

With random graphs $g_b(s,t)$ is the probability that the best path exists in a realisation of G. A more appropriate measure could be the probability that at least one path exists between s and t. This measure is closely related to the theory of *network reliability* [24], and the desired measure

$$g_R(s,t) = \Pr(H : H \subset G, H \text{ contains an } s\text{–}t\text{-path}), \tag{19}$$

where H is a random instantiation of the uncertain graph G, is the *two-terminal network reliability* of G with terminals s and t. (The connected parties are called terminals in the reliability literature.)

4.4 Estimation of Link Significance

We eventually want to measure how strongly two given vertices s and t are related in graph G. Link goodness measures, such as those discussed in Section 4, allow ranking of links but their values may be difficult to put into perspective. For example, assume we have $f(s,t) = 0.4$ for some goodness measure f. Is this particular value of f high or low? This obviously depends on the data and the specific instances of s and t.

We can estimate the statistical significance of the link by using the goodness value $f(s,t)$ as a test statistic. Returning to the previous example this tells us how likely it is to obtain a link with goodness 0.4 or better by chance. There are multiple meaningful null hypotheses:

N1. Vertices s and t of types $\tau_s \in T_v$ and $\tau_t \in T_v$ are not more strongly connected than randomly chosen vertices s' and t' of types τ_s and τ_t.
N2. Vertex s of type $\tau \in T_v$ is not more strongly connected to vertex t than a randomly chosen vertex s' of type τ.
N3. Vertices s and t are not more strongly connected in the given graph G than in random graph H with edge weights $w' : E(H) \mapsto \mathbb{R}$ generated by model $\mathcal{H}$ similar to the (unknown) model which generated G and w.

The last null hypothesis N3 is clearly the most complicated one: it is not easy to come up with model $\mathcal{H}$ that generates random graphs that are sufficiently similar to the observed graph. The choice from the first two null hypotheses depends on what we are testing. In a symmetrical case, for example when testing the significance of connection between two candidate genes, N1 is appropriate. If the roles of the vertices are asymmetric, as in testing for the connection from a set of candidate genes to a single phenotype, N2 should be used.

Under null hypothesis N1 we can estimate p-value for the test statistic $f(s,t)$ by randomly sampling N pairs of vertices (s',t') from V. Let us denote the sample by $S = \{(s_1,t_1),\ldots,(s_N,t_N)\}$. To obtain an empirical null distribution we compute the value of test statistic $f(s_i, t_i)$ for each $(s_i, t_i) \in S$, and let $S_+ = \{(s_i, t_i) \in S : f(s_i, t_i) \geq f(s, t)\}$. Then the estimated p-value $\tilde{p}$ is simply

$$\tilde{p} = \frac{|S_+|}{N}. \tag{20}$$

The same procedure can be used under null hypothesis N2 by sampling single vertices $S = \{t_1, \ldots, t_n\}$ and letting $S_+ = \{t_i \in S : f(s, t_i) \geq f(s, t)\}$.

Because vertices of the same type may have wildly varying degrees one should sample vertices s' and t' that have degrees similar to s and t, respectively. If several hypotheses are tested (several candidate genes, for example), the resulting p-values should be adjusted accordingly to account for multiple testing.

5 Related Work

Concurrently with the development of Biomine, several other data integration systems have been proposed in the literature. Of these, most similar to our approach are ONDEX [25] and Biozon [26], which both collect the data from various sources under a single data store. They also use a graph data schema. In both systems, the data model is a graph with typed nodes and edges, allowing for the incorporation of arbitrary data sources. In addition to curated data derived from the source databases, both ONDEX and Biozon include in-house data such as similarity links computed from sequence similarity of proteins and predicted links derived by text mining. Biozon provides several types of queries, most interestingly searching by graph topology and ranking of nodes by importance defined by the graph structure. In ONDEX, the integrated data is accessed by a pipeline, in which individual filtering and graph layout operations may be combined to process the graph in application-specific ways. BioWarehouse [27] aims to provide generic tools for enabling users to build their own combinations of biological data sources. Their data management approach is rather similar to ONDEX and Biozon, but the data is stored in a relational database with a dedicated table for each data type instead of a generic graph structure. This approach allows database access through standard SQL queries, and is not directly suitable for graph-oriented queries.

6 Conclusion

We presented Biomine, a system that integrates data from a number of heterogenous sources into a single, graph-structured index. The current implementation of Biomine contains over 1 million entities and over 8 million relations between them, with focus on human genetics. The index can be queried using a public web interface[2], and results are visualized graphically. Biomine in its current form is a functional proof of concept, covering only part of the available data and with a limited focus on human genetics. Initial experimental results indicate that Biomine and other similar approaches have strong potential for predicting links and annotations.

[2] `biomine.cs.helsinki.fi`

Acknowledgements. We thank the Biomine team for co-operation. This work has been supported by the Algorithmic Data Analysis (Algodan) Centre of Excellence of the Academy of Finland (Grant 118653). We also thank the European Commission under the 7th Framework Programme FP7-ICT-2007-C FET-Open, contract BISON-211898.

References

1. Kötter, T., Berthold, M.R.: From Information Networks to Bisociative Information Networks. In: Berthold, M.R. (ed.) Bisociative Knowledge Discovery. LNCS (LNAI), vol. 7250, pp. 33–50. Springer, Heidelberg (2012)
2. Dubitzky, W., Kötter, T., Schmidt, O., Berthold, M.R.: Towards Creative Information Exploration Based on Koestler's Concept of Bisociation. In: Berthold, M.R. (ed.) Bisociative Knowledge Discovery. LNCS (LNAI), vol. 7250, pp. 11–32. Springer, Heidelberg (2012)
3. Sevon, P., Eronen, L., Hintsanen, P., Kulovesi, K., Toivonen, H.: Link discovery in graphs derived from biological databases. In: Proceedings of Data Integration in the Life Sciences, Third International Workshop, pp. 35–49 (2006)
4. Getoor, L., Diehl, C.P.: Link mining: A survey. SIGKDD Explorations 7, 3–12 (2005)
5. Maglott, D., Ostell, J., Pruitt, K.D., Tatusova, T.: Entrez Gene: gene-centered information at NCBI. Nucleic Acids Research 35, D26–D31 (2007)
6. Sayers, E.W., Barrett, T., Benson, D.A., Bolton, E., Bryant, S.H., Canese, K., Chetvernin, V., Church, D.M., DiCuccio, M., Federhen, S., Feolo, M., Geer, L.Y., Helmberg, W., Kapustin, Y., Landsman, D., Lipman, D.J., Lu, Z., Madden, T.L., Madej, T., Maglott, D.R., Marchler-Bauer, A., Miller, V., Mizrachi, I., Ostell, J., Panchenko, A., Pruitt, K.D., Schuler, G.D., Sequeira, E., Sherry, S.T., Shumway, M., Sirotkin, K., Slotta, D., Souvorov, A., Starchenko, G., Tatusova, T.A., Wagner, L., Wang, Y., Wilbur, W.J., Yaschenko, E., Ye, J.: Database resources of the National Center for Biotechnology Information. Nucleic Acids Research 38, 5–16 (2010)
7. The Uniprot Consortium: The Universal Protein Resource (UniProt) in 2010. Nucleic Acids Research 38, D142–D148 (2010)
8. Hunter, S., Apweiler, R., Attwood, T.K., Bairoch, A., Bateman, A., Binns, D., Bork, P., Das, U., Daugherty, L., Duquenne, L., Finn, R.D., Gough, J., Haft, D., Hulo, N., Kahn, D., Kelly, E., Laugraud, A., Letunic, I., Lonsdale, D., Lopez, R., Madera, M., Maslen, J., McAnulla, C., McDowall, J., Mistry, J., Mitchell, A., Mulder, N., Natale, D., Orengo, C., Quinn, A.F., Selengut, J.D., Sigrist, C.J.A., Thimma, M., Thomas, P.D., Valentin, F., Wilson, D., Wu, C.H., Yeats, C.: InterPro: the integrative protein signature database. Nucleic Acids Research 37, D211–D215 (2009)
9. Jensen, L.J., Kuhn, M., Stark, M., Chaffron, S., Creevey, C., Muller, J., Doerks, T., Julien, P., Roth, A., Simonovic, M., Bork, P., von Mering, C.: STRING 8—a global view on proteins and their functional interactions in 630 organisms. Nucleic Acids Research 37, D412–D416 (2009)

10. The Gene Ontology Consortium: Gene ontology: tool for the unification of biology. Nature Genetics 25(1), 25–29 (2000)
11. Kanehisa, M., Goto, S., Furumichi, M., Tanabe, M., Hirakawa, M.: KEGG for representation and analysis of molecular networks involving diseases and drugs. Nucleic Acids Research 38, D355–D360 (January 2010)
12. Gerhard, D.S., et al.: The status, quality, and expansion of the NIH full-length cDNA project: The Mammalian Gene Collection (MGC). Genome Research 14, 2121–2127 (2004), full list of authors http://dx.doi.org/10.1101/gr.2596504
13. Liben-Nowell, D., Kleinberg, J.: The link prediction problem for social networks. Journal of the American Society for Information Science and Technology 58(7), 1019–1031 (2007)
14. Newman, M.E.J.: Clustering and preferential attachment in growing networks. Physical Review E 64(2), 025102 (2001)
15. Adamic, L.A., Adar, E.: Friends and neighbors on the Web. Social Networks 25(3), 211–230 (2003)
16. Faloutsos, C., McCurley, K.S., Tomkins, A.: Fast discovery of connection subgraphs. In: Proceedings of the Tenth ACM SIGKDD International Conference on Knowledge Discovery and Data Mining, pp. 118–127 (2004)
17. Koren, Y., North, S.C., Volinsky, C.: Measuring and extracting proximity graphs in networks. In: Proceedings of the 12th ACM SIGKDD International Conference on Knowledge Discovery and Data Mining, pp. 245–255 (2006)
18. Doyle, P.G., Snell, J.L.: Random walks and electric networks (January 2000), http://arxiv.org/abs/math.PR/0001057
19. Brandes, U., Fleischer, D.: Centrality Measures Based on Current Flow. In: Diekert, V., Durand, B. (eds.) STACS 2005. LNCS, vol. 3404, pp. 533–544. Springer, Heidelberg (2005)
20. Katz, L.: A new status index derived from sociometric analysis. Psychometrika 18(1), 39–43 (1953)
21. Chen, H., Zhang, F.: The expected hitting times for finite Markov chains. Linear Algebra and its Applications 428(11-12), 2730–2749 (2008)
22. Jeh, G., Widom, J.: SimRank: a measure of structural-context similarity. In: Proceedings of the Eighth ACM SIGKDD International Conference on Knowledge Discovery and Data Mining, Edmonton, Canada, pp. 538–543. ACM (July 2002)
23. Brin, S., Page, L.: The anatomy of a large-scale hypertextual web search engine. Computer Networks and ISDN Systems 30(1-7), 107–117 (1998)
24. Colbourn, C.J.: The Combinatorics of Network Reliability. Oxford University Press (1987)
25. Köhler, J., Baumbach, J., Taubert, J., Specht, M., Skusa, A., Rüegg, A., Rawlings, C., Verrier, P., Philippi, S.: Graph-based analysis and visualization of experimental results with ONDEX. Bioinformatics 22(11), 1383–1390 (2006)
26. Birkland, A., Yona, G.: BIOZON: a system for unification, management and analysis of heterogeneous biological data. BMC Bioinformatics 7(1), 70 (2006)
27. Lee, T., Pouliot, Y., Wagner, V., Gupta, P., Stringer-Calvert, D., Tenenbaum, J., Karp, P.: BioWarehouse: a bioinformatics database warehouse toolkit. BMC Bioinformatics 7(1), 170 (2006)

Semantic Subgroup Discovery and Cross-Context Linking for Microarray Data Analysis

Igor Mozetič[1], Nada Lavrač[1,2], Vid Podpečan[1], Petra Kralj Novak[1], Helena Motaln[3], Marko Petek[3], Kristina Gruden[3], Hannu Toivonen[4], and Kimmo Kulovesi[4]

[1] Jožef Stefan Institute, Jamova 39, Ljubljana, Slovenia
{igor.mozetic,nada.lavrac,vid.podpecan,petra.kralj.novak}@ijs.si
[2] University of Nova Gorica, Vipavska 13, Nova Gorica, Slovenia
[3] National Institute of Biology, Večna pot 111, Ljubljana, Slovenia
{helena.motaln,marko.petek,kristina.gruden}@nib.si
[4] Department of Computer Science, University of Helsinki, Finland
{hannu.toivonen,kimmo.kulovesi}@cs.helsinki.fi

Abstract. The article presents an approach to computational knowledge discovery through the mechanism of *bisociation*. Bisociative reasoning is at the heart of creative, accidental discovery (e.g., serendipity), and is focused on finding unexpected links by crossing contexts. Contextualization and linking between highly diverse and distributed data and knowledge sources is therefore crucial for the implementation of bisociative reasoning. In the article we explore these ideas on the problem of analysis of microarray data. We show how enriched gene sets are found by using ontology information as background knowledge in semantic subgroup discovery. These genes are then contextualized by the computation of probabilistic links to diverse bioinformatics resources. Preliminary experiments with microarray data illustrate the approach.

1 Introduction

Systems biology studies and models complex interactions in biological systems with the goal of understanding the underlying mechanisms. Biologists collect large quantities of data from wet lab experiments and high-throughput platforms. Public biological databases, like Gene Ontology and Kyoto Encyclopedia of Genes and Genomes, are sources of biological knowledge. Since the growing amounts of available knowledge and data exceed human analytical capabilities, technologies that help analyzing and extracting useful information from such large amounts of data need to be developed and used.

The concept of association is at the heart of many of today's ICT technologies such as information retrieval and data mining (for example, association rule learning is an established data mining technology, [1]). However, scientific discovery requires creative thinking to connect seemingly unrelated information, for example, by using metaphors or analogies between concepts from different domains. These modes of thinking allow the mixing of conceptual categories and

M.R. Berthold (Ed.): Bisociative Knowledge Discovery, LNAI 7250, pp. 379–389, 2012.

contexts, which are normally separated. One of the functional basis for these modes is the idea of *bisociation*, coined by Artur Koestler half a century ago [8]:

> "The pattern ... is the perceiving of a situation or idea, L, in two self-consistent but habitually incompatible frames of reference, M_1 and M_2. The event L, in which the two intersect, is made to vibrate simultaneously on two different wavelengths, as it were. While this unusual situation lasts, L is not merely linked to one associative context but *bisociated* with two."

Koestler found bisociation to be the basis for human creativity in seemingly diverse human endeavors, such as humor, science, and arts. The concept of bisociation in science is discussed in depth in [2]. Here we take a more restricted and focused view (illustrated in Figure 1).

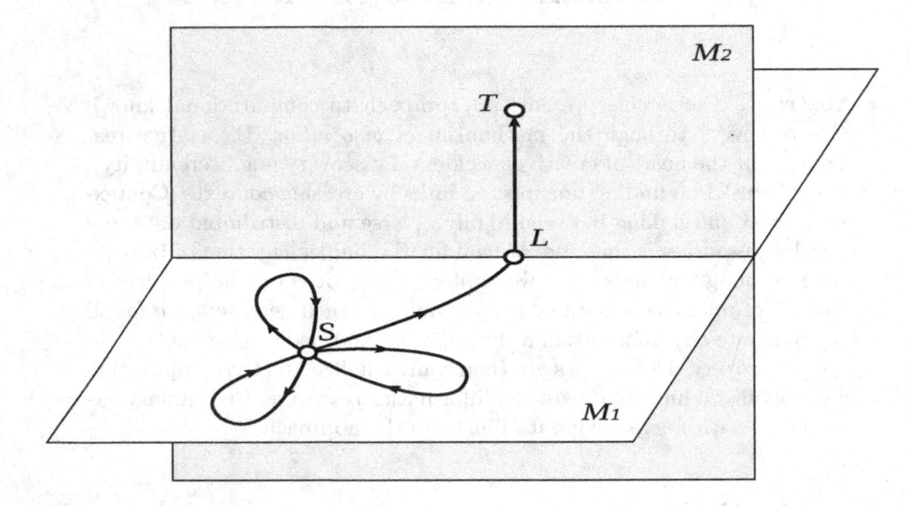

Fig. 1. Koestler's schema of bisociative discovery in science ([8], p. 107)

We are interested in creative discoveries in science, and in particular in computational support for knowledge discovery from large and diverse sources of data and knowledge. The computational realization of bisociative reasoning is based on the following, somewhat simplified, assumptions:

- A bisociative information network (named BisoNet) can be created from available resources. BisoNet is a large graph, where nodes are concepts and edges are probabilistic relations. Unlike semantic nets or ontologies, the graph is easy to construct automatically since it carries little semantics. To a large extent it encodes just circumstantial evidence that concepts are somehow related through edges with some probability.
- Different subgraphs can be assigned to different contexts (frames of reference).

- Graph analysis algorithms can be used to compute links between distant nodes and subgraphs in a BisoNet.
- A bisociative link is a link between nodes (or subgraphs) from different contexts.

In this article we thus explore one specific pattern of bisociation: long-range links between nodes (or subgraph) which belong to different contexts. A long-range link is a special case of a bridging graph [2] since it has the form of a path between two nodes. More precisely, we say that two concepts are bisociated if:

- there is no direct, obvious evidence linking them,
- one has to cross contexts to find the link, and
- this new link provides some novel insight into the problem domain.

We have to emphasize that context crossing is subjective, since the user has to move from his 'normal' context (frame of reference) to an *habitually incompatible context* to find the bisociative link [2]. In Koestler's terms (Figure 1), a habitual frame of reference (plane M_1) corresponds to a BisoNet subgraph as defined by a user or his profile. The rest of the BisoNet represents different, habitually incompatible contexts (in general, there may be several planes M_2). The creative act here is to find links (from S to target T) which lead 'out-of-the-plane' via intermediate, bridging concepts (L). Thus, contextualization and link discovery are two of the fundamental mechanisms in bisociative reasoning.

Finding links between seemingly unrelated concepts from texts was already addressed by Swanson [12]. The Swanson's approach implements *closed discovery*, the so-called A-B-C process, where A and C are given and one searches for intermediate B concepts. On the other hand, in *open discovery* [18], only A is given. One approach to open discovery, RaJoLink [9], is based on the idea to find C via B terms which are rare (and therefore potentially interesting) in conjunction with A. Rarity might therefore be one of the criteria to select links which lead out of the habitual context (around A) to known, but non-obviously related concepts C via B.

In this article we present an approach to bisociative discovery and contextualization of genes which helps in the analysis of microarray data. The approach is based on semantic subgroup discovery (by using ontologies as background knowledge in microarray data analysis), and the linking of various publicly available bioinformatics databases. This is an ongoing work, where some elements of bisociative reasoning are already implemented: creation of the BisoNet graph, identification of relevant nodes in a BisoNet, and computation of links to indirectly related concepts. Currently, we are expanding the BisoNet with textual resources from PubMed, and implementing open discovery from texts through BisoNet graph mining. We envision that the open discovery process will identify potentially interesting concepts from different contexts which will act as the target nodes for the link discovery algorithms. Links discovered in this way, crossing contexts, might provide instances of bisociative discoveries.

The currently implemented steps of bisociative reasoning are the following. The *semantic subgroup discovery* step is implemented by the SEGS system [16].

SEGS uses as background knowledge data from three publicly available, semantically annotated biological data repositories, GO, KEGG and Entrez. Based on the background knowledge, it automatically formulates biological hypotheses: rules which define groups of differentially expressed genes. Finally, it estimates the relevance (or significance) of the automatically formulated hypotheses on experimental microarray data. The *BisoNet creation* and the *link discovery* steps are implemented by the Biomine system [3,11]. Biomine weakly integrates a large number of biomedical resources, and computes most probable links between elements of diverse sources. It thus complements the semantic subgroup discovery technology, due to the explanatory potential of additional link discovery and Biomine graph visualization. While this link discovery process is already implemented, our current work is devoted to the contextualization of Biomine nodes for bisociative link discovery.

The article is structured as follows. Section 2 gives an overview of five steps in exploratory analysis of gene expression data. Section 3 describes an approach to the analysis of microarray data, using semantic subgroup discovery in the context of gene set enrichment. A novel approach, a first attempt at bisociative discovery through contextualization, composed of using SEGS and Biomine (SegMine, for short) is in Section 4. An ongoing experimental case study is presented in Section 5. We conclude in Section 6 with plans for future work.

2 Exploratory Gene Analytics

This section describes the steps which support bisociative discovery, targeted at the analysis of differentially expressed gene sets: gene ranking, the SEGS method for enriched gene set construction, linking of the discovered gene set to related biomedical databases, and finally visualization in Biomine. The schematic overview is in Figure 2.

The proposed method consists of the following five steps:

1. **Ranking of genes.** In the first step, class-labeled microarray data is processed and analyzed, resulting in a list of genes, ranked according to differential expression.
2. **Ontology information fusion.** A unified database, consisting of GO[1] (biological processes, functions and components), KEGG[2] (biological pathways) and Entrez[3] (gene-gene interactions) terms and relationships is constructed by a set of scripts, enabling easy updating of the integrated database (details are discussed by [14]).
3. **Discovering groups of differentially expressed genes.** The ranked list of genes is used as input to the SEGS algorithm [16], an upgrade of the RSD relational subgroup discovery algorithm [4,5,15], specially adapted to microarray data analysis. The result is a list of most relevant gene groups that

[1] http://www.geneontology.org/
[2] http://www.genome.jp/kegg/
[3] ftp://ftp.ncbi.nlm.nih.gov/gene/GeneRIF/interaction_sources

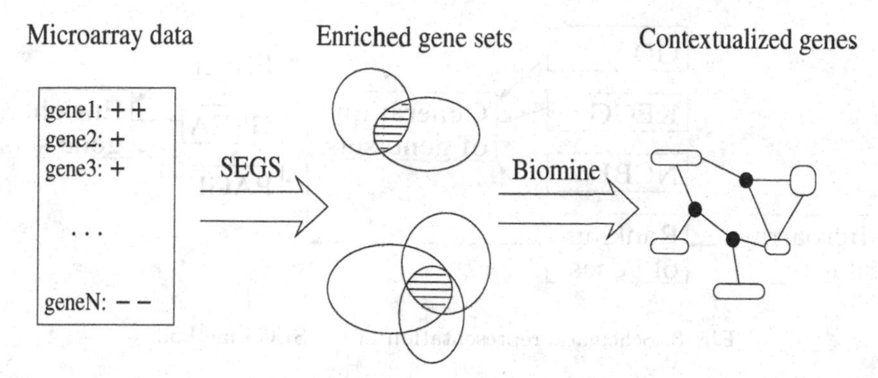

Fig. 2. Microarray gene analytics proceeds by first finding candidate enriched gene sets, expressed as intersections of GO, KEGG and Entrez gene-gene interaction sets. Selected enriched genes are then put in the context of different bioinformatic resources, as computed by the Biomine link discovery engine. The '+' and '-' signs under Microarray data indicate over- and under-expression values of genes, respectively.

semantically explain differential gene expression in terms of gene functions, components, processes, and pathways as annotated in biological ontologies.

4. **Finding links between gene group elements.** The elements of the discovered gene groups (GO and KEGG terms or individual genes) are used to formulate queries for the Biomine link discovery engine. Biomine then computes most probable links between these elements and entities from a number of public biological databases. These links help the experts to uncover unexpected relations and biological mechanisms potentially characteristic for the underlying biological system.
5. **Gene group visualization.** Finally, in order to help in explaining the discovered out-of-the-context links, the discovered gene relations are visualized using the Biomine visualization tools.

3 SEGS: Search for Enriched Gene Sets

The goal of the gene set enrichment analysis is to find gene sets which form coherent groups and are different from the remaining genes. More precisely, a gene set is *enriched* if the member genes are semantically coherent and statistically significantly differentially expressed as compared to the rest of the genes. Two methods for testing the enrichment of gene sets were developed: Gene set enrichment analysis (GSEA) [13] and Parametric analysis of gene set enrichment (PAGE) [7]. Originally, these methods take individual terms from GO and KEGG (which annotate gene sets), and test whether the genes that are annotated by a specific term are statistically significantly differentially expressed in the given microarray dataset.

The novelty of the SEGS method, developed by Trajkovski et al. [14,16] and used in this study, is that the method does not only test existing gene sets for

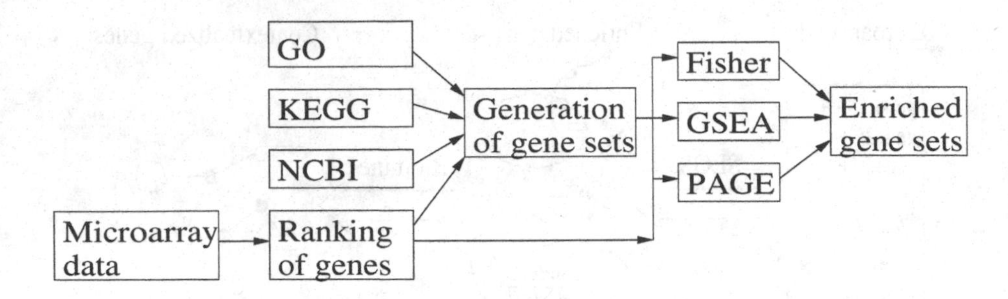

Fig. 3. Schematic representation of the SEGS method

differential expression but it also generates new gene sets that represent novel biological hypotheses. In short, in addition to testing the enrichment of individual GO and KEGG terms, this method tests the enrichment of newly defined gene sets constructed by the intersection of GO terms, KEGG terms and gene sets defined by taking into account also the gene-gene interaction data from Entrez.

The SEGS method has four main components:

- the background knowledge (the GO, KEGG and Entrez databases),
- the SEGS hypothesis language (the GO, KEGG and interaction terms, and their conjunctions),
- the SEGS hypothesis generation procedure (generated hypotheses in the SEGS language correspond to gene sets), and
- the hypothesis evaluation procedure (the Fisher, GSEA and PAGE tests).

The schematic workflow of the SEGS method is shown in Figure 3.

4 SegMine: Contextualization of genes

We made an attempt at exploiting bisociative discoveries within the biomedical domain by explicit contextualization of enriched gene sets. We applied two methods that use publicly available background knowledge for supporting the work of biologists: the SEGS method for searching for enriched gene sets [16] and the Biomine method for contextualization by finding links between genes and other biomedical databases [3,11]. We combined the two methods in a novel way. We used SEGS for hypothesis generation in the form of interesting gene sets, which are constructed as intersections of terms from different ontologies (different contexts). Queries are then formulated to Biomine for out-of-the-context link discovery and visualization (see Figure 4). We believe that this combination provides an easier interpretation of the biological mechanisms underlying differential gene expression for biologists.

In the Biomine[4] project [3,11], data from several publicly available databases were merged into a large graph, a BisoNet, and a method for link discovery

[4] `http://biomine.cs.helsinki.fi/`

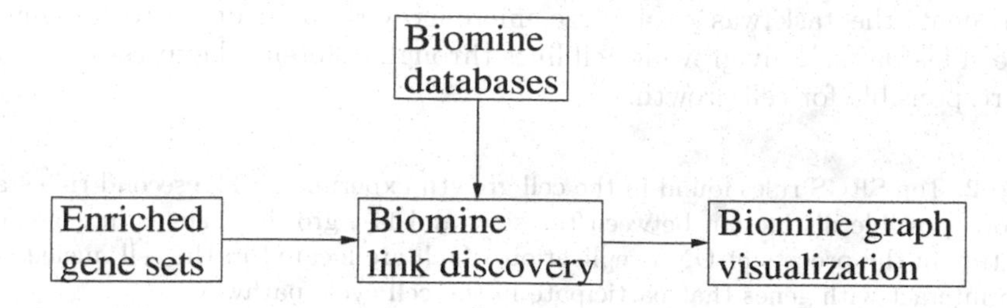

Fig. 4. SegMine workflow

between entities in queries was developed. In the Biomine framework nodes correspond to entities and concepts (e.g., genes, proteins, GO terms), and edges represent known, probabilistic relationships between nodes. A link (a relation between two entities) is manifested as a path or a subgraph connecting the corresponding nodes.

Table 1. Databases included in the Biomine snapshot used in the experiments

Vertex Type	Source Database	Nodes	Degree
Article	PubMed	330,970	6.92
Biological process	GO	10,744	6.76
Cellular component	GO	1,807	16.21
Molecular function	GO	7,922	7.28
Conserved domain	ENTREZ Domains	15,727	99.82
Structural property	ENTREZ Structure	26,425	3.33
Gene Entrez	Gene	395,611	6.09
Gene cluster	UniGene	362,155	2.36
Homology group	HomoloGene	35,478	14.68
OMIM entry	OMIM	15,253	34.35
Protein Entrez	Protein	741,856	5.36
Total		1,968,951	

The Biomine graph data model consists of various biological entities and annotated relations between them. Large, annotated biological data sets can be readily acquired from several public databases and imported into the graph model in a relatively straightforward manner. Some of the databases used in Biomine are summarized in Table 1. The snapshot of Biomine we use consists of a total of 1,968,951 nodes and 7,008,607 edges. This particular collection of data sets is not meant to be complete, but it certainly is sufficiently large and versatile for real link discovery.

5 A Case Study

In the systems biology domain, our goal is to computationally help the experts to find a creative interpretation of wet lab experiment results. In the particular experiment, the task was to analyze microarray data in order to distinguish between fast and slowly growing cell lines through differential expression of gene sets, responsible for cell growth.

Table 2. Top SEGS rules found in the cell growth experiment. The second rule states that one possible distinction between the slow and fast growing cells is in genes participating in the process of DNA replication which are located in the cell nucleus and which interact with genes that participate in the cell cycle pathway.

Enriched Gene Sets
1. SLOW-vs-FAST ← GO_Proc('DNA metabolic process') & INTERACT(GO_Comp('cyclin-dep. protein kinase holoenzyme complex'))
2. SLOW-vs-FAST ← GO_Proc('DNA replication') & GO_Comp('nucleus') & INTERACT(KEGG_Path('Cell cycle'))
3. SLOW-vs-FAST ← . . .

Table 2 gives the top rules resulting from the SEGS search for enriched gene sets. For each rule, there is a corresponding set of over expressed genes from the experimental data. Figure 5 shows a part of the Biomine graph which links a selected subset of enriched gene set to the rest of the nodes in the Biomine graph.

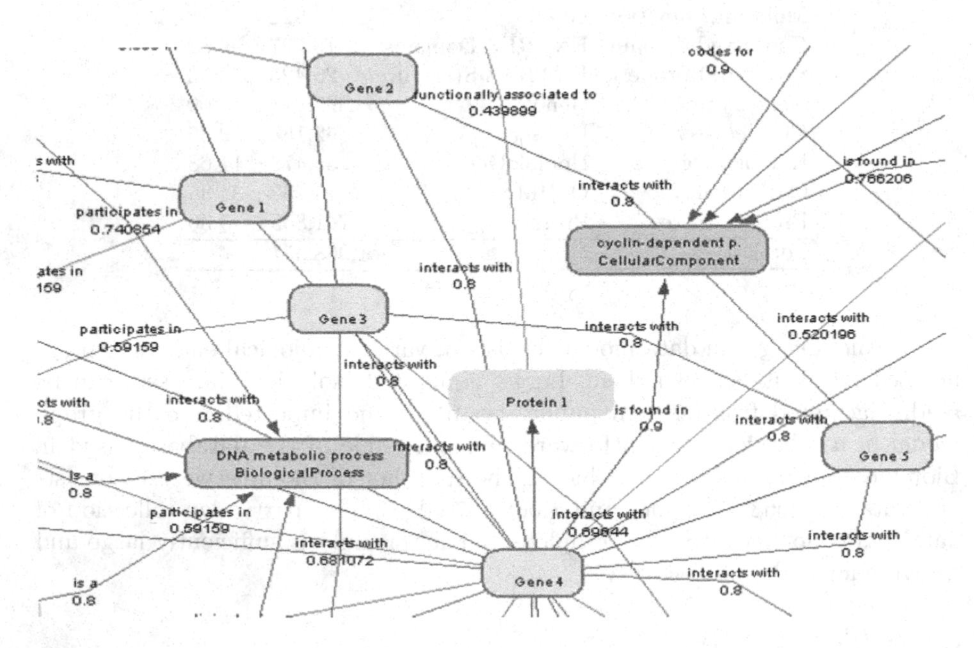

Fig. 5. Biomine subgraph related to five genes from the enriched gene set produced by SEGS. Note that the gene and protein names are not explicitly presented, due to the preliminary nature of these results.

The wet lab scientists have assessed that SegMine, SEGS in combination with Biomine, provides additional hints on what to focus on when comparing the expression data of cells. In subsequent analysis of senescence in human stem cells, the use of SegMine resulted in formulation of three novel research hypotheses which could improve understanding of the underlying mechanisms and identification of candidate marker genes [10].

In principle, such an in-silico analysis can considerably lower the costs of in-vitro experiments with which the researchers in the wet lab are trying to get a hint of a novel process or phenomena observed. This may be especially true for situations when one cannot find deeper explanation for drug effects, organ functions, or diseases from surface observations only. Namely, the gross, yet important characteristics of the cells (organ function) are not directly accessible (since they do not affect visual morphology) or could not be identified soon enough. An initial requirement for this approach is wide accessibility and low costs of high throughput microarray analysis which generate appropriate data for in-silico analysis.

6 Conclusions

We presented SegMine, a bisociation discovery system for exploratory gene analytics. It is based on the non-trivial steps of subgroup discovery (SEGS) and link discovery (Biomine). The goal of SegMine is to enhance the creation of novel biological hypotheses about sets of genes. An implementation of the gene analytics software, which enhances SEGS and creates links to Biomine queries and graphs, is available as a set of workflows in the Orange4WS[5] service-oriented platform at `http://segmine.ijs.si/`.

In the future we plan to enhance the contextualization of genes with contexts discovered by biomedical literature mining. We will add PubMed articles data into the BisoNet graph structure. In particular, we already have a preliminary implementation of software, called Texas [6], which creates a probabilistic network (BisoNet, compatible to Biomine) from textual sources. By focusing on different types of links between terms (e.g., frequent and rare co-ocurances) we expect to get hints at some unexpected relations between concepts from different contexts.

Our long term goal is to help biologists better understand inter-contextual links between genes and their role in explaining (at least qualitatively) underlying mechanisms which regulate gene expression. The proposed approach is considered a first step at computational realization of bisociative reasoning for creative knowledge discovery in systems biology.

Acknowledgements. The work presented in this article was supported by the European Commission under the 7th Framework Programme FP7-ICT-2007-C FET-Open project BISON-211898, by the Slovenian Research Agency grants P2-0103, J4-2228, P4-0165, and by the Algorithmic Data Analysis (Algodan)

[5] `http://orange4ws.ijs.si/`

Centre of Excellence of the Academy of Finland. We thank Igor Trajkovski for his previous work on SEGS, and Filip Železný and Jakub Tolar for their earlier contributions leading to SEGS.

References

1. Agrawal, R., Srikant, R.: Fast algorithms for mining association rules in large databases. In: Proc. 20th Intl. Conf. on Very Large Data Bases, VLDB, Santiago, Chile, pp. 487–499 (1994)
2. Dubitzky, W., Kötter, T., Schmidt, O., Berthold, M.R.: Towards Creative Information Exploration Based on Koestler's Concept of Bisociation. In: Berthold, M.R. (ed.) Bisociative Knowledge Discovery. LNCS (LNAI), vol. 7250, pp. 11–32. Springer, Heidelberg (2012)
3. Eronen, L., Hintsanen, P., Toivonen, H.: Biomine: A Network-Structured Resource of Biological Entities for Link Prediction. In: Berthold, M.R. (ed.) Bisociative Knowledge Discovery. LNCS (LNAI), vol. 7250, pp. 364–378. Springer, Heidelberg (2012)
4. Gamberger, D., Lavrač, N.: Expert-guided subgroup discovery: Methodology and application. Journal of Artificial Intelligence Research 17, 501–527 (2002)
5. Gamberger, D., Lavrač, N., Železný, F., Tolar, J.: Induction of comprehensible models for gene expression datasets by the subgroup discovery methodology. Journal of Biomedical Informatics 37, 269–284 (2004)
6. Juršič, M., Lavrač, N., Mozetič, I., Podpečan, V., Toivonen., H.: Constructing information networks from text documents. In: ECML/PKDD 2009 Workshop on Explorative Analytics of Information Networks, Bled, Slovenia (2009)
7. Kim, S.Y., Volsky, D.J.: PAGE: Parametric Analysis of Gene Set Enrichment. BMC Bioinformatics 6, 144 (2005)
8. Koestler, A.: The Act of Creation. The Macmillan Co., New York (1964)
9. Petrič, I., Urbančičc, T., Cestnik, B., Macedoni-Lukšič, M.: Literature mining method RaJoLink for uncovering relations between biomedical concepts. Journal of Biomedical Informatics 42(2), 219–227 (2009)
10. Podpečan, V., Lavrač, N., Mozetič, I., Kralj Novak, P., Trajkovski, I., Langohr, L., Kulovesi, K., Toivonen, H., Petek, M., Motaln, H., Gruden, K.: SegMine workflows for semantic microarray data analysis in Orange4WS. BMC Bioinformatics 12, 416 (2011)
11. Sevon, P., Eronen, L., Hintsanen, P., Kulovesi, K., Toivonen, H.: Link Discovery in Graphs Derived from Biological Databases. In: Leser, U., Naumann, F., Eckman, B. (eds.) DILS 2006. LNCS (LNBI), vol. 4075, pp. 35–49. Springer, Heidelberg (2006)
12. Swanson, D.R., Smalheiser, N.R., Torvik, V.I.: Ranking indirect connections in literature-based discovery: The role of Medical Subject Headings (MeSH). JASIST 57(11), 1427–1439 (2006)

13. Subramanian, P., Tamayo, P., Mootha, V.K., Mukherjee, S., Ebert, B.L., Gillette, M.A.: Gene set enrichment analysis: A knowledge based approach for interpreting genome-wide expression profiles. Proc. of the National Academy of Science, USA 102(43), 15545–15550 (2005)
14. Trajkovski., I.: Functional interpretation of gene expression data. Ph.D. Thesis, Jozef Stefan International Postgraduate School, Ljubljana, Slovenia (2007)
15. Trajkovski, I., Železny, F., Lavrač, N., Tolar, J.: Learning relational descriptions of differentially expressed gene groups. IEEE Transactions of Systems, Man and Cybernetics C, Special Issue on Intelligent Computation for Bioinformatics 38(1), 16–25 (2008)
16. Trajkovski, I., Lavrač, N., Tolar, J.: SEGS: Search for enriched gene sets in microarray data. Journal of Biomedical Informatics 41(4), 588–601 (2008)
17. Železny, F., Lavrač, N.: Propositionalization-based relational subgroup discovery with RSD. Machine Learning 62(1-2), 33–63 (2007)
18. Weeber, M., Klein, H., de Jong-van den Berg, L.T.W., Vos, R.: Using concepts in literature-based discovery: Simulating Swanson's Raynaud-fish oil and migraine-magnesium discoveries. J. Am. Soc. Inf. Sci. Tech. 52(7), 548–557 (2001)

Contrast Mining from Interesting Subgroups

Laura Langohr[1], Vid Podpečan[2], Marko Petek[3], Igor Mozetič[2], and Kristina Gruden[3]

[1] Department of Computer Science and Helsinki Institute for Information Technology (HIIT), University of Helsinki, Finland
laura.langohr@cs.helsinki.fi

[2] Department of Knowledge Technologies, Jožef Stefan Institute, Ljubljana, Slovenia
{vid.podpecan,igor.mozetic}@ijs.si

[3] Department of Biotechnology and Systems Biology, National Institute of Biology, Ljubljana, Slovenia
{marko.petek,kristina.gruden}@nib.si

Abstract. Subgroup discovery methods find interesting subsets of objects of a given class. We propose to extend subgroup discovery by a second subgroup discovery step to find interesting subgroups of objects specific for a class in one or more contrast classes. First, a subgroup discovery method is applied. Then, contrast classes of objects are defined by using set theoretic functions on the discovered subgroups of objects. Finally, subgroup discovery is performed to find interesting subgroups within the two contrast classes, pointing out differences between the characteristics of the two. This has various application areas, one being biology, where finding interesting subgroups has been addressed widely for gene-expression data. There, our method finds enriched gene sets which are common to samples in a class (e.g., differential expression in virus infected versus non-infected) and at the same time specific for one or more class attributes (e.g., time points or genotypes). We report on experimental results on a time-series data set for virus infected potato plants. The results present a comprehensive overview of potato's response to virus infection and reveal new research hypotheses for plant biologists.

1 Introduction

Subgroup discovery is a classical task in data mining for finding interesting subsets of objects. We extend subgroup discovery by a second subgroup discovery step to find interesting subgroups of objects of a specific class in one or more contrast classes. Contrast classes can represent, for example, different time points or genotypes. Their exact definition depends on the interest of the user. We build on a generic assumption that objects are grouped into classes and described by features (e.g., terms). Often several terms can be summarized under a more

M.R. Berthold (Ed.): Bisociative Knowledge Discovery, LNAI 7250, pp. 390–406, 2012.

general term. We use hierarchies to incorporate such background knowledge about terms. We are not concerned whether objects represent individuals, genes, or something else, and neither what features, classes, and hierarchies represent. Consider the following examples.

In bioinformatics a common problem is that high-throughput techniques and simple statistical tests produce rankings of thousands of genes. Life-scientists have to choose few genes for further (often expensive and time consuming) experiments. Genes can be annotated, for example, by molecular functions or biological processes, which are organized as hierarchies. A life-scientist might be interested in studying an organism in virus infected and non-infected condition (classes) at different time points after infection (contrast classes). In this context, subgroup discovery is known as gene set enrichment, where genes represent features and the aim is to find subgroups of features. In contrast, to fit the retrieval of gene sets into the general subgroup discovery context, we consider genes as objects, their ranking values and their annotations as features. See Table 1 for a line-up of the terms used in the two communities.

We report on experimental results on a time-series data set for virus infected *Solanum tuberosum* (potato) plants. As *S. tuberosum* has only sparsely biological annotations, we use bisociations. Bisociations are concepts that are bridging two domains which are connected only very sparsely or not at all [1]. In our experiments we transfer knowledge from the well studied model plant *A. thaliana* to *S. tuberosum*, our plant under investigation.

Table 1. Synonyms from different communities

Subgroup Discovery	Bioinformatics
object or instance	gene
feature or attribute value, e.g., a term in a hierarchy	annotation or biological concept, e.g., a GO term
class attribute	gene expression under a specific experimental condition such as a specific time point or genotype
class (or class attribute value), e.g., positive/negative	differential/non-differential gene expression
subgroup of objects	gene set
interesting subgroup	enriched gene set

In sociology objects are individuals which are described by different features. For example, bank customers can be described by their occupation, location, loan and insurance type. An economist then might be interested in comparing bank customers who are big spender (classes) and those who are not, before and after the financial crisis (contrast classes). Consider, as a toy example, bank customers in Table 2 and four background hierarchies in Fig. 1. The economist might know that before the financial crisis there were more big spenders than afterwards. Other, perhaps less obvious subgroups, can be more interesting. For example, the economist might not expect that the subgroup described by the term Ljubljana is statistically significant for a contrast class "after financial crisis" in comparison to the contrast class "before the financial crisis".

While subgroup discovery has been addressed in different applications before (see Section 2 for related work). We propose and formulate the problem of subgroup discovery from interesting subgroups and describe how well-known algorithms can be combined to solve the problem (Section 3). In Section 4 we show how these definitions can be applied to find interesting subgroups of genes. We report on experimental results on a time-series data set for virus infected potato plants in Section 5. In Section 6 we conclude with some notes about the results and future work.

Table 2. Bank customers described by features: occupation (OCC), location (LOC), loan (LOAN), insurance (INS) (adapted from Kralj Novak et al. [2]). Different classes are big spender (BSP) as well as before/after financial crisis.

	Before financial crisis					After financial crisis				
ID	OCC	LOC	LOAN	INS	BSP	OCC	LOC	LOAN	INS	BSP
1	private	Maribor	flat	yes	yes	private	Maribor	flat	yes	yes
2	private	Piran	no	no	yes	private	Ljubljana	no	no	yes
3	private	Ljubljana	flat	no	yes	private	Ljubljana	no	no	yes
4	public	Ljubljana	flat	yes	yes	private	Ljubljana	no	no	yes
5	public	Maribor	no	yes	yes	private	Maribor	no	yes	yes
6	private	Maribor	no	no	yes	unemployed	Maribor	no	no	no
7	private	Ljubljana	car	no	yes	unemployed	Ljubljana	car	no	no
8	public	Maribor	no	no	yes	unemployed	Maribor	no	no	no
9	unemployed	Maribor	no	no	yes	unemployed	Ljubljana	no	no	no
10	private	Ljubljana	no	yes	no	private	Ljubljana	no	yes	no
11	private	Piran	no	no	no	unemployed	Piran	no	no	no
12	public	Piran	car	yes	no	public	Piran	car	yes	no
13	unemployed	Piran	no	no	no	unemployed	Piran	no	no	no
14	unemployed	Ljubljana	flat	no	no	unemployed	Ljubljana	no	no	no
15	unemployed	Piran	car	no	no	unemployed	Ljubljana	car	no	no

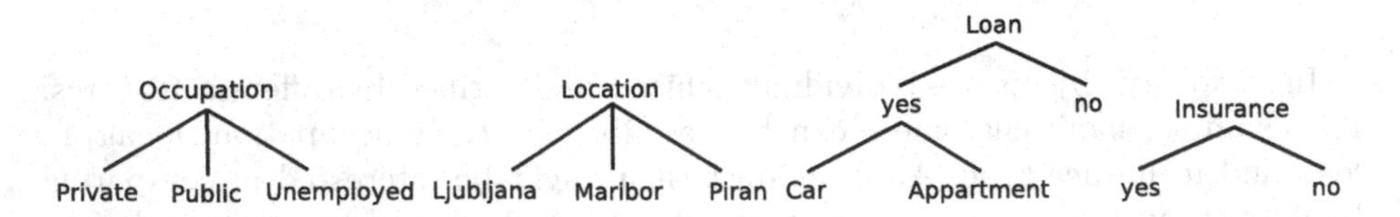

Fig. 1. Bank account feature ontologies (adapted from Kralj Novak et al. [2])

2 Related Work

Discovering patterns in data is a classical problem in data mining and machine learning [3,4]. To represent patterns in an explanatory form they are described by rules (or logical implications) *Condition* $\mapsto$ *Subgroup*, where the antecedent *Condition* is a conjunction of attributes (e.g., terms) and the consequent *Subgroup* is a set of objects.

Subgroup discovery methods find interesting subgroups of objects of a specific class compared to a complementary class. A subgroup of objects is interesting, when the feature values within the subgroup differ statistical significant from the feature values of the other objects. To analyze the constructed subgroups we use Fisher's exact test [5] and a simple test of significance. Alternatively, other statistical tests, like χ^2 test can be used.

Various application areas exist: sociology [6,7], marketing [8], vegetation data [9] or transcriptomics [10] amongst others. In sociology objects typically represent individuals and the aim is to find interesting subgroups of individuals.

In bioinformatics subgroup discovery is known as gene set enrichment. There, objects represent genes and the aim is to find subgroups of genes. A gene set is interesting (or enriched) if the differential expression of the genes of that gene set are statistically significant compared to the rest of the genes. The expression values of several samples are transformed into one feature value, called differential expression, and the genes are partitioned into two classes: differentially and not differentially expressed. Then, subgroup discovery methods find enriched gene sets. Alternatively, gene set enrichment analysis (GSEA) [11] or parametric analysis of gene set enrichment (PAGE) [12] can be used to analyze whether a subgroup is interesting (a gene set is enriched) or not. Both methods use not a partitioning of the genes into two classes, but a ranking of differential expressions instead.

Subgroup discovery differs from typical *time series analysis* where one observation per time point is given. Recently, different approaches have been described which split time series into shorter time-windows to be clustered in separated groups [13] or to find interesting subgroups [14,15]. However, subgroup discovery is not restricted to time series. In addition to time points it can also compare other types of classes, for example, healthy individuals compared to virus infected ones.

Contrast set mining aims to understand the differences between contrasting groups [16]. It is a special case of rule discovery [17] that can be effectively solved by subgroup discovery [18]. It is thus a generalization of subgroup discovery, in which two contrast classes are defined, in contrast subgroup discovery, where one class and it's complement are used.

Association rules describe associations like Y tends to be in the database if X is in it, where X and Y are item sets (sets of terms) [19]. *Exception rules* are association rules which differ from a highly frequent association rule [20]. Alike in our approach they aim to find unexpected rules. Their approach differs from the one presented here, as we are not only interested in finding subgroups in one specific class, but in set theoretic combinations like intersections or set differences of subgroups found by a first subgroup discovery instance.

Frequent item set mining aims to find item sets describing a set of transactions (a subgroup) that are frequent [21]. Similar to the approach presented here, some methods intersect transactions to find closed frequent item sets [22,23,24].

Descriptive induction algorithms aim to discover individual rules defining interesting patterns in the data. This includes association rule learning [19], clausal discovery [25], contrast set mining [16] and subgroup discovery [6,7] amongst others. In contrast, *predictive induction* aims to construct rules to be used for classification and/or prediction [4]. We will focus on descriptive induction, even though our proposed approach could be adapted for predictive induction.

Semantic data mining denotes data mining methods which use background knowledge to improve pattern discovery and interpretation by using semantic annotations of objects as features [2]. Michalski [4] describes different types of background knowledge which can be subsumed under the term *ontology*. An ontology is a representation of a conceptualization and is often represented by a hierarchy, where nodes represent concepts and edges a subsumption relations [26]. Several ontologies can be modeled by a single ontology [27].

In biology commonly used ontologies include Gene Ontology (GO)[1] [28] and Kyoto Encyclopedia of Genes and Genomes (KEGG) Orthology (KO)[2] [29]. GoMapMan[3] is an extension of the MapMan [30] ontology for plants used in our experiments. These ontologies are hierarchical vocabularies of gene annotations (semantic descriptors) organized as a directed acyclic graphs. Nodes represent molecular functions, biological processes or cellular components in GO, molecular pathways in KEGG and plant's molecular functions or biological processes in GoMapMan. Edges represent "is a" or "part of" relationships between the concepts (nodes).

Ontologies are extensively used in gene set enrichment [11,12]. Other application areas include association rule mining [27,31], where the transactions are either extended [27] or frequent item sets are generated one level at at a time [31]. Here, we use the subgroup construction method by Trajkovski et al. [32], which combines terms from the same level as well as from different levels.

3 Contrast Mining from Interesting Subgroups

Given a set of objects described by features and different classes of objects, the goal is to find interesting subgroups of objects of a specific class in one or more contrast classes. That is, for example, to find interesting subgroups specific for big spenders (class) after the financial crisis (contrast class).

Our approach finds such subgroups by dividing the task into three steps: First, interesting subgroups are found by a subgroup discovery method. Second, contrast classes on those subgroups are defined by set theoretic functions. Third,

[1] `http://www.geneontology.org/`
[2] `http://www.genome.jp/kegg/ko.html`
[3] `http://www.gomapman.org/`

subgroup discovery finds interesting subgroups in the contrast classes. Next, we will describe each step in detail.

3.1 Subgroup Discovery (Step 1)

To find interesting subgroups, we use search for enriched gene set (SEGS) [32], a method developed for gene set enrichment analysis, but not restricted to this application area [2].

First, all subgroups that contain at least a minimal number of objects are constructed by a depth-first traversal [32]. Afterwards, the constructed subgroups are analyzed if they are statistically significant for the class of interest.

Construction of Subgroups. We use hierarchies of terms as background knowledge to construct subgroups that contain at least a minimal number of objects. Subgroups are constructed by individual terms and logical conjunctions of terms.

Subgroup Construction by Individual Terms. Let S be the set of all objects and T the union of all terms of n background knowledges. Each term $t \in T$ defines a subgroup $S_t \subset S$ that consists of all objects s where feature value t is true, that is, are annotated by term t:

$$S_t = \{s \mid s \text{ is annotated by } t\}. \quad (1)$$

Subgroup Construction with Logical Conjunctions. Subgroups can be constructed by intersections, which are described by logical conjunctions of terms. Let $S_1, \ldots, S_k$ be k subgroups described by terms $t_1, \ldots, t_k$. Then, the logical conjunction of k terms defines the intersection of k subgroups:

$$t_1 \wedge t_2 \wedge \ldots \wedge t_k \mapsto S_1 \cap S_2 \cap \ldots \cap S_k . \quad (2)$$

Example 1. In Table 2, before the financial crisis, the conjunction *Ljubljana* $\wedge$ $\neg$*Insurance* defines a subgroup of three bank customers $\{3, 7, 14\}$.

A subgroup description can be seen as the condition part of a rule *Condition* $\mapsto$ *Subgroup* [33]. If an object is annotated by several terms, it is a member of several subgroups. A subgroup might be a subset of another subgroup. In particular, consider the example hierarchies in Figure 1. Then, an object that is annotated by a term t is also annotated by its ancestors.

To construct all possible subgroups one ontology is used, where the root has n children, one for each ontology. We start with the root term and recursively replace each term by each of its children. We are not interested in constructing all possible subgroups, but only those representing at least a minimal number of objects. Therefore, we extend a condition only if the subgroup defined by it contains more than a minimum number of objects. If a condition defines the same group of objects as a more general condition, the more general condition is deleted. Furthermore, in each recursion we add another term to the rule to obtain intersections of two or more subgroups and test if the intersection represents at least a minimal number of objects.

Analysis of Constructed Subgroups. Statistical tests can be used to analyze if the constructed subgroups are interesting, that is, the feature values within the subgroup differ statistically significant from the feature values of the other objects with respect to given classes A and B. For each subgroup $S_t \subset S$ the data is arranged in a table:

	A	B
S_t	n_{11}	n_{12}
$S \setminus S_t$	n_{21}	n_{22}

where $n = |S| = n_{11} + n_{12} + n_{21} + n_{22}$, n_{11} is the number of objects in S_t that are annotated by A, n_{12} is the number of objects in S_t that are annotated by B, n_{21} is the number of objects in $S \setminus S_t$ that are annotated by A, and n_{21} is the number of objects in $S \setminus S_t$ that are annotated by B.

Fisher's Exact Test. Fisher's exact test evaluates if the equal proportions and the observed difference is within what is expected by chance alone or not [5]. The probability of observing each possible table configuration is calculated by

$$P(X = n_{11}) = \binom{n_{11}+n_{12}}{n_{11}}\binom{n_{21}+n_{22}}{n_{21}} / \binom{n}{n_{11}+n_{21}} . \tag{3}$$

The p-value is then the sum of all probabilities for the observed or more extreme (that is, $X < n_{11}$) observations:

$$p = \sum_{i=0}^{n_{11}} P(X = i) . \tag{4}$$

Example 2. Consider the bank customers in Table 2, the condition *Maribor* and the class big spender versus not big spender and a significance level α. There are five bank customers in Maribor: $S_t = \{1, 5, 6, 8, 9\}$, which are all big spenders. Hence, the p-value is $p \approx 0.043956$.

Test of Significance. To address the multiple testing problem, that is, that subgroups might have occurred by chance alone, we correct the p-values. Therefore, we randomly permute the genes and calculate the p-value for each subgroup. We repeat this first step for $1,000$ permutations, create a histogram by the p-values of each permutation's best subgroup, and estimate the (corrected) p-value using the histogram: The corrected p-value is the reciprocal of the permutations in which the p-value obtained by Fisher's exact test is smaller than all p-values obtained from the permutations. For example, if the p-value obtained by Fisher's exact test is in all permutations smaller, then the corrected p-value is $p = 0.001$. If the corrected p-value is smaller than the given significance level α then the feature values within the subgroup differ statistical significantly from the feature values from the other objects and we call the subgroup interesting and the subgroup is called interesting.

3.2 Construction of Contrast Classes (Step 2)

Let $S_1, \ldots, S_n$ denote the interesting subgroups found for n classes. Then, two contrast classes S_f and S_g are defined by two set theoretic functions f and g:

$$f(S_1, \ldots, S_n) = S_f \subseteq \bigcup_i S_i \,. \tag{5}$$

and $g(S_1, \ldots, S_n)$ is defined as the complement. If $g(\cdot)$ is defined as something else than the complement, the next step is contrast set mining rather than subgroup discovery (see [33] for a line-up of both approaches).

Which set theoretic functions should be used depends on the objective. For example, if we aim to find interesting subgroups which are common to all classes, then $f(\cdot)$ is defined as the set of objects occurring in at least one interesting subgroup of each class:

$$f(S_1, \ldots, S_n) = \bigcap_{i \in \{1,\ldots,n\}} S_i \,. \tag{6}$$

Hence, every object of S_f occurred in each class in at least one interesting subgroup.

Alternatively, if the aim is to find interesting subgroups which are specific for class k, then $f(\cdot)$ is defined as the set of objects only occurring in interesting subgroups found for k^{th} class:

$$f(S_1, \ldots, S_n) = S_k \setminus \bigcup_{\substack{i \in \{1,\ldots,n\}, \\ i \neq k}} S_i \,. \tag{7}$$

Hence, every object in S_f occurred in one or more interesting subgroups of class S_k, but not in a single one of the other classes.

Example 3. Consider again the bank customers in Table 2, subgroups with at least four bank customers and $\alpha = 0.3$ (for sake of simplicity we consider a relatively high significance level in this toy example). For the"before financial crisis" class we obtain four subgroups: *Maribor*, *Maribor* $\wedge$ *¬Loan*, *Piran*, and *Unemployed*. The set of bank customers described by at least one of them is $S_1 = \{1, 5, 6, 8, 11, \ldots, 15\}$. For the "after financial crisis" class we obtain two subgroups: *Private* and *Unemployed* and the set $S_2 = \{1, 2, 3, 6, \ldots, 11, 13, 14, 15\}$. Then the sets $S_f = S_2 \setminus S_1 = \{2, 3, 7, 10\}$ and $S_g = S_1$ specify contrast classes.

3.3 Subgroup Discovery (Step 3)

We find interesting subgroups in contrasting classes by a second subgroup discovery instance, where the two classes are now the sets S_f and S_g. The p-values are calculated by (3) and (4), followed by a test of significance.

Example 4. Given the contrast classes (sets) of bank customers $S_f = \{2, 3, 7, 10\}$ and $S_g = \{1, 5, 6, 8, 9, 11, \ldots, 15\}$ we analyze the statistical significance of subgroups with respect to these contrast classes. The condition *Ljubljana* has after the financial crisis eight bank customers $\{2,3,4,7,9,10,14,15\}$, from which four are in S_f and three in S_g. Hence, we obtain a p-value of $p = 0.0699301$. Next, we test the p-value for significance to assure we did not obtain the subgroup by chance alone. In the first subgroup discovery instance, we did not not obtain *Ljubljana* as logical condition. When compared to the contrast class "before financial crisis", and assuming it passed the significance test, *Ljubljana* is found to be statistically significant for the contrast class "after financial crisis".

4 An Instance of Our Method: Gene Set Enrichment from Enriched Gene Sets

Next, we will discuss how our proposed method can be applied in the area of gene set enrichment. In gene-expression experiments objects are genes and features are their annotations by, for example, GO and KEGG terms. Here, our aim is to find enriched gene sets of a specific class (e.g., virus infected plants) in one or more other classes (e.g., different time points). Next, we describe measures used for transforming the expression values of several samples (e.g., different individuals). into a feature value, called differential expression, and how the constructed gene sets are analyzed for statistical significance.

Measures for Differential Expression. After preprocessing the data (including microarray image analysis and normalization) the genes can be ranked according to their gene expression.

Fold change (FC) is a metric for comparing the expression level of a gene g between two distinct experimental conditions (classes) A and B [10]. It is the log ratio of the average gene-expression levels with respect to two conditions [34]. However, FC values do not indicate the level of confidence in the designation of genes as differently expressed or not.

The *t-test statistic* is a statistical test to determine the statistically significant difference of gene g between two classes A and B [10]. Though, the probability that a real effect can be identified by the t-test is low if the sample size is small [34]. A Bayesian t-test is advantageous if few (that is, two or three) replicates are used only, but no advantage is gained if more replicated are used [35]. In our experiments we used four replicates and therefore will use the simple t-test.

Analysis of Gene Set's Enrichment. For the enrichment analysis of gene sets statistical tests like Fisher's exact test [5] can be used. Alternatively, GSEA and PAGE can be used. We next describe each of them.

Fisher's Exact Test. In the gene set enrichment setting S_t is the gene set analyzed and $S \setminus S_t$ is the gene set consisting of all other genes. The two classes are differential expression and non-differential expression. To divide the genes into two classes a cut off is set in the gene ranking: genes in the upper part are defined as differentially expressed and the genes in the lower part are defined as not differentially expressed genes. Then the p-values are calculated and tested for significance.

Gene Set Enrichment Analysis (GSEA) [11]. Given a list $L = \{g_1, \ldots, g_n\}$ of n ranked genes, their expression levels $e_1, \ldots, e_n$, and a gene set S_t , GSEA evaluates whether S_t's objects are randomly distributed throughout L or primarily found at the top or bottom [36]. An enrichment score (ES) is calculated, which is the maximum deviation from zero of the fraction of genes in the set S_t weighted by their correlation and the fraction of genes not in the set:

$$ES(S_t) = \max_{i \in \{1,\ldots,n\}} \left| \sum_{\substack{g_j \in S_t \\ j \leq i}} \frac{|e_j|^p}{n_w} - \sum_{\substack{g_j \notin S_t \\ j \leq i}} \frac{1}{n - n_w} \right| \tag{8}$$

where $n_w = \sum_{g_j \in S_t} |e_j|^p$. If the enrichment score is small, then S_t is randomly distributed across L. If it is high, then the genes of S_t are concentrated in the beginning or end of the list L. The exponent p controls the weight of each step. $ES(S_t)$ reduces to the standard Kolmogorov-Smirnov statistic if $p = 0$:

$$ES(S) = \max_{i \in \{1,\ldots,n\}} \left| \sum_{\substack{g_j \in S_t \\ j \leq i}} \frac{1}{|S_t|} - \sum_{\substack{g_j \notin S_t \\ j \leq i}} \frac{1}{|S| - |S_t|} \right|. \tag{9}$$

The significance of $ES(S_t)$ is then estimated by permutating the sample labels, reordering the genes, and re-computing $ES(S_t)$. From 1,000 permutations a histogram is created and the nominal p-value for S_t is estimated by using the positive (or the negative) portion if $ES(S_t) > 0$ (or $ES(S_t) < 0$, respectively).

Parametric Analysis of Gene Set Enrichment (PAGE). PAGE is a gene set enrichment analysis method based on a parametric statistical analysis model [12]. For each gene set S_t a Z-score is calculated, which is the fraction of mean deviation to the standard deviation of the ranking score values:

$$Z(S_t) = (\mu_{S_t} - \mu)\frac{1}{\sigma}\sqrt{|S_t|} \tag{10}$$

where σ is the standard deviation and μ and μ_{S_t} are the means of the score values for all genes and for the genes in set S_t, respectively. The Z-score is high if the deviation of the score values is small or if the means largely differ between the gene set and all genes. As gene sets may vary in size, the fraction is scaled by the square root of the set size. However, because of this scaling the Z-score is also high if S_t is very large. Assuming a normal distribution, a p-value for each gene set is calculated. Finally, the p-values are corrected by a test of significance.

Using normal distributions for statistical inference makes PAGE computationally lighter than GSEA which requires permutations. On the other hand, GSEA makes no assumptions about the variability and can be used if the distribution is not normal or unknown. Kim and Volsky [12] studied different data sets for which PAGE generally detected a larger number of significant gene sets than GSEA. Trajkovski et al. [32] used the sum of GSEA's and PAGE's p-values, weighted by percentages (e.g., one third of GSEA's and two third of PAGE's or half of both). Hence, gene sets with small *p*-values for GSEA and PAGE are output as enriched gene sets.

In the second gene set enrichment analysis instance, we want to analyze subgroups with respect to the constructed contrast classes, and not with respect to to the differential expression. Now, we have two classes, but not a ranking and thus GSEA and PAGE cannot be used for analyzing the constructed gene sets. Statistical test for categorical analysis can still be used. We use Fisher's exact test to compare the two classes S_f and S_g against each other.

5 Experiments

For our experiments we use a *Solanum tuberosum* (potato) time course gene-expression data set for virus infected and non-infected plants. The data set consists of three time points: one, three and six days after virus infection when the viral infected leaves as well as leaves from non-infected plants were collected. The aim is to find enriched gene sets which are common to virus infected samples compared to non-infected samples (classes in subgroup discovery of Step 1), and at the same time specific for one or all time points (classes in subgroup discovery of Step 3).

Test Setting. Recently, *S. tuberosum*'s genome has been completely sequenced [37], but only few GO or KEGG annotations of *S. tuberosum* genes exist. However, plenty GO and KEGG annotations exist for the well studied model plant *Arabidopsis thaliana*. We use homologs between *S. tuberosum* and *A. thaliana* to make gene set enrichment analysis for *S. tuberosum* possible. There are more than 26.000 homologs provided by the POCI consortium [38] for more than 42.000 *S. tuberosum* genes. We consider only the best (with respect to the e-value) in case there are several homologs. Gene set enrichment analysis is performed based on expression values in the dataset, the gene IDs of the *A. thaliana* homologs, and GO and KEGG annotations for *A. thaliana*.

In parallel, we built potato ontologies independently using Blast2GO[4] to obtain homologue sequences in NCBI (BLASTX with high scoring segment pair (HSP) length 33 and e-value $1e-15$) and their GO annotations (GO weight 5, cutoff 55 and e-value $1e-15$). In this case, enrichment analysis is performed using the gene IDs and expression values of *S. tuberosum*, and the GO and KEGG annotations obtained with Blast2GO.

For both approaches we carried out gene set enrichment experiments in an Orange4WS[5] workflow [39]. We restricted gene sets to contain at minimum ten genes, the gene set description to contain at maximum four terms, and the *p*-value to be 0.05 or smaller. For analyzing the constructed gene sets in Step 1 we used Fisher's exact test, GSEA, PAGE and the combined GSEA and PAGE (equal percentages).

[4] `http://www.blast2go.org/`
[5] `http://orange4ws.ijs.si/`

We consider two types of contrast classes for gene set enrichment (Step 2): genes that are common to all classes compared to the genes occurring in some gene sets, but not in all (obtained by (6) and genes that are specific for one class compared to the genes of the gene sets of the other classes (obtained by (7). Fisher's exact test is used to analyze gene set enrichment in Step 3 for both approaches.

Results. Several subgroup descriptions that are known to relate to potato's response to virus infection were found. That is, our method reveals molecular functions, biological processes and pathways that have a central role in it. We are interested in assisting the biologist in generating new research hypotheses. Therefore, we evaluate our results by counting the number of gene set descriptions which were unexpected to a plant biologist to relate to potato's response to virus infection. In this context, "unexpected" means that the knowledge was contained in GO, KEGG or GoMapMan, but it was not shown previously to be related to experimental conditions studied (here, related to the response of potato to viral infection).

The amount of enriched gene sets found for the *A. thaliana* homologs approach are shown in Table 3. and for the GO ontologies for potato genes approach in Table 4. For both approaches, both subgroup discoveries (Step 1 and 3) found few rules if any at all for the first and third day, whereas for the sixth day several rules are found. This matches well with the biological knowledge about potato's respond on virus infection: In the first days the potato activates the defense response, but the full effect can be witnessed only on day six.

The quantities of unexpected enriched gene sets found for the *A. thaliana* homologs approach are shown in Table 5. and for the GO ontologies for potato genes approach in Table 6. Few enriched gene sets are found in the first stage when using GSEA or the combination of GSEA and PAGE for analyzing the gene sets of the first stage. Hence, few enriched gene sets (if any at all) are found in Step 3. When using either Fisher's exact test or PAGE instead, more enriched gene sets are found, from which several are of interest to a plant biologist, suggesting one of these methods should be preferred.

The subgroups discovered in Step 3 revealed some enriched gene sets for the intersection, but none of them was more specific in comparison to the enriched gene sets found in Step 1 or even unexpected for the biologist. This is most likely due to the characteristic of a defense response: The gene expression of the first days (when activating the defense response) differs from the gene expression on day six (when the defense response is active) and therefore the intersection reveals only few enriched gene sets that are active at all time points.

For the set differences we obtain new and more specific gene sets. Some of them we did not find in the first stage, some other are more specific than in Step 1, both of interest for biologists. Hence, this shows that our proposed method reveals new enriched gene sets if the set theoretic functions are selected appropriately for the experiment and user's objective.

Table 3. Quantities of enriched gene sets found for the *A. thaliana* homologs approach for Fisher (F), GSEA (G), PAGE (P), and the combined approach of GSEA and PAGE with equal percentages (G+P)

		F	G	P	G+P
Step 1	first day	6	4	5	1
	third day	7	4	16	5
	sixth day	14	5	12	5
Step 3	first day set difference	9	0	7	1
	third day set difference	7	0	7	6
	sixth day set difference	21	5	16	6
	intersection	4	4	16	4

Table 4. Quantities of enriched gene sets found for the GO ontologies for potato genes approach for Fisher (F), GSEA (G), PAGE (P), and the combined approach of GSEA and PAGE with equal percentages (G+P)

		F	G	P	G+P
Step 1	first day	1	0	4	0
	third day	1	1	5	0
	sixth day	25	21	33	16
Step 3	first day set difference	15	0	7	0
	third day set difference	5	1	10	0
	sixth day set difference	42	2	34	3
	intersection	0	0	1	0

Table 5. Quantities of unexpected enriched gene sets found for the *A. thaliana* homologs approach for Fisher (F), GSEA (G), PAGE (P), and the combined approach of GSEA and PAGE with equal percentages (G+P). In Step 3 only unexpected enriched gene sets are counted which were new or more specific in comparison to Step 1.

		F	G	P	G+P
Step 1	first day	2	2	0	0
	third day	4	2	4	4
	sixth day	14	5	12	5
Step 3	first day set difference	1	0	4	0
	third day set difference	1	0	1	0
	sixth day set difference	11	1	4	1
	intersection	0	0	0	0

Table 6. Quantities of unexpected enriched gene sets found for the GO ontologies approach for Fisher (F), GSEA (G), PAGE (P), and the combined approach of GSEA and PAGE with equal percentages (G+P). In Step 3 only unexpected enriched gene sets are counted which were new or more specific in comparison to Step 1.

		F	G	P	G+P
Step 1	first day	0	0	1	0
	third day	1	1	2	0
	sixth day	24	21	28	16
Step 3	first day set difference	4	0	0	0
	third day set difference	0	0	2	0
	sixth day set difference	15	0	13	0
	intersection	0	0	0	0

6 Conclusion

We addressed the problem of subgroup discovery from interesting subgroups. After reviewing subgroup discovery we introduced the construction of contrast classes on the discovered subgroups. Subgroup discovery then finds interesting subgroups in those contrast classes. Thereby, we allow the user to specify contrast classes she is interested in, for example, she can choose to contrast several time points.

We showed how our approach works on an example of bank customers and applied it to a gene set enrichment application, a time-series data set for virus infected potato plants. The results indicate that our proposed approach reveals new research hypotheses for biologists.

Further experimental evaluation is planned, including experiments on other data sets and with more complex set theoretic functions. A careful interpretation of our results is needed as the subgroup discovery of the first step reduced the number of genes (objects) and hence Fisher was applied (in the third step) on a relatively small number of genes. Furthermore, gene set descriptions were often biologically redundant which we will address in future, for example, by clustering or filtering the obtained gene sets.

We will carry out a more extensive evaluation by analyzing the quality of gene sets descriptions which are unknown to relate to potato's virus response and visualize the gene sets and their relations with the enrichment map tool. We will evaluate quantity and quality of the genes of the unknown gene sets with Biomine, a search engine for visualization and discovery of non-trivial connections between biological entities, such as genes. Finally, some genes will be selected for wet-lab experiments, which may further the understanding of the biological mechanisms of virus response, particularly that of potatoes.

Acknowledgments. We would like to thank Kamil Witek, Ana Rotter, and Špela Baebler for the test data and the help with interpreting the results and

Nada Lavrač and Hannu Toivonen for their valuable comments and suggestions on the chapter.

This work has been supported by the European Commission under the 7th Framework Programme FP7-ICT-2007-C FET-Open, contract no. BISON-211898, by the Algorithmic Data Analysis (Algodan) Centre of Excellence of the Academy of Finland and by the Slovenian Research Agency grants P2-0103, J4-2228 and P4-0165.

References

1. Berthold, M.R. (ed.): Bisociative Knowledge Discovery. LNCS (LNAI), vol. 7250. Springer, Heidelberg (2012)
2. Kralj Novak, P., Vavpetič, A., Trajkovski, I., Lavrač, N.: Towards Semantic Data Mining with g-SEGS. In: SiKDD 2010 (2010)
3. Bruner, J., Goodnow, J., Austin, G.: A Study of Thinking. Wiley (1956)
4. Michalski, R.: A Theory and Methodology of Inductive Learning. Artificial Intelligence 20(2), 111–161 (1983)
5. van Belle, G., Fisher, L., Heagerty, P., Lumley, T.: Biostatistics: A Methodology for the Health Sciences, 2nd edn. Wiley series in probability and statistics. Wiley-Interscience (1993)
6. Klösgen, W.: Explora: a Multipattern and Multistrategy Discovery Assistant. In: Fayyad, U., Piatetsky-Shapiro, G., Smyth, P., Uthurusamy, R. (eds.) Advances in Knowledge Discovery and Data Mining, pp. 249–271. AAAI (1996)
7. Wrobel, S.: An Algorithm for Multi-Relational Discovery of Subgroups. In: Komorowski, J., Żytkow, J.M. (eds.) PKDD 1997. LNCS, vol. 1263, pp. 78–87. Springer, Heidelberg (1997)
8. del Jesus, M., Gonzalez, P., Herrera, F., Mesonero, M.: Evolutionary Fuzzy Rule Induction Process for Subgroup Discovery: A Case Study in Marketing. Transactions on Fuzzy Systems 15, 578–592 (2007)
9. May, M., Ragia, L.: Spatial Subgroup Discovery Applied to the Analysis of Vegetation Data. In: Karagiannis, D., Reimer, U. (eds.) PAKM 2002. LNCS (LNAI), vol. 2569, pp. 49–61. Springer, Heidelberg (2002)
10. Allison, D., Cui, X., Page, G., Sabripour, M.: Microarray Data Analysis: from Disarray to Consolidation and Consensus. Nature Reviews, Genetics 5, 55–65 (2006)
11. Mootha, V., Lindgren, C., Eriksson, K.F., Subramanian, A., Sihag, S., Lehar, J., Puigserver, P., Carlsson, E., Ridderstrale, M., Laurila, E., Houstis, N., Daly, M., Patterson, N., Mesirov, J., Golub, T., Tamayo, P., Spiegelman, B., Lander, E., Hirschhorn, J., Altshuler, D., Groop, L.: PGC-1α-responsive Genes Involved in Oxidative Phosphorylation are Coordinately Downregulated in Human Diabetes. Nature Genetics 34(3), 267–273 (2003)
12. Kim, S.Y., Volsky, D.: PAGE: Parametric Analysis of Gene Set Enrichment. BMC Bioinformatics 6(1), 144 (2005)

13. Antoniotti, M., Ramakrishnan, N., Mishra, B.: GOALIE, A Common Lisp Application to Discover Kripke Models: Redescribing Biological Processes from Time-Course Data. In: ILC 2005 (2005)
14. Antoniotti, M., Carreras, M., Farinaccio, A., Mauri, G., Merico, D., Zoppis, I.: An Application of Kernel Methods to Gene Cluster Temporal Meta-Analysis. Computers & Operations Research 37(8), 1361–1368 (2010)
15. Zoppis, I., Merico, D., Antoniotti, M., Mishra, B., Mauri, G.: Discovering Relations Among GO-Annotated Clusters by Graph Kernel Methods. In: Măndoiu, I.I., Zelikovsky, A. (eds.) ISBRA 2007. LNCS (LNBI), vol. 4463, pp. 158–169. Springer, Heidelberg (2007)
16. Bay, S., Pazzani, M.: Detecting Group Differences: Mining Contrast Sets. Data Mining and Knowledge Discovery 5, 213–246 (2001)
17. Webb, G., Butler, S., Newlands, D.: On Detecting Differences between Groups. In: KDD 2003, pp. 256–265. ACM (2003)
18. Kralj Novak, P., Lavrač, N., Gamberger, D., Krstacic, A.: CSM-SD: Methodology for Contrast Set Mining through Subgroup Discovery. Journal of Biomedical Informatics 42(1), 113–122 (2009)
19. Agrawal, R., Mannila, H., Srikant, R., Toivonen, H., Verkamo, A.: Fast Discovery of Association Rules. In: Fayyad, U., Piatetsky-Shapiro, G., Smyth, P., Uthurusamy, R. (eds.) Advances in Knowledge Discovery and Data Mining, pp. 307–328. AAAI (1996)
20. Suzuki, E.: Autonomous Discovery of Reliable Exception Rules. In: KDD 1997 (1997)
21. Agrawal, R., Imieliński, T., Swami, A.: Mining Association Rules Between Sets of Items in Large Databases. In: SIGMOD 1993, pp. 207–216. ACM (1993)
22. Mielikäinen, T.: Intersecting Data to Closed Sets with Constraints. In: FIMI 2003 (2003)
23. Pan, F., Cong, G., Tung, A., Yang, J., Zaki, M.: Carpenter: Finding Closed Patterns in Long Biological Datasets. In: KDD 2003, pp. 637–642. ACM (2003)
24. Borgelt, C., Yang, X., Nogales-Cadenas, R., Carmona-Saez, P., Pascual-Montano, A.: Finding Closed Frequent Item Sets by Intersecting Transactions. In: EDBT/ICDT 2011, pp. 367–376. ACM (2011)
25. De Raedt, L., Dehaspe, L.: Clausal Discovery. Machine Learning 26, 99–146 (1997)
26. Gruber, T.: Toward principles for the design of ontologies used for knowledge sharing. International Journal of Human-Computer Studies 43, 907–928 (1995)
27. Srikant, R., Agrawal, R.: Mining Generalized Association Rules. In: VLDB 1995, pp. 407–419 (1995)
28. Khatri, P., Drăghici, S.: Ontological Analysis of Gene Expression Data: Current Tools, Limitations, and Open Problems. Bioinformatics 21(18), 3587–3595 (2005)
29. Aoki-Kinoshita, K., Kanehisa, M.: Gene Annotation and Pathway Mapping in KEGG. In: Walker, J.M., Bergman, N.H. (eds.) Comparative Genomics, vol. 396, pp. 71–91. Humana Press (2007)
30. Thimm, O., Bläsing, O., Gibon, Y., Nagel, A., Meyer, S., Krüger, P., Selbig, J., Müller, L., Rhee, S., Stitt, M.: MapMan: a User-driven Tool to Display Genomics Data Sets Onto Diagrams of Metabolic Pathways and Other Biological Processes. The Plant Journal 37(6), 914–939 (2004)
31. Han, J., Fu, Y.: Discovery of Multiple-Level Association Rules from Large Databases. In: VLDB 1995, pp. 420–431. Morgan Kaufmann Publishers Inc. (1995)
32. Trajkovski, I., Lavrač, N., Tolar, J.: SEGS: Search for enriched gene sets in microarray data. Journal of Biomedical Informatics 41(4), 588–601 (2008)

33. Kralj Novak, P., Lavrač, N., Webb, G.: Supervised Descriptive Rule Discovery: A Unifying Survey of Contrast Set, Emerging Pattern and Subgroup Mining. Journal of Machine Learning Research 10, 377–403 (2009)
34. Cui, X., Churchill, G.: Statistical Tests for Differential Expression in cDNA Microarray Experiments. Genome Biology 4(4), 210.1–210.10 (2003)
35. Baldi, P., Long, A.: A Bayesian Framework for the Analysis of Microarray Expression Data: Regularized *t*-test and Statistical Inferences of Gene Changes. Bioinformatics 17(6), 509–519 (2001)
36. Subramanian, A., Tamayo, P., Mootha, V., Mukherjee, S., Ebert, B., Gillette, M., Paulovich, A., Pomeroy, S., Golub, T., Lander, E., Mesirov, J.: Gene Set Enrichment Analysis: A Knowledge-based Approach for Interpreting Genome-wide Expression Profiles. PNAS 102(43), 15545–15550 (2005)
37. The Potato Genome Sequencing Consortium: Genome sequence and analysis of the tuber crop potato. Nature 475, 189–195 (2011)
38. Bioinformatics @ IPK Gatersleben: BLASTX against Arabidopsis, `http://pgrc-35.ipk-gatersleben.de/pls/htmldb_pgrc/f?p=194:5:941167238168085::NO` (visited on March 2011)
39. Podpečan, V., Lavrač, N., Mozetič, I., Kralj Novak, P., Trajkovski, I., Langohr, L., Kulovesi, K., Toivonen, H., Petek, M., Motaln, H., Gruden, K.: SegMine Workflows for Semantic Microarray Data Analysis in Orange4WS. BMC Bioinformatics 12, 416 (2011)

Link and Node Prediction in Metabolic Networks with Probabilistic Logic

Angelika Kimmig[1] and Fabrizio Costa[2,*]

[1] Departement Computerwetenschappen, K.U. Leuven
Celestijnenlaan 200A - bus 2402, B-3001 Heverlee, Belgium
angelika.kimmig@cs.kuleuven.be

[2] Institut für Informatik, Albert-Ludwigs-Universität,
Georges-Koehler-Allee, Geb 106, D-79110 Freiburg, Germany
costa@informatik.uni-freiburg.de

Abstract. Information on metabolic processes for hundreds of organisms is available in public databases. However, this information is often incomplete or affected by uncertainty. Systems capable to perform automatic curation of these databases and capable to suggest pathway-holes fillings are therefore needed. To this end such systems should exploit data available from related organisms and cope with heterogeneous sources of information (e.g. phylogenetic relations). Here we start to investigate two fundamental problems concerning automatic metabolic networks curation, namely *link prediction* and *node prediction* using ProbLog, a simple yet powerful extension of the logic programming language Prolog with independent random variables.

1 Introduction

Living organisms rely on a large interconnected set of biochemical reactions to provide the requirements of mass and energy for the cellular processes to take place. This complex set of reactions constitute the organism's metabolic network [1]. Highly specialized proteins, called enzymes, are used to regulate the time and place for the various processes as most of the reactions taking place in organisms would be too slow without them. Enzymes control in practice which parts of the overall metabolic network is active in a given cell region in a given cellular phase. A large quantity of information about these networks accumulated through years of research, and is nowadays stored and organized in databases allowing researchers to develop network based approaches to study organisms metabolism. There exist collections of metabolic networks for several hundreds of organisms (e.g., the Kyoto Encyclopedia of Genes and Genomes (KEGG) [2] or the BioCyc database [3]) where relations between genes, enzymes, reactions and chemical compounds are available and organized in collections called "pathways". The knowledge that we have of these relations is however incomplete (most annotation efforts fail to assign functions to 40-60% of the protein sequences [4]) and is affected by uncertainty (wrong catalytic function assignment,

* FC was a postdoctoral fellow at K.U. Leuven while this work was initiated.

M.R. Berthold (Ed.): Bisociative Knowledge Discovery, LNAI 7250, pp. 407–426, 2012.

incomplete annotation (e.g., only one function of a multi-domain protein) or non-specific assignment (e.g., to a protein family)). Systems capable to perform automatic curation of these databases and capable to suggest pathway-holes fillings are therefore in dear need. However, in order to overcome the limitations of homology searches, it is paramount to make use of information from heterogeneous sources and to therefore encode all the available data into complex relational data bases (i.e., BisoNets [5]). Finally, to leverage the different amount of coverage for different organisms (i.e., there is more information regarding humans than for other vertebrates), a case-based approach that uses information on related organisms should also be employed. All these requirements raise the problem of how to integrate heterogeneous and uncertain sources of information in a principled way.

Although systems for reconstructing pathways from relevant gene sets [6] and filling pathway-holes [7] are known in literature, they do not offer sufficient flexibility when new additional sources of information become available or, more importantly, in case one needs to change the set of queries involved in the solution of a specific task.

We study an approach that satisfies these flexibility requirements by representing metabolic networks in the probabilistic logical framework ProbLog [8], a simple yet powerful extension of the logic programming language Prolog with independent random variables in the form of *probabilistic facts*. This allows us to easily include background knowledge affected by uncertainty, and to obtain an answer to several key questions by performing probabilistic inference in a principled manner.

In this work, we start to investigate some fundamental problems concerning automatic metabolic networks curation, namely: 1) *link prediction*, i.e., estimation of the degree of belief in a link between a gene and an enzyme, and 2) *node prediction*, that is, whether the existence of a certain enzyme (and its link to an unknown gene) has to be hypothesized in order to maintain the contiguity of a pathway. For both tasks, the key components of our probabilistic model are (1) a preliminary estimate of the degree of belief for an association between a gene G and an enzyme E in an organism O, (2) background knowledge BK on organisms related to O obtained from the KEGG database, and (3) a linear model that predicts the probability of the gene-enzyme relation $G - E$ for the organism O given the dataset BK. The features employed in the linear model are complex queries and the associated values correspond to the probability of the query in BK including the preliminary estimate. The parameters of the model encode the relevance of the query for the specific pair gene-enzyme. The core idea is to leverage the flexibility of ProbLog to define meaningful queries at a conveniently abstract level. We finally compute the probability of the gene-enzyme relation $G - E$ based on the queries that are satisfied with high probability and that are predicted to be relevant for $G - E$.

The chapter is organized as follows: in Section 2 we introduce the probabilistic logic framework ProbLog; in Section 3 we describe how we model the knowledge associated with the metabolic reconstruction tasks and how we query this model for prediction; finally in Section 4 we present some initial empirical results on a specific pathway in yeast.

2 The Probabilistic Logic Environment: ProbLog

Our work uses ProbLog to model data and queries. ProbLog is a probabilistic extension of the logic programming language Prolog. It thus combines the expressivity of a first order modeling language with the ability to reason under uncertainty. In contrast to propositional graphical models (such as Bayesian Networks), connections between random variables in ProbLog can be specified on the first order level, thus avoiding the need of explicitly grounding all information a priori. This results in a higher level of abstraction and more flexibility in the specification of queries. In this section, we briefly illustrate the basic ideas of ProbLog by means of an example; for more details, we refer to [8].

The following ProbLog program[1] models a tiny fraction of the type of network considered in this chapter:

```
0.8 :: ortholog(g1, g2).   0.7 :: ortholog(g1, g3).
0.6 :: function(g1, e1).   0.9 :: function(g2, e1).   0.5 :: function(g3, e1).
```

With probability 0.8, genes **g1** and **g2** are orthologs, with probability 0.6, the enzymatic function of **g1** is **e1**, and so forth. One can now add background knowledge to the program to define more complex relations. For instance,

```
edge(X, Y) :- ortholog(X, Y).
edge(X, Y) :- function(X, Y).
connected(X, Y) :- edge(X, Y).
connected(X, Y) :- edge(X, Z), connected(Z, Y).
```

defines a simple general path relation in terms of the edges present in the network, whereas

```
connected_via_ortholog(X, Y) :- ortholog(X, Z), function(Z, Y).
```

defines a specific type of connection from a gene via an ortholog gene to an enzymatic function.

More formally, a *ProbLog program* T consists of a set of labeled facts $p_i :: f_i$ together with a set of definite clauses encoding *background knowledge* (BK).[2] Each ground instance of such a fact f_i is true with probability p_i, that is, corresponds to a random variable with probability p_i. All random variables are assumed

[1] We use standard Prolog notation, that is, arguments starting with lower case letters are constants, those starting with upper case letters are variables, and a definite clause $\mathtt{h} :- \mathtt{b_1}, \ldots, \mathtt{b_n}$ is read as "if the $\mathtt{b_i}$ are all true, $\mathtt{h}$ is true as well".

[2] Uncertain clauses can be modeled by adding a probabilistic fact to the clause body.

to be mutually independent. The program thus naturally defines a probability distribution

$$P^T(L) = \prod_{f_i \in L} p_i \prod_{f_i \in L^T \setminus L} (1 - p_i)$$

over logic programs $L \subseteq L^T = \{f_1, \cdots, f_n\}$. The *success probability* of query q is then defined as

$$P_s^T(q) = \sum_{L \subseteq L^T : L \cup BK \models q} P^T(L). \quad (1)$$

It thus corresponds to the probability that q is *provable* in a randomly sampled logic program.

Given the example program above, one could now ask for the probability of a connection between `g1` and `e1`, that is, for the success probability of query `connected(g1, e1)`. As enumerating all possible programs (subgraphs in the example) is infeasible in most cases, ProbLog instead calculates success probabilities using all proofs of a query. The query `connected(g1, e1)` has three proofs in our example: one direct connection, and two connections involving an additional gene each, with probabilities 0.6, $0.8 \cdot 0.9 = 0.72$ and $0.7 \cdot 0.5 = 0.35$, respectively. As there are several subgraphs that contain more than one of these connections, we cannot simply sum the probabilities of proofs. This problem is also known as the *disjoint-sum-problem* or the two-terminal network reliability problem, which is #P-complete [9]. When calculating success probabilities from proofs, one has to take care to address this problem and to remove the overlap between proofs. In the example, this could be done by explicitly stating that proofs only add information if none of the previous ones are true. That is, the second proof via `g2` only adds to the probability if the direct connection is not present, and its contribution therefore needs to be reduced to $0.8 \cdot 0.9 \cdot (1 - 0.6)$. Similarly, the third proof only adds information if neither the first nor the second are true, resulting in an overall probability of

$$\begin{aligned} P_s^T(\texttt{connected}(\texttt{g1}, \texttt{e1})) &= 0.6 + 0.8 \cdot 0.9 \cdot (1 - 0.6) && (2) \\ &+ 0.7 \cdot 0.5 \cdot (1 - 0.6) \cdot (1 - 0.8) && (3) \\ &+ 0.7 \cdot 0.5 \cdot (1 - 0.6) \cdot 0.8 \cdot (1 - 0.9) && (4) \\ &= 0.9272 \end{aligned}$$

Here, (2) lists the contributions of the first and second proof as explained above, (3) and (4) that of the third proof, split into the two possible causes for the second proof being invalidated, that is, `ortholog(g1, g2)` being false, or `ortholog(g1, g2)` being true, but `function(g2, e1)` being false.

While this disjoining approach is sound for any order of the proofs, it does not scale very well. In practice, ProbLog therefore represents all proofs of the query as a propositional formula, and then uses advanced data structures to calculate the probability of this formula; we refer to [8] for the technical details.

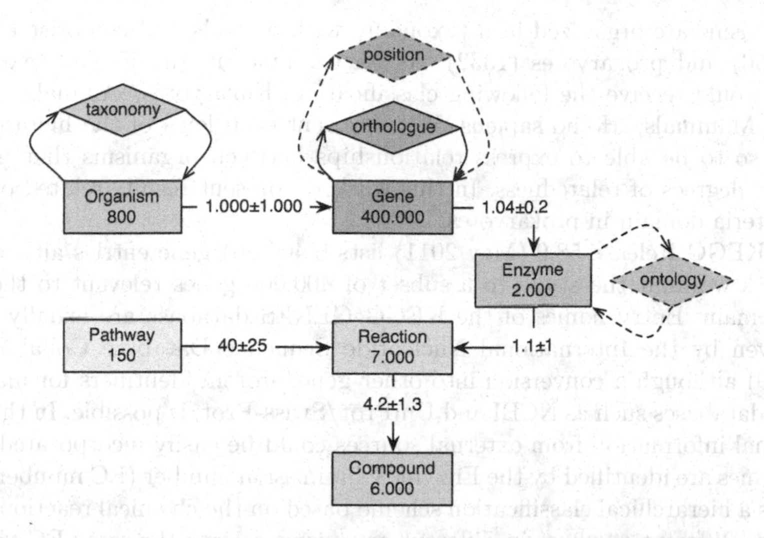

Fig. 1. Part of KEGG metabolic network used. The number in the node shape is the cardinality of the element set. The number on the edge is the average ± standard deviation number of relations between the element at the starting endpoint and the elements at the final endpoint of the edge. Dashed elements represent information present in KEGG but not currently used.

3 Method

We first discuss the information modeled in the background knowledge and then introduce the structural queries used in the prediction models.

3.1 Metabolic Network Representation

We represent the knowledge about metabolic networks in a probabilistic logical framework. To this end, we identify the main entities involved in the problem and encode all relations between them quantifying the uncertainty of each relation with an associated probability value. The entities that we consider (and that are represented as vertices in the global network) are: organisms, genes, enzymes, reactions, compounds (also called metabolites) and pathways (see Fig. 1).

Informally, a metabolic network contains information on the set of genes that belong to specific organisms and how these code for proteins, called enzymes, that are responsible for specific reactions involving the transformation of one compound into another. An organism is thus capable to perform certain related sets of reactions (semantically grouped under a single pathway concept) in order to produce and transform sets of metabolites, only if the organism can express the enzymes needed to catalyze those reactions.

We derive all data from the Kyoto Encyclopedia of Genes and Genomes (KEGG) [2].

Organisms are organized in a taxonomy with 5 levels and comprise eukaryotes (256) and prokaryotes (1332). As an example, in the KEGG taxonomy human would receive the following classification: Eukaryotes/ Animals/ Vertebrates/ Mammals/ Homo sapiens. We represent each level of the hierarchy as a node so to be able to express relationships between organisms that involve different degrees of relatedness. In this work we present results related only to the bacteria domain in prokaryotes.

The KEGG Release 58.0 (May 2011) lists 6,405,661 gene entries although in this work we limit the study to a subset of 400,000 genes relevant to the bacteria domain. Entry names of the KEGG GENES database are usually *locus-tags* given by the International Nucleotide Sequence Database Collaboration (INSDC) although a conversion into other gene/protein identifiers for main sequence databases such as NCBI and UniProt/Swiss-Prot, is possible. In this way additional information from external sources could be easily incorporated.

Enzymes are identified by the Enzyme Commission number (EC number) [10], which is a hierarchical classification scheme based on the chemical reactions they catalyze. Different enzymes in different organisms receive the same EC number if they catalyze the same reaction. Every enzyme code consists of the letters "EC" followed by four numbers separated by periods. Those numbers represent a progressively finer classification of the enzyme and induce a functional hierarchy. For example, the tripeptide aminopeptidases have the code "EC 3.4.11.4", whose components indicate the following groups of enzymes: EC 3 enzymes are hydrolases (enzymes that use water to break up some other molecule); EC 3.4 are hydrolases that act on peptide bonds; EC 3.4.11 are those hydrolases that cleave off the amino-terminal amino acid from a polypeptide; EC 3.4.11.4 are those that cleave off the amino-terminal end from a tripeptide.

The compounds involved in the metabolic transformations are a collection of small molecules, biopolymers, and other chemical substances that are relevant to biological systems. We consider 6000 unique compounds.

Enzyme mediated reactions between specific compounds are uniquely identified. The compounds involved in the reaction are distinguished into substrates and products. Note however that the reaction is considered to be bidirectional as we do not make use of more complex (and less reliable) reaction rate information.

Finally, the concept of *pathways* is used to express and organize our knowledge on metabolic processes occurring in a cell. A pathway is a set of related chemical reactions where a principal substance is modified by a series of chemical processes. Given the many compounds ("metabolites") and co-factors that are involved, single metabolic pathways can be quite complex. Moreover the separation in pathways is induced by human knowledge rather than being defined in a natural and uncontroversial way. Finally, the metabolic output of one pathway is the input for another, which implies that all the pathways are interconnected into the global complex metabolic network (see Fig. 2[3]).

All the aforementioned entities constitute vertices in our relational representation and are connected by several types of relations: at the highest level, the

[3] Image source: `http:\\commons.wikimedia.org\wiki\File:Metabolism_790px.svg`

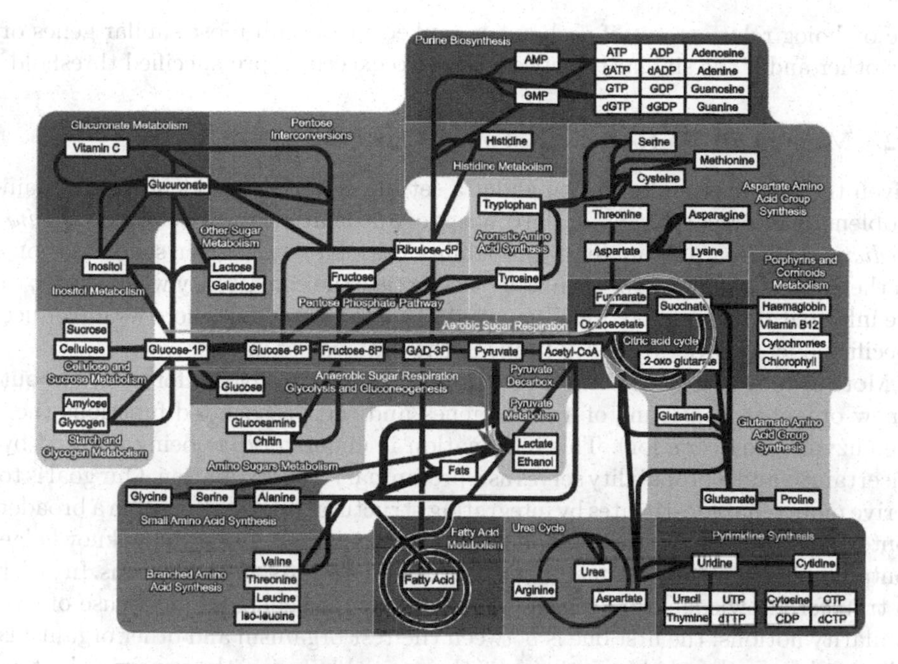

Fig. 2. Graphical representation of the major known pathways

various organisms are phylogenetically related to each other; genes are related to the organisms they are part of and they are related to each other via the ortholog relationship (see further in the text); enzymes are organized in a hierarchy following the Enzyme Commission number system; reactions are related to the compounds they require as substrate and to those they produce; genes are related to the enzymatic function of the protein that they code for; enzymes are related to the reactions they catalyze; and finally pathways are collections of related reactions. Our current model only treats the gene-enzyme relation probabilistically while all the other relations are assumed to be known with certainty. Note that in principle all relations are of the type many-to-many although in practice a gene is almost always associated to a single enzyme, which in turn catalyzes almost always a single reaction (see Fig. 1).

While the majority of these relations are intuitive, the ortholog relationship deserves some further detail. Orthologs, or orthologous genes, are genes in different species that are similar to each other because they descended from a single gene of the last common ancestor. Information about ortholog genes is available in KEGG and is obtained via a heuristic method that determines an ortholog cluster identifier in a bottom-up approach [11]. In this method, each gene subgroup is considered as a representative gene and the correspondence is computed using bi-directional best hit (BBH) relations obtained from the KEGG SSDB database which stores all-vs-all Smith-Waterman similarity scores. For efficiency reasons, the similarity score is thresholded and binarized: two genes are linked via the ortholog relation only if each one is ranked in the top most similar genes of the other and if the similarity between the two exceeds a pre-specified threshold.

3.2 Models for Automatic Network Curation

Given the metabolic information about a set of organisms we identify two main problems of interest relevant for the concept of automatic network curation: 1) *link prediction*, where we estimate the probability associated to a given set of relations on the basis of an initial guess, in order to increase the consistency with respect to the information on related organisms; and 2) *node prediction*, where we introduce specific nodes in order to best fill gaps in the pathway of interest.

More in detail, we work in the following setting. We are given information about a new organism consisting of a set of genes and their associated functions (i.e., the enzyme they code for). This information is understood as being affected by uncertainty, and a probability serves as a preliminary approximation. Our goal is to derive more reliable estimates by integrating structural information from a broader context based on this first set of probabilities. The available background knowledge contains information on the metabolic network for a large set of organisms. In order to transfer knowledge from related organisms and/or genes we make use of two similarity notions: the first one is between the test organism and other organisms (obtained from the phylogenetic tree), the second between the genes in the test organism and genes in other organisms (via the ortholog relationship).

In principle we prefer evidence that is consistent across multiple sources as noise is likely to affect each source in an uncorrelated way. In practice, it is at times hard to propagate information from multiple sources because of the partial knowledge that we have of the metabolic network. In particular: a) not all genes of a test organism have an initial associated function; b) not all genes have known orthologs; c) not all reactions are known in a given pathway.

Another source of troubles in propagating evidence is to be found in the topological properties of the reaction network itself, known as the "small world" property [12]. A network is said to exhibit a small world property if there exist paths (reaction chains) of short length that can be followed to connect any two vertices (metabolites). This apparently surprising property of real metabolic networks can be explained by the presence of so called "currency" or "commodities" compounds [13], i.e., substances that occur commonly in any chemical process and that are assumed to be present in any needed quantity at any time in the cell environment. Common examples of such substances are water and ADP. Saying that two unrelated metabolites are connected because water is present in different reactions that involve them is therefore just an artifact of the data representation that has to be dealt with in an ad-hoc way. The problem is made non-trivial by the fact that there is no consensus on how to identify these substances. In this work we make use of the flexibility offered by the ProbLog language and specify a list of "accepted" (and "forbidden") compounds that can (cannot) be part of the path definition used to propagate information. Here we create such lists based on the frequency of the compounds in different reactions but expert knowledge can be as easily incorporated.

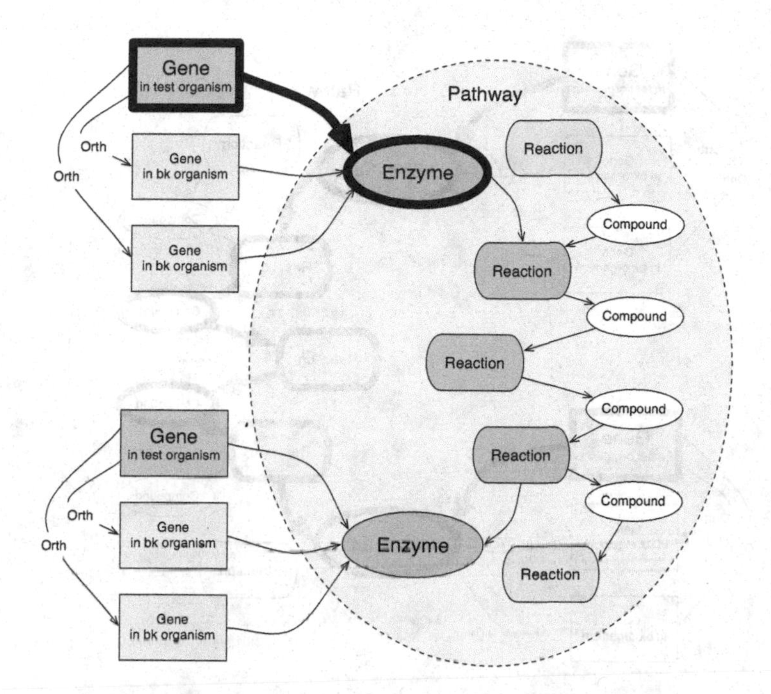

Fig. 3. Graphical representation of the portion of metabolic network used to obtain evidence for the link prediction task. The single gene-enzyme edge marked in bold corresponds to the substructures of type (1) used to obtain evidence for the link prediction task.

To summarize, the key idea of our prediction models is to use structural queries of increasing complexity to combine different forms of evidence. In the following, we discuss the queries we use for link prediction, their adaptation for node prediction, and the linear model that combines the success probabilities of the individual queries. Prediction then corresponds to a call to the ProbLog inference engine to compute the associated probability value.

Link Prediction Task. Figure 3 shows the part of the background knowledge queried to obtain support in link prediction. We use three types of queries of increasing complexity, illustrated in Figures 3, 4 and 5:

1. an estimate of the degree of belief for a gene-enzyme relation, either given a-priori or estimated by an external predictive system;
2. support coming from paths that contain the probabilistic gene-enzyme link under consideration; and
3. support coming from more complex subgraphs, that is, network portions that involve both the probabilistic gene-enzyme link and links to ortholog genes in related organisms.

For all queries, we only consider enzymes linked to a reaction in the pathway of interest. In particular, we require (2) to be a path that traverses in order the

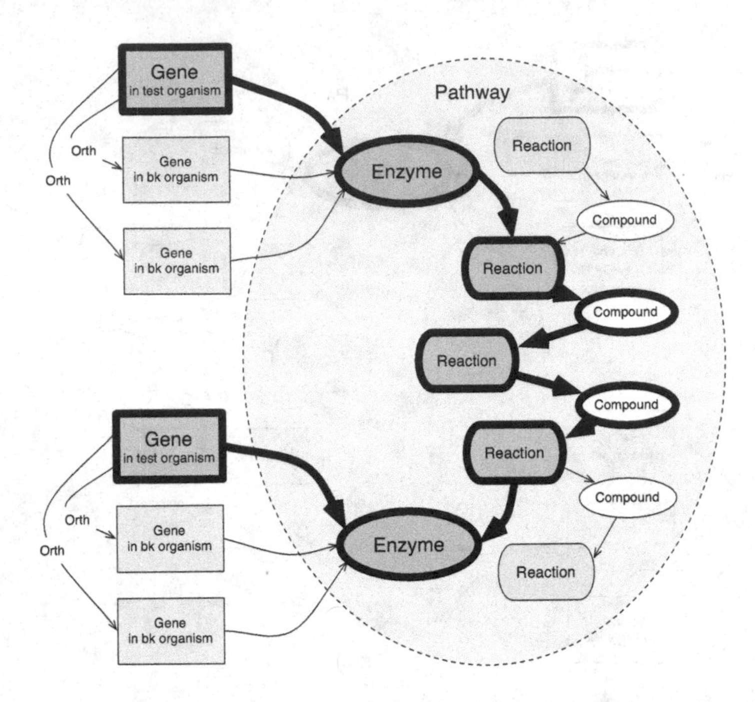

Fig. 4. Graphical representation of the substructures of type (2) used to obtain evidence for the link prediction task (marked in bold): path between two genes

following selected types of entities: gene, enzyme, reaction, compound, (reaction-compound)*, reaction, enzyme, gene. The intended meaning of the star notation here is that the path is only allowed to follow further reaction-compound links if the current reaction does not have an enzyme associated in the database. This latter condition is motivated by both computational efficiency issues (i.e., we do not consider all possible paths but only the shortest ones) and the desire to favor paths that make use of information relevant to the test organism. In words: we consider linear chains that originate in one gene of the test organism and end up in another gene of the same organism traversing the enzyme-reaction network relevant to a specific pathway. The subgraph for case (3) is obtained considering paths of type (2) with the addition of two extra paths at both ends. These provide additional links between the genes and enzymes at the end of the path via ortholog genes. The ratio here is to prefer evidence that is consistent with the information on similar genes in different organisms.

ProbLog allows us to specify the characteristics of these substructures at an intensional level. The network links are encoded using a set of (possibly probabilistic) predicates. Facts of the form `reaction_compound_reaction(r1,c,r2)` represent connections between reactions `r1` and `r2` via compound `c`. The list of compounds that may be traversed in queries is given as facts of the form `accept_compound(c)`. `ortholog(g1,g2)` facts list pairs of ortholog genes `g1` and `g2`, whereas `function(g,e)` facts link genes `g` to their enzymatic functions `e`.

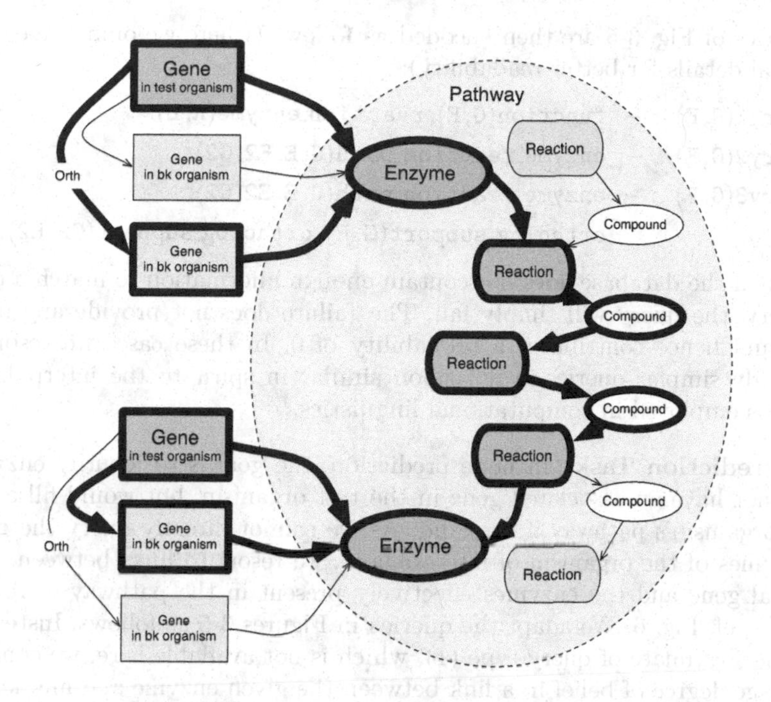

Fig. 5. Graphical representation of the substructures of type (3) used to obtain evidence for the link prediction task (marked in bold): subgraph involving ortholog genes

Finally, `reaction_enzyme(r,e)` facts connect reactions `r` in the background network to enzymes `e`. The background knowledge then defines additional relations and subgraph structures.

```
reaction_reaction(R1,R2) :- reaction_compound_reaction(R1,C,R2),
                            accept_compound(C).
```

restricts the reaction network to those links connected via accepted compounds as defined by the user.

```
enzyme_reaction_path(G1,E1,E2,G2) :- function(G1,E1),
                                     reaction_enzyme(R1,E1),
                                     reaction_reaction(R1,R2),
                                     reaction_enzyme(R2,E2),
                                     function(G2,E2).
```

corresponds to the second query (modulo the star part), but making the gene and enzyme at the other end explicit, which is used in the third query to extend the query towards ortholog genes using

```
ortholog_support(G,E) :- ortholog(G,G2),function(G2,E).
```

The queries of Fig. 3-5 are then encoded as follows (where we omit some computational details for better readability):

```
query1(G,E) :- function(G,E),reaction_enzyme(R,E).
query2(G,E) :- enzyme_reaction_path(G,E,E2,G2).
query3(G,E) :- enzyme_reaction_path(G,E,E2,G2),
               ortholog_support(G,E),ortholog_support(G2,E2).
```

Note that if the database does not contain enough information to match a complex query, the query will simply fail. The failure does not provide any information and hence contributes a probability of 0. In these cases we resort to increasingly simpler queries in a fashion similar in spirit to the interpolation techniques employed in computational linguistics.[4]

Node Prediction Task. In node prediction, the goal is to identify enzymes that do not have an associated gene in the test organism, but would fill a hole in that organism's pathway if they did. As we cannot directly query the genes and enzymes of the organism of interest here, we resort to links between a hypothetical gene and the enzymes effectively present in the pathway of related organisms, cf. Fig. 6. We adapt the queries in Figures 3-5 as follows. Instead of the a-priori estimate of query type (1), which is not available here, we consider the average degree of belief in a link between the given enzyme and any known gene present in related organisms. For queries of types (2) and (3), we replace the test organism's gene at the top by a gene in some other related organism, but still require the path to end in a gene that is known to belong to the test organism.

Model. In both the link and node prediction setting, we estimate degrees of belief for our target relation by calculating the success probability (cf. Equation (1)) for each of the three types of supporting queries in the given model. We combine those results to answer the two main questions: 1) what is the probability of a specific gene of a test organism to be associated to a specific enzyme in the pathway? and 2) what is the probability of some unknown gene of a test organism to be associated to a specific enzyme in the pathway?

The combination is done via a linear model whose weights encode the reliability for each type of query.[5] Let $Q_i(G,E)$ be the success probability of the query of type i that relates the gene G with the enzyme E. The probability $p(G,E)$ that the gene effectively encodes the function E is computed as a convex combination of the success probability of each type of query, that is:

$$p(G,E) = \sum_{i=1,2,3} w_i(E) Q_i(G,E)$$

[4] When employing *n-gram* models, a common practice is to assess the probability of complex n-grams using the frequency counts of smaller n-grams that are more likely to occur in (small) datasets.

[5] Technically, the linear model is itself encoded as a ProbLog query and inference thus done in a single step without obtaining the individual success probabilities.

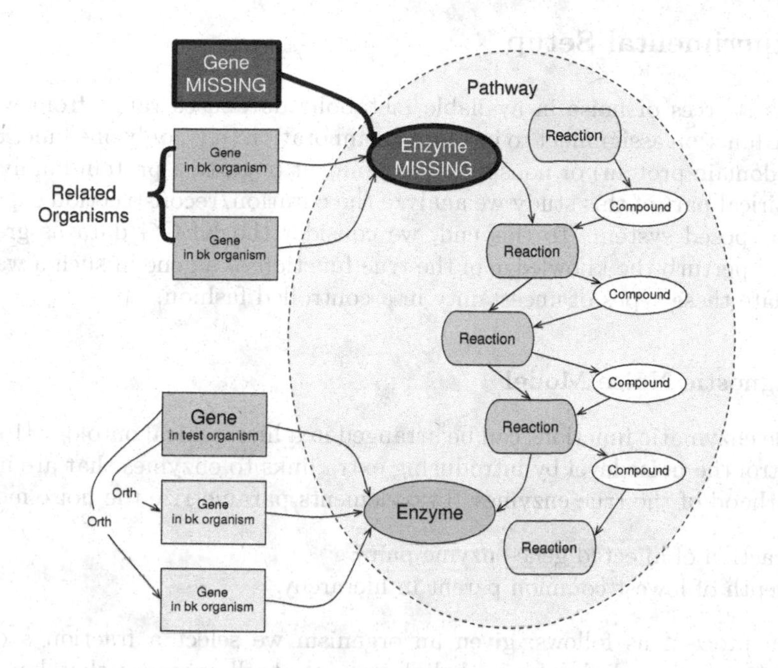

Fig. 6. Graphical representation of the portion of metabolic network used to obtain evidence for the node prediction task

where for each enzyme E, $\sum_{i=1,2,3} w_i(E) = 1$. We consider two variants of this model: one with enzyme-specific weights $w_i(E)$, and a global model that uses identical $w_i(E)$ for all enzymes.

The idea behind the linear model is to adapt to the level of missing information in the network: when assessing the degree of belief for an enzyme that is embedded in a network region where few reactions are known, it is better to trust the prior estimate with respect to more complex queries since they will mainly fail over the poorly connected reaction network; analogously when ortholog genes are known for a given enzyme, the evidence from the more complex queries becomes compelling. In summary, we adapt to the unknown local quality of the network by estimating the relative reliability of each query for the final answer on related organisms known in a background knowledge base.

In this work we explore two ways to induce the weights:

Frequency estimation: for each query type and enzyme, we count the number of proofs obtained for both positive and negative examples and obtain first estimates as $p/(p+n)$; these are then normalized over the three query types. Parameters for the global model, which does not model the dependency on enzymes, are obtained by summing counts over all enzymes before calculating frequencies.

Machine learning estimation: the weights are learned with ProbLog's gradient-descent approach to parameter learning [14]. Given a set of queries with associated target probabilities, this method uses standard gradient descent to minimize the *mean squared error* (MSE) on the training data.

4 Experimental Setup

Common sources of noise in available metabolic databases range from wrong catalytic function assignment to incomplete annotation (e.g., only one function of a multi-domain protein) or nonspecific assignment (e.g., to a protein family). In the empirical part of this study we analyze the curation/reconstruction capacity of the proposed system. To this end, we consider the KEGG data as ground truth and perturb the knowledge of the true function of a gene in such a way as to simulate these types of uncertainty in a controlled fashion.

4.1 Agnostic Noise Model

Since the enzymatic functions can be arranged in a hierarchical ontology [10], we can control the noise level by introducing extra links to enzymes that are in the neighborhood of the true enzymes. Two elements parametrize the noise model:

1. s: fraction of affected gene-enzyme pairs;
2. d: depth of lowest common parent in hierarchy.

We then proceed as follows: given an organism we select a fraction s of its known gene-enzyme links; for each link, we select all enzymes that have the lowest common parent with the link's enzyme at depth d in the hierarchy and that appear in the background knowledge network of the pathway of interest. We then introduce a uniform distribution over the set of gene-enzyme links resulting from the original gene and the selected enzymes.

General Setting. In the experiments reported here, we focus on the Pyruvate metabolism pathway (cf. Fig. 7) and organisms from subfamilies of proteobacteria, cf. Fig. 8. Pyruvate is an important intermediate in the fermentative metabolism of sugars by yeasts and is located at a major junction of assimilatory and dissimilatory reactions as well as at the branch-point between respiratory dissimilation of sugars and alcoholic fermentation.

A total of 40 organisms are picked uniformly at random, ensuring that all organisms of the smallest three subfamilies are included. For each such organism, we construct the background knowledge network by superimposing the networks of all organisms of the other five subfamilies, thus leaving out the most closely related organisms.

We create six different noise settings by perturbing the true relationships for $s = 1/5/10\%$ of gene-enzyme links, using $d = 2$ and $d = 3$, and use the linear model to rank candidate instances of the target relationship in each setting. For efficiency reasons, the linear model parameters are computed using the simple frequency estimate.

Experimental Results: Link Prediction. In the link prediction setting, positive examples are the test organism's real gene-enzyme links, while negative ones are the ones added by the noise model. The linear model uses the three queries

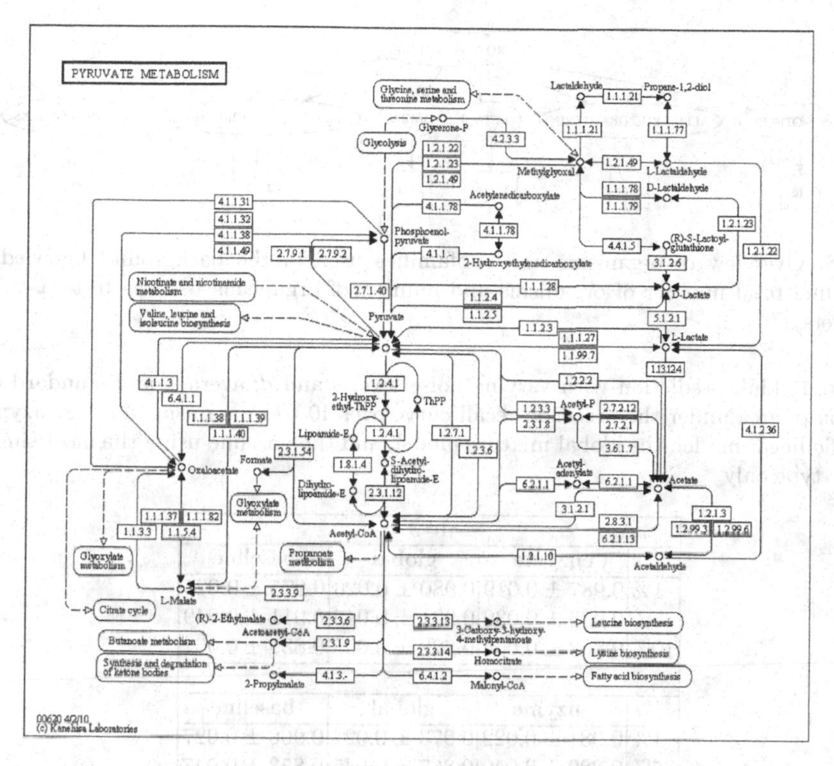

Fig. 7. Pyruvate metabolism pathway

depicted in Fig. 3-5. As the data is unbalanced, we report the area under the precision-recall curve as a performance measure. Results are summarized in Table 1 for the enzyme-specific linear model, the global mixture model, and the baseline using the most simple query type only. With increasing noise levels, the enzyme-based mixture model clearly improves over the baseline that does not take into account background information, and also over the less flexible global mixture model.

Experimental Results: Node Prediction. In the node prediction setting, examples are pairs of organisms and enzymes from the background knowledge. If the enzyme occurs in the organism's network, such an example is considered positive, and negative otherwise. We adopt an enzyme level leave-one-out design among those enzymes in the background knowledge that are not associated to any gene in the test organism. We remove these enzymes in turn and we measure the precision at one, that is, the fraction of times that the missing enzyme is ranked in first position as the most probable among all the missing enzymes.

The linear model uses the queries described in Section 3. Results are summarized in Table 2. While both mixture models significantly improve over the random ranking of all background enzymes, there is no significant difference between the global model (which doesn't take into account enzyme-specific in-

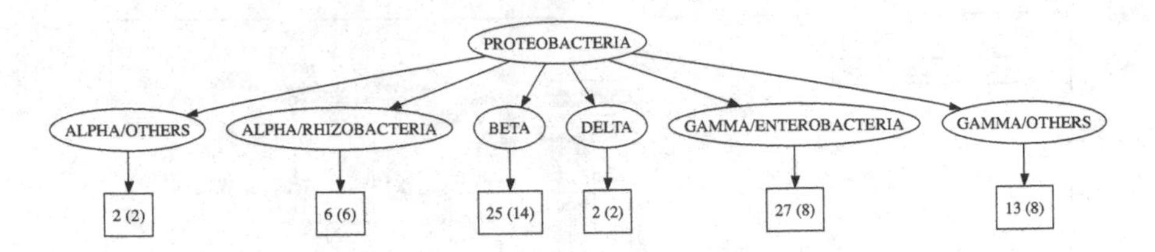

Fig. 8. Overview of organisms and subfamilies used in the background knowledge, including total number of organisms and number of organisms used as test cases (in brackets)

Table 1. Link prediction with varying noise level s and d: average and standard deviation of area under the precision-recall curve over 40 test organisms for the enzyme-specific linear model, the global mixture model, and the baseline using the most simple query type only

d=3			
s	enzyme	global	baseline
1%	0.987 ± 0.019	0.980 ± 0.026	0.975 ± 0.025
5%	0.935 ± 0.039	0.921 ± 0.045	0.911 ± 0.049
10%	0.863 ± 0.065	0.828 ± 0.068	0.831 ± 0.062
d=2			
s	enzyme	global	baseline
1%	0.981 ± 0.022	0.973 ± 0.027	0.966 ± 0.027
5%	0.889 ± 0.040	0.867 ± 0.045	0.853 ± 0.047
10%	0.775 ± 0.059	0.721 ± 0.064	0.743 ± 0.052

Table 2. Node prediction with varying noise level s and d: average and standard deviation of precision at one over 40 test organisms for the enzyme-specific linear model, the global mixture model, and the baseline using a random ranking

d=3			
s	enzyme	global	baseline
1%	0.218 ± 0.111	0.271 ± 0.143	0.020 ± 0.000
5%	0.217 ± 0.091	0.340 ± 0.124	0.020 ± 0.000
10%	0.198 ± 0.082	0.325 ± 0.144	0.020 ± 0.000
d=2			
s	enzyme	global	baseline
1%	0.223 ± 0.107	0.290 ± 0.165	0.020 ± 0.000
5%	0.224 ± 0.045	0.386 ± 0.081	0.019 ± 0.005
10%	0.152 ± 0.035	0.262 ± 0.080	0.011 ± 0.010

formation) and the enzyme-specific model. We conjecture that averaging the performance over "easy" and "hard" to predict enzymes yields a too coarse result and that a more detailed analysis is needed to identify the conditions that favour the enzyme-specific vs. the global model.

4.2 Noise Model for Unreliable Predictions

In this scenario, we assume that a predictor (i.e., a machine learning algorithm) is available and that it can compute the enzymatic function of a gene with a certain reliability. Instead of working with a specific predictor here we perturb the knowledge of the true function of a gene in order to simulate different degrees of reliability. Once again we make use of the fact that the enzymatic functions can be arranged in a hierarchical ontology [10]. Under this assumption we relate the topological distance in the ontology tree to the functional distance, i.e., the closer two enzyme nodes are in the hierarchy the more similar their functions. Under this assumption we build a noise model described by the following parameters:

1. s: fraction of affected genes;
2. k: number of noisy gene-enzyme links added per gene;
3. σ_{EC}: parameter controlling the size of the neighborhood where to randomly sample the additional noisy gene-enzyme links;
4. σ_N: parameter controlling the quantity of noise added to the gene-enzyme relationship probability estimate.

We then proceed as follows (see Fig. 9). Given an organism, we select a fraction s of its genes. For each selected gene, we add k extra links to randomly sampled *nearby* enzymes. Sampling selects enzymes using a normal distribution $N(0, \sigma_{EC})$ over their topological distance induced by the ontology, i.e., the length of the shortest path between the leafs containing the actual and the sampled enzyme in the tree structured ontology. Finally, we obtain the degree of belief for the link between the gene and the randomly selected enzyme as the probability of selecting the enzyme plus additional $N(0, \sigma_N)$ noise. In this way enzymes that are less related to (i.e., more distant from) the true enzymatic function of the original gene receive on average a smaller probability.

Experimental Results. In the experiments reported here, we focus on the Pyruvate metabolism pathway for the Escherichia coli UTI89 test organism. We perturb the true relationships with k=5 extra links for s = 50% of genes. The probability estimate of the gene-enzyme relationship receives additional noise from $N(0, \frac{1}{8})$.

The linear model parameters are computed using ProbLog's gradient-descent approach to parameter learning [14]. We use default settings in our experiments and run learning for at most 50 iterations, stopping earlier if the MSE on the training data does not change between two successive iterations. Training data is generated from the other organisms with the same parent in the organism hierarchy as the test organism, and target probabilities are set to 1.0 for positive and 0.0 for negative examples, respectively.

In the link prediction setting, positive examples are real gene-enzyme links, while negative ones are the ones added by the noise model where no real one is known between these entities. We use the three queries depicted in Fig. 3-5. We measure the area under the precision-recall curve.

When using the initial (perturbed) estimate for the gene-enzyme link we achieve an AUCPR of 0.69. If we use only the most complex query (type (3))

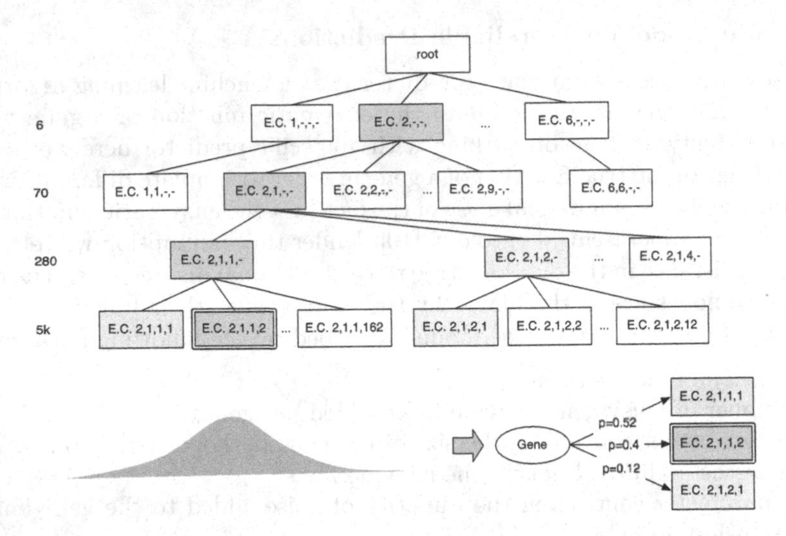

Fig. 9. Noise model: the E.C. hierarchy induced metric notion (i.e., topological distance between nodes) is used for the perturbed enzymatic function. The hypothetical true enzyme is marked with a double line. In the example a gene is associated to an incorrect enzymatic activity with probability 0.52 and to the correct one with probability 0.4.

we increase to 0.74, but when we learn the linear model over all queries we achieve 0.80. Note that simply learning a fixed mixture of experts for the whole organism (i.e., not modeling the dependency on the enzyme) we do not improve over the initial 0.69 result, as for this particular test organism, it is better to resort on average to the most simple query.

In the node prediction experiment, we follow the same scheme as above. That is, we adopt an enzyme level leave-one-out design among those enzymes in the background knowledge that are not associated to any gene in the test organism, remove these enzymes in turn and measure the precision at one.

The set of training examples is the set of all pairs of training organisms (as before) and enzymes appearing in the pathway for organisms different from the test organism. Such a pair is considered positive if the enzyme appears in the organism's pathway, and negative else.

We use the query described in Section 3 both with and without ortholog information, as well as a basic query that predicts each enzyme with the average probability of a gene-enzyme link involving this enzyme in one of the training organisms. In this experiment we achieve a precision at one of 0.66 over 35 possible enzymes (i.e., the baseline random guessing precision at one would be 0.03).

5 Conclusions

We have started tackling the problem of automatic network curation by employing the ProbLog probabilistic logic framework. To overcome the limitations of homology searches, we have made use of information from heterogeneous sources, encoding all available data into a large BisoNet. To leverage the different quantity and quality of information available for different organisms, we have used a case-based approach linking information on related organisms. The use of a probabilistic logic framework has allowed us to: a) represent the knowledge about the metabolic network even when affected by uncertainty, and b) express complex queries to extract support for the presence of missing links or missing nodes in an abstract and flexible way. Initial experimental evidence shows that we can partially recover missing information and correct inconsistent information. Future work includes the integration of gene function predictor and the development of novel queries that make use of additional sources of information such as the gene position in the genome or the co-expression of genes in the same pathway from medical literature abstract analysis.

Acknowledgments. A. Kimmig is supported by the Research Foundation Flanders (FWO Vlaanderen). F. Costa was supported by the GOA project 2008/08 Probabilistic Logic Learning and by the European Commission under the 7th Framework Programme, contract no. BISON-211898.

References

1. Alm, E., Arkin, A.: Biological networks. Current Opinion in Structural Biology (13), 193–202 (January 2003)
2. Kanehisa, M., Goto, S., Furumichi, M., Tanabe, M., Hirakawa, M.: KEGG for representation and analysis of molecular networks involving diseases and drugs. Nucleic Acids Research 38(Database issue), D355–D360 (2010)
3. Karp, P., Ouzounis, C., Moore-Kochlacs, C., Goldovsky, L., Kaipa, P., Ahren, D., Tsoka, S., Darzentas, N., Kunin, V., Lopez-Bigas, N.: Expansion of the BioCyc collection of pathway/genome databases to 160 genomes. Nucleic Acids Research 19, 6083–6089 (2005)
4. Pouliot, Y., Karp, P.: A survey of orphan enzyme activities. BMC Bioinformatics 8(1), 244 (2007)
5. Kötter, T., Berthold, M.R.: From Information Networks to Bisociative Information Networks. In: Berthold, M.R. (ed.) Bisociative Knowledge Discovery. LNCS (LNAI), vol. 7250, pp. 33–50. Springer, Heidelberg (2012)
6. Moriya, Y., Itoh, M., Okuda, S., Yoshizawa, A., Kanehisa, M.: KAAS: an automatic genome annotation and pathway reconstruction server. Nucleic Acids Research 35(Web Server issue), W182–W185 (2007)
7. Green, M., Karp, P.: A Bayesian method for identifying missing enzymes in predicted metabolic pathway databases. BMC Bioinformatics 5(1), 76 (2004)
8. Kimmig, A., Demoen, B., De Raedt, L., Santos Costa, V., Rocha, R.: On the implementation of the probabilistic logic programming language ProbLog. Theory and Practice of Logic Programming (TPLP) 11, 235–262 (2011)

9. Valiant, L.G.: The complexity of enumeration and reliability problems. SIAM Journal on Computing 8(3), 410–421 (1979)
10. Webb, E.C.: Enzyme nomenclature 1992: recommendations of the Nomenclature Committee of the International Union of Biochemistry and Molecular Biology on the nomenclature and classification of enzymes. Published for the International Union of Biochemistry and Molecular Biology by Academic Press, San Diego (1992)
11. Moriya, Y., Katayama, T., Nakaya, A., Itoh, M., Yoshizawa, A., Okuda, S., Kanehisa, M.: Automatic generation of KEGG OC (Ortholog Cluster) and its assignment to draft genomes. In: International Conference on Genome Informatics (2004)
12. Wagner, A., Fell, D.A.: The small world inside large metabolic networks. Proceedings of the Royal Society B: Biological Sciences 268(1478), 1803–1810 (2001)
13. Huss, M., Holme, P.: Currency and commodity metabolites: their identification and relation to the modularity of metabolic networks. IET Systems Biology 1, 280–285 (2007)
14. Gutmann, B., Kimmig, A., Kersting, K., De Raedt, L.: Parameter Learning in Probabilistic Databases: A Least Squares Approach. In: Daelemans, W., Goethals, B., Morik, K. (eds.) ECML PKDD 2008, Part I. LNCS (LNAI), vol. 5211, pp. 473–488. Springer, Heidelberg (2008)

Modelling a Biological System: Network Creation by Triplet Extraction from Biological Literature

Dragana Miljkovic[1,*], Vid Podpečan[1], Miha Grčar[1], Kristina Gruden[3], Tjaša Stare[3], Marko Petek[3], Igor Mozetič[1], and Nada Lavrač[1,2]

[1] Jožef Stefan Institute, Ljubljana, Slovenia
{dragana.miljkovic,vid.podpecan,miha.grcar}@ijs.si,
{igor.mozetic,nada.lavrac}@ijs.si
[2] University of Nova Gorica, Nova Gorica, Slovenia
[3] National Institute of Biology, Ljubljana, Slovenia
{kristina.gruden,tjasa.stare,marko.petek}@nib.si

Abstract. The chapter proposes an approach to support modelling of plant defence response to pathogen attacks. Such models are currently built manually from expert knowledge, experimental results, and literature search, which is a very time consuming process. Manual model construction can be effectively complemented by automated model extraction from biological literature. This work focuses on the construction of triplets in the form of subject-predicate-object extracted from scientific papers, which are used by the Biomine automated graph construction and visualisation engine to create the biological model. The approach was evaluated by comparing the automatically generated graph with a manually developed Petri net model of plant defence. This approach to automated model creation was explored also in a bisociative setting. The emphasis is not on creative knowledge discovery, but rather on specifying and crossing the boundaries of knowledge of individual scientists. This could be used to model the expertise of virtual scientific consortia.

1 Introduction

The mechanism of a plant's defence response to virus attacks is a hot topic of current biological research. Despite a vivid interest in creating a holistic model of plant defence, only partial and oversimplified models of the entire defence system are created so far.

The motivation of biologists to develop a more comprehensive model of the entire defence response is twofold. Firstly, it will provide a better understanding of the complex defence response mechanism in plants by highlighting important connections between biological molecules and understanding how the mechanism operates. Secondly, prediction of experimental results through simulation saves time and indicates further research directions to biological scientists. The development of a more comprehensive model of plant defence for simulation purposes raises three research questions:

* Corresponding author.

M.R. Berthold (Ed.): Bisociative Knowledge Discovery, LNAI 7250, pp. 427–437, 2012.

- What is the most appropriate formalism for representing the plant defence model?
- How to extract the model, i.e. network structure; more precisely, how to retrieve relevant molecules and relations between them?
- How to determine network parameters such as initial molecules values, types and speeds of the reactions, threshold values, etc.?

Having studied different representation formalisms, we have decided to represent the model of the given biological network in the form of a graph. This chapter addresses the second research question, i.e. automated extraction of the graph structure through information retrieval and natural language processing techniques. We propose a methodology to support modelling of plant defence response to pathogen attacks, and present its implementation in a workflow which combines open source natural language processing tools, data from publicly available databases, and hand-crafted knowledge. The evaluation of the approach is carried out using a manually crafted Petri net model which was developed by fusing expert knowledge and the results of manual literature search.

The structure of the chapter is as follows. Section 2 presents existing approaches to modelling plant defence and discusses their advantages and shortcomings. Section 3 briefly presents our manually crafted Petri net model, followed by Section 4 which proposes a methodology used for automated model extraction from the biological literature. Section 5 explores the results of model extraction in a bisociative setting, where extracted knowledge of different scientists is combined. Section 6 concludes the chapter and proposes directions for further work.

2 Related Work

Due to the complexity of the plant defence mechanism, the challenge of building a general model for simulation purposes is still not fully addressed. Early attempts to accomplish simulation by means of Boolean formalism from experimental microarray data [5] have already indicated the complexity of defence response mechanisms, and highlighted many crosstalk connections. However, several of the interconnecting molecules were not considered in the model presented in that work. These intermediate molecules as well as the discovery of new connections between them are of particular interest for biological scientists.

Other existing approaches, such as the MoVisPP tool [6], attempt to automatically retrieve information from databases and transfer the pathways into the Petri net formalism. MoVisPP is an online tool which automatically produces Petri net models from KEGG and BRENDA pathways. However, not all pathways are accessible, and the signalling pathways for plant defence do not exist in the databases.

Tools for data extraction and graphical representation are also related to our work as they are used to help experts to understand the underlying biological principles. They can be roughly grouped according to their information sources: databases (Biomine [14][4], Cytoscape [15], ProteoLens [8], VisAnt [7], PATIKA [2]), databases and experimental data (ONDEX [9], BiologicalNetworks [1]), and literature (TexFlame [10]). More general approaches, such as visualisation of arbitrary textual data through triplets [13] are also relevant. However, such general systems have to be adapted in order to produce domain-specific models.

3 Manually Constructed Petri Net Model of Plant Defence Response

This section presents a part of the manually crafted Petri net model using the Cell Illustrator software [12]. We briefly describe the development cycle of the model and show some simulation results.

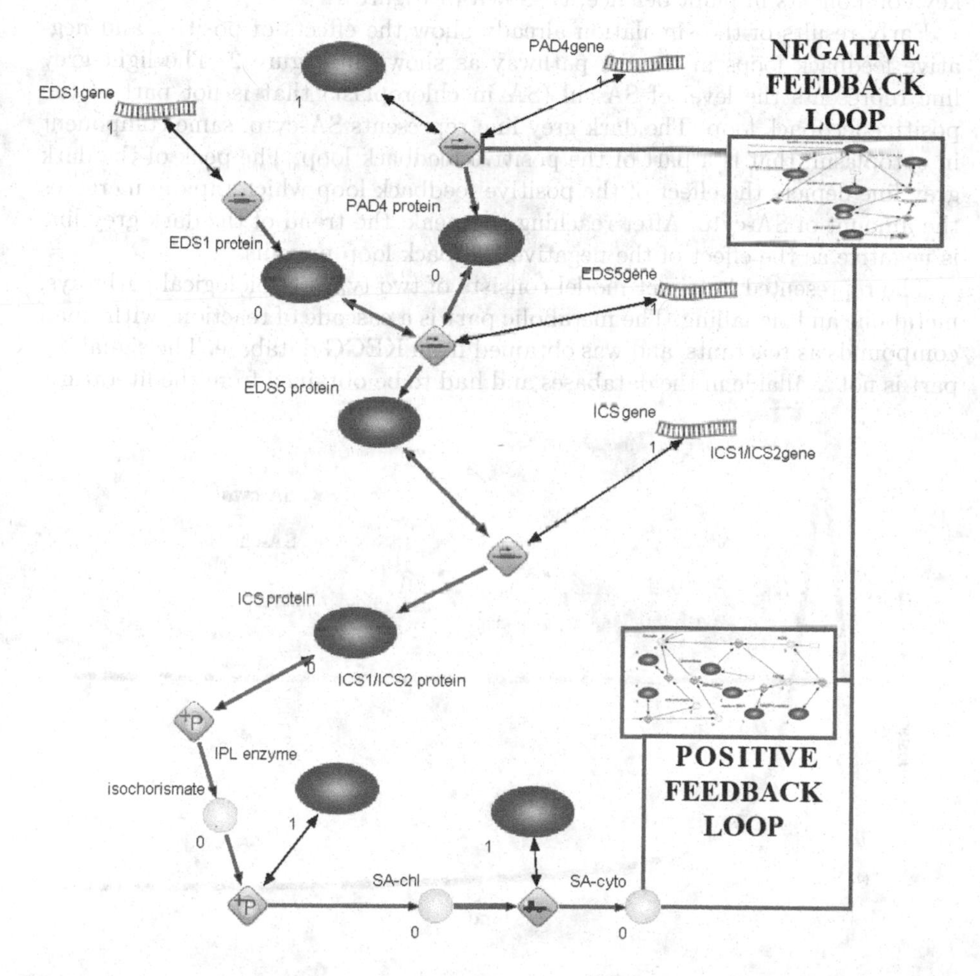

Fig. 1. A partial and simplified Petri net model of SA biosynthesis and signalling pathway in plants. Biological molecules SA-chl and SA-cyto represent SA located in different parts of the cell. Both SA-chl and SA-cyto of this figure correspond to node SA in the graph of Figure 4.

A Petri net is a bipartite graph with two types of nodes: places and transitions. Standard Petri net models are discrete in terms of variables and sequence of events, but their various extensions can represent both qualitative and quantitative models. The Cell Illustrator software implements an extension of Petri nets, called hybrid functional Petri net, which was used in our study. In the hybrid functional Petri net formalism, the speed of transitions depends on the amount of input components and both, discrete and continuous places, are supported.

Our manually crafted Petri net model of plant defence currently represents a complex network where molecules and reactions, according to the Petri net formalism, correspond to places and transitions, respectively. A part of the model of salicylic acid (SA) biosynthesis and signalling pathway, which is one of the key components in plant defence, is shown in Figure 1.

Early results of the simulation already show the effects of positive and negative feedback loops in the SA pathway as shown in Figure 2. The light grey line represents the level of SA-chl (SA in chloroplast) that is not part of the positive feedback loop. The dark grey line represents SA-cyto, same component in cytoplasm, that is a part of the positive feedback loop. The peak of the dark grey line depicts the effect of the positive feedback loop which rapidly increases the amount of SA-cyto. After reaching the peak, the trend of the dark grey line is negative as the effect of the negative feedback loop prevails.

The represented Petri net model consists of two types of biological pathways: metabolic and signalling. The metabolic part is a cascade of reactions with small compounds as reactants, and was obtained from KEGG database. The signalling part is not available in the databases and had to be obtained from the literature.

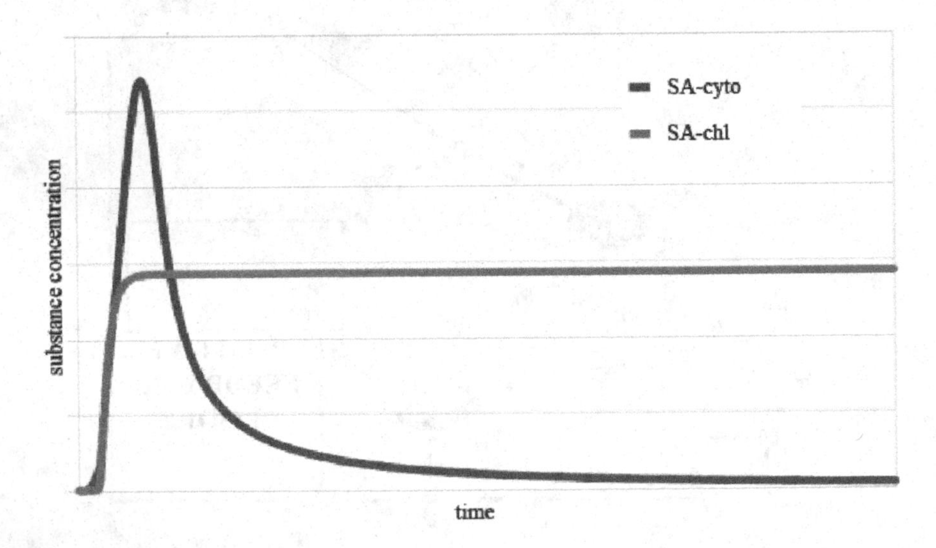

Fig. 2. Simulation results of the Petri net model of SA pathway. The light grey line represents the level of SA-chl, i.e. SA in chloroplast that is not part of the positive feedback loop. The dark grey line represents the same component in cytoplasm, SA-cyto, which is in the positive feedback loop.

The biological scientists have manually extracted relevant information related to this pathway within a period of approximately two months. Keeping in mind that the SA pathway is only one out of three pathways involved in plant defence response, it is clear that a purely manual approach would be very time-consuming.

4 Automated Extraction of Plant Defence Response Model from Biological Literature

The process of fusing expert knowledge and manually obtained information from the literature as presented in the previous section turns out to be time consuming and error-prone. Therefore, we suggest the automated extraction of information from the scientific literature relevant to the construction and curation of such models. The proposed methodology consists of a series of text mining and information retrieval steps, which offer reusability and repeatability, and can be easily extended with additional components. For natural language processing we employed functions from the NLTK library [11], which were transformed into web services and used in the proposed triplet extraction, graph construction, visualisation and exploration workflow. Additionally, the GENIA tagger [16] for biological domains was used to perform part-of-speech tagging and shallow parsing. The data was extracted from PubMed[1] and Web of Science[2] using web-service enabled access.

The methodology for information retrieval from public databases to support modelling of plant defence is shown in Figure 3. Computer-assisted creation of plant defence models from textual data, is performed by using following services:

1. PubMed web service and Web of Science search to extract the article data,
2. PDF-to-text converter service, which is based on Poppler[3], an open source PDF rendering library,
3. natural language processing web services based on NLTK: tokenizer, shallow parser (chunker), sentence splitter,
4. the GENIA tagger,
5. filtering components, e.g. negation removal, etc.

The goal of this study is to extract sets of triplet in the form:

$$(Subject, Predicate, Object)$$

from biological texts which are freely available. The defence response related information is obtained by employing the vocabulary which we have manually developed for this specific field. *Subject* and *Object* are biological molecules such as proteins or small compounds, and their names with synonyms are built

[1] PubMed is a free database that comprises biomedical journals and online books.

[2] Web of Science is an online academic citation index constructed to access multiple databases and explore cross-disciplinary studies and specialized subfields within a scientific domain.

[3] `http://poppler.freedesktop.org/`

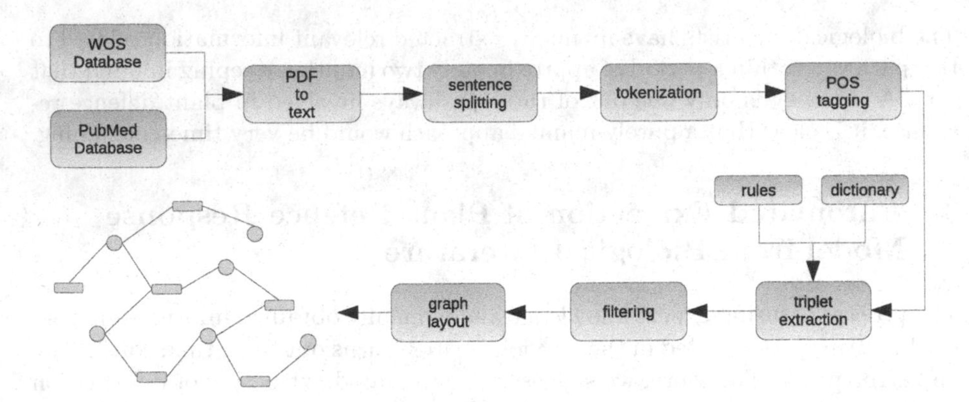

Fig. 3. Methodology for information retrieval from public databases to support modelling of plant defence response

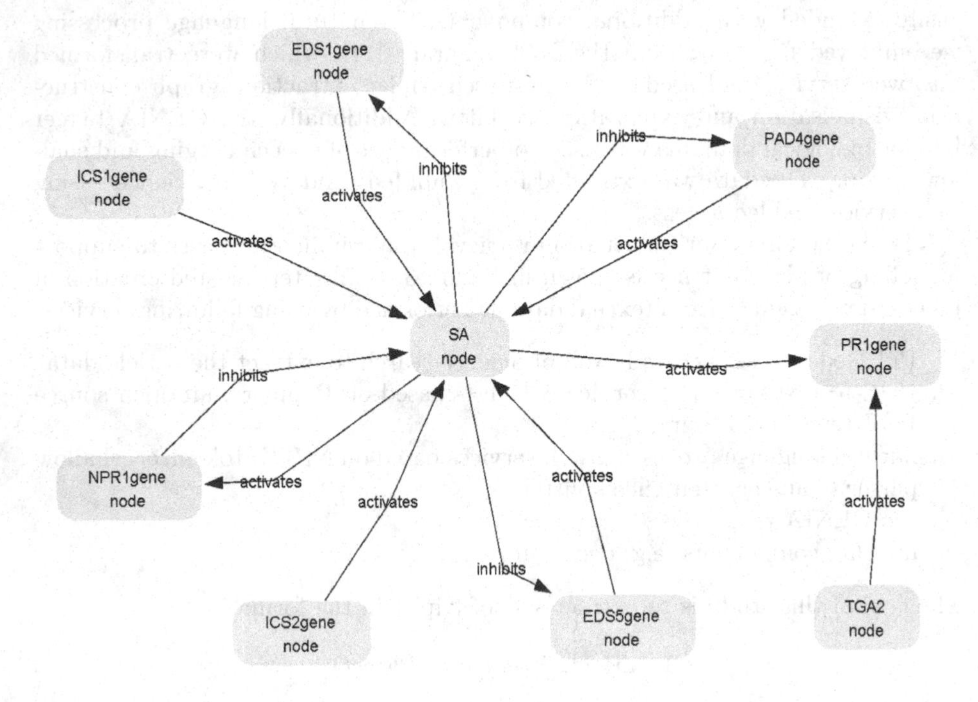

Fig. 4. A graph constructed from a set of triplets, extracted from ten documents, visualised using the Biomine visualisation engine

into the vocabulary. *Predicate* represents the relation or interaction between the molecules. We have defined three types of reactions, i.e. *activation*, *inhibition* and *binding*, and the synonyms for these reactions are also part of the vocabulary. An example of such a triplet is shown below:

$$(PAD4\,protein,\ activates,\ EDS5\,gene)$$

Such triplets, if automatically found in text, composed and visualised as a graph, can assist the development of the plant defence Petri net model. Triplet extraction is performed by employing simple rules to find the last noun of the first phrase as *Subject. Predicate* is a part of a verb phase located between the noun phrases. *Object* is then detected as a part of the first noun phrase after the verb phrase. In addition to these rules, pattern matching from the dictionary is performed to search for more complex phrases in text to enhance the information extraction. The graph is then constructed and visualised using the Biomine graph construction and visualisation engine [14]. An example of such a graph is shown in Figure 4.

While such automatically extracted knowledge currently cannot compete — in terms of details and correctness — with the manually crafted Petri net model, it can be used to assist the expert in the process of building and curation of the model. Also, it can provide novel and relevant information to the biological scientist.

5 Two Modelling Scenarios

5.1 An Illustrative Example

Consultation with biological scientists resulted in the first round of experiments performed on a set of ten most relevant articles from the field which were published after 2005. Figure 4 shows the extracted triplets, visualised using the Biomine graph visualiser.

SA appears to be the central component in the graph, which confirms the biological fact that SA is indeed one of the three main components in plant defence. The information contained in the graph of Figure 4 is similar to the initial knowledge obtained from biological scientists by manual information retrieval from the literature[4]. Such a graph, however, cannot provide the cascade network type which is closer to reality (and to the manually crafted Petri net model).

The feedback from the biologists was positive. Even though this approach cannot completely substitute human experts, biologists consider it a helpful tool in speeding up information acquisition from the literature. The presented results indicate the usefulness of the proposed approach but also the necessity to further improve the quality of information extraction.

5.2 Crossing the Boundaries of Individual Readers

The goal of the second experiment is to elicit differences in knowledge and interests between different scientists. We take a simplifying assumption that each

[4] It is worth noting that before the start of joint collaboration between the computer scientists and biologists, the collaborating biological scientists have tried to manually extract knowledge from scientific articles in the form of a graph, and have succeeded to build a simple graph representation of the SA biosynthesis and signalling pathway.

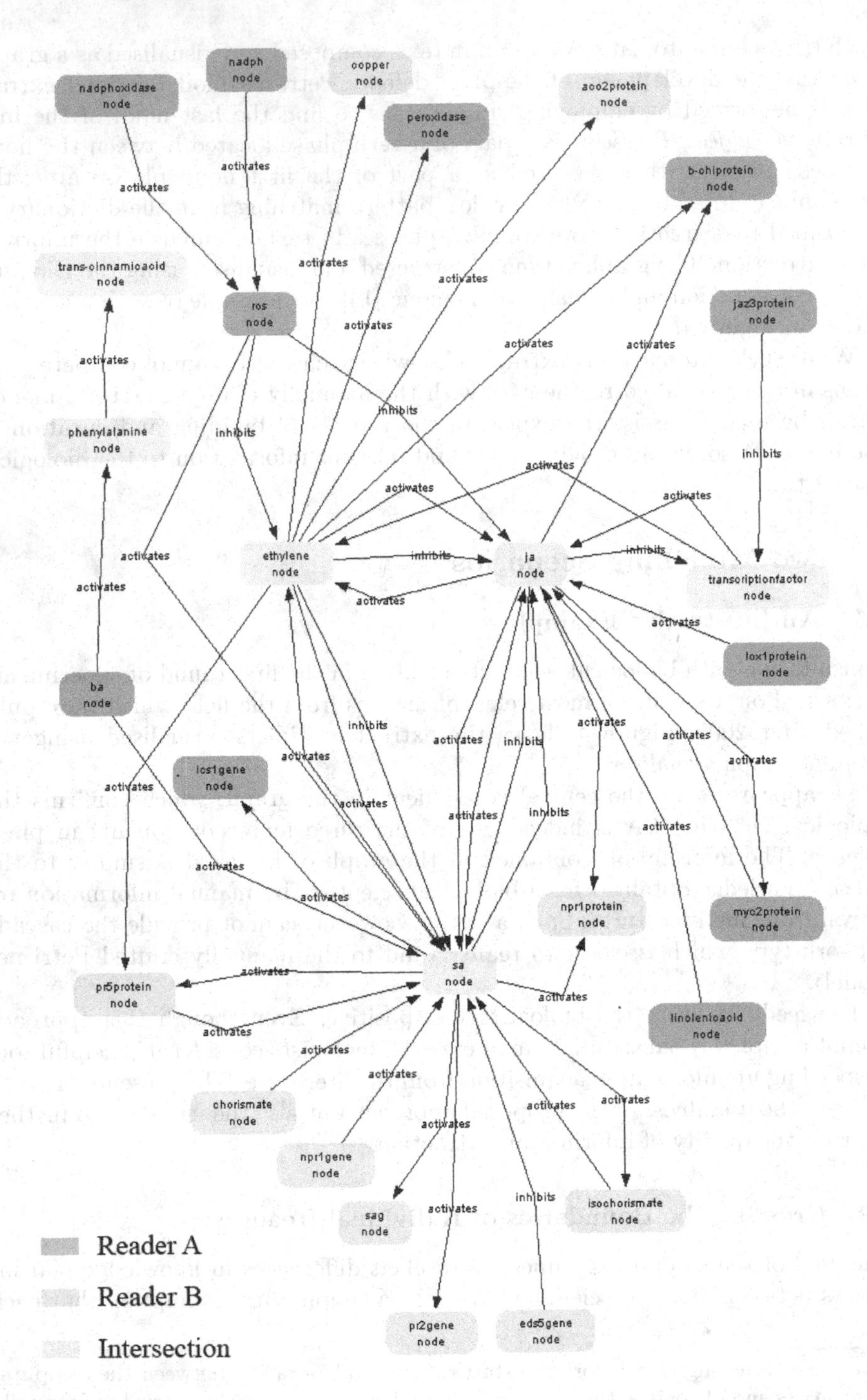

Fig. 5. A model constructed from a set of triplets extracted from 122 documents, read by two different readers and displayed using the Biomine graph visualisation engine

scientists' knowledge corresponds to a set of papers it read. The extracted triplets and subgraph thus model her/his subjective, habitual knowledge [3]. By combining subjective knowledge bases we obtain a join BisoNet where the intersecting subgraph represents a bridging graph pattern of bisociation. In particular, we extracted triplets from a set of 122 documents, read by two biology experts:

Reader A: Reader A (colored dark grey) has read 91 papers, of which 13 unique triplets were extracted automatically.
Reader B: Reader B (colored medium grey) has read 31 papers, of which 21 unique triplets were extracted automatically.
Intersections: Eight common triplets, extracted from 91 publications read by reader A and from 31 publications read by reader B, were colored in light grey colour.

Figure 5 shows the model extracted from 122 articles read by the two readers (two biological scientists). Besides supporting the automatic model construction, there are other benefits from visualising knowledge of different domain experts as illustrated in Figure 5. For instance, one can clearly see which nodes are in the intersection of interest of the two experts (coloured light grey in Figure 5).

This could indicate the areas of joint interest which the two experts might want to investigate jointly in more detail, e.g., to get answers to some yet unexplored research question in the intersection of their domains of expertise. On the other hand, this visualisation enables to see also who has some unique expertise in the field, with no intersection with other experts (coloured dark and medium grey in Figure 5). If applied to modelling the knowledge of larger consortia of readers, this type of information could be used to determine the complementarities of research groups.

The proposed approach to modelling and visualisation of knowledge extracted from the literature could be used also for modelling the know-how of large project consortia where it is hard to track the expertise of all project participants. Consequently, the proposed approach to cross-context modelling may be viewed as a step towards creating virtual laboratory knowledge models.

6 Conclusion

In this chapter we presented a methodology which supports the domain expert in the process of creation, curation, and evaluation of plant defence response models by combining publicly available databases, natural language processing tools, and hand-crafted knowledge. The methodology was implemented as a reusable workflow of software services, and evaluated using a hand crafted Petri net model. This Petri net model has been developed by fusing expert knowledge, experimental results and biological literature, and serves as a baseline for evaluation of automatically extracted plant defence response knowledge, but it also enables computer simulation and prediction.

This chapter presented also an approach to modelling the knowledge of different domain experts, by visualising the intersections as well as complementarities of their expertise, with a potential of providing a global overview of the expertise of consortia members. This type of modelling can be used to analyze and monitor knowledge of larger groups of experts to establish how their knowledge grows and evolves in terms of time and research topics.

Acknowledgments. This work has been supported by the European Commission under the 7th Framework Programme FP7-ICT-2007-C FET-Open, contract no. BISON-211898, the European Community 7th frame-work program ICT-2007.4.4 under grant number 231519 e-Lico: An e-Laboratory for Interdisciplinary Collaborative Research in Data Mining and Data Intensive Science, AD Futura scholarship and the Slovenian Research Agency grants P2-0103 and J4-2228. We are grateful to Claudiu Mihăilă for the initial implementation of the triplet extraction engine, and Lorand Dali and Delia Rusu for constructive discussions and suggestions.

References

1. Baitaluk, M., Sedova, M., Ray, A., Gupta, A.: BiologicalNetworks: visualization and analysis tool for systems biology. Nucl. Acids Res. 34(suppl. 2), W466–W471 (2006)
2. Demir, E., Babur, O., Dogrusoz, U., Gursoy, A., Nisanci, G., Cetin-Atalay, R., Ozturk, M.: PATIKA: An integrated visual environment for collaborative construction and analysis of cellular pathways. Bioinformatics 18(7), 996–1003 (2002)
3. Dubitzky, W., Kötter, T., Schmidt, O., Berthold, M.R.: Towards Creative Information Exploration Based on Koestler's Concept of Bisociation. In: Berthold, M.R. (ed.) Bisociative Knowledge Discovery. LNCS (LNAI), vol. 7250, pp. 11–32. Springer, Heidelberg (2012)
4. Eronen, L., Hintsanen, P., Toivonen, H.: Biomine: A Network-Structured Resource of Biological Entities for Link Prediction. In: Berthold, M.R. (ed.) Bisociative Knowledge Discovery. LNCS (LNAI), vol. 7250, pp. 364–378. Springer, Heidelberg (2012)
5. Genoud, T., Trevino Santa Cruz, M.B., Metraux, J.-P.: Numeric Simulation of Plant Signaling Networks. Plant Physiology 126(4), 1430–1437 (2001)
6. Hariharaputran, S., Hofestädt, R., Kormeier, B., Spangardt, S.: Petri net models for the semi-automatic construction of large scale biological networks. Springer Science and Business. Natural Computing (2009)
7. Hu, Z., Mellor, J., Wu, J., DeLisi, C.: VisANT: data-integrating visual framework for biological networks and modules. Nucleic Acids Research 33, W352–W357 (2005)
8. Huan, T., Sivachenko, A.Y., Harrison, S.H., Chen, J.Y.: ProteoLens: a visual analytic tool for multi-scale database-driven biological network data mining. BMC Bioinformatics 9(suppl. 9), S5 (2008)

9. Köhler, J., Baumbach, J., Taubert, J., Specht, M., Skusa, A., Röuegg, A., Rawlings, C., Verrier, P., Philippi, S.: Graph-based analysis and visualization of experimental results with Ondex. Bioinformatics 22(11), 1383–1390 (2006)
10. Le Novère, N., Hucka, M., Mi, H., Moodie, S., Schreiber, F., Sorokin, A., Demir, E., Wegner, K., Aladjem, M.I., Wimalaratne, S.M., Bergman, F.T., Gauges, R., Ghazal, P., Kawaji, H., Li, L., Matsuoka, Y., Villéger, A., Boyd, S.E., Calzone, L., Courtot, M., Dogrusoz, U., Freeman, T.C., Funahashi, A., Ghosh, S., Jouraku, A., Kim, S., Kolpakov, F., Luna, A., Sahle, S., Schmidt, E., Watterson, S., Wu, G., Goryanin, I., Kell, D.B., Sander, C., Sauro, H., Snoep, J.L., Kohn, K., Kitano, H.: The Systems Biology Graphical Notation. Nature Biotechnology 27(8), 735–741 (2009)
11. Loper, E., Bird, S.: NLTK: The Natural Language Toolkit. In: Proceedings of the ACL Workshop on Effective Tools and Methodologies for Teaching Natural Language Processing and Computational Linguistics, pp. 62–69. Association for Computational Linguistics, Philadelphia (2002)
12. Matsuno, H., Fujita, S., Doi, A., Nagasaki, M., Miyano, S.: Towards Biopathway Modeling and Simulation. In: van der Aalst, W.M.P., Best, E. (eds.) ICATPN 2003. LNCS, vol. 2679, pp. 3–22. Springer, Heidelberg (2003)
13. Rusu, D., Fortuna, B., Mladenić, D., Grobelnik, M., Sipoš, R.: Document Visualization Based on Semantic Graphs. In: Proceedings of the 13th International Conference Information Visualisation, pp. 292–297 (2009)
14. Sevon, P., Eronen, L., Hintsanen, P., Kulovesi, K., Toivonen, H.: Link Discovery in Graphs Derived from Biological Databases. In: Leser, U., Naumann, F., Eckman, B. (eds.) DILS 2006. LNCS (LNBI), vol. 4075, pp. 35–49. Springer, Heidelberg (2006)
15. Shannon, P., Markiel, A., Ozier, O., Baliga, N.S., Wang, J.T., Ramage, D., Amin, N., Schwikowski, B., Ideker, T.: Cytoscape: A software environment for integrated models of biomolecular interaction networks. Genome Research 13, 2498–2504 (2003)
16. Tsuruoka, Y., Tateishi, Y., Kim, J.-D., Ohta, T., McNaught, J., Ananiadou, S., Tsujii, J.: Developing a Robust Part-of-Speech Tagger for Biomedical Text. In: Bozanis, P., Houstis, E.N. (eds.) PCI 2005. LNCS, vol. 3746, pp. 382–392. Springer, Heidelberg (2005)

Bisociative Exploration of Biological and Financial Literature Using Clustering

Oliver Schmidt[1], Janez Kranjc[2], Igor Mozetič[2], Paul Thompson[1], and Werner Dubitzky[1]

[1] University of Ulster, Northern Ireland, UK
schmidt-o1@email.ulster.ac.uk,
[2] Jozef Stefan Institute, Ljubljana, Slovenia

Abstract. The *bile acid and xenobiotic system* describes a biological network or system that facilitates detoxification and removal from the body of harmful xenobiotic and endobiotic compounds. While life scientists have developed a relatively comprehensive understanding of this system, many mechanistic details are yet to be discovered. Critical mechanisms are those which are likely to significantly further our understanding of the fundamental components and the interaction patterns that govern this systems gene expression and the identification of potential regulatory nodes. Our working assumption is that a creative information exploration of available bile acid and xenobiotic system information could support the development (and testing) of novel hypotheses about this system. To explore this we have set up an information space consisting of information from biology and finance, which we consider to be two semantically distant knowledge domains and therefore have a high potential for interesting bisociations. Using a cross-context clustering approach and outlier detection, we identify bisociations and evaluate their value in terms of their potential as novel biological hypotheses.

Keywords: Clustering, outlier detection, bisociative information exploration.

1 Introduction

Bisociative information exploration is based on the assumption that the pooling of information from different domains could facilitate the discovery of new knowledge. In this study we explore bisociative information discovery based on literature from molecular biology and finance. Our hypothesis is that the bisociative approach may help life scientists interested in the bile acid and xenobiotic system to generate (and possibly test) novel hypotheses which will ultimately support the discovery of biological mechanisms.

The presented approach is based on the work by Petrič et al. [10] who developed methods to investigate the role of outliers in literature-based knowledge discovery. Their approach rests upon the assumption that cluster outliers of two document sets with known classification can be used to discover new, useful

M.R. Berthold (Ed.): Bisociative Knowledge Discovery, LNAI 7250, pp. 438–451, 2012.

knowledge. In this context we define outliers as domain-labeled documents that are further away from the centroid of their knowledge domain then the majority of documents from its domain.

The work by Petrič et al.[10], which focuses on the domains of biology and medicine, differs from our approach in the way that we consider selected documents from two *unrelated* domains, namely, finance and biology. With *unrelated* domains we mean knowledge domains or domain theories (as defined in *Part I: Bisociation [2]*) that share less concepts than the knowledge domains of biology and medicine for instance. Therefore we expect to find less documents than between *related* domains, which in turn enables us to have a more detailed semi-automatic analysis of those documents.

We investigate the cluster outliers and their opposite-domain neighborhood in order to identify bisociations between biology and finance. In particular, we are looking for shared features in scientific abstracts across the two domains. Such features might be common terms or even sets of relationships within one domain which have correspondences in the other domain.

2 The Bile Acid and Xenobiotic System

The bile acid and xenobiotic system (BAXS) defines a biological network that facilitates two distinct but intimately overlapping physiological processes. The enterohepatic circulation and maintenance of bile acid concentrations (Fig. 1) and the detoxification and removal from the body of harmful xenobiotic (e.g. drugs, pesticides) and endobiotic compounds (e.g., steroid hormones) [8]. The system involves the coordination of several levels of gene activity, including control of mRNA and protein expression and regulation of metabolizing enzyme and transporter protein function in tissues such as liver, intestine/colon and kidney. Bile acids are necessary for the emulsification and absorption of dietary fats and are therefore valuable compounds, however as their build-up can cause harm, their concentrations need to be appropriately regulated and recycled. Similarly there is a requirement for a system that can "sense" the accumulation of xenobiotic and endobiotic compounds and facilitate their detoxification and removal from the body. The BAXS accomplishes this and maintains enterohepatic circulation (the circulation of biliary acids from the liver as depicted in Fig. 1) through a complex network of sensors in the form of nuclear receptors that function as ligand-activated transcription factors (see molecular interaction network depicted in Fig. 2). They serve to detect fluctuations in concentration of many compounds and initiate a physiological response by regulating the BAXS.

Transcriptional regulation by nuclear receptors[1] involves both activating and repressive effects upon specific "sets" of genes. There is considerable overlap exhibited between nuclear receptors in the genes they target and also the ligands

[1] Nuclear receptors are a class of proteins within the interior of cells responsible for sensing the presence of steroid and thyroid hormones and certain other molecules. In response, these receptors work in concert with other proteins to regulate the expression of specific genes, thereby controlling the development, homeostasis, and metabolism of the organism.

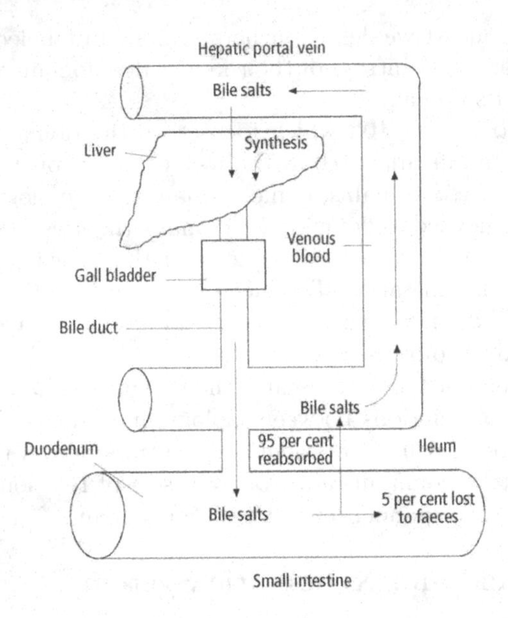

Fig. 1. The enterohepatic circulation system of the BAXS

that bind to and activate them, i.e. each gene has multiple functions within this system depending on the tissue it is expressed. It is these factors that contribute, for example, to the phenomenon of drug-drug interactions, e.g. between St. John's Wort and Cyclosporine or St. John's Wort and oral contraceptive [1,7].

The goal of the BAXS application within the BISON project is to support the discovery of hitherto unknown but important biological mechanisms in the BAXS. Critical mechanisms are those which are likely to significantly further our understanding of the fundamental components and the interaction patterns that govern BAXS gene expression and the identification of potential regulatory nodes. It has been established that the overall flux of the BAXS is achieved through a regulatory transcriptional network mediated through the activities of members of the nuclear receptors (such as FXR, LXR, PXR, CAR) and nuclear factors (such HNF1α, HNF4α). However, given the overall complexity of the bile acid/xenobiotic system it is difficult to assess the exact importance of each receptor and modulatory factor with respect to BAXS activity in different tissues. One of the key issues in the understanding of the BAXS is to decipher the components and the interaction patterns that govern BAXS gene expression and the identification of potential regulatory nodes. This understanding is essential to identify targets for treatment regimes, to understand the components impacting drug-drug interactions, to provide a framework for the design of large-scale, integrated prediction studies, and to aid in the definition of high-quality "gold standards" or research frameworks for future systems biology studies.

To investigate the potential of bisociative exploration of the BAXS, we are pooling two groups of information resources from the biological and financial domains respectively.

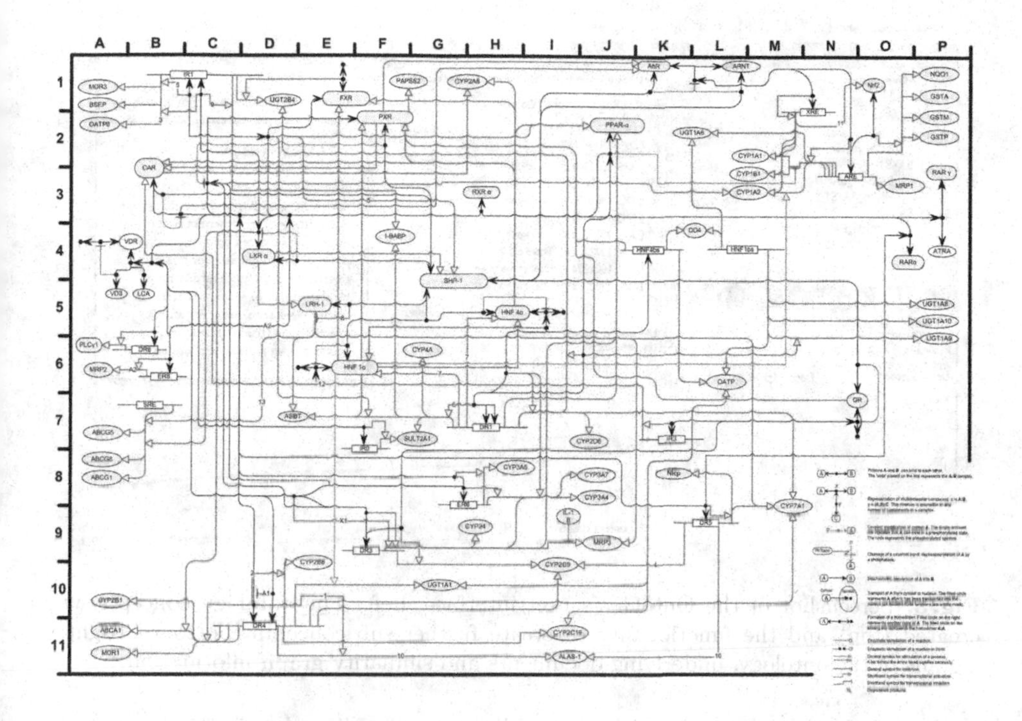

Fig. 2. The gene network and molecular interactions of the BAXS

3 Materials and Methods

OntoGen[2] is a semi-automatic and data-driven ontology editor facilitating the editing of topic ontologies (a hierarchy of topics connected with the subtopic_of relation). The editor uses text mining and machine learning techniques to help create ontologies from a set of text documents. For this the tool adopts k-means clustering and latent semantic indexing techniques(LSI) [6]. A screenshot of the tool is shown in Fig. 3.

In OntoGen each text document is represented as a term vector which is generated by the standard *bag of words* approach and the assignment of the *term frequency / inverse document frequency* (*TFIDF*)[3] measure to each term. The similarity between term vectors is calculated using Cosine similarity and shown in the center of Fig. 3. The similarity values are recalculated when the document selection changes. For further mentioning of the similarity between documents we refer to these similarity values calculated by OntoGen.

[2] `http://ontogen.ijs.si/`

[3] Elements of vectors are weighted with the TFIDF weights as follows [3]: the ith element of the vector containing frequency of the ith term is multiplied with $IDF_i = log(N/df_i)$, where N represents the total number of documents and df_i is document frequency of the ith term (i.e. the number of documents from the whole corpus in which the ith term appears).

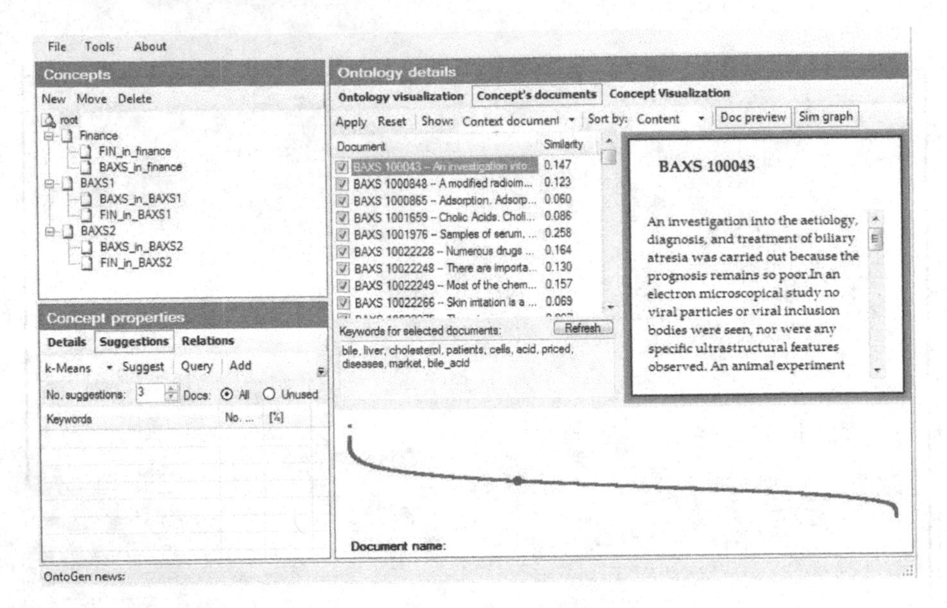

Fig. 3. Screenshot of the OntoGen's user interface. Left: The ontology concepts we created (top) and the functionality to create further sub concepts (bottom). Right: Details of the ontology, underlying documents and similarity graph information.

One of OntoGen's powerful features allows to visualize the similarity among a set of selected documents, which is called *Concept visualization*(as shown in Fig. 4). This visualization is created using dimensionality reduction techniques by combining linear subspace and multidimensional scaling methods. For a detailed description and explanation of OntoGen's main components and functionality we refer to work by Fortuna and colleagues [3,6,5,4].

In the following we describe how we generated and explored the document outliers across two domains: BAXS and finance. Petrič et al. [10] outline procedures that facilitate the identification of cross-context outliers with OntoGen. The main steps comprise the k-means clustering of documents with two different labels, the further subdivision of each cluster according to the labels and the outlier analysis of these misclassified documents in contrast to the clusters. Before going into the details of outlier detection, we describe the retrieval of the scientific documents for both domains. A *document* in this study refers to the *abstract* and associated *keywords* of a published scientific article.

The document corpus for the biological domain (as relevant to the BAXS) consists of documents from PubMed[4]. PubMed is a free resource containing over 20 million biomedical article citations and articles.

To compile a corpus of BAXS-relevant PubMed abstracts, we used keywords and phrases that reflect important concepts in relation to BAXS research. Furthermore, we restricted the search to articles that discuss these concepts in the context of human biology (ignoring other species). We used the following

[4] www.pubmed.gov

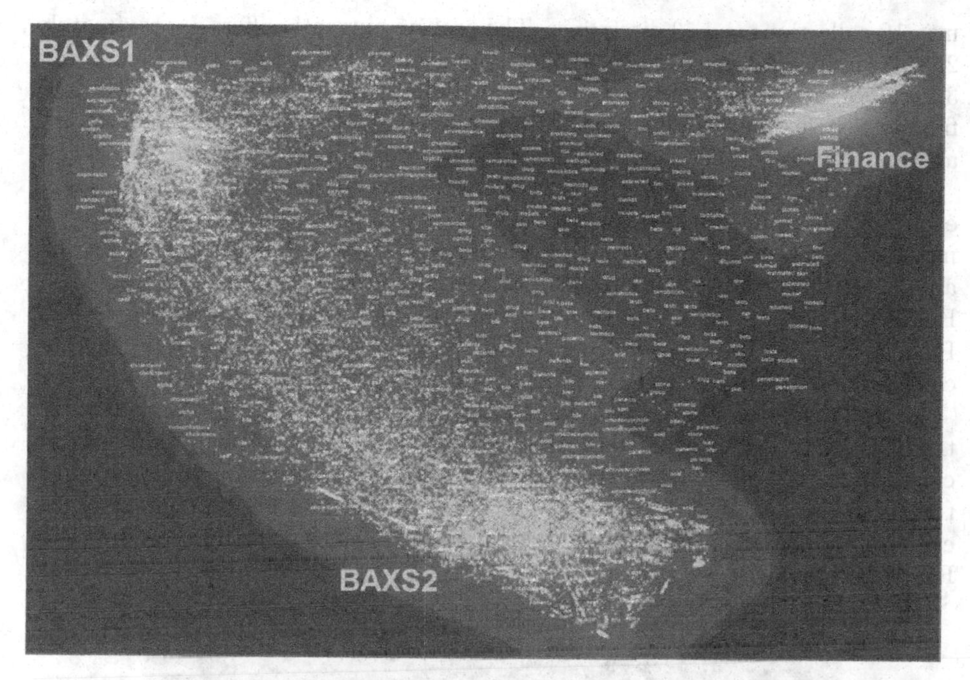

Fig. 4. OntoGen's concept visualization of all documents. Yellow crosses denote documents and the white labels depict document terms. The 3 potential clusters are labeled accordingly

PubMed query to select the articles:

```
(''bile acids and salts'' [MeSH Terms] OR
(''bile'' [All Fields] AND ''acids'' [All Fields] AND
''salts'' [All Fields] ) OR ''bile acids and salts''
[All Fields] OR (''bile'' [All Fields] AND
''acids'' [All Fields]) OR ''bile acids'' [All Fields]) OR
(''xenobiotics'' [MeSH Terms] OR
''xenobiotics'' [All Fields] ) AND ''humans'' [MeSH Terms]
```

The query resulted in 21 565 articles of which 16 106 had an abstract. In addition to the abstracts, we retrieved all articles with the MeSH terms provided by PubMed, i.e., we included also articles with MeSH[5] terms only. With this approach we compiled 21 276 documents containing either abstracts, MeSH terms or both.

The information resources from the financial domain are abstracts from the financial literature. We obtained these from the Journal Storage[6] (JSTOR). Currently, JSTOR contains approximately 1224 journals, which are categorized

[5] Medical Subject Headings (MeSH) provides a vocabulary of ca. 25 000 terms used to characterize the content of biomedical document such as articles in scientific journals.

[6] `http://dfr.jstor.org/`

into 19 collections and 53 disciplines. The archive currently offers approximately 295 000 individual journal issues and about 6.4 million articles of which about 3.2 million articles are full-text articles available online. Out of 27 613 available finance articles that we retrieved, 7674 provided an abstract. In addition to the abstracts, we retrieved a set of JSTOR-generated keywords for each abstract.

We created a simple text file for the BAXS and finance articles, respectively, each file containing all the abstracts in that domain. Within each file the documents were organized as a sequence of lines, each line representing a document, and the first word in the line is used as the title of the document. A list of 532 English stop words was used to filter out the "low-content" words followed by Porter stemming to reduce inflected (or derived) words to their stem [9]. For the construction of the similarity graph the document labels were ignored.

After some experimentation and visual analysis of the clusters, we generated the final set of clusters using k-means clustering with $k = 3$. Viewing the resulting clustering of the document vectors (as shown in Fig. 4), one can easily recognize the three clusters. In order to get an idea about the information content of the clusters and the area between them, a more detailed view on Fig. 4 (shown in Fig. 5) is provided.

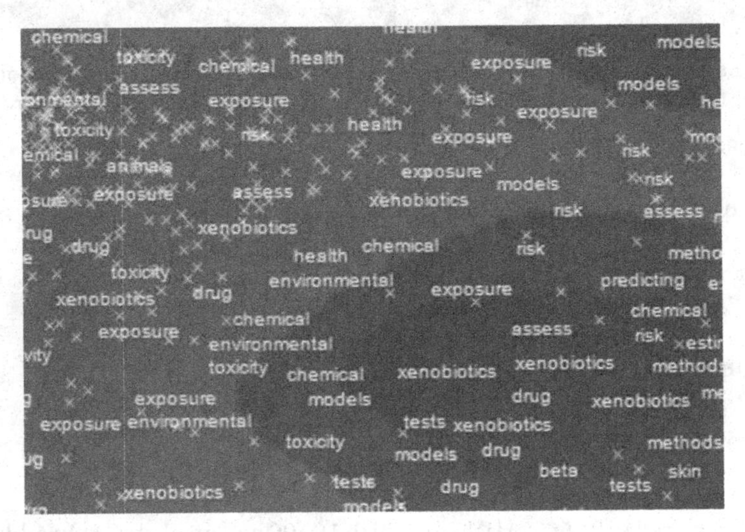

Fig. 5. Zoomed in detail from the top centre of Figure 4 (area between ***BAXS1*** and ***Finance cluster***).

Another interesting feature of OntoGen is its ability to determine the most common keywords for each cluster. OntoGen uses two methods to extract the keywords (1) using centroid vectors to get descriptive keywords and (2) Support Vector Machine classification to extract distinctive keywords. In this study we considered keywords extracted from method (1) only. The keywords provide users with an idea of the meaning of the clusters. According to this we characterized the three clusters as follows:

1. The ***Finance Cluster*** characterized by the keywords *priced, market, stocks, returned, firm,trading, investments, models, portfolio* and *rate.*
2. The ***BAXS1 Cluster*** with the keywords *cells, activity, protein, drug, transport, receptor, enzyme, xenobiotics, expression* and *gene.*
3. The ***BAXS2 Cluster*** with the keywords *bile, cholesterol, patients, liver, acid, diseases, age,* bile acid, *biliary* and *ursodeoxycholic.*

In order to determine the outliers for each cluster, we separated the documents of each of the three clusters according to the assigned document labels BAXS and FIN. The resulting topic ontology is depicted in Fig. 6.

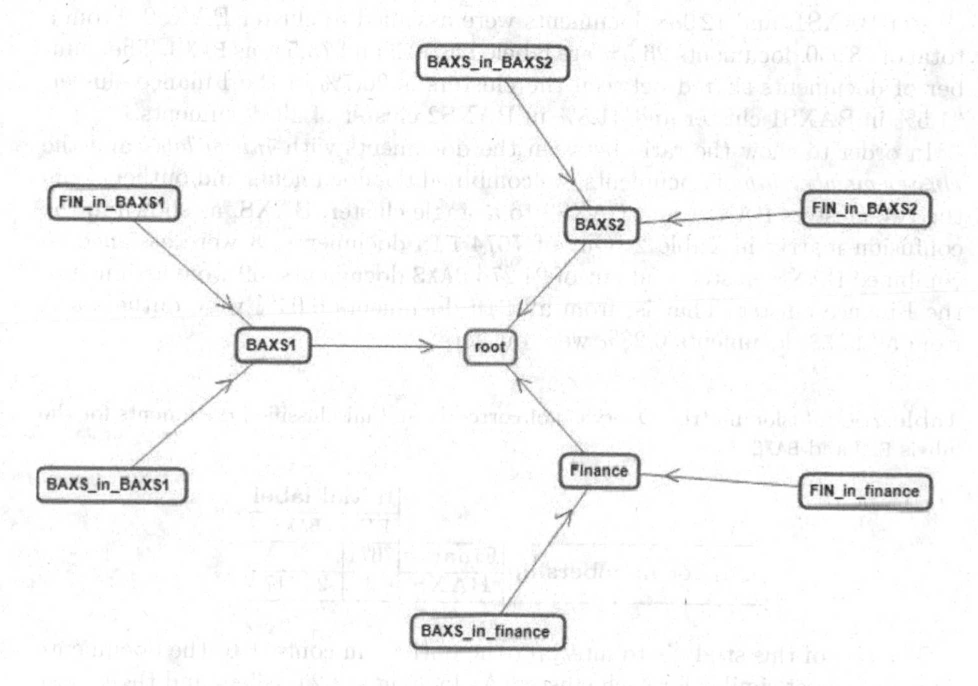

Fig. 6. Visualization of the topic ontology with the root in the middle and its three ascending nodes/clusters (*Finance*, *BAXS1*, *BAXS2*) of which each has two ascending nodes/clusters corresponding to either labels BAXS or FIN.

Fig. 6 suggests that each cluster contains outliers, i.e., *BAXS_in_Finance* for BAXS-labeled documents in the Finance cluster, *FIN_in_BAXS1* for FIN-labeled documents in the BAXS1 cluster, and *FIN_in_BAXS2* for FIN-labeled documents in the BAXS2 cluster. A detailed breakdown of the distribution of outliers over the clusters is provided in Table 1 as a contingency table.

The rows in the table described the number of documents falling into one of the three clusters, and the columns describe the number of documents labeled BAXS and FIN respectively. Out of 7674 FIN-labeled documents, 7671 were assigned to cluster Finance, 2 documents were assigned to cluster BAXS1, and

Table 1. Contingency table: Overview of document distribution over the clusters *Finance*, *BAXS1* and *BAXS2*

Cluster/Label	FIN	BAXS	Total
Finance	7671	59	7730
BAXS1	2	9129	9131
BAXS2	1	12 088	12 089
Total	7674	21 276	28 950

1 document was assigned to cluster BAXS2. Out of 21 276 BAXS-labeled documents, 59 were assigned to cluster Finance, 9129 documents were assigned to cluster BAXS1, and 12 088 documents were assigned to cluster BAXS2. From a total of 28 950 documents 26,5% are labeled as FIN and 73,5% as BAXS. The number of documents shared between the clusters is 26,7% in the Finance cluster, 31,5% in BAXS1 cluster and 41,8% in BAXS2 cluster of all documents.

In order to show the ratio between the documents with *initial label* and the *cluster membership* of documents, we combined the documents and outliers from the two clusters BAXS1 and BAXS2 to a single cluster, BAXS, as shown in the confusion matrix in Table 2. Out of 7674 FIN documents, 3 were assigned to combined BAXS cluster, and out of 21 276 BAXS documents, 59 were assigned to the Finance cluster. That is, from all FIN documents 0.04% were outliers and from all BAXS documents 0.28% were outliers.

Table 2. Confusion matrix: Overview of correctly and misclassified documents for the labels FIN and BAXS

		Initial label	
		FIN	BAXS
Cluster membership	**Finance**	7671	59
	BAXS	3	21 217

The aim of this study is to interpret the outliers in context to the documents which are most similar for each cluster. As looking at 62 outliers and their most similar neighbours for each cluster would be manually not feasible we reduced the amount of BAXS outliers within the Finance cluster.

From the 59 BAXS outliers within the Finance cluster, we decided to consider 13 of them to be relevant for this study. The decision making process was accomplished by experts analysing the 59 BAXS documents and filtering them. We selected the outliers that were most relevant to BAXS but also most promising to find relationships in finance. The list of outliers and the topic covered is shown in Table 3.

As next step we investigated the most similar documents for each outlier document for each cluster. We considered the 5 most similar documents for the BAXS clusters (BAXS1 and BAXS2) and the 6 most similar documents for the Finance cluster. The reason for considering one more document from the Finance cluster was that FIN-labeled documents were generally much shorter than BAX

Table 3. General topic of outlier documents

Outlier	Topic
BO_01	Data analysis using statistical models
BO_02	Study about how paracetamol dissolves in different conditions
BO_03	A new model in pharmacokinetics is introduced
BO_04	Paper discusses legal criteria and judicial precedents related to hormesis
BO_05	How to calculate reference doses for toxic substances
BO_06	Nonlinear system to model concentration of hormones important for menstrual cycle
BO_07	Paper about volume of distribution and mean residence time of pharmaceutics in the body
BO_08	Book about chronic kidney disease and future perspective for Japan
BO_09	Application of data mining methods for biomonitoring and usage of xenobiotics
BO_10	Xenobiotics in the air close to industrial centres affect mechanisms of inheritance
BO_11	Data analysis about colorectal polyp prevention
BO_12	Statistical data analysis for Alzheimer's disease
BO_13	Costs and effectiveness of anti-fungal agents for patients are assessed
FO_01	Analysis about how new drug launches affect life expectancy
FO_02	Analysis of Russian's investment in transport infrastructure
FO_03	Relationship between choice of treatment for illness and getting a job

documents, and were therefore considered to offer less information than the BAXS documents. The set of abstracts we retrieved contained on average 8 sentences for PubMed abstracts and 4 for JSTOR.

As OntoGen wasn't designed or intended to be used in such a way we adopted the following procedure to obtain the documents that are most similar to an outlier. First, we selected the outlier of interest and deselected all other documents. Then, using OntoGen, we recalculated the similarity of all documents for this outlier. The most similar documents to the outlier are then listed below the outlier.

In order to find the most similar documents from a different cluster, one had to assign the considered outlier to the other cluster. This was achieved by the selection of the outlier and selecting OntoGen'S *move* function (shown in the top left corner in Fig. 7). After the outlier was moved to one of the other clusters, the procedure was repeated to obtain the most similar neighborhood documents.

Petrič et al. [10] considered only neighborhood documents within the same cluster for the interpretation of outliers. Our approach extends this approach by taking into account the neighborhood documents from the other clusters. Thus, possible relationships between the clusters can be assessed, where the outliers serve as link between the most similar neighborhood documents. Table 4 lists the common bridging terms between the outliers and their neighborhood documents that we found.

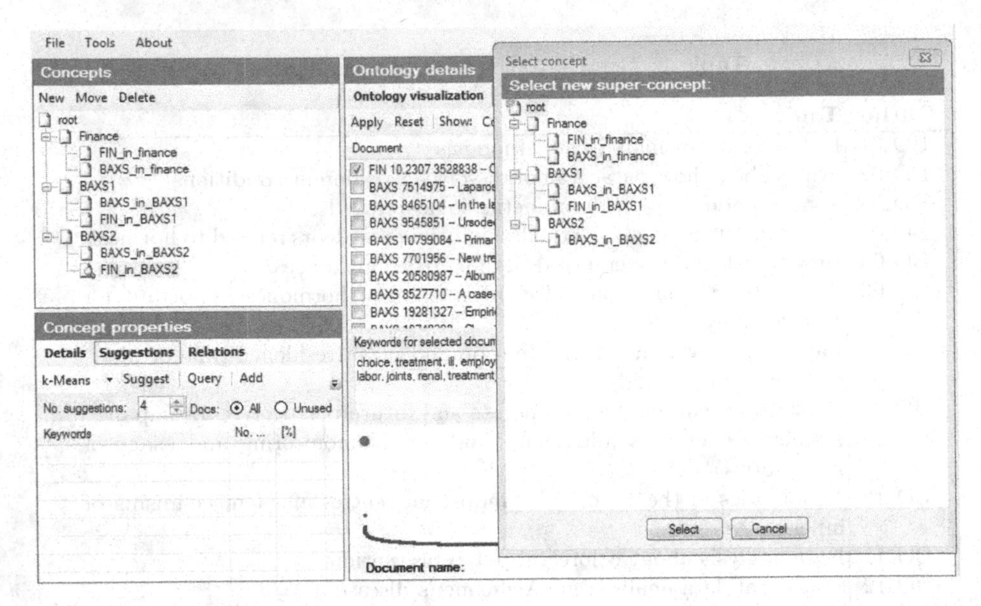

Fig. 7. Screenshot showing how the outlier document FIN 10.2307 3528838 in cluster *BAXS2* can be moved to different clusters

4 Results and Discussion

Initially, we wanted to look at all 16 outliers and how they might be linked or related to the topic they cover. For this we first looked at the selected BAXS outlier documents and how they cover the topics of all `BAXS` outliers within the Finance cluster. As shown in Fig. 8, the 13 outliers (the dark labels) seem to cover most of the topic space of all 59 outliers. We realized that the outliers are not evenly distributed over this space, which is due to the bias created by manual selection by the experts. In fact, there are more outliers close to each other on the left side of the diagram then on the right side.

If we look at the topics covered by each outlier (as shown in Table 3), we see that most of them cover topics related to clinical studies, statistical analysis or models related to BAXS in one way or another.

The `FIN` outliers, on the other hand, do not seem to be similar to each other. There do not seem to be any obvious relationships between the life expectancy affected by drug launches (FO 01), how Russians invest in their transport infrastructure (FO 02), or how the choice of treatment for illness is related to getting a job (FO 03).

Therefore, we analyzed the outliers in detail and determined the terms they share with their most similar neighbor documents. The results of this analysis are summarized in Table 4. The table lists the most frequent bridging terms (b-terms) between each outlier and their neighbors.

Then we looked at the bridging terms for the BAXS outliers within each cluster. The b-terms between the 13 BAXS outliers and the documents in the

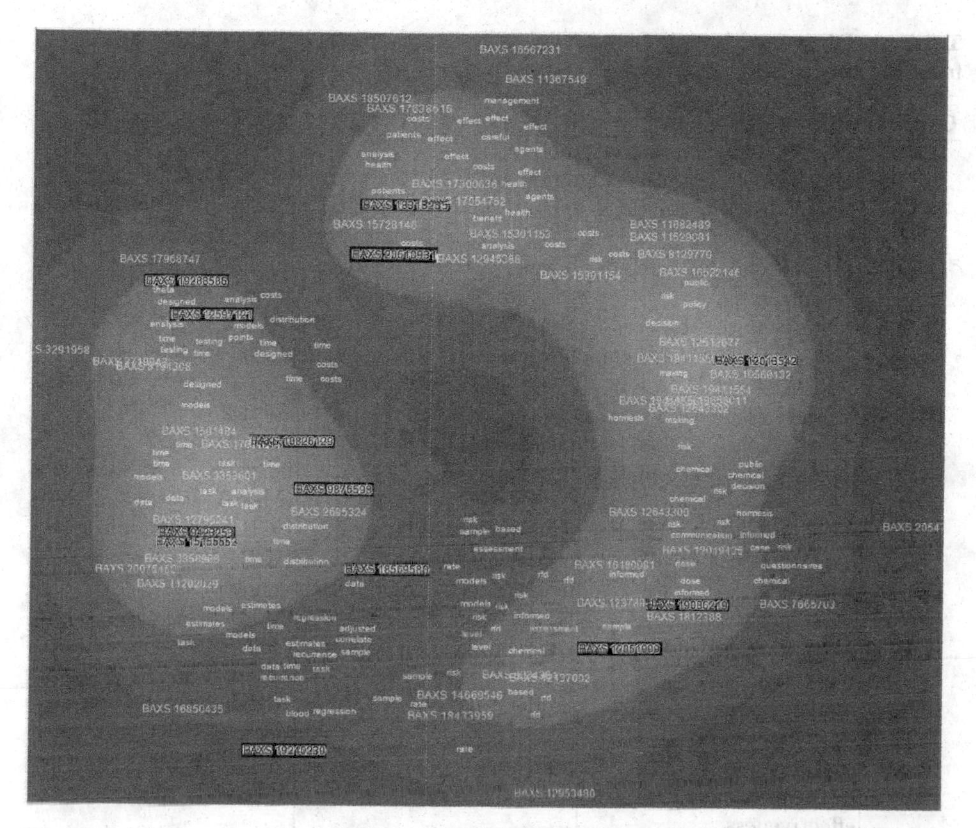

Fig. 8. Concept visualization of all BAXS outlier documents within the Finance cluster. The relevant outliers under investigation are highlighted in black

Finance cluster are usually general terms such as *model, event, predict, estimate* or *risk*. The b-terms within the *BAXS1* cluster and *BAXS2* appear to be more specific, i.e., biological or medical terms such as *hygiene, hormone, pharmacokinetic, colon cancer*. The b-terms between the 3 *Finance* outliers within *BAXS1* and *BAXS2* do not provide any new insights compared to Table 3

Meaningful bisociative relationships between the outliers and their neighbor documents could not be found. As this neglects the neighbor documents within the other clusters, we decided to search for relationships between the neighbor documents from the clusters. For this we picked the outliers BO 05 and FO 02 (see Table 4) due to their common b-terms in each cluster.

The BO 05 outlier relates finance models concerned with the risk of investment to abstracts which analyze the risk of cancer depending on different factors. The most promising outlier for this was FO 02, as it relates BAXS abstracts to transporter pathways within and across cells with the transporting infrastructure within and across countries.

Table 4. Discovered bridging terms via outliers with most similar neighbours from the clusters *Finance*, *BAXS1* and *BAXS2*.

Outlier	Finance	BAXS1	BAXS2
BO 01	event, model, simulation	model	regression
BO 02	model, predict, statistical	model, dissolution	dissolution
BO 03	curve, distribution, long-term, model	model, pharmacokinetic	model, pharmacokinetic (PK), kinetic
BO 04	decision-making, regulatory, agency	hormesis, toxic, regulator	humans (Mesh)
BO 05	risk, estimate, observe	NOAEL, BMD	risk
BO 06	cycles, non-linear systems, model	hormone	cycle, women, hormone
BO 07	volume, estimate	clearance, estimate, pharmacokinetic	volume
BO 08	japanese, assume, analyse	filtration	humans, middle aged (Mesh)
BO 09	decision-tree, predict, model	environmental	PCB, biomonitoring, HCB
BO 10	air, risk	hygiene, air, xenobiotics (Mesh)	xenobiotics
BO 11	estimation, method	colorectal, colon cancer	recurrence, prevention, polyp
BO 12	model, predict	model, analysis, PLSDA	Alzheimer's disease, patient
BO 13	cost, hospitals	antifungal	cost, leukemia
FO 01	health, expenditure, cost-effectiveness	drug, database	drug
FO 02	Russia, transport, transit	transport	transit, transport
FO 03	choice	renal	treatment, disease, chronic, cholestasis

5 Conclusions

In this case study we investigated the potential bisociations between the finance and BAXS domain based on document outliers as determined by a cross-context clustering approach.

One main issue in this study is the asymmetric nature of knowledge, i.e., we have more knowledge about the BAXS than about the finance domain. Another issue is the asymmetric nature of the data sources (73% `BAXS` documents and 27% `Finance` documents). Both put a strong bias on the discovered outliers in this study and therefore reduce the quality of the results. Based on this study it would appear that the most promising method to find potential bisociations is to look at the neighbor documents from the different clusters related to one outlier. The use of scientific abstracts only could be another reason for the lack of finding interesting relationships between the domains. More work is needed to explore this approach to bisociative information discovery but the approach presented here shows promise in the discovery of novel connections between domains

References

1. Barone, G., Gurley, B., Ketel, B., Lightfoot, M., Abul-Ezz, S.: Drug interaction between St. John's wort and cyclosporine. The Annals of Pharmacotherapy 34(9), 1013–1016 (2000)
2. Dubitzky, W., Kötter, T., Schmidt, O., Berthold, M.R.: Towards Creative Information Exploration Based on Koestler's Concept of Bisociation. In: Berthold, M.R. (ed.) Bisociative Knowledge Discovery. LNCS (LNAI), vol. 7250, pp. 11–32. Springer, Heidelberg (2012)
3. Fortuna, B., Grobelnik, M., Mladenič, D.: Visualization of text document corpus (2005)
4. Fortuna, B., Grobelnik, M., Mladenič, D.: Semi-automatic data-driven ontology construction system (2006)
5. Fortuna, B., Grobelnik, M., Mladenic, D.: OntoGen: Semi-automatic Ontology Editor. In: Smith, M.J., Salvendy, G. (eds.) HCII 2007, Part II. LNCS, vol. 4558, pp. 309–318. Springer, Heidelberg (2007)
6. Fortuna, B., Mladenič, D., Grobelnik, M.: Semi-automatic Construction of Topic Ontologies. In: Ackermann, M., Berendt, B., Grobelnik, M., Hotho, A., Mladenič, D., Semeraro, G., Spiliopoulou, M., Stumme, G., Svátek, V., van Someren, M. (eds.) EWMF 2005 and KDO 2005. LNCS (LNAI), vol. 4289, pp. 121–131. Springer, Heidelberg (2006)
7. Hall, S.D., Wang, Z., Huang, S., Hamman, M.A., Vasavada, N., Adigun, A.Q., Hilligoss, J.K., Miller, M., Gorski, J.C.: The interaction between St. John's wort and an oral contraceptive[ast]. Clinical Pharmacology and Therapeutics 74(6), 525–535 (2003)
8. Kliewer, S.A.: The nuclear pregnane X receptor regulates xenobiotic detoxification. The Journal of Nutrition 133(7), 2444S–2447S (2003)
9. Natarajan, J., Berrar, D., Hack, C.J., Dubitzky, W.: Knowledge discovery in biology and biotechnology texts: A review of techniques, evaluation strategies, and applications. Critical Reviews in Biotechnology 25(1-2), 31–52 (2005)
10. Petrič, I., Cestnik, B., Lavrač, N., Urbančič, T.: Outlier detection in cross-context link discovery for creative literature mining. The Computer Journal (2010), doi:10.1093/comjnl/bxq074

Bisociative Discovery in Business Process Models

Trevor Martin[1,2] and Hongmei He[1,3]

[1] Artificial Intelligence Group, University of Bristol BS8 1UB UK
firstname.lastname@bristol.ac.uk
[2] BT Innovate and Design, Adastral Park, Ipswich IP5 3RE, UK
[3] Current Address : School of Computing and Intelligent Systems,
University of Ulster, Magee Campus, BT48 7JL, UK
h.he@ulster.ac.uk

Abstract. Bisociative knowledge discovery - finding useful, previously unknown links between concepts - is a vital tool in unlocking the economic and social value of the vast range of networked data and services that is now available. An important application for bisociative knowledge discovery is business process analysis, where bisociation could lead to improvements in one domain being disseminated to other domains. We identify two forms of bisociation, based on structural similarity, that are applicable to business processes, and present examples using real-world data to show how bisociative reasoning can be applied.

1 Introduction

Business Intelligence has been defined as a broad category of "applications and technologies for gathering, storing, analyzing, and providing access to data to help users and automated systems make fact-driven business decisions[1] " Business process analysis is a subfield, arising from the need for companies to learn more about how their processes operate in the real world. According to Andersen and Fagerhaug [1], a business process is "a logical series of related transactions that converts input to results or output" In particular, business process analysis involves aspects such as

- discovery of process models, based on event logs
- process conformance (do event logs follow prescribed paths through process models)
- process analysis (e.g. are there significant bottlenecks, can process instances be grouped on the basis of different paths through the process model)
- extension of process models, in cases where the actual process execution is not properly reflected in the model.

A business process can be represented naturally as a graph, and hence business processes form suitable inputs for the bisociation operations described in [2] – particularly

[1] This definition was taken from www.oracle.com/us/solutions/sap/database/sapbocmtsizing-352636.pdf; there are many similarly phrased descriptions on the web.

M.R. Berthold (Ed.): Bisociative Knowledge Discovery, LNAI 7250, pp. 452–471, 2012.

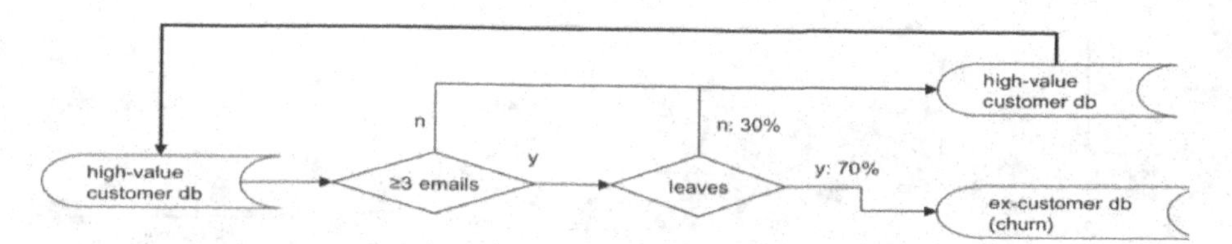

Fig. 1. Simplified process diagram showing 70% of high value customers leave after sending 3 or more emails to a support centre

the notion of graph (structural) similarity. There are, however, some features which distinguish process graphs from most of the other (document- and graph- based) demonstrators mentioned in [3]. In particular, the sequential nature of most business processes is fundamental - there is a specific order required for the steps within a process. In contrast, measures of similarity which depend on counting the number of occurrences (or co-occurrences) of words, nodes, etc. do not require a specific order of occurrence. Additionally, the BisoNet representation assumes that numerical edge labels reflect the probability or strength of a link. In process graphs, edges can be labelled by the time taken to move from one process stage to the next. Notwithstanding these differences, the Bison framework has been used to generate useful suggestions for domain experts, showing its versatility and potential.

As discussed in [4], creative knowledge discovery can be distinguished from "standard" knowledge discovery by defining the latter as the search for explanatory and/or predictive patterns and rules in large volume data within a specific domain. For example, a knowledge discovery process might examine an ISP (internet service provider)'s customer database and determine that people who have a high monthly spend and who send more than three emails to the support centre in a single month are very likely to change to a different provider in the following month. Such knowledge is implicit within the data but is useful in predicting and understanding behaviour. Figure 1 illustrates this as a summary process diagram (the actual set of process instances would be a more complex graph).

By contrast, creative knowledge discovery is more concerned with "thinking the unthought-of" and looking for new links, new perspectives, etc. Such links are often found by drawing parallels between different domains and looking to see how well those parallels hold - for example, compare the ISP example mentioned above to a hotel chain finding that regular guests who report dissatisfaction with two or more stays often cease to be regular guests unless they are tempted back by special treatment (such as complimentary room upgrades), as illustrated in Fig. 2. This is a simple illustration of similar problems (losing customers) in different domains. There is a structural similarity, and a solution in one domain (complimentary upgrades) could inspire a solution in the second (e.g. a higher download allowance at the same price). Of course, such analogies may break down when probed too far but they often provide the creative insight necessary to spark a new solution through a new way of looking at a problem. In many cases, this inspiration is referred to as "serendipity", or accidental discovery.

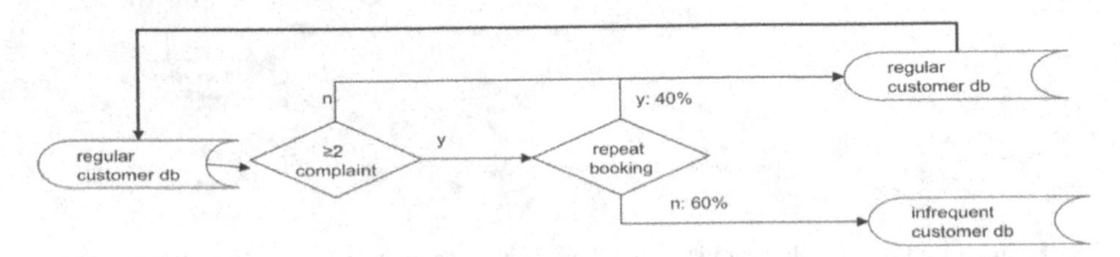

Fig. 2. (a) Simplified process showing 60% of regular customers do not re-book after making 2 or more complaints

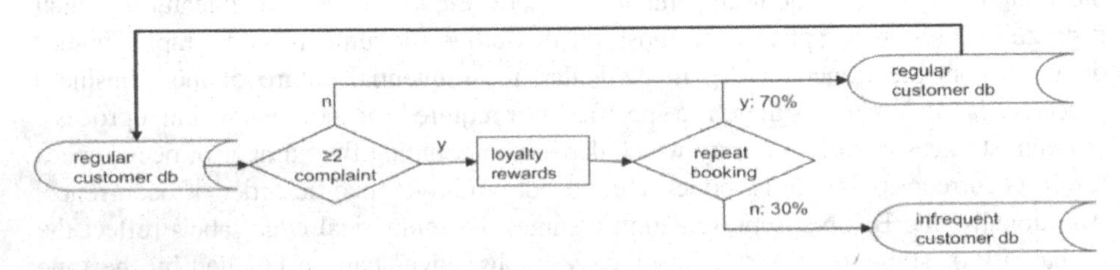

Fig. 2. (b) Updated process introducing loyalty rewards after which only 30% of regular customers do not re-book after making 2 or more complaints

The core of the Bison project is the automation of creativity, in this sense of making novel connections between previously unrelated concepts. For networks representing business processes, we have investigated two possible modes of bisociation:

(i) structural bisociation between two process networks from different domains, with respect to a specific mapping between node types in the two networks. The processes in one domain are assumed to need improvement. We look for sections of the two networks where there is high similarity between most of the nodes and links, with a small segment exhibiting low similarity. The operation of bisociation swaps the (mapped) low similarity segments, as illustrated in the ISP/hotel chain example above.

(ii) conceptual bisociation, which is a form of structural bisociation, requiring one process network and a generalisation / specialisation hierarchy on the node types. Similar process graphs for use in bisociative combination can be generated using the generalisation/ specialisation hierarchy. The origin of this approach is explained below.

Sherwood [5] proposed a systematic method, in which a situation or artefact is represented as an object with multiple attributes, and the consequences of changing attributes, removing constraints, etc are progressively explored. For example, given an old style reel-to-reel tape recorder as starting point, Sherwood's approach is to list some of its essential attributes, substitute plausible alternatives for a number of these attributes, and evaluate the resulting conceptual design or solution. Table 1 shows how this could have led to the Sony Walkman in the late 70s [5]. Again, with the benefit of hindsight the reader should be able to see that by changing magnetic tape to a hard disk and by also considering the way music is purchased and distributed, the same method could (retrospectively, at least) lead one to invent the iPod. Of course, having the vision to choose new attributes and the knowledge and foresight to evaluate the result is the hard part - and the creative steps are usually only obvious with hindsight.

This systematic approach is ideally suited to handling data which is held in an object-attribute-value format, with a taxonomy defined on the domain of (at least) one attribute. This provides a means of changing/generalising attribute values, so that "sensible" changes can be made (e.g. *mains electricity, battery* are possible values for a *power* attribute). Representing an object O as a set of attribute-value pairs

$$\{(a_i, v_i) | \text{attribute } a_i \text{ of object } O \text{ has value } v_i\}$$

we generate a new "design"

$$O^* = \{(a_i, T(v_i))\}$$

by changing one or more values using T, a non-deterministic transformation of a value to another value from the same taxonomy. Given sufficient time, this would simply enumerate all possible combinations of attribute values. We can reduce the search space by looking at the solution to an analogous problem in a different domain, as in the structural bisociation method, so that analogies can be found. This requires tools for taxonomy matching e.g. [6], for converting the data into object-attribute-value form (or extending the number of attributes), and for detecting structure arising from the attribute patterns. The latter two are covered in the next section.

Table 1. Attributes of two music players (taken from [4])

Conventional tape recorder	Sony Walkman
big	small
clumsy	neat
records	does not record
plays back	plays back
uses magnetic tape	uses magnetic tape
tape is on reels	tape is in cassette
speakers in cabinet	speakers in headphones
mains electricity	battery

2 Tools Used for Pre-processing Data

Two sets of process data were examined, as described in section 3. In order to identify taxonomic elations within the data, we used fuzzy formal concept analysis. One dataset contained short text sequences at each process step, and fuzzy grammars were used to extract key features from the text, prior to the concept analysis. For completeness, both methods (fuzzy grammars and fuzzy formal concept analysis) are briefly described in this section.

2.1 Fuzzy Grammars

A text fragment is a sequence of symbols, and it is common to use shallow processing (e.g. presence / absence of keywords) to find attributes of the text. This operation can be viewed as a way to label sub-sequences of symbols with different tags indicating the nature of the text fragment. For example, we could have a schema for the process step "arranging to call back a customer", including attributes such as the time, which party requested the call-back, reason for the call-back, etc. It is not necessary to extract every attribute from the text, and it is possible that information may not be recognised due to unexpected ways of expressing the information or abbreviations, mis-spelling etc. The latter case can be handled by extending the matching process to include fuzzy matches, that is sequences of symbols that almost conform to the pattern and are sufficiently close to be recognisable as examples of the pattern.

It is often not possible to define simple patterns (such as regular expressions) which can reliably identify the information structure. The key contribution of the fuzzy grammar approach is the definition and use of approximate grammars, where a degree of support is calculated for the matching process between an approximate grammar and a sequence of symbols that may not precisely conform to the grammar.

For example, the following fragments are all examples where a call back has been arranged:

spoke with Michelle the cust partner need a call after 2hrs.
cust need a call tomorrow.
cust is going to think about charge and call back if needs to.
eu is going to call back in a minute on a different phone

(*cust* is an abbreviation for *customer,* and *eu* is an abbreviation for *"end user"* i.e. customer). It is difficult to anticipate all possible forms (including abbreviations) in a regular expression. Full details of the fuzzy grammar approach are given in [7, 8]

2.2 Fuzzy Formal Concept Analysis

Formal concept analysis (FCA) [9, 10] is a way of extracting hidden structure (specifically, lattice-based structure) from a dataset that is represented in object-attribute-value form. In its simplest form, FCA considers a binary-valued relation on a set of objects O and attributes A

$$R \subseteq O \times A$$

The structure *(O, A, R)* is a formal context. Given X, a subset of objects and Y, a subset of the attributes,

$$X \subseteq O$$
$$Y \subseteq A$$

the operators $\uparrow$ and $\downarrow$ are defined as follows:

$$X^{\uparrow} = \{ y \in Y \mid \forall x \in X : (x, y) \in R \} \quad (1)$$

$$Y^{\downarrow} = \{ x \in X \mid \forall y \in Y : (x, y) \in R \} \quad (2)$$

Any pair (X, Y) such that $X^{\uparrow}=Y$ and $Y^{\downarrow}=X$ is a formal concept.

Table 2. A simple formal context

	a1	a2	a3
o1	1	0	1
o2	1	1	0
o3	0	0	1

For example, Table 2 shows the relation between three objects *o1, o2, o3* and attributes *a1, a2* and *a3*. The resulting concepts are

({o2}, {a1, a2})
({o1}, {a1, a3})
({o1, o2}, {a1})
({o1, o3}, {a3})

i.e. the object *o2* is the only one to have both attributes *a1* and *a2*, objects *o1* and *o2* are the only objects to have attribute *a1* etc. In larger tables, this is less obvious to inspection.

A partial order, $\leq$, is defined on concepts such that

$$(X1, Y1) \leq (X2, Y2)$$

means $X1 \subseteq X2$ and $Y2 \subseteq Y1$ i.e. the higher concept contains more objects and fewer conditions (attributes that must be true) than the lower concept. This gives rise to a lattice, enabling us to discover relations between concepts - for example, in Fig 3 we see that attributes *a2* and *a3* in Table 2 are mutually exclusive, since no object has both attributes i.e. the least upper bound is equal to the top element of the lattice and the greatest lower bound is equal to the bottom. Each node drawn as a large circle represents an object (or objects), and each object has all the attributes attached to its node and all higher nodes linked directly or indirectly to it. The software draws a node with a black lower half if it represents an object (or set of objects), and with a blue upper half if it is the highest node corresponding to an attribute; this convention allows the diagram to be simplified by omitting labels.

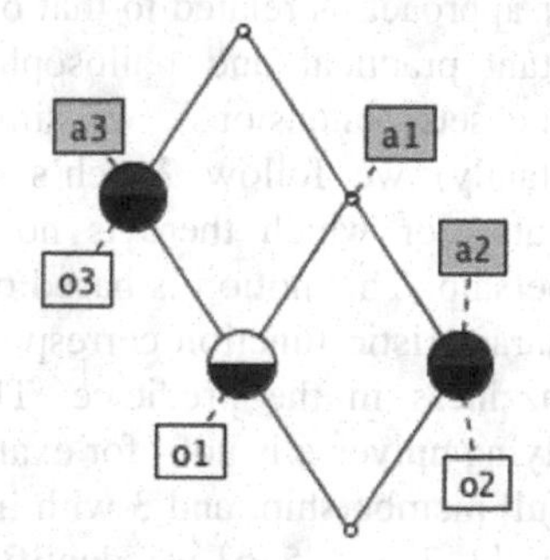

Fig. 3. Concept lattice corresponding to the formal context in Table 2. The lattice is drawn using the *conexp* tool (conexp.sourceforge.net).

2.2.1 Conceptual Scaling

For attributes which take a range of values (rather than true / false as above), the idea of "conceptual scaling" is introduced [11]. This transforms a many-valued attribute (e.g. a number) into a symbolic attribute - for example, an attribute such as "*time in seconds*", given an integer or real value between 0 and 200 could be transformed to attributes "*timeLessthan50*", "*timeFrom50to99*", etc. These derived attributes have true/false values and can thus be treated within the framework described above.

2.2.2 Fuzzy FCA

Clearly the idea of conceptual scaling is ideally suited to a fuzzy treatment, which reduces artefacts introduced by having to draw crisp lines between the categories. The idea of a binary-valued relation is easily generalised to a fuzzy relation, in which an object can have an attribute to a degree. Instead of defining the relation

$$R \subseteq O \times A$$

as

$$R : O \times A \rightarrow \{0,1\}$$

we define it as a fuzzy relation

$$R : O \times A \rightarrow [0,1]$$

where each tuple of R, $(o,a) \in R$ has a membership value in [0, 1].

We define a fuzzy formal concept as a pair X, Y where X is a fuzzy set of objects and Y is a crisp set of attributes such that $X^{\uparrow}$=Y and $Y^{\downarrow}$=X where

$$X^{\uparrow} = \{y \in Y \mid \forall x \in X : \mu_R(x,y) \geq \mu_X(x)\} \quad (3)$$

$$Y^{\downarrow} = \left\{ x / \mu_X(x) \middle| \mu_X(x) = \min_{y \in Y}(\mu_R(x,y)) \right\} \quad (4)$$

It is also possible to define crisp sets of objects and fuzzy sets of attributes, using a dual of these operators.

Our approach is related to that of Belohlavek (e.g. [12], [13]) but differs in some important practical and philosophical respects. By restricting ourselves to crisp attribute sets (intensions), we simplify the calculation of the closures but (more importantly) we follow Zadeh's original motivation for fuzzy sets - modelling predicates for which there is no clear boundary between membership and non-membership. This notion is based on a universe of discourse, and a generalisation of the characteristic function corresponding to the set defined by a predicate, reflecting the fuzziness in the predicate. The extension of the predicate is fuzzy but the underlying universe is not - for example, the set of *small* dice values could be 1 and 2 with full membership, and 3 with intermediate membership. From the set of possible values, {1, 2, 3, 4, 5, 6} we identify a fuzzy subset of *small* values. Given the value 1, we can say that it definitely has the attribute *small*, whereas given the value 3 we can say that it only has the attribute *small* to an intermediate degree. If we are told that a single dice roll has resulted in a *small* value, we can model it as a possibility distribution using the same membership function. We have a crisp event (small dice roll) with a fuzzy attribute (value displayed on the dice).

In contrast, methods based on residuated implication allow both intension and extension to be fuzzy.

Methods based on the alpha-cut are essentially crisp, once the choice of a threshold is made; changing the threshold is equivalent to defining a different conceptual scaling.

3 Process Data

Access to a number of process datasets was provided by an industrial partner, BT Innovation and Design. The datasets were taken from real operations, and were anonymised by removal of obvious personal details; in order to ensure commercial and customer confidentiality, the datasets were not taken offsite. Two datasets were selected for study:

1. Repair Data - a dataset of approximately 55000 process instances, stored in an XML format. Each process instance represented a single case of fault analysis and repair, and contained data on the tasks carried out in that case, the length of time taken for each task, the location, etc. Process and task names were structured but not necessarily understandable - for example, task names (or WorkFlowModelElement, using the XML tag) mostly consisted of a 5 character code (e.g. UK106) representing the centre at which the task was carried out, followed by a three character identifier (e.g. TS2) representing the actual task. Process instances varied in length from 3 up to 440 (including start and end) e.g.

start BB450GAB end

Figure 4 shows the distribution of path lengths. Over 30 centre identifiers were included in the data, representing a wide range of repair functions within the company. Python and Unix scripts, plus custom java modules were used with KNIME to convert the data into BisoNet form.

2. Call-Centre Data - a dataset of call-centre interactions, related to different business units within the company. Each process instance involved a number of questions designed to elicit information (customer details, problem symptoms, etc) and find a solution (including an immediate fix, additional tests, or appointment for an engineer to visit). These questions were a mixture of scripted and free-form text. Each step in the process had a unique identifier; additional data included an identifier for each process instance, the customer and call-centre agent, date/time and duration of the step, and information about the handling department and ultimate resolution of the problem. The data was recorded automatically, with scripted questions provided by the system and unscripted questions plus responses entered by the call centre agent. The free-form nature of unscripted questions and the number of abbreviations, mis-spellings and grammatical shortcuts taken when these questions are typed added an additional complication to the dataset.

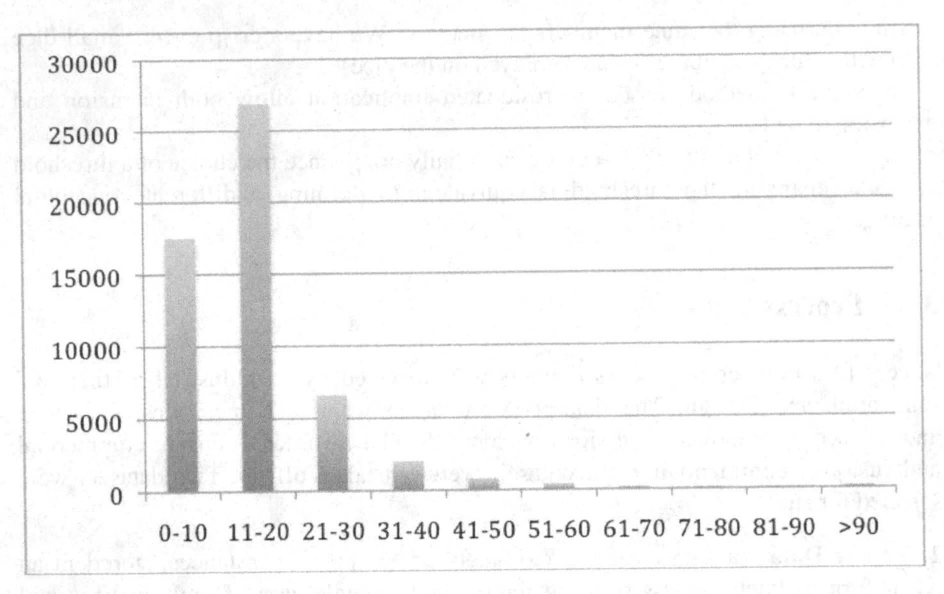

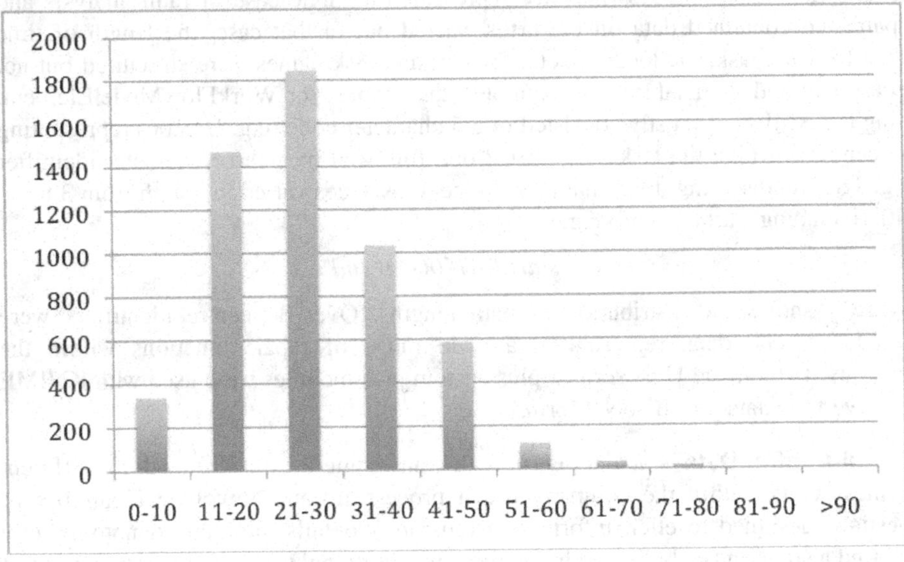

Fig. 4. Distribution of path lengths in dataset 1 (top) and dataset 2 (bottom)

The dataset consisted of around 5500 process instances and a total of over 65000 steps. The process data was in the form of a series of questions (and answers) plus time taken, and identifiers for caller and call-centre agent, date/time and other data. The complete set of attributes (with brief description) was

CASE_ID	a unique identifier for this process instance
USER_RESPONSE_ID	unique identifier for this process step

AGENT	call centre agent
CONTACT_CREATED	timestamp for one (or more) steps
CUSTOMER	identifier for customer
QUESTION	text of scripted or unscripted question / other notes
RESPONSE	text of answer / summary of test result
DURATION	System-generated time taken by this process step
EXIT_POINT1, 2, 3	internal data
CASESTATUS	boolean indicating whether process has finished
DEPARTMENT	name of dept that handled this process

Figure 4 shows the distribution of path lengths (note that dataset 1 contains approximately 10 times as many instances as dataset 2). Table 4 shows a small part of an interaction; the "question" field was used to record scripted questions and notes made by the agent. Each sequence of questions as a process instance, represented as a directed graph.

Because there was so much flexibility in the question/answer patterns, we pre-processed the text to extract key features, using fuzzy grammar tagging [7] to add attributes. This went beyond a simple keyword-recognition approach (which was found to be inadequate) and was able to cope with the non-standard language employed in the questions. Table 4 shows examples of the tags added; these were used as node labels in the directed graphs.

Subsequent to the tagging, a combination of Unix scripts and customised Java / KNIME workflows were used to convert the data into Bisonet form.

Table 3. Example of call centre interaction (a single process instance)

Question	Response	Duration
What is the call about?	New fault	11
What type of fault?	No incoming calls	30
Is the Customer calling from home?	Yes, using line with the fault	4
Is this an early life Customer?	No	3
Are there any Open orders or billing restrictions (including issues from high value accounts) on the account which could be causing this problem?	No problems	4
Ask the Customer if they are reporting an issue with a BB talk hub phone, a VOIP phone or an internet phone.	No	5
Is there an open SR with a valid test within the last 30 minutes?	No Open SR	2
Start the test to check settings on the asset in OneView, use the Incoming calls option.	OK	34
…	…	…
System Test Line Test	Green Test Result	3
Have all line/equipment checks been completed for this problem?	Yes	2
cust will do more checking	Progress Saved	147

Table 4. Tags applicable to example shown in Table 3

Question	Fuzzy Tag(s)
What is the call about?	<g1FindCustomerProblem />
What type of fault?	<g1FindProblemDetails />
Is the Customer calling from home?	<g1FindProblemDetails />
Is this an early life Customer?	<g1FindProblemDetails />
Are there any Open orders or billing restrictions (including issues from high value accounts) on the account which could be causing this problem?	<g1FindAccountDetails />
Ask the Customer if they are reporting an issue with a BB talk hub phone, a VOIP phone or an internet phone.	<g1ProblemFeature />
Is there an open SR with a valid test within the last 30 minutes?	<g1SystemTest />
Start the test to check settings on the asset in OneView, use the Incoming calls option.	<g1SystemTest />
…	…
System Test Line Test	<g1CheckTestResult />
Have all line/equipment checks been completed for this problem?	<unknown />
cust will do more checking	<g1EndCall />

The process instances were derived from different call centres, dealing with different business areas (business and consumer phone services, broadband, cable broadcasting, etc). This was indicated to some extent by the "Department" field, and we took this as a starting point for finding different Bison domains within the data. Departments whose processes involved similar sequences of questions were grouped together using fuzzy FCA; we also divided each of these domains into good and bad process instances. The characteristics of a good process instance are

- it does not involve multiple interactions,
- it does not contain loops, and
- it is completed in a reasonably short time.

4 Bisociative Knowledge Discovery in Business Processes

4.1 Illustrative Example

We first provide a simple illustration to show how the fuzzy FCA approach can aid in conceptual bisociation for creative knowledge discovery. The data used to create the examples shown in Figs 1 and 2 leads to the concept lattices shown in Figs 5 and 6. Note that this is a "toy" example and the similarity between lattices is obvious in this case. We have developed methods which facilitate this matching by comparing lattices [14] and by finding fuzzy association confidences in fuzzy class hierarchies [15].

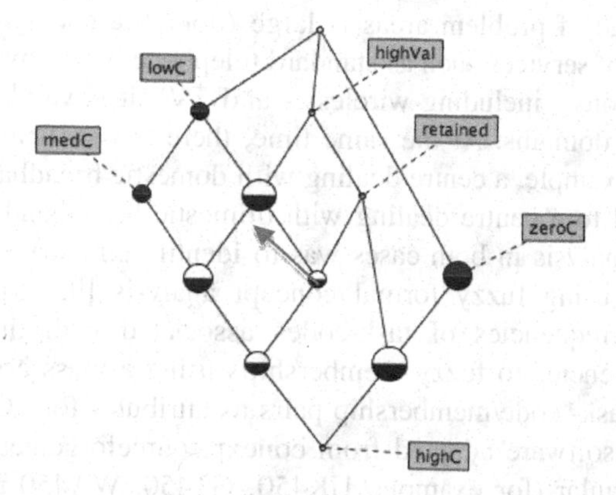

Fig. 5. Concept lattice corresponding to the process shown in Fig 1. The arrow indicates nodes used in calculating the association confidence (key performance indicator)

The attribute *highVal* indicates that a customer is a member of the set of high valued customers at the beginning of a specified period, time point t_0; the attribute *retained* indicates membership at the end of the period, time point t_1, and *zeroC, lowC, medC, highC* show the number of complaints (respectively, zero, low, medium, high) made in the period. Note that every object with membership in *medC* is also in *lowC* because of the overlapping nature of these sets. In this dataset, there are no customers who have made a high number of complaints, so the concept labelled *highC* is at the bottom of the lattice with no elements,.

Figure 5 shows the concept lattice for the ISP example of Fig. 1. The arrow indicates the association rule between the set of high value customers who complained a non-zero (*low* or *medium)* number of times and the subset who also satisfy the *retained* attribute. In this case, the confidence is 40% and this forms a key performance indicator for the process.

Figure 6 shows concept lattices corresponding to the hotel example of Fig. 2. The introduction of the *reward* attribute makes a major difference to the key performance indicator, raising it from 30% to 70%. Because the lattice is isomorphic to Fig. 5, the automated creative knowledge discovery process suggests that introduction of "something like" a rewards programme could also benefit the ISP in retaining high value customers Although the parallels are obvious here, practical examples require considerable search to find the best match between lattices. Work in this area is ongoing, outside the Bison project.

4.2 Business Process Example - Definition of Domains

Our second application looks for structural bisociations, and we start by defining domains. In both cases (datasets 1 and 2), data was gathered during a specific time interval, and was not balanced across different business units. Since the business units are (effectively) anonymised, the first step was to group processes from different units into domains for bisociation.

Because the range of problem areas is large (domestic and businesses customers using a wide range of services such as standard telephone, voice over IP, all aspects of broadband connections - including wireless - and TV), it is valid to regard different centres as different domains. At the same time, there is significant overlap between some centres - for example, a centre dealing with domestic broadband in one region is effectively identical to a centre dealing with domestic broadband in another region. The first stage of analysis in both cases was to identify similarities between centres; this was achieved using fuzzy formal concept analysis [9, 15] In dataset 1, we extracted relative frequencies of task-codes associated with the various centres, converted the frequencies to fuzzy memberships using a mass assignment approach [16] and used the task- code/membership pairs as attributes for FCA. The result (Fig 7, displayed using software adapted from conexp.sourceforge.net) shows that some centres are very similar (for example, UK450, GT450, WA450 near the top of the diagram), that there is a generalisation hierarchy (the UK450, GT450, WA450 cluster is more general than BB450, in terms of tasks performed), and that there are dissimilarities (e.g. UK107, UK106 near the bottom left have no overlap). The opinion of a domain expert was that these groupings were a realistic reflection of the functions.

In dataset 2, we used the fuzzy tags assigned by fuzzy grammar analysis as attributes, leading to the concept lattice in Fig 7(b). Here, it is possible to assess the groupings by inspection of the centre names - for example, it is not surprising to see the strong connection between centres dealing with businesses (six connected nodes on right hand side), with vision products (three nodes on left), etc.

4.3 Bisociations

There are a number of indicators for "good" and "bad" process execution. Reaching a satisfactory end point in a relatively short time, with no unnecessary loops is an ideal situation; cases which require repeated work, suffer from long delays and/or incorrect execution paths are not ideal.

Multiple Domains in a Single Dataset

Having defined different domains within each dataset, we looked for possible overlapping concepts between the domains. We first combined all process instances within a domain by adding universal start-process and finish-process nodes, and combining common paths from / to these universal nodes (using a modification of the DAWG algorithm in [17]).

In dataset 2, we used the fuzzy tags assigned by fuzzy grammar analysis as attributes, Three variants were initially produced for each set of instances. The first retained loops, but unrolled them so that each vertex had indegree and outdegree of 1 (other than the start-process and finish-process nodes). Second and third variants were produced, in which a node representing the loop (including the source/sink node of the loop) was given a derived identifier or given the same arbitrary identifier as all loops. Figure 8 shows an example of a single process with a loop from dataset 1.

Bisociations were sought by looking for structural similarity between domains. This was interpreted as finding a consistent mapping from the set of nodes in one graph to the set of nodes in the second graph, such that paths (i.e. process instances) are preserved (NB timing data for process steps was ignored here). For two domains (V_1, E_1) and (V_2, E_2) we search for a mapping

$$f : V_1 \rightarrow V_2$$

such that for each process instance from domain 1

$$P_{1i} = (v_{i1}, v_{i2}, \ \ldots \ , v_{in}) \text{ where each } v_{1i} \in V_1$$

there is a corresponding process

$$P_{2k} = f(P_{1j}) = (f(v_{j1}), f(v_{j2}), \ \ldots \ , f(v_{jn}))$$

in domain 2.

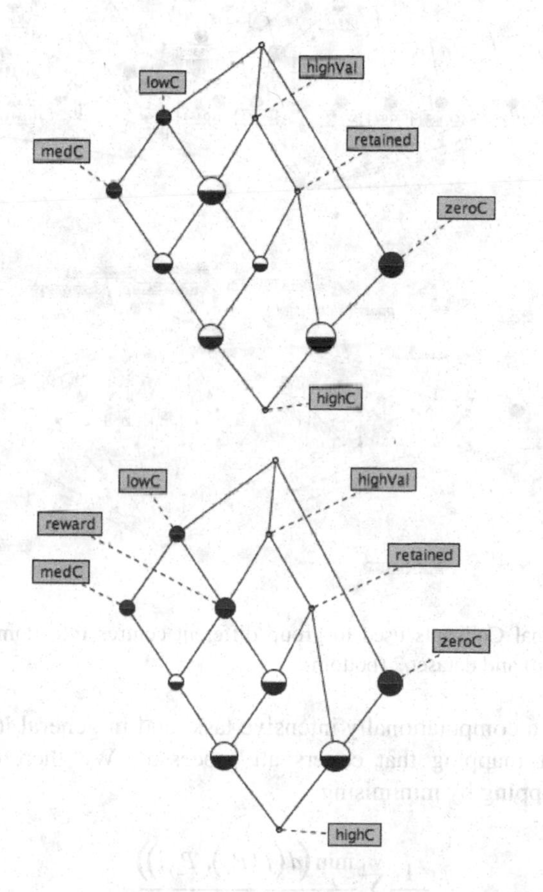

Fig. 6. Concept lattices corresponding to the processes shown in Fig 2. The key performance indicator is not shown, but improves from 40% to 70% . The similarity to Fig 5 is clear, and the suggestion to add an attribute corresponding to "reward" is obvious, once the parallel between contexts is seen.

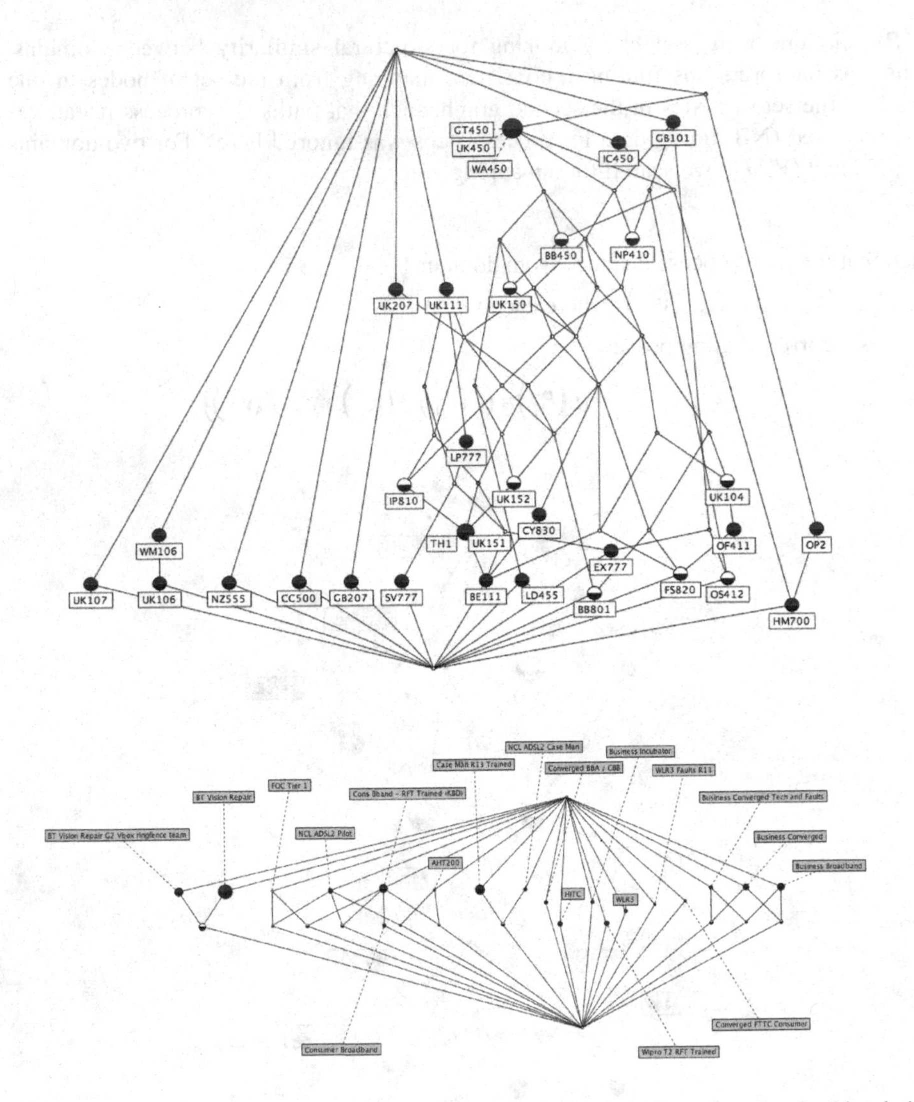

Fig. 7. Fuzzy Formal Concepts used to group different centres into domains for bisociation within dataset 1 (top) and dataset 2 (bottom)

Clearly this is a computationally intensive task, and in general it is not possible to find a consistent mapping that covers all processes. We therefore measured the goodness of a mapping by minimising

$$\frac{1}{N_{P1}}\sum_{i=1}^{N_{P1}}\frac{\min\limits_{j}\left(d\left(f\left(P_{1i}\right),\ P_{2j},\right)\right)}{length\left(P_{1i}\right)} \tag{5}$$

where d is the edit distance between the two sequences, *length* measures the number of steps in a process instance and N_{P1} is the number of process instances in domain 1 and j indexes processes in domain 2. The value of (5) ranges between 0 (every mapped process instance in domain 1 is identical to a process instance in domain 2) and 1 (no overlap at all). Subtracting the value of (5) from 1 gives an indication of the degree of overlap.

A number of heuristics were used to guide the search for a good mapping, based on the frequencies of nodes and node pairs.

Obviously if there is an exact mapping, there is an equivalence between the process domains and the only contribution from bisociative reasoning would be to suggest that improvements in one domain might also be made in the other. In cases where there is a short distance between a process instance and its image in the target domain, bisociative reasoning might suggest process modification - for example if

$$d\left(f\left(P_{1i}\right), P_{2k}\right)=1$$

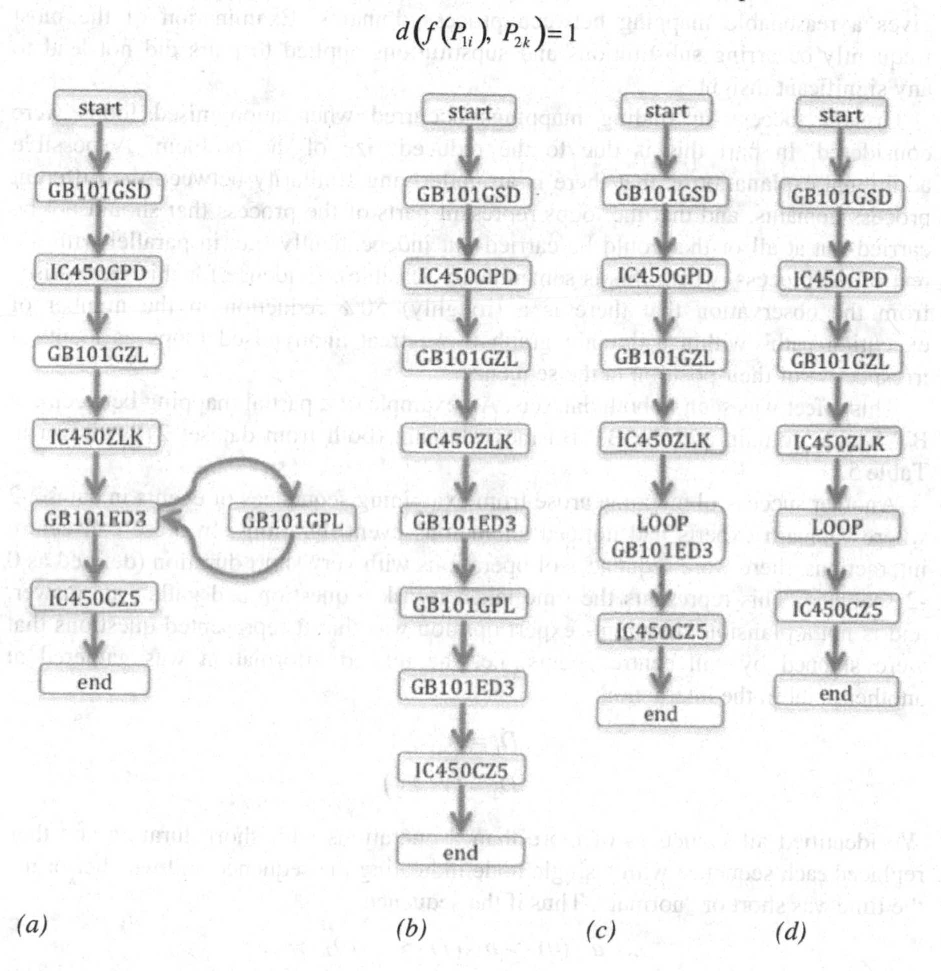

Fig. 8. Different options for treating loops in processes
(a) original process with a loop
(b) unrolled loop
(c) all loop nodes replaced by a single node, named by its start/end node
(d) all loop nodes replaced by an anonymous *loop* node

for some process P_{2k} then there is one node where the processes differ. Bisociation would suggest replacing this node by the inverse image of its counterpart in D_2. That is, if

$$P_{2k} = \left(f\left(v_{j1}\right), f\left(v_{j2}\right), \ldots, v_{kP}^{*} \quad \ldots, f\left(v_{jn}\right)\right)$$

then we should change the first process to

$$P_{1j} = \left(v_{j1}, v_{j2}, \ldots, f^{-1}\left(v_{kl}^{*}\right), \quad \ldots, v_{jn}\right)$$

This is a limited interpretation of bisociation, and - in the cases studied here - meets with little success, not least, because of the difficulty in finding a possible f which gives a reasonable mapping between process domains. Examination of the most frequently occurring substitutions and substitutions applied to pairs did not lead to any significant insight.

Greater success in finding mappings occurred when anonymised loops were considered. In part this is due to the reduced size of the problem. A possible additional explanation is that there is an underlying similarity between the different process domains, and that the loops represent parts of the process that should not be carried out at all or that could be carried out independently (i.e. in parallel with the rest of the process, where this is semantically feasible). Evidence for this view arises from the observation that there is a (roughly) 50% reduction in the number of execution paths within a domain graph if we treat anonymised loops as identical irrespective of their position in the sequence.

This effect was seen in both datasets. An example of a partial mapping between the BT Vision domain and the BT Business domain (both from dataset 2) is shown in Table 5.

Another successful outcome arose from examining sequences of events in dataset 2 where domain experts had noticed anomalous event durations. In these call centre interactions, there were sequences of operations with very short duration (defined as 0 -2 seconds). This represents the time taken to ask a question and gather an answer, and is not a plausible duration - expert opinion was that it represented questions that were skipped by call centre agents, i.e. the related information was gathered at another point in the interaction.

$$D_1 = \left(V_1, E_1\right)$$
$$D_2 = \left(V_2, E_2\right)$$

We identified all sequences of more than 2 operations with short duration and then replaced each sequence with a single node indicating the sequence and whether or not the time was short or "normal". Thus if the sequence

... *a - (0) -> b - (1) -> c - (0) ->* ...

was found, then all sequences *a-b-c* were replaced by a single node *ABC-short* or *ABC-normal.* The two domains for bisociation were defined as (i) processes containing one or more nodes denoting a short sequence and (ii) processes containing

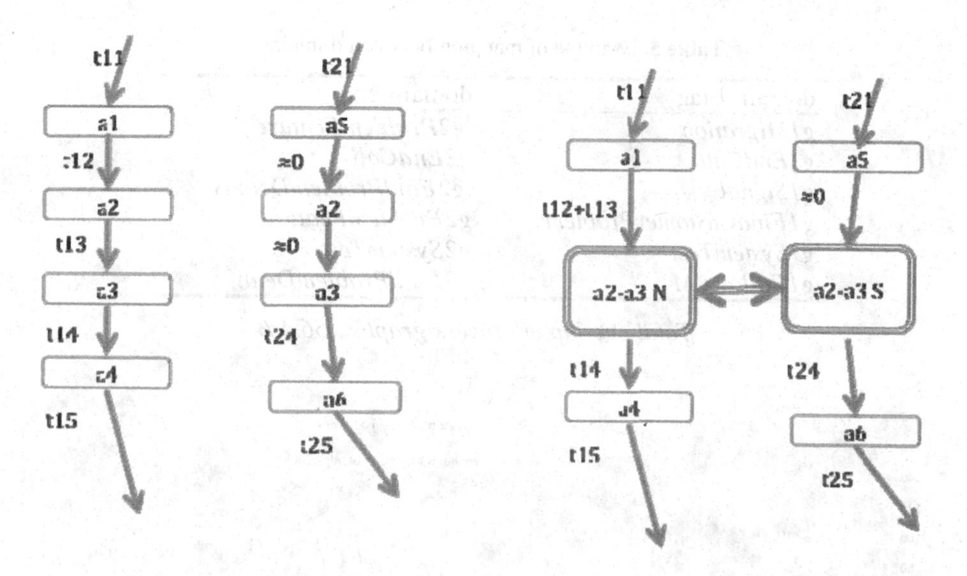

Fig. 9. Schematic illustration of bisociation between short duration sequences and normal duration sequences. The two graphs on the left both contain the sequence *-a2-a3-* , in the first case with "normal" duration and in the second with abnormally short duration. The sequences are concatenated to *a2-a3-normal (N)* and *-short (S),* and the durations of adjacent nodes are compared - see Fig. 10.

one or more nodes denoting a normal sequence. The replacement nodes were treated as bridging concepts (e.g. *ABC-short* in one domain was matched with *ABC-normal* in the second domain). Process time was examined in the joined graphs, since it was key to the bridging concepts, and we found that there was a significant increase in process step time for the immediate predecessors / successors of the bridging nodes (see Fig. 10). This suggests that although questions were skipped, the related information was gathered during preceding or succeeding questions. In turn, this means that the sequences could be moved e.g. they could be asked whilst waiting for another part of the process to complete. Such delays can happen when tests are run on the line, for instance, but further work would be required to test the feasibility of the suggestion.

The final example of bisociation was reached by comparing all of dataset 1 with all of dataset 2. Within each dataset, all processes were combined into a single large graph (with universal start-process and finish-process nodes). Based on the previous investigations, we chose as bridging concepts the loops in dataset 1 and the short-duration sequences in dataset 2. These were used to derive a mapping between nodes from domain 1 and domain 2, and the overlap in process graphs arising from the mapping was estimated by (5). Note that we used relative frequencies of process paths, since there are approximately 10 times more process instances in dataset 1 than in dataset 2. The resultant mapping between domains was deemed to be relatively high quality, since it led to high similarity between the mapped domain 1 and domain 2.

Table 5. Example of mapping between domains

domain 1 tag	domain 2 tag
g1Migration	*g2ProblemFeature*
g1EndCall	*g2EndCall*
g1Signal	*g2FindProblemDetails*
g1FindCustomerProblem	*g2ProblemFeature*
g1SystemTest	*g2SystemTest*
g1FollowKM	*g2FindProblemDetail*

Total overlap in process graphs : 56.4%

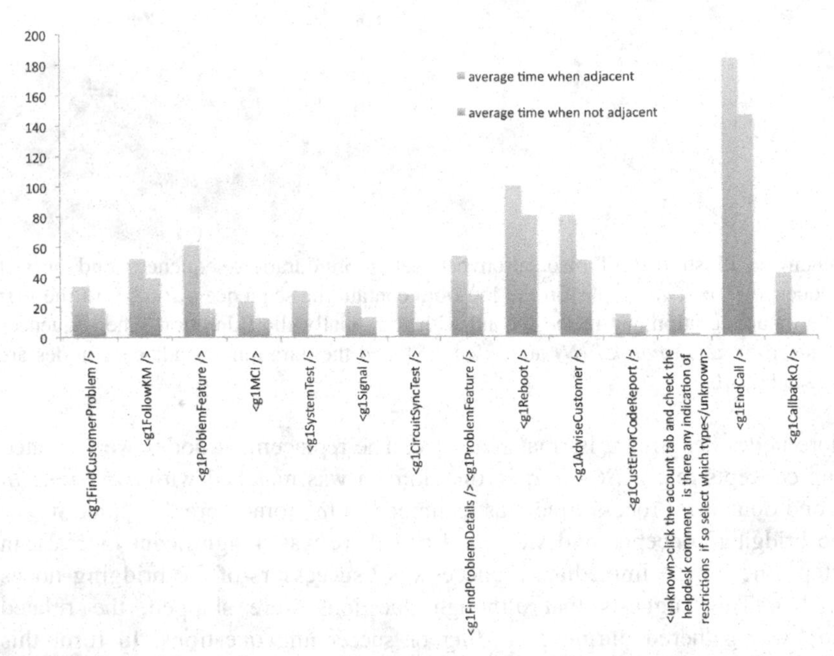

Fig. 10. Comparison of process durations for nodes adjacent to abnormally short duration sequences. The most frequent nodes are shown. Left (blue) column denotes the average duration when adjacent to an abnormal sequence, the right (red) column shows the average duration when not adjacent to an abnormal sequence. The difference may be due to additional information being gathered in adjacent nodes

5 Summary

Application of bisociation analysis to the task of creative process engineering has generated novel insight into the underlying data and into possible improvements - in particular, by suggesting parts of processes that could be performed at different points in the process sequence. The results of this study are sufficiently encouraging to warrant further investigation. Areas for future work include better presentation and visualisation of results, particularly with large data sets, the need to handle matching in edges as well as within the node structure, and issues relating to the non-static nature of process data (relevant links that may emerge and change with further data).

Acknowledgment. This work was partly funded by BT Innovate and Design and by the FP7 BISON (Bisociation Networks for Creative Information Discovery) project, number 211898

References

[1] Andersen, B., Fagerhaug, T.: Advantages and disadvantages of using predefined process models. Strategic Manufacturing: IFIP WG5 (2001)
[2] Kotter, T., Berthold, M.R.: From Information Networks to Bisociative Information Networks. In: Berthold, M.R. (ed.) Bisociative Knowledge Discovery. LNCS (LNAI), vol. 7250, pp. 33–50. Springer, Heidelberg (2012)
[3] Berthold, M.R. (ed.): Bisociative Knowledge Discovery. LNCS (LNAI), vol. 7250. Springer, Heidelberg (2012)
[4] Berthold, M.R. (ed.): Bisociative Knowledge Discovery. LNCS (LNAI), vol. 7250. Springer, Heidelberg (2012)
[5] Sherwood, D.: Koestler's Law: The Act of Discovering Creativity-And How to Apply It in Your Law Practice. Law Practice 32 (2006)
[6] Martin, T.P., Shen, Y.: Fuzzy Association Rules to Summarise Multiple Taxonomies in Large Databases. In: Laurent, A., Lesot, M.-J. (eds.) Scalable Fuzzy Algorithms for Data Management and Analysis: Methods and Design, pp. 273–301. IGI-Global (2009)
[7] Martin, T.P., Shen, Y., Azvine, B.: Incremental Evolution of Fuzzy Grammar Fragments to Enhance Instance Matching and Text Mining. IEEE Transactions on Fuzzy Systems 16, 1425–1438 (2008)
[8] Sharef, N.M., Martin, T.P.: Incremental Evolving Fuzzy Grammar for Semi-structured Text Representation. Evolving Systems (2011) (to appear)
[9] Ganter, B., Wille, R.: Formal Concept Analysis: Mathematical Foundations. Springer (1998)
[10] Priss, U.: Formal Concept Analysis in Information Science. Annual Review of Information Science and Technology 40, 521–543 (2006)
[11] Prediger, S.: Logical Scaling in Formal Concept Analysis. In: Delugach, H.S., Keeler, M.A., Searle, L., Lukose, D., Sowa, J.F. (eds.) ICCS 1997. LNCS, vol. 1257, pp. 332–341. Springer, Heidelberg (1997)
[12] Belohlavek, R.: Fuzzy Relational Systems. Springer (2002)
[13] Belohlavek, R., Sklenar, V., Zacpal, J.: Crisply Generated Fuzzy Concepts. In: Albrecht, A.A., Jung, H., Mehlhorn, K. (eds.) Parallel Algorithms and Architectures. LNCS, vol. 269, pp. 269–284. Springer, Heidelberg (1987)
[14] Martin, T.P., Majidian, A.: Dynamic Fuzzy Concept Hierarchies (2011) (to appear)
[15] Martin, T., Shen, Y., Majidian, A.: Discovery of time-varying relations using fuzzy formal concept analysis and associations. International Journal of Intelligent Systems 25, 1217–1248 (2010)
[16] Baldwin, J.F.: The Management of Fuzzy and Probabilistic Uncertainties for Knowledge Based Systems. In: Shapiro, S.A. (ed.) Encyclopedia of AI, 2nd edn., pp. 528–537. John Wiley (1992)
[17] Sgarbas, K.N., Fakotakis, N.D., Kokkinakis, G.K.: Optimal Insertion in Deterministic DAWGs. Theoretical Computer Science 301, 103–117 (2003)

Bisociative Music Discovery and Recommendation

Sebastian Stober, Stefan Haun, and Andreas Nürnberger

Data & Knowledge Engineering Group, Faculty of Computer Science,
Otto-von-Guericke-University Magdeburg, D-39106 Magdeburg, Germany
{sebastian.stober,stefan.haun,andreas.nuernberger}@ovgu.de

Abstract. Surprising a user with unexpected and fortunate recommendations is a key challenge for recommender systems. Motivated by the concept of bisociations, we propose ways to create an environment where such serendipitous recommendations become more likely. As application domain we focus on music recommendation using MusicGalaxy, an adaptive user-interface for exploring music collections. It leverages a non-linear multi-focus distortion technique that adaptively highlights related music tracks in a projection-based collection visualization depending on the current region of interest. While originally developed to alleviate the impact of inevitable projection errors, it can also adapt according to user-preferences. We discuss how using this technique beyond its original purpose can create distortions of the visualization that facilitate bisociative music discovery.

1 Introduction

One of the big challenges of computer science in the 21st century is the digital media explosion. Online music stores already contain several millions of music tracks and steadily growing hard-drives are filled with personal music collections of which a large portion is almost never used. Music recommender systems aim to help us cope with this amount of data and find new interesting music or rediscover once loved pieces we have forgotten about – a task also called "recomindation" [22]. One common problem that many recommender systems face is that their recommendations are often too obvious and thus not particularly useful when it comes to discovering new music. Especially, collaborative filtering approaches are prone to a strong popularity bias [2]. In fact, McNee et al. argue that there is too much focus on improving the accuracy of recommender systems. They identify several important aspects of human-recommender interaction of which serendipity is specifically related to the above phenomenon [17]. A serendipitous recommendation is unexpected and fortunate – something that is particularly hard to grasp and evaluate.

We recently conducted a user study to assess the usability and usefulness of a visualization technique for the exploration of large multimedia collections. One task was to find photographs of lizards in a collection of photos taken in Western Australia. The user-interface was supposed to support the participants

M.R. Berthold (Ed.): Bisociative Knowledge Discovery, LNAI 7250, pp. 472–483, 2012.

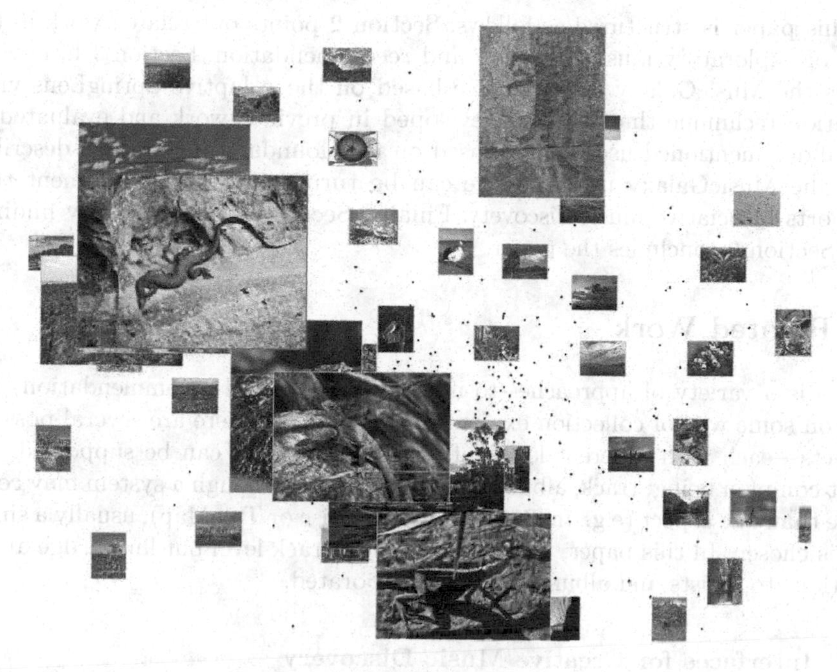

Fig. 1. Serendipitous encounter with a rock painting of a lizard when looking for photographs of a lizard (using the Adaptive SpringLens visualization for exploring multimedia collections [26])

by pointing out possibly relevant photos for a seed photo. As it happened, one of the participants encountered a funny incident: While looking for photographs showing a lizard, he selected an image of a monitor lizard as seed. To his surprise, the system retrieved an image showing the rock painting of a lizard (Figure 1). Interestingly, rock paintings were actually another topic to find photos for and the relevant photos were a lot harder to make out in the collection than the lizards. Bearing in mind that according to Isaac Asimov "the most exciting phrase to hear in science, the one that heralds new discoveries, is not 'Eureka!' (I found it!) but 'That's funny ...' ", we decided to further investigate this phenomenon. What the participant encountered is called a *bisociation* – a bridging element between the two distinct domains: animals and rock paintings. While most associations are found between concepts of one domain, there are certain paths which either bridge two different domains or connect concepts by incorporating another domain. In his book *The Act of Creation*, Arthur Köstler, an Austrian publisher, coined the term *bisociation* for these types of associations and as it turns out, many scientific discoveries are in some way bisociations [9].

Admittedly, no one expects scientific discoveries from a music recommender application. However, the question persists whether we can leverage the effect of bisociations and create an environment where serendipitous recommendations become more likely. After all, the concept of bisociation is much easier to grasp than serendipity and can even be formalized by means of graph theory [10].

This paper is structured as follows: Section 2 points out related work in the field of exploratory music discovery and recommendation. Section 3 briefly reviews the MusicGalaxy user-interface based on the Adaptive SpringLens visualization technique that we have developed in previous work and evaluated in the above mentioned user study. Based on this foundation, Section 4 describes how the MusicGalaxy user-interface can be turned into an environment that supports bisociative music discovery. Finally, Section 5 discusses early findings and Section 6 concludes the paper.

2 Related Work

There is a variety of approaches to music discovery and recommendation that rely on some way of collection exploration. Generally, there are several possible aspects—each with different levels of abstraction—that can be supported, the most common being: track, album, artist and genre. Though a system may cover more than one aspect (e.g., in [31] visualized as disc or TreeMap), usually a single one is chosen. In this paper, the focus is on the track level but knowledge about relations to artists and albums is also incorporated.

2.1 Interfaces for Creative Music Discovery

MusicRainbow [19] is an interface to explore music collections at the artist level. Using a traveling salesman algorithm, similar artists are mapped near each other on a circular rainbow where the colors of the rainbow reflect the genres. Audio-based similarity is combined with words extracted from web pages related to the artists. The words are used to label the rainbow and describe the artists.

MusicSun [20] applies a similar concept to discover artists. Recommendations are based on one or more artists that are selected by the user and displayed in the center of a sun. The sun rays (triangles) represent words that describe these seed artists. The size of a ray's base reflects how well the respective word fits to the artist and its length is proportional to the number of artists in the collection that can also be described by that word. Selecting a ray, a list of recommended artists is generated. Similarly to the work presented in this paper, users can also adapt the impact of three different aspects of music similarity that are combined.

Musicream [7] facilitates active, flexible, and unexpected encounters with musical pieces by extending the common concept of query by example: Several tubs provide streams of music pieces (visualized as discs) that the user can grab and drop into the playback region of the interface or use as a magnet to filter similar pieces from the streams. The interface also provides enhanced playback functions such as building playlists of playlists or going back to any previous point in the play history.

The MusicExplorer FX[1] takes a different approach: Built upon the EchoNest API[2], it displays a local similarity graph, connecting an artist with the most

[1] http://musicexplorerfx.citytechinc.com/

[2] http://developer.echonest.com/

similar ones. The interface also shows a navigation history containing the previously visited artists. A similar approach is taken by the Relational Artist Map RAMA [24] that additionally displays labels as graph overlay. However, both lack a global overview of the whole artist space and users need to specify a seed artist to start with. In contrast to this, the Last.fm artist map[3] displays the whole graph (based on the Last.fm API[4]). As this results in a lot of clutter caused by crossing edges, it is hard to navigate and explore the graph. Consequently, it is rather suited to map a user's listening preferences.

2.2 Projection of a Similarity Space

In contrast to the already described works, the visualization approach taken here is primarily based on a projection of a similarity space. This is a very common method to create an overview of a collection. Popular dimensionality reduction techniques applied are self-organizing maps (SOM) [21,8,18,15,27] principal component analysis (PCA) [13] and multidimensional scaling (MDS) or similar force-based approaches [12,4,14]. Mapping the collection from high-dimensional feature/similarity space onto display space, it is usually impossibly to correctly preserve all distances (independent of the method used). Some objects will appear closer than they actually are and on the other side, some objects that are distant in the projection may in fact be neighbors in the original space.[5] Only a small number of approaches tries to additionally visualize such properties of the projection itself: The MusicMiner [18] draws mountain ranges between songs that are displayed close to each other but are dissimilar. The SoniXplorer [15] uses the same geographical metaphor but in a 3D virtual environment that the user can navigate with a game pad. The "Islands of Music" [21] and its related approaches [8,4] use the third dimension the other way around: Here, islands or mountains refer to regions of similar songs (with high density). Both ways, local properties of the projection are visualized – neighborhoods of either dissimilar or similar songs. Soundbite [14], on the other hand, attempts to visualize properties of the projection that are not locally confined: For selected objects in the (MDS) projection, edges are drawn additionally that connect them to their nearest neighbors – according to the underlying similarity and not the distance in the projection. We take a similar approach, interpreting connections between neighbors that are distant in the projection as "wormholes" through the high-dimensional feature space in analogy to the concept in astrophysics.

2.3 User-Adaption during the Exploration Process

Additionally, our goal is to support user-adaptation during the exploration process by means of weighting aspects of music similarity. Of the above approaches,

[3] `http://sixdegrees.hu/last.fm/interactive_map.html`

[4] `http://www.last.fm/api`

[5] Note that it is impossible to fix these problems without causing damage elsewhere as the projection is in general already optimal with respect to the projection technique applied.

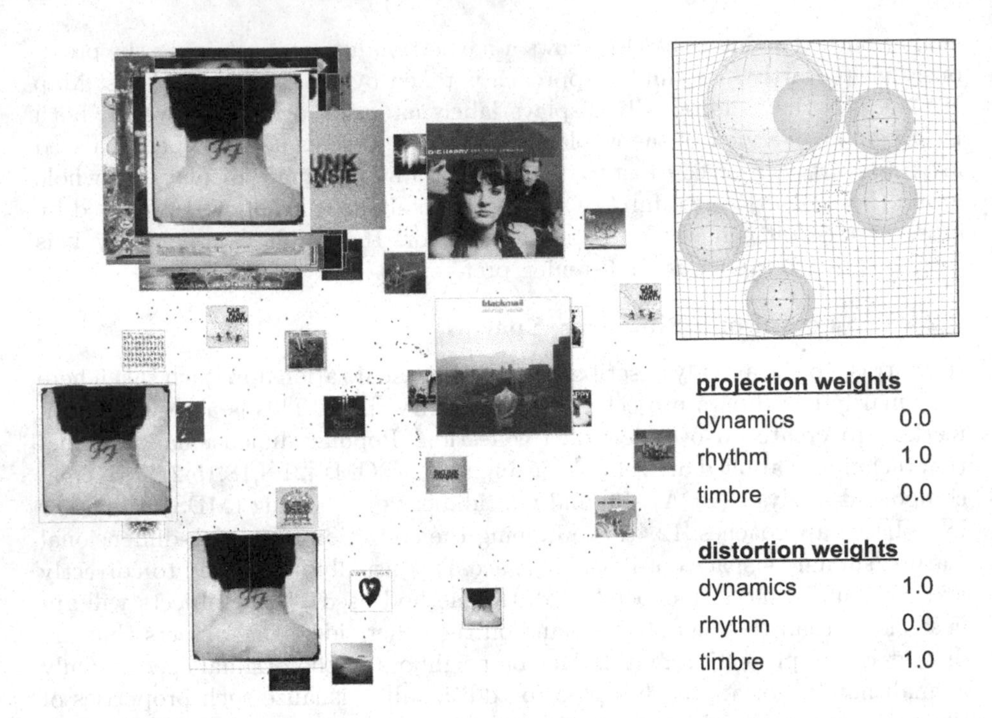

Fig. 2. Left: MusicGalaxy visualization. Top right: corresponding SpringLens distortion resulting from primary focus (red) and 5 secondary lenses (blue). Bottom right: facet weights for the projection and distortion distance measures.

only the revised SoniXplorer [15], MusicBox [13], MusicSun [20] and our original SOM-based prototype [27] allow automatic adaptation of the view on the collection through interaction. Apart from this, there exist systems that also adapt a similarity measure but not to change the way the collection is presented in an overview but to directly generate playlists (e.g., [1,32]). In contrast to these systems that purely focus on the task of playlist generation, we pursuit a more general goal in providing an adaptive overview of the collection that can then be used to easily generate playlists as, e.g., already shown in [8] or [13].

3 The MusicGalaxy Visualization

In previous work [28,29], we have developed an interface for exploring large music collections using a galaxy metaphor that addresses the problem of distorted neighborhoods. Figure 2 shows a screenshot of the interface visualizing a music collection.[6] Each track is displayed as a star, i.e., a point, with its brightness and—to some extend—its hue depending on a predefined importance measure (here a play count obtained from last.fm – other measures such as a general popularity or ratings are possible). A spatially well distributed subset of the

[6] A demo video is available at `http://www.dke-research.de/aucoma`

collection (specified by filters) is additionally displayed as an album cover for orientation. The arrangement of the stars is computed using multi-dimensional scaling (MDS) [11] relying on a set of descriptive features to be extracted beforehand.[7] MDS is a popular neighborhood-preserving projection technique that attempts to preserve the distances (dissimilarities) between the objects in the projection. The result of the MDS is optimal with respect to the minimization of the overall distance distortions. Thus, fixing one distorted neighborhood is not possible without damaging others. However, if the user shows interest in a specific neighborhood, this one can get a higher priority and be temporarily fixed (to some extend) at the cost of the other neighborhoods. To this end, an adaptive distortion technique called SpringLens [5] is applied that is guided by the user's focus of interest. The SpringLens is a complex overlay of multiple fish-eye lenses divided into primary and secondary focus. The primary focus is a single large fish-eye lens used to zoom into regions of interest. At the same time, it compacts the surrounding space but does not hide it from the user to preserve overview. While the user can control the primary focus, the secondary focus is automatically adapted. It consists of a varying number of smaller fish-eye lenses. When the primary focus changes, a neighbor index is queried with the object closest to the center of focus. If nearest neighbors are returned that are not in the primary focus, secondary lenses are added at the respective positions. As a result, the overall distortion of the visualization temporarily brings the distant nearest neighbors back closer to the focused region of interest. This way, distorted distances introduced by the projection can to some extend be compensated.

The user-interface has been evaluated in a study as reported in [26]. In the study, 30 participants had to solve an exploratory image retrieval task[8]: Each participant was asked to find representative images for five non-overlapping topics in a collection containing 350 photographs. This was repeated on three different collections – each one with different topics and with varying possibilities for interaction, comparing the fish-eye with traditional panning and zooming and a combination thereof. In total, each participant spent between 30 and 60 minutes using the system. The participants clearly preferred the fish-eye and the combined interface over the traditional panning and zooming in terms of helpfulness, simplicity and intuitivity. Further, gaze information recorded with an eye-tracker showed extensive use of the secondary focus to find more relevant images belonging to the same topic as the one in primary focus. As anticipated, some participants used the primary lens to skim through the photo collection in a rather continuous fashion. But surprisingly, there was also a group that browsed the collection mostly by moving (in a single click) the primary focus to some (previously) secondary focus region step-by-step – much like navigating an invisible neighborhood graph. Thus, it can

[7] Alternatively, feature information may also be annotated manually or collected from external sources.

[8] Images were used instead of tracks because (1) it could be guaranteed that the collection is unknown to all users and (2) visual similarity and relevance are much quicker assessed.

be concluded that the multi-focus SpringLens technique is very well suited for exploratory recommendation scenarios.

An aspect not addressed in the user study is that MusicGalaxy additionally allows to adapt the underlying music similarity. To this end, music similarity is represented as a distance measure that is a weighted linear combination of facet distances. Each facet covers a specific aspect of music similarity such as melody, harmony, rhythm, dynamics or lyrics and is defined by one ore more representative features and an appropriate distance measure. The importance of the individual facets can be adapted by changing their weights for the linear aggregation. To this end, the user interface has a control panel (not shown in the screenshot) with respective sliders. As we have shown in recent experiments with simulated user-interaction [30], it is also possible to adapt the weights automatically based on (relative) preferences derived from user actions such as judging two objects to be more similar with respect to a third one. Using adaptation, it becomes possible to personalize the music similarity measure used for recommendations.

4 Bisociative Lens Distortions

How can MusicGalaxy be turned into an environment that supports bisociative music discovery? The general idea is to combine two distinct domain views into one visualization by using the secondary focus to highlight connections to nearest neighbors *in a different domain* than the one used for projection: The "primary domain" is directly visualized by the projection and contains the displayed tracks connected by neighborhood relations that are implicitly induced between each track and its neighbors in the projection.[9] Additionally the "secondary domain"—which is used to identify nearest neighbors for the secondary focus distortion—is not directly visible to the user. A bisociation occurs in this setting if two tracks are not neighbors in the projection domain, i.e., close to each other in the display, but are connected in the secondary domain. In this case, the secondary focus will highlight this connection by focusing on the bisociated track – or similar image with respect to another domain as shown in Figure 1.

4.1 Orthogonal Similarity Measures

The simplest way to create such a setting is to use orthogonal similarity measures, i.e., defined on non-overlapping facet sets, for the two domains by choosing the facet weights accordingly. E.g., in Figure 2 the tracks in secondary focus are very different in rhythm (large distance in projection) but very similar in dynamics and timbre with respect to the track in primary focus. This approach could also be used in different applications. To illustrate the possibilities, imagine a user wants to explore a collection of world-music as, e.g., addressed by

[9] This is rather an artificial mental model a user perceives as no connections are explicitly visualized. Due to possible distortions introduced by dimensionality reduction, it only approximates the one derived from the actual distances in the original space.

mHashup [16]. In such applications, a straightforward way for the arrangement of the tracks would be according to their geographical origin, i.e., mapping the tracks on a common world map. Using this primary domain instantly gives the user an overview of the geographic distribution of the tracks in the collection. With the primary fish-eye lens, the user could magnify a region he is interested in. This would allow to display the local distribution of tracks in more detail and differentiate smaller (sub)regions. Note that in this special case, the arrangement of the tracks is perfect in the sense that all distances can be displayed distortion-free (except for the neglectible mapping of the earth's surface to a plane) because there is no dimensionality reduction involved. The secondary focus in its original setting would be unnecessary here anyway and it could therefore be freely used to highlight regions with nearest neighbors with respect to other aspects addressed by the secondary domain – e.g., acoustic similarity as a combination of its respective facets. Further, analyzing the interaction with the user, the system can—over time—learn which (acoustic) facets (of the secondary domain) are particularly important for the user and personalize the similarity measure for nearest neighbor retrieval accordingly. This has already been described and evaluated in [30].

4.2 Generalization to Domain Graphs

The above example uses an orthogonal similarity measure for the secondary domain. This is, however, only a very special case. Generally, the secondary domain might be any graph that contains at least the tracks as concepts (nodes) and allows to find neighboring tracks by some way of traversing relations between the concepts. An orthogonal similarity measure as described above induces such a graph: In this case, the graph contains only the tracks as concepts plus relations between tracks that are nearest neighbors and finding nearest neighbors for a track means simply returning all directly related tracks. An alternative way to construct such a sparse neighborhood graph for the secondary domain is to use any (black-box) system that recommends similar tracks for a seed track or even a combination of several such systems. However, the graph does not need to be confined to tracks. In fact, it may be arbitrarily complex – e.g., contain also artists, albums plus respective relations and possibly allowing multiple paths between tracks. For instance, from the freely available data from MusicBrainz[10], a user maintained community music meta-data base, a large graph can be constructed containing more than 10M tracks, 740K albums, 600K artists and 48K labels.[11] Between these entities, common relationships exist that, e.g., link tracks to artists and albums as well as albums to artists and labels. Apart from this, a large variety of advanced relationships links (ARL) exists. They are particularly interesting as they go beyond trivial information, such as links from tracks and albums to mastering and recording engineers, producers and studios (in total more than 281K artist-album and 786K artist-recording ARLs), how artists are

[10] `http://musicbrainz.org/`

[11] Figures as of January 2011 when the graph was created.

related with each other (more than 135K ARLs), or which tracks contain samples of others (more than 44K recording-recording ARLs).[12]

Nearest neighbors for a track in primary focus can be found by traversing the MusicBrainz graph in breadth-first order collecting paths to other tracks. Graph traversal stops when either the traversal depth or the number of reached track nodes exceeds a predefined threshold. As only the most relevant tracks can be highlighted by the secondary focus, some relevance measure is required to rank the retrieved tracks. Because increasing serendipity is the main objective, the relevance measure should capture how likely a track will be a lucky surprise for the user. This is however all but trivial. Possible simple heuristics are:

- Prefer tracks that are projected far away from the primary focus (and thus most likely sound very different).
- Prefer tracks that the user has not listened to a lot or for a long time (and probably is no longer aware of).
- Prefer tracks of different artists and/or albums.

The result of using either heuristic or a combination thereof will most likely surprise the user but at the same time the risk is high that the connection to the primary focus is too far fetched. Therefore, paths need to be judged according to their interestingness. Platt [23] defines discrete edge distances depending on the type of relationships for a similar graph created on a dataset from the All Music Guide [3]. Similar weightings can be applied here. Alternatively, weights could be assigned to common path patterns instead – possibly penalizing longer paths. For instance, some path patterns are straightforward such as *track-artist-track* (same artist) or *track-album-track* (same album) where the latter is more interesting in terms of serendipity because it could be a compilation that also contains tracks of other artists. Both weighting approaches require empirical tuning of the respective weights. Another option is to count the frequencies of occurring path patterns and boost infrequent and thus remarkable patters which can be interpreted as analogy to the idf weights used in text retrieval. This favors patterns containing ARLs. If multiple paths between two tracks are found, their weights can be aggregated, e.g., using the maximum, minimum or average. More sophisticated methods like those described in [25] are currently developed to facilitate bisociations on text collections and could also be applied here to further increase the chances of bisociative recommendations from complex domain graphs. This is currently studied more thoroughly as the impact of the different heuristics and the values of their respective parameters are not yet fully clear.

5 Discussion

This research in the field of bisociative music collection exploration is still in an early stage and clearly leaves several options for elaboration. For instance, it would be possible to extend the domain graph beyond MusicBrainz by incorporating information from other sources such as last.fm, The EchoNest or Myspace

[12] Full list available at: `http://wiki.musicbrainz.org/Category:Relationship_Type`

(see Section 2 for some graphs created from artist-similarity relations that can be obtained from these resources).

The user-interface needs to better integrate the graph information – possibly displaying (single) interesting connections. It can also be important to point out why a specific track is highlighted by the secondary focus. Such explanations would make the recommendation more understandable and less ambiguous. Currently, a user can only recognize tracks of the same album (because of the same cover) and to some extend tracks of the same artists (given he can associate the album covers with the respective artists). Looking at the screenshot of MusicGalaxy shown in Figure 2, four tracks from the same album can be seen in secondary focus. This is in fact because of a strong album effect (the album contains jazz cover versions of Beatles songs) captured entirely only by acoustic facets and without knowledge of track-album or track-artist relations. However, a similar result could have been produced by using the MusicBrainz graph as secondary domain. There is currently no visual clue to differentiate one from the other. A deeper analysis of the relationship graph could lead to more sophisticated ways of judging the interestingness of paths to related tracks. In order to personalize recommendations and increase the chance of surprises, additional information from a user-profile could be incorporated. Finally, it is necessary to test the proposed approach in another user study. However, it still remains an open question how to objectively judge the quality of recommendations in terms of serendipity.[13]

6 Conclusions

This paper described an approach to increase the chance of serendipitous recommendations in an exploratory music retrieval scenario. Instead of addressing serendipity directly, we proposed to exploit the related concept of bisociations that can be formalized by means of graph theory. We demonstrated how separating the underlying similarity measures for projection and distortion in the MusicGalaxy interface makes is possible to link two distinct domain views on a music collection – creating a setting that promotes bisociations where serendipitous recommendations become more likely. We hope that this paper can contribute to the ongoing discussion of improving the serendipity of recommendations and at the same time spreads the awareness of the bisociation concept.

Acknowledgments. The authors would like to thank the members of the EU BISON project for many fruitful discussions on the topic of bisociation. This work was supported in part by the German National Merit Foundation, the German Research Foundation (DFG) project AUCOMA and the European Commission under FP7-ICT-2007-C FET-Open, contract no. BISON-211898.

[13] For a general discussion about the evaluation of exploratory user interfaces see [6].

References

1. Baumann, S., Halloran, J.: An ecological approach to multimodal subjective music similarity perception. In: Proc. of 1st Conf. on Interdisc. Musicology, CIM 2004 (2004)
2. Celma, O., Cano, P.: From hits to niches?: or how popular artists can bias music recommendation and discovery. In: Proc. of the 2nd KDD Workshop on Large-Scale Recommender Systems and the Netflix Prize Competition, NETFLIX 2008 (2008)
3. Ellis, D.P.W., Whitman, B., Berenzweig, A., Lawrence, S.: The quest for ground truth in musical artist similarity. In: Proc. of 3rd Int. Conf. on Music Information Retrieval, ISMIR 2002 (2002)
4. Gasser, M., Flexer, A.: Fm4 soundpark audio-based music recommendation in everyday use. In: Proc. of the 6th Sound and Music Computing Conference, SMC 2009 (2009)
5. Germer, T., Götzelmann, T., Spindler, M., Strothotte, T.: Springlens: Distributed nonlinear magnifications. In: Eurographics 2006 - Short Papers, pp. 123–126. Eurographics Association, Aire-la-Ville, Switzerland (2006)
6. Gossen, T., Nitsche, M., Haun, S., Nürnberger, A.: Data exploration for bisociative knowledge discovery: A brief overview of tools and evaluation methods. In: Berthold, M.R. (ed.) Bisociative Knowledge Discovery. LNCS (LNAI), vol. 7250, pp. 287–300. Springer, Heidelberg (2012)
7. Goto, M., Goto, T.: Musicream: New music playback interface for streaming, sticking, sorting, and recalling musical pieces. In: Proc. of 6th Int. Conf. on Music Information Retrieval, ISMIR 2005 (2005)
8. Knees, P., Pohle, T., Schedl, M., Widmer, G.: Exploring Music Collections in Virtual Landscapes. IEEE MultiMedia 14(3), 46–54 (2007)
9. Köstler, A.: The Act of Creation. Macmillan (1964)
10. Kötter, T., Thiel, K., Berthold, M.: Domain bridging associations support creativity. In: Proc. 1st Int. Conf. on Computational Creativity (ICCC 2010), Lisbon (2010)
11. Kruskal, J., Wish, M.: Multidimensional scaling. Sage (1986)
12. Lamere, P., Eck, D.: Using 3D visualizations to explore and discover music. In: Proc. of 8th Int. Conf. on Music Information Retrieval (ISMIR 2007), pp. 173–174 (2007)
13. Lillie, A.S.: Musicbox: Navigating the space of your music. Master's thesis, MIT (2008)
14. Lloyd, S.: Automatic playlist generation and music library visualisation with timbral similarity measures. Master's thesis, Queen Mary University of London (2009)
15. Lübbers, D., Jarke, M.: Adaptive multimodal exploration of music collections. In: Proc. of 10th Int. Conf. on Music Information Retrieval, ISMIR 2009 (2009)
16. Magas, M., Casey, M., Rhodes, C.: mhashup: fast visual music discovery via locality sensitive hashing. In: SIGGRAPH 2008: ACM SIGGRAPH 2008 New Tech Demos, p. 1. ACM, New York (2008)
17. McNee, S.M., Riedl, J., Konstan, J.A.: Making recommendations better: an analytic model for human-recommender interaction. In: CHI 2006 Extended Abstracts on Human Factors in Computing Systems (2006)
18. Mörchen, F., Ultsch, A., Nöcker, M., Stamm, C.: Databionic visualization of music collections according to perceptual distance. In: Proc. of 6th Int. Conf. on Music Information Retrieval, ISMIR 2005 (2005)

19. Pampalk, E., Goto, M.: Musicrainbow: A new user interface to discover artists using audio-based similarity and web-based labeling. In: Proc. of 7th Int. Conf. on Music Information Retrieval (ISMIR 2006), pp. 367–370 (2006)
20. Pampalk, E., Goto, M.: Musicsun: A new approach to artist recommendation. In: Proc. of 8th Int. Conf. on Music Information Retrieval (ISMIR 2007), pp. 101–104 (2007)
21. Pampalk, E., Rauber, A., Merkl, D.: Content-based organization and visualization of music archives. In: Proc. of ACM MULTIMEDIA 2002 (2002)
22. Plate, C., Basselin, N., Kröner, A., Schneider, M., Baldes, S., Dimitrova, V., Jameson, A.: Recomindation: New Functions for Augmented Memories. In: Wade, V.P., Ashman, H., Smyth, B. (eds.) AH 2006. LNCS, vol. 4018, pp. 141–150. Springer, Heidelberg (2006)
23. Platt, J.C.: Fast embedding of sparse music similarity graphs. In: Advances in Neural Information Processing Systems (NIPS 2003), vol. 16 (2004)
24. Sarmento, L., Gouyon, F., Costa, B., Oliveira, E.: Visualizing networks of music artists with rama. In: Int. Conf. on Web Information Systems and Technologies, Lisbon (2009)
25. Segond, M., Borgelt, C.: Selecting the links in bisoNets generated from document collections. In: Berthold, M.R. (ed.) Bisociative Knowledge Discovery. LNCS (LNAI), vol. 7250, pp. 56–67. Springer, Heidelberg (2012)
26. Stober, S., Hentschel, C., Nürnberger, A.: Evaluation of adaptive springlens – a multi-focus interface for exploring multimedia collections. In: Proc. of 6th Nordic Conf. on Human-Computer Interaction, NordiCHI 2010 (2010)
27. Stober, S., Nürnberger, A.: Towards User-Adaptive Structuring and Organization of Music Collections. In: Detyniecki, M., Leiner, U., Nürnberger, A. (eds.) AMR 2008. LNCS, vol. 5811, pp. 53–65. Springer, Heidelberg (2010)
28. Stober, S., Nürnberger, A.: A multi-focus zoomable interface for multi-facet exploration of music collections. In: Proc. of 7th Int. Symposium on Computer Music Modeling and Retrieval, CMMR 2010 (2010)
29. Stober, S., Nürnberger, A.: MusicGalaxy – an adaptive user-interface for exploratory music retrieval. In: Proc. of 7th Sound and Music Computing Conference, SMC 2010 (2010)
30. Stober, S., Nürnberger, A.: Similarity Adaptation in an Exploratory Retrieval Scenario. In: Detyniecki, M., Knees, P., Nürnberger, A., Schedl, M., Stober, S. (eds.) AMR 2010. LNCS, vol. 6817, pp. 144–158. Springer, Heidelberg (2012)
31. Torrens, M., Hertzog, P., Arcos, J.L.: Visualizing and exploring personal music libraries. In: Proc. of 5th Int. Conf. on Music Information Retrieval, ISMIR 2004 (2004)
32. Vignoli, F., Pauws, S.: A music retrieval system based on user driven similarity and its evaluation. In: Proc. of 6th Int. Conf. on Music Information Retrieval, ISMIR 2005 (2005)

Author Index

Author Index